Alien Introgression in Wheat

Márta Molnár-Láng • Carla Ceoloni
Jaroslav Doležel
Editors

Alien Introgression in Wheat

Cytogenetics, Molecular Biology, and Genomics

Springer

Editors
Márta Molnár-Láng
Agricultural Institute
Centre for Agricultural Research
Hungarian Academy of Sciences
Martonvásár, Hungary

Carla Ceoloni
Department of Agriculture, Forestry,
Nature and Energy
University of Tuscia
Viterbo, Italy

Jaroslav Doležel
Institute of Experimental Botany
Centre of the Region Hana
for Biotechnological & Agricultural
Research
Olomouc, Czech Republic

ISBN 978-3-319-23493-9 ISBN 978-3-319-23494-6 (eBook)
DOI 10.1007/978-3-319-23494-6

Library of Congress Control Number: 2015954645

Springer Cham Heidelberg New York Dordrecht London

Printed on acid-free paper

Springer International Publishing AG Switzerland is part of Springer Science+Business Media (www.springer.com)

Preface

Humankind will face unprecedented challenges in the twenty-first century, including growing pressure on agricultural production, in particular, major staple crops such as wheat. In order to cope with a suite of climatic, environmental and socio-economic changes and achieve food security in a sustainable manner, it will be critical to grow varieties of wheat with increased resistance to diseases, pests and adverse environmental conditions and with improved yield, as well as to reduce post-harvest losses and employ improved agronomic practices. However, the development of suitable varieties may be limited by the narrow genetic diversity of wheat germplasm, which is the result of selection pressure imposed during wheat domestication and many years of breeding. An approach that could play an important role in breeding novel wheat cultivars relies on the use of alien germplasm, including relatively distant gene pools, which represent an attractive source of value-added traits. The need to counteract wheat genetic erosion was highlighted by the intensive selection pressures encountered in the twentieth century's Green Revolution.

The idea of interspecific hybridization attracted researchers and breeders long ago, and the first successful crosses were already reported in the eighteenth century. The production of a wheat–rye hybrid in 1876 by the Scottish botanist A. Stephen Wilson was a milestone in the efforts to hybridize wheat with related species. Although the hybrid was sterile, intensive programs were set up in many countries with the hope of incorporating useful traits of rye into wheat and soon led to the development of fertile wheat–rye hybrids. A notable case of the early efforts to transfer alien genes from related species into wheat was carried out by Edgar S. McFadden, a USDA agronomist at the Texas Agricultural Experiment Station, and Ernest R. Sears, a USDA geneticist at the University of Missouri, Columbia. During a long-lasting, fruitful collaboration initiated in 1938, McFadden's determination to use wide crosses in wheat breeding could be realized thanks to the methods and cytogenetic tools developed by Sears. Unfortunately, the initial efforts in "radical breeding" were inefficient due to hybrid sterility and linkages between desirable and undesirable trait loci. Nonetheless, their work not only contributed to the breeding, cytology, genetics, phylogeny, taxonomy and evolution of wheat but also opened the way for the targeted exploitation of the genetic diversity in the Triticeae tribe for wheat improvement, for which Sears coined the term "chromosome engineering" in 1972.

The species related to wheat exhibit very great genetic diversity, which can be exploited by breeders. A large proportion of wheat relatives from the Triticeae tribe can be crossed with wheat, and chromosome segments carrying genes coding for specific traits can be incorporated into wheat by means of backcrossing or the more sophisticated cytogenetic methodologies of chromosome engineering, depending on the degree of relatedness between the genomes of wheat and of the alien species concerned. To date, numerous genes, particularly for resistance to diseases, have been transferred into wheat from rye and from various species in the *Triticum, Aegilops* and *Thinopyrum* genera and other perennial Triticeae, as well as from the *Dasypyrum* species. However, considering the extant genetic diversity, only a fraction has been used, and a huge reservoir of unexploited genes and alleles for a wide range of traits remains to be tapped. Another poorly explored resource is that of existing wheat cultivars known to benefit from alien introgessions for various favourable traits, in which, however, the nature of the introgressed chromatin and the genes underlying the alien traits are still unknown.

Each new technique developed in the twentieth century gave renewed momentum to work on interspecific and intergeneric crosses. From the 1920s onwards, the elaboration of cytogenetic methods for chromosome staining enabled the chromosome number of hybrids to be determined and chromosome pairing to be traced. Improvements in in vitro tissue culture methods, including embryo rescue, enabled new, previously unsuccessful, hybrid combinations to be developed. Methods became available to control pairing between only partially homologous (homoeologous) chromosomes of wheat and of the related species, and to induce chromosome rearrangements. The Giemsa chromosome banding technique, developed in the 1970s, solved the problem of identifying individual chromosomes of wheat, rye and, later, of an increasing number of Triticeae species, enabling the presence of alien chromosome(s) in hybrids and in their progeny to be traced. The application of molecular cytogenetics (fluorescence in situ hybridization [FISH] and genomic in situ hybridization [GISH]) allowed the more accurate identification of chromosomes and chromosome segments from different species and the pinpointing of wheat-alien translocation breakpoints. The recent progress in molecular biology and genomics makes the development of impressive numbers of DNA markers possible, thus facilitating the detailed characterization of genetic diversity, the mapping of genes of interest, the identification and characterization of introgressed chromatin, and its traceability in cross progenies.

Despite these advances, alien gene transfer and introgression breeding remain empirical to a large extent, and more research is needed to allow the targeted transfer of genes and alleles of interest and the efficient production of improved wheat lines, without compromising other traits. Poor knowledge of the genome structure of wild relatives may result in unexpected and undesirable outcomes of hybridization and recombination between wheat and alien chromosomes. The mechanisms controlling meiotic recombination and the determination of the position of recombinations along chromosomes remain obscure, making some chromosome regions inaccessible to recombination. Little is known about the changes in genome structure and epigenetic changes induced by interspecific hybridization, and almost nothing is known on the function of introgressed genes and how the alien genes and wheat genome interact.

This book provides an overview of the major results achieved in the field of alien gene introgression into wheat. Separate chapters are devoted to introgressions originating from various genera of the Triticeae tribe, including *Triticum, Aegilops, Thinopyrum* and other perennial genera, as well as *Hordeum* and *Secale*. The book includes an introduction to the taxonomy of the Triticeae tribe, while other chapters discuss the evolution of wheat and related Triticeae species, as well as the current tasks facing wheat breeders. A detailed description is given of crossability genes, of genes controlling homoeologous pairing, and of gametocidal genes, while the last chapter outlines the potential of genomics to facilitate the alien introgression breeding of wheat.

We hope that this book will stimulate more intensive characterization of the genetic diversity of wild wheat relatives, their genome structure and their exploitation in wheat breeding. Wild wheat relatives and related cultivated species represent a so far poorly explored treasure trove of genes and alleles for wheat breeding. The methods developed and knowledge gained in the past, together with new molecular genetics and genomics techniques, should make the exploitation of wheat relatives more efficient and the breeding of improved wheat cultivars with novel traits a realistic goal.

Martonvásár, Hungary — Márta Molnár-Láng
Viterbo, Italy — Carla Ceoloni
Olomouc, Czech Republic — Jaroslav Doležel
July 2015

Contents

Chapter 1
Taxonomic Treatments of Triticeae and the Wheat Genus *Triticum*

Nadine Bernhardt

1.1 The Tribe Triticeae

The economically important grass tribe Triticeae Dumort. consists, depending on taxonomic treatment, of about 360 species and several subspecies in 20–30 genera, adding up to 500 taxa. Triticeae occur in temperate regions all over the Earth and harbor the important cereals, bread wheat (*Triticum aestivum*), barley (*Hordeum vulgare*), rye (*Secale cereale*) and their wild relatives, many forage grasses, species crucial for soil stabilization, and important elements of diverse plant communities. Morphologically they are characterized by the possession of open leaf sheath margins on culm leaves, membranous ligules, inflorescences having sessile spikelets, and ovaries with a hairy top. The tribe comprises annual as well as perennial species, with the annual members being most abundant in Western Asia and the Mediterranean, while the highest species diversity for the perennial members is reported in Central (and East) Asia, particularly China (Seberg and Frederiksen 2001; Barkworth and von Bothmer 2009). There are self-pollinating as well as cross-pollinating species (Escobar et al. 2010). Within Triticeae, all taxa have a chromosome base number of $x=7$ and 2n chromosome numbers are 14 or multiples of 14. Supernumerary or B chromosomes are reported in only a few genera like *Secale* and seem to be a result of major karyotype reorganizations (Martis et al. 2012). The genomes vary in size, but are generally very large. For instance, the genome size of *Triticum aestivum* amounts to more than 100 times of the genome size of *Arabidopsis thaliana*.

The tribe is of interest to many areas of research, including archaeology, domestication research, crop breeding, and evolutionary biology. Approximately 12,000 years ago, barley (*Hordeum vulgare*), einkorn wheat (*Triticum monococcum*), and emmer wheat (*Triticum turgidum*) were domesticated in the Fertile Crescent in

N. Bernhardt (✉)
Leibniz Institute of Plant Genetics and Crop Plant Research (IPK), Taxonomy & Evolutionary Biology, Corrensstr. 3, D-06466 Gatersleben, Germany
e-mail: bernhardt@ipk-gatersleben.de

M. Molnár-Láng et al. (eds.), *Alien Introgression in Wheat*,
DOI 10.1007/978-3-319-23494-6_1

Southwest Asia, marking the transition of human society from hunter-gatherer to a sedentary lifestyle (Kilian et al. 2010). Triticeae species are also of research interest due to their potential for crop improvement by understanding and utilization of the genetic diversity of wild relatives to ensure global food security (Kellogg et al. 1996; Mochida and Shinozaki 2013). Furthermore, the tribe exhibits a complex evolutionary history that is not well understood and adds to its appeal for evolutionary biologists.

Approximately 80 species of the tribe are diploid, while the majority of species are allopolyploid, i.e., combining parental genomes from different diploid species or genera of the tribe. Polyploid species are likely to have originated repeatedly, involving genetically different parent species and thus resulting in genetically diverse polyploid species-complexes (Soltis and Soltis 1999; Mason-Gamer 2004; Jakob and Blattner 2010; Brassac et al. 2012). *Triticum aestivum* is the most prominent allopolyploid species, formed by recurrent hybrid speciation involving three diploid species of the tribe and thereby combing three related genomes (named **A**, **B**, and **D**) (Petersen et al. 2006; Marcussen et al. 2014). In Triticeae and in many other crops, such genomes were defined through cytogenetic characterization of chromosomes together with the analysis of their pairing behavior in interspecific and intergeneric crosses. In comparison to other grass tribes, Triticeae show a low barrier against hybridization and other introgressive events (i.e., gene flow between different species). This, together with large-scale and small-scale genome duplications and incomplete lineage sorting of ancestral polymorphisms, results in diverse genealogical histories for different parts of the genome.

The tribe Triticeae belongs to the Poaceae family and the subfamily of Pooideae. It is generally accepted to be monophyletic (Watson et al. 1985; Kellogg 1989; Soreng et al. 1990; Seberg and Frederiksen 2001), this means all taxa of the tribe are derived from a most recent common ancestor. However, the taxonomic treatment of the tribe's members continues to be under long-standing debate. Because of its economic importance together with the worldwide distribution, it was always of interest to many researchers. Different opinions of taxonomists about the relationships of taxa, as well as changing methods and perceptions for good classificatory systems, resulted in various taxonomic treatments that in several cases led to various correct scientific names for the same taxon. The circumscription of several of the tribe's genera is also under debate (e.g., to include the genus *Aegilops* into *Triticum* or to split it into several genera). The most recent description of Triticeae genera and a discussion of their level of acceptance can be found in Barkworth and von Bothmer (2009), and is summarized in Table 1.1. Today, there is common consent that a good classification of a taxonomic unit like Triticeae starts with a thorough evaluation of phylogenetic relationships, based on analyses of different portions of the genome and including taxa of all genera or genomic groups and thereby covering their whole distribution area (Barkworth 2000; Mason-Gamer 2005; Petersen et al. 2006). So far, there is no study available that resolves the tribe's evolutionary history in such a manner that it can be used as a basis for a conclusive taxonomic treatment. Published studies largely agree, however, that *Bromus,* the only genus in the tribe Bromeae, is the sister group to all Triticeae (Kellogg 1992; Schneider et al. 2009)

Table 1.1 List of genera and genomes in Triticeae (modified from Barkworth and von Bothmer 2009)

Genus	Genomic composition
Aegilops L.	**B, C, D, N, M, U, X, BU, CU, CD, DN, DM, MU, DDM, DBM, DMU**
Agropyron Gaertn.	**P**
Amblyopyrum (Jaub. & Spach) Eig	**T**
Anthosachne Steud.	**StHW**
Australopyrum (Tzvelev) A.Löve	**W**
Crithopsis Jaub. & Spach	**K**
Dasypyrum (Coss. & Durieu) T.Durand	**V**
Douglasdeweya C.Yen, J.L.Yang, & B.R.Baum	**StP**
Elymus L.	**St** plus at least one of **H, W, Y**
*Elytrigia** Desv.	**St, E, H, N**
Eremium Seberg & Linde-Laursen	**Ns**
Eremopyrum (Ledeb.) Jaub. & Spach	**F, Xe**
Festucopsis (C.E.Hubb.) Melderis	**L**
Henrardia C.E.Hubb.	**O**
Heteranthelium Hochst.	**Q**
Hordelymus (Jess.) Harz	**Ns**
Hordeum L.	**H, I, Xa, Xu, HXa**
*Hystrix** Moench	**StH or Ns**
Kengyilia C.Yen & J.L.Yang	**StPY**
Leymus Hochst.	**Ns**
Pascopyrum A.Löve	**St, H, N**
Peridictyon Seberg, Fred. & Baden	**Xp**
Psathyrostachys Nevski	**Ns**
Pseudoroegneria (Nevski) A.Löve	**St**
*Roegneria** K.Koch	**St, Y**
Secale L.	**R**
Stenostachys Turcz.	**HW**
Taeniatherum Nevski	**Ta**
Thinopyrum A.Löve incl. *Psammopyrum* A.Löve	**E** sometimes with **P, St,** or **L**
Triticum L.	**A, AB, AAB, ABD**

Generic names marked with an asterisk were not accepted by Barkworth and von Bothmer (2009) for diverse reasons: The type species of *Elytrigia* and *Hystrix* were moved to *Elymus*, hence the generic names are no longer valid following this taxonomic view. Genome designations for *Eremopyrum* follow Wang et al. (1994) and Seberg and Frederiksen (2001). *Leymus* genome designation is given following Fan et al. (2014). The genus *Roegneria* was not accepted because of e.g., a lack or morphological information needed to distinguish it from *Elymus*. *Eremium* resembles *Leymus* genomically, and some taxa do so also morphologically. *Hordelymus* is widely accepted but often included in *Leymus*. Löve's (1984) usage of the '**N**' genome for *Leymus* had been replaced by '**Ns**'. The '**N**' genome in *Aegilops* was designated '**L**' by Löve (Yen and Yang 2009). All known genome combinations for *Aegilops* and *Triticum* are given only in alphabetical order. The **B** genome in *Aegilops* might be referred to **S** in other treatments. The **B** genome in *Triticum* might also be treated as the two different genomes **B** and **G** and that the genera *Psathyrostachys* and *Hordeum* diverged early on from the rest of Triticeae. *Aegilops* and *Triticum* are closely related and are of rather recent origin (Kellogg et al. 1996; Petersen and Seberg 1997; Mason-Gamer and Kellogg 2000; Escobar et al. 2011). Furthermore, it can be assumed that diploid species and monogenomic taxa (i.e., taxa possessing a single genome type in diploids and polyploids) are the basic units within Triticeae and that heterogenomic polyploids (mostly intergeneric allopolyploidsthat combine different genomes in various combinations) form a second level of taxonomic entities (Kellogg 1989; Seberg and Frederiksen 2001).

The aim of this chapter is to give a short review on the important taxonomic treatments of Triticeae through time (and in more detail on *Triticum*), thereby providing guidance through the multitude of classificatory systems.

1.2 Short Introduction to Taxonomy, Phylogenetics, and Nomenclature

Taxonomy in a broad sense is the theory and practice of identifying, describing, and classifying organisms into a hierarchical system of taxa (e.g., genera, species, subspecies), including phylogenetics that uncovers evolutionary relationships among organisms, and nomenclature, which is how to correctly name an organism. Taxonomy is not static and its treatments might change whenever new knowledge is gained that more appropriately reflects the evolutionary history of taxa. The rational behind the objective to anchor taxonomy in the evolutionary history of the organisms is the assumption that this, in the long-term, will result in a stable system, as all taxa should have originated through a single evolutionary process. There is general agreement that taxonomic decisions should therefore be made by integrating all available knowledge from different scientific disciplines including morphological, cytogenetic, and molecular data. This combination of conceptual and methodological achievements was described as integrative taxonomy (Padial et al. 2010), but there are no formal rules to follow on how to classify tribes into lower-ranked taxa (Barkworth and von Bothmer 2009). Taxonomist working on the same group of taxa may disagree on which kind of data, and to what extent that data, should be taken into account for taxonomic decisions, and if few large or many smaller taxonomic units should be preferred. Taxonomists may even differ in their opinion regarding what a species actually is. There are several contemporary and partially conflicting species concepts. Some might prioritize the phylogenetic species concept in that a species is the smallest unit for which a monophyly (i.e., a clade comprising an ancestor and all of its descendants, which is inferred by the possession of shared derived characteristics, so called synapomorphies) can be found (Rosen 1979; Cracraft 1983). Recently, de Queiroz (2007) proposed a unifying species concept, based on the common denominator of all concepts. He stated that species are evolutionary lineages independent of other such lineages and that, for instance, crossing behavior, ecology, and phylogeny are tools to infer such independence of lineages.

Although the main objective of the International Code of Nomenclature for algae, fungi, and plant (ICN; McNeill et al. 2012) is to ensure that there is only one correct scientific name for every taxon, the existence of several different and incompatible taxonomic treatments result in the existence of many formally correct names in use at the same time. This is especially true for taxa within Triticeae, since a reliable system of evolutionary relationships that can be used as a base for decision-making is still lacking. Moreover, the priority rule of the ICN states "for any taxon from family to genus inclusive, the correct name is the earliest legitimate one with the same rank." Although intended to result in taxonomic stability, it may lead to changes of names even for taxonomic ranks not under debate if older legitimate names are detected. A striking example in wheat grasses is the recent confusion about the tribe's name: Reveal (2004, 2011) stated that Martinov designated the name Hordeeae (as "Hordeaceae") at the rank of tribe in 1820, implicating that Triticeae would be a younger and therefore illegitimate name given by Dumortier in 1824. This resulted in a gradual renaming of the tribe from Triticeae into Hordeeae till a recent survey by Welker et al. (2014) revealed a misreading of Martinov's work, as in it no specific taxonomic rank was assigned to Hordeeae. Hence, Triticeae Dumortier (1824) remains the valid name for the tribe.

1.3 Taxonomic Treatments in Triticeae from Linnaeus (1753) to Tzvelev (1976)

Since the beginning of the Triticeae taxonomy, many considerably different classificatory systems have been proposed by diverse authors. The treatments reflect different aims of taxonomists, the state of the art of classificatory concepts, employed methods, and recognized taxa of the time in which they were published. When Linnaeus published his *Species Plantarum* in 1753, it was the start of modern biological (binomial) nomenclature, providing the names for plant classifications. Early taxonomists like him aimed for a classification that allowed taxa to be easily recognized morphologically. They grouped species being similar to each other and different to others into genera, thereby following a typological taxonomic concept (Linnaeus 1753; Bentham 1882).

With the development of new species concepts and analysis techniques, taxonomists aimed for their classification to reflect the evolutionary history of the tribe. Nevski (1934) was the first to propose a classification for Triticeae that reflects the tribe's phylogeny. Thus it was largely different from earlier generic treatments (e.g., Bentham 1882) in the number of accepted genera and their description. Nevski was well aware of the fact that morphological traits might evolve independently and hence, might not necessarily reflect evolution correctly. In addition to morphological data, he included phytogeographic and cytogenetic data in his proposal. Noteworthy is the taxon sampling, which covered the center of diversity of perennial Triticeae. His work was adopted in the Flora of the USSR, but its application in the West was delayed by several years. Taxonomic treatments in North America were mainly based on Hitchcock (1935, 1951), which in turn were based on Bentham (1882) and Hackel (1887). On the basis of observations made studying crossing-relationships in the tribe, Stebbins (1956) argued that all taxa of the tribe might be subsumed into a single genus *Triticum*. His proposal was made under the influence of the biological species concept (Mayr 1942) that defines a species as members of populations that can produce fertile offspring, and that are reproductively isolated from other such populations. Although a perfectly defensible proposal, it was not adopted and was considered impractical to describe the diverse Triticeae. The treatment of the Russian taxonomist Tzvelev (1976) considered several additional annual and perennial genera in comparison to Hitchcock (1951) and was quickly adopted in the USSR (for a review see Barkworth 2000).

1.4 Löve and Dewey's Genomic System of Generic Classification (1984)

Until the 1980's, several classificatory systems for members of the Triticeae had been proposed. They differed in goals, taxonomic and regional scope, and were adopted differently in different parts of the world. Furthermore, since the start of genome analysis (Kihara 1930), a large amount of cytogenetic data had been accumulated. To overcome existing taxonomical inconsistencies, Löve (1984) and Dewey (1984) independently proposed a generic classification for Triticeae that was solely based on the recognition of so called *genomes* (i.e., the monoploid set of chromosomes). They analyzed the pairing behavior of homologous chromosomes in metaphase I of meiosis in interspecific hybrids and recorded the number of chiasmata in a particular number of cells (often 50). Homologous chromosomes showing near complete pairing are defined as belonging to the same genome. Both authors considered hybrids sharing the same genome or genome combination to be congeneric, i.e., to be placed in the same genus. In contrast, hybrids showing little or no pairing of homologous chromosomes have different genomes and were considered intergeneric. Distinct genomes were depicted by different bold uppercase letters, while genomes considered related but not identical are depicted with the same capital letter, extended by lowercase letters or a superscript extension (Baum et al. 1987).

Dewey's revision treated perennial Triticeae, while Löve's work was based on the whole tribe. Thereby, Löve summarized extremely valuable information on all synonyms he was aware of, together with their original citation, including chromosome number and genomic constitution. He also provided descriptions for all genera and subgenera (Barkworth and von Bothmer 2005). Löve recognized 37 genera, 13 of which are treated as *Aegilops* by most taxonomists today. He defined one genus per genome combination (**A**, **AB**, **ABD**) or ploidy level, respectively, for what is unified in *Triticum* s.str. today (van Slageren 1994). Although Löve was not completely consistent (e.g., *Elymus* consisted of three different genome constitutions; Dewey 1984) and genus affiliations were partly inferred from morphological similarities (as at that time genome composition was not analyzed for all taxa), this treatment was the start of the modern taxonomy of Triticeae (Barkworth and von Bothmer 2009).

1.5 Criticism on the Genomic System

The genomic concept was sharply criticized, (1) mainly because the pairing behavior of homologous chromosomes was interpreted as single-character classification by most taxonomists. Löve and Dewey, themselves interpreted pairing behavior as an indication of overall genetic similarity along the chromosomes, and therefore as a result of multiple characters (Barkworth and von Bothmer 2005). Severe disapproval was also mentioned, when (2) interpreting the extent of chromosomal pairing as an indicator for similarity between the genomes. This was considered a random division of a continuum, and therefore no reliable taxonomic trait. Further criticism (3) was based on the fact that the failure of chromosomes to pair does not necessarily indicate dissimilarity. At this time, it was already known that a number of loci control the pairing behavior and recombination of chromosomes. It was suggested that 50 loci might have an impact on the chromosomal pairing in *Pisum* (Gottschalk 1973; Farooq et al. 1990; Seberg and Petersen 1998). Today, *Ph1* is the most extensively studied locus involved in this trait (Moore 2009). Additionally, (4) the system was considered overall unstable, meaning each time new genomes or combinations of them are recognized, species need to be transferred between genera, and (5) impractical in the field, when each time a new species is discovered, an appropriate name can only be given after the genome compositions have been identified (Baum et al. 1987; Kellogg 1989; Frederiksen and Petersen 1998; Barkworth 2000; Barkworth and von Bothmer 2005).

1.6 Impact of the Genomic System

After publication of the genomic system of generic classification, taxonomists split into two different directions: On the one hand, the group rejecting the genomic concept, due to the lack of complete understanding of pairing behavior in meiosis, and on the other hand, the group acknowledging the additional information it provides, e.g., considering the possible convergent evolution of morphological and physiological traits. In any case, Löve's revision was apprehended as a valuable and comprehensive collection of nomenclature names of members of Triticeae, and the information about pairing behavior of chromosomes, i.e., information about the potential of species to exchange genetic information, which is of use to plant breeders.

Although criticized heavily, the genomic system contributed in the clarification and understanding of evolutionary relationships within the tribe. Kellogg (1989) argued that the genomic designations are good indicators of phylogenetic relatedness, since Löve treated them as discrete character, i.e., as an "all-or-nothing phenomenon" of complete pairing versus nearly complete failure to pair. However, genome data should never be used solely, but combined with additional data from other disciplines to overcome differences in genetic regulation of chromosome pairing found among some taxa. In this manner, species groups that share a single genome and are also morphologically uniform can be considered monophyletic, and in several cases were found to belong to such groups through later phylogenetic analyses (e.g., Blattner 2004; Petersen et al. 2004; Blattner 2009). In contrast, heterogenomic species (i.e., species that combine two or more different genomes) are clearly the result of hybridization. Therefore, they are in conflict with the monophyly criterion of the phylogenetic species concept as more than one parental clade was involved in their formation. Heterogenomic taxa can, however, be treated as units of their own in plant groups with a high extent of hybrid formation like Triticeae. Heterogenomic taxa can be included as reticulations in cladistic analyses, for review see Kellogg (1989). Löve and Dewey's work showed that evolutionary relationships are not necessarily strictly based on bifurcating ancestral lineages and hence, cannot be covered by a classificatory system that allows only for hierarchical relationships.

1.7 Taxonomic Treatments of the Genus *Triticum*

There are two different interpretations of the wheat genus "*Triticum*". Depending on the taxonomic treatment under use, *Triticum* either also includes taxa that are otherwise described under *Amblyopyrum* and *Aegilops* (i.e., *Triticum* s.l.) or is restricted to species having the **A**genome or one of the genome combinations **AB**, **AAB**, or **ABD** (Table 1.1; Barkworth and von Bothmer 2009). In the past, *Aegilops* and *Triticum* were rarely subsumed into one genus (Bowden 1959; Dvořák and Zhang 1992; Yen et al. 2005) as proposed by Stebbins (1956). However, treating them separately makes *Aegilops* paraphyletic, as not all descending lineages are included in the same genus, and *Triticum* polyphyletic, since *Aegilops* played a key role in the formation of tetraploid and hexaploid wheat. Thus, *Aegilops tauschii* contributed the **D**genome to *T. aestivum,* the **B** (also referred to as **S** in the parental species, or **G** in other polyploid species) genome stems from *Aegilops speltoides* (Petersen et al. 2006). Agreement on the congeneric status of *Aegilops* and *Triticum* would solve this nomenclatural problem. But there is a long tradition in keeping both genera separate because of their clear morphological distinctions. Of 153 inspected floristic treatments that were published between 1753 and 1994, approximately 86% considered them as different genera (van Slageren 1994). In the following parts, I adopt the traditional view and treat *Triticum* as separate from *Aegilops* (also following Barkworth and von Bothmer 2009).

1.8 Pros and Cons of Available *Triticum* Treatments: Dorofeev et al. vs. Mac Key/van Slageren

On top of two different interpretations of the genus *Triticum* (i.e., keeping *Aegilops* and *Triticum* separate vs. their unification in one genus), there exist different and incongruent taxonomic treatments for *Triticum* s.str. leading to confusion for the international research community. A good classificatory system for *Triticum* seems pivotal not only to unravel its origin but also to efficiently preserve its morphological and genetic diversity. It was argued that, therefore, an especially detailed taxonomic system is needed that also meets the needs for future wheat breeding (Goncharov 2002; Goncharov 2011). Additionally, the lack of consistent rules of how to treat natural and artificial hybrids resulted in many unnamed artificial hybrids as well as hybrids with invalid names (Goncharov 2011).

There are two main opposing classificatory systems, and both take into account genomic data. On the one hand, there is the latest monograph from Dorofeev et al. (1979) that is based on a very fine discrimination of morphological characters, but was also influenced by Flaksberger's (1935) treatment founded on the division of taxa into groups by ploidy level (diploids, tetraploids, and hexaploids). Dorofeev et al. recognized 27 *Triticum* species in two subgenera, and no less than 1054 infraspecific taxa. The subgenera are divided by the possessionof different version of the Agenome ($\mathbf{A}^b$ vs. $\mathbf{A}^u$, Table 1.2). Species rank was given also to cultivated wheat. This also accounts for cultivated wheat with only relict or locally restricted importance, e.g., *T. compactum*, *T. macha*, and *T. spelta* (van Slageren 1994). Although well known, since it was written in Russian, the monograph was of limited access to the global research community. A translation into English is expected to become available soon (pers. comm. H. Knüpffer, Gatersleben).

On the other hand, Mac Key (1966, 1977, 1989, 2005) formulated his concept based on classical (Mendelian) genetics, i.e., observations made on a small number of genes that play a role in the development of distinct morphological characters in wheat (Goncharov 2011). In his most recent treatment, Mac Key (2005) recognized ten species and 20 infraspecific taxa in four subdivisions. He defined the subdivision *Triticosecale* for three intergeneric hybrid species of *Triticum* and *Secale*. The treatment was acknowledged as an open system that allows adjustment (van Slageren 1994) or as a "plausible and workable concept" (Hammer et al. 2011). But it was also criticized because of its simplicity. It was argued that it tends to overlook and thereby exclude morphologically distinct entities from genetic resources (Goncharov 2011; Hammer et al. 2011). Nevertheless, treatments based on Mac Key have been widely accepted (e.g., Petersen et al. 2006; Feldman and Kislev 2007; Matsuoka 2011). A detailed comparison of the nomenclatural relationships between Dorofeev et al. (1979) and Mac Key (2005) can be found in Hammer et al. (2011). This publication also refers to other studies that applied a meticulous morphological treatment or influenced Dorofeev et al. 1979 (e.g., Körnicke 1885; Percival 1921; Vavilov 1925; Mansfeld 1951).

Table 1.2 Overview of the *Triticum* classification after Dorofeev et al. (1979)

Subgenus	Section	Species	Genome
Triticum	*Urartu* Dorof. & A.Filat.	*T. urartu* Thum. ex Gandil.	A^u
	Dicoccoides Flaksb.	*T. dicoccoides* (Koern. ex Aschers. & Graeb.) Schweif.	A^uB
		T. dicoccum (Schrank) Schuebl.	
		T. karamyschevii Nevski	
		T. ispahanicum Heslot	
	Triticum	*T. turgidum* L.	A^uB
		T. jakubzineri Udacz. & Schachm.	
		T. durum Desf.	
		T. turanicum Jakubz.	
		T. polonicum L.	
		T. aethiopicum Jakubz.	
		T. carthlicum Nevski (syn. *T. persicum* Vav.)	
		T. macha Dekapr. & Menabde	A^uBD
		T. spelta L.	
		T. vavilovii (Thum.) Jakubz.	
		T. compactum Host	
		T. aestivum L.	
		T. sphaerococcum Perciv.	
		T. petropavlovskyi Udacz. & Migusch.	
Boeoticum	*Monococcon* Dumort.	*T. boeoticum* Boiss.	A^b
Migusch. & Dorof.		*T. monococcum* L.	
		T. sinskajae A.Filat. & Kurk.	
	Timopheevii A.Filat. & Dorof.	*T. araraticum* Jakubz.	A^bG
		T. timopheevii (Zhuk.) Zhuk.	
		T. zhukovskyi Menabde & Erizjan	A^bA^bG
		T. militinae Zhuk. & Migusch.	A^bG
	Kiharae Dorof. & Migusch.	*T. kiharae* Dorof. & Migusch.	A^bGD

Species were divided into subgenera by the possession of different variants of the **A** genome (**A^u** vs. **A^b**)

Van Slageren (1994) followed Mac Key (1977, 1989) in that taxa at the species and subspecies level should only contain “commercially” cultivated wheat and their closest wild relatives. Artificial autoploids or amphiploids should be included in a nothogenus and excluded from *Triticum*; the same accounts for interspecific crosses that should be subsumed into nothospecies. He formulated three sections, a total of six species, and 17 subspecies. Goncharov et al. (2009) published the most recent taxonomic revision for *Triticum*. They mainly followed the principles of Dorofeev

Table 1.3 Revision of Mac Key's *Triticum* treatment by van Slageren (1994)

Section	Species	Subspecies	Genome
Monococcon	*T. monococcum* L.	subsp. *monococcum*	A
Dumort.		subsp. *aegilopoides* (Link) Thell.	
	T. urartu Tumanian ex Gandilyan		A
Dicoccoidea	*T. turgidum* L.	subsp. *turgidum*	BA
Flaksb.		subsp. *carthlicum* (Nevski) A.Löve & D.Löve	
		subsp. *dicoccum* (Schrank ex Schübl.) Thell.	
		subsp. *durum* (Desf.) Husn.	
		subsp. *paleocolchicum* (Menabde) A.Löve & D.Löve	
		subsp. *polonicum* (L.) Thell.	
		subsp. *turanicum* (Jakubz.) A.Löve & D.Löve	
		subsp. *dicoccoides* (Körn. ex Asch. & Graebn.) Thell.	
	T. timopheevii	subsp. *timopheevii*	GA
	(Zhuk.) Zhuk.	subsp. *armeniacum* (Jakubz.) van Slageren	
Triticum	*T. aestivum* L.	subsp. *aestivum*	BAD
		subsp. *compactum* (Host) Mac Key	
		subsp. *macha* (Dekapr. & Menabde) Mac Key	
		subsp. *spelta* (L.) Thell.	
		subsp. *sphaerococcum* (Percival) Mac Key	
	T. zhukovskyi Menabde & Ericz.		GAA

Genome designations are given as "female parent x male parent". Subspecies are arranged by autonym first, followed by cultivated ones in alphabetical order, followed by wild forms. The usage of *T. turgidum* subsp. *dicoccum* (Schrank ex Schübl.) Thell. was chosen over T. *turgidum* subsp. *dicoccon* (Schrank) Thell. following van Slageren and Payne (2013)

et al. (1979), but also included some molecular genetic data (Golovnina et al. 2007, 2009). Goncharov et al. (2009) recognized 29 *Triticum* species in five sections and criticized the recognition of *T. kiharae* by Dorofeev et al. (1979) and Mac Key (2005), as the only human-made wheat. Therefore, the section "*Compositum*" was erected for all possible genome combinations of artificial amphiploids, which enables their effective collection and preservation. Van Slageren (1994, 2013), Mac Key (2005), and Goncharov (2011) disagreed on whether or not, or if so at which point, single spontaneous or induced mutations should be given species rank. Hence, they disagreed on the acceptance of some species described by Dorofeev et al. (1979) like e.g., *T. sinskajae*. Tables 1.2 and 1.3 depict the taxonomic treatments for

Triticum s.str. according to Dorofeev et al. (1979) and the revised treatment of Mac Key by van Slageren (1994). In this work van Slagerens revision was chosen over Mac Key (2005) since it is the most consistent and most used treatment (e.g., Matsuoka 2011; Feldman and Levy 2012; Zohary et al. 2012).

1.9 The Era of Molecular Phylogenies: What Have We Learned?

The taxonomy within the tribe continued to evolve, particularly as improved phylogenetic analyses started to shed light on species relationships. Different kinds of molecular data were generated using isozymes (McIntyre 1988; Jarvie and Barkworth 1990), DNA/DNA hybridizations, and restriction-site analyses (Ogihara and Tsunewaki 1988; Monte et al. 1993; Mason-Gamer and Kellogg 1996). In particular, DNA sequencing resulted in fast advances in molecular phylogenies within the tribe. In the last few decades, a multitude of molecular markers have been employed using coding and non-coding, nuclear and plastid, single-copy or low-copy number, and repetitive loci (Hsiao et al. 1995; Kellogg et al. 1996; de Bustos and Jouve 2002; Mason-Gamer et al. 2002; Blattner 2004; Petersen et al. 2006; Golovnina et al. 2007; Mason-Gamer 2008; Jakob and Blattner 2010; Petersen et al. 2011; Bordbar et al. 2011; Yan et al. 2011; Brassac et al. 2012; Yan and Sun 2012; Mason-Gamer 2013). Studies have mainly aimed for a better understanding within single genera (e.g., *Hordeum*, *Elymus*) of the tribe and/or a small subset of related genera with the main focus being *Hordeum*, *Pseudoroegneria,* and *Elymus*, or the *Aegilops-Triticum* complex. Petersen et al. (2006) used two single-copy nuclear genes and one plastid marker including tetraploid and hexaploids *Aegilops* and *Triticum* species, as well as species that represent all diploid genomes of Triticeae. The comparison to other molecular phylogenies was severely hampered by the differences in taxon sampling among studies. Petersen et al. (2006) suggested that neither *Aegilops* and *Triticum* separately, nor the combination of both genera, can be assumed monophyletic.

Generally, the comparison of studies dealing with the same genus rarely revealed similar phylogenetic relationships. Possible reasons for incongruences between different studies are (1) true but different evolutionary histories of distinct genes caused by hybridization, introgressions, incomplete lineage sorting, and/or selection, (2) differently conducted analyses, and (3) limited, biased, and different taxon sampling (Kellogg et al. 1996; Petersen et al. 2006; Blattner 2009). There is a consensus that a good taxonomic treatment of the tribe's members has to start with a robust phylogeny that is based on a combination of data from different parts of the genome (Mason-Gamer 2005), including all representative genera to provide conclusive results as well as several accessions per species to test for monophyly within species (Petersen et al. 2006). So far, only a few molecular phylogenies have been published trying to understand the tribe's history as a whole, leaving many questions unanswered. These studies considered only diploid taxa, with rarely more than one accession per species and only a small part or biased fractions of the genome (Hsiao et al. 1995; Petersen and Seberg 1997; Petersen and Seberg 2000; Petersen 2002; Escobar et al. 2011). Although not including all diploid genera, Escobar et al. (2011) have to date published the most comprehensive multigenic data set including one chloroplast and 26 nuclear loci, most of which are probably located on a single chromosome. This study resulted in the recognition of up to five major clades in diploid Triticeae and agrees with other studies, for instance in the basal positions of *Psathyrostachys* and *Hordeum* in the phylogenetic trees. Furthermore, it pinpoints clades that are particularly affected by reticulate evolution or incomplete lineage sorting (e.g., *Aegilops*, *Dasyprum*, *Secale*, and *Triticum*).

Recently, the advent of methods that retrieve sequence information from many different portions of the genome for many accessions and species group(s) under study at the same time (e.g., methods that make use of restriction-site-associated differences or aim for the targeted enrichment of genomic parts; for review see Lemmon and Lemmon 2013) coupled with high-throughput sequencing platforms (e.g., Illumina or Ion Torrent) allow new analysis approaches (Brassac and Blattner 2015). This offers new possibilities for a taxonomically complex group, like that of the grass tribe Triticeae, and will hopefully result in more robust phylogenetic data to be used for taxonomic decision-making in the near future.

1.10 Triticeae Taxonomy: Is It Intractable?

After several decades of intense research into Triticeae taxonomy, there is still no general agreement on a good classificatory system for the tribe. The reasons for this are multifaceted. Once a good understanding for the evolutionary history among monogenomic units is obtained, this information can be used to unravel the relationships among heterogenomic taxa. A conclusive taxonomic treatment for monogenomic and heterogenomic taxa should be based on integrating all available data from different science disciplines as well as sequence information from different portions of the genome for a representative sampling of taxa. Though still quite cost-intensive, newly available DNA sequencing techniques allow for the generation of such data. Nevertheless, the handling and interpretation of this data will be very demanding due to the highly reticulate evolutionary processes in Triticeae together with the presence of paralogous loci caused by large- and small-scale duplications within genomes. Many incongruent gene trees, that depict the truly different histories of the genomic fractions under study, are to be expected. An average phylogeny, true for most parts of the genomes, and associated alternative phylogenies, will be necessary to represent relationships within the tribe. New insights into the tribe's evolutionary history can mark the starting point for a generally accepted taxonomic treatment. Still, assuming all taxonomists working on the tribe could be convinced to follow the same treatment, the achievement of reaching an agreement on a single accepted name for a certain taxon would be extremely laborious. It requires extensive and careful literature search to ensure that only valid names (in agreement with the priority rule of the ICN) will be given to taxa. However, once established, a single and conclusive taxonomic treatment that is generally used would greatly simplify the understanding of researchers working on Triticeae taxa. The same is true for the usage of the genomic symbols. Generally, genome designations offer an efficient and abbreviated way to depict relationships between allopolyploids and their progenitors. For this reason, they are used in nearly every study dealing with polyploids in Triticeae. Nonetheless, since Löve's (1984) revision of the tribe, several suggestions have already been made to rename some of the genomes (Table 1.1). Often, these suggestions have been followed by only some scientists, which led to another level of confusion. Depending on the publication at hand, it can be difficult to track back the change of a genome name and the reasons for it.

With respect to the genus *Triticum*, first, a consensus on its interpretation (in a broad or narrow sense) is desirable. The separate generic status of *Aegilops* and *Triticum* s.str. can be favored because of clear morphological distinctions and the long tradition in keeping the taxa separate (Barkworth and von Bothmer 2009). Additionally, there are taxonomists that suggest the general acceptance of paraphyletic taxa (e.g., Brummitt 2006), as this reflects the course of evolutionary history. The situation within *Triticum* s.str. is more complex, since two different taxonomic treatments are adopted in different parts of the world. The treatment of Dorofeev et al. (1979) is mainly used in Eastern countries, whereas Mac Key (1977, 1989) or the revision made by van Slageren (1994) are used in Western countries. Hence, there is no long tradition for one or the other of the treatments. Although the two main classificatory systems consider genomic information, it is then partially interpreted in different ways (i.e., the grouping of species into subgenera depending on the possession of the $\mathbf{A}^b$ or $\mathbf{A}^u$genome by Dorofeev et al. (1979). Both treatments are mainly based on different data and differ in the acceptance of the species rank for "commercially" cultivated and artificial hybrid wheat. So far, there is no treatment that sufficiently incorporates molecular data. Finally, a taxonomic treatment for *Triticum* (s.l. or s.str.) needs an open system offering a flexible solution to cope with new taxa that need to be included because of, for instance, success in breeding or the retention of new spontaneous or induced mutations (e.g., van Slageren 1994). This seems essential to establish a stable taxonomic treatment for the genus. Moreover, a classificatory system that reflects the progenitors of a taxon already in its name is both self-informative and easy to learn. To avoid loss of information related to biodiversity and genetic resources, morphological distinctions can be made under lower taxonomic ranks, e.g., varieties in case of wild taxa. In case of species or lower taxonomic rank brought into cultivation, a cultivar name can be given according to the rules of the International Code of Nomenclature for Cultivated Plants (ICNCP, Article 20; Brickell et al. 2009). The naming of hybrid taxa is also regulated by the ICN (Appendix I; McNeill et al. 2012).

However, a future classificatory system for Triticeae and/or *Triticum* can only decrease current confusion in the taxonomy, if it experiences general attention and acceptance by scientists working with these taxa. Therefore, one central database could provide an easily accessible repository for valid names. The establishment of such a database is not feasible for a single or few collaborating taxonomists. The literature that needs to be considered is at least partially difficult to access and written in different languages, such as Chinese, English, German, and Russian. Taxonomists working on Triticeae would be asked to share their knowledge by completing a database like the International Plant Name Index database (www.ipni.org). This would greatly facilitate the tracking and understanding of taxonomic changes made in the past including access to all available information regarding a given taxon.

Acknowledgements I thank Frank Blattner, Jonathan Brassac, Klaus Pistrick, and Hart Poskar for valuable remarks and suggestions on the manuscript. The Deutsche Forschungsgemeinschaft (DFG) financially supported this work through grant number BL 462/10.

References

Barkworth ME (2000) Changing perceptions of the Triticeae. In: Jacobs SWL, Everett J (eds) Grasses: Systematics and evolution. CSIRO Publishing, Collingwood, VIC, pp 110–120

Barkworth ME, von Bothmer R (2005) Twenty-one years later: The impact of Löve and Dewey's genomic classification proposal. Czech J Genet Plant Breed 41:3–9

Barkworth ME, von Bothmer R (2009) Scientific names in the Triticeae. In: Feuillet C, Muehlbauer GJ (eds) Genetics and genomics of Triticeae. Springer, New York, NY, pp 3–30

Baum BR, Estes JR, Gupta PK (1987) Assessment of the genomic system of classification in the Triticeae. Am J Bot 74:1388–1395

Bentham G (1882) Notes on Gramineae. J Linn Soc Lond Bot 19:14–134. doi:10.1111/j.1095-8339.1881.tb00355.x

Blattner FR (2004) Phylogenetic analysis of *Hordeum* (Poaceae) as inferred by nuclear rDNA ITS sequences. Mol Phylogenet Evol 33:289–299. doi:10.1016/j.ympev.2004.05.012

Blattner FR (2009) Progress in phylogenetic analysis and a new infrageneric classification of the barley genus *Hordeum* (Poaceae: Triticeae). Breed Sci 59:471–480

Bordbar F, Rahiminejad MR, Saeidi H, Blattner FR (2011) Phylogeny and genetic diversity of D-genome species of *Aegilops* and *Triticum* (Triticeae, Poaceae) from Iran based on microsatellites, ITS, and *trn*L-F. Plant Syst Evol 291:117–131. doi:10.1007/s00606-010-0375-1

Bowden WM (1959) The taxonomy and nomenclature of the wheats, barleys, and ryes and their wild relatives. Can J Bot 37:657–684

Brassac J, Blattner FR (2015) Species-level phylogeny and polyploid relationships in *Hordeum* (Poaceae) inferred by next-generation sequencing and *in silico* cloning of multiple nuclear loci. Syst Biol 64:792–808. doi: 10.1093/sysbio/syv035

Brassac J, Jakob SS, Blattner FR (2012) Progenitor-derivative relationships of *Hordeum* polyploids (Poaceae, Triticeae) inferred from sequences of *TOPO6*, a nuclear low-copy gene region. PLoS ONE 7, e33808. doi:10.1371/journal.pone.0033808

Brickell CD, Alexander C, David JC et al. (2009) International code of nomenclature for cultivated plants. International Society for Horticultural Science, Korbeek-Lo

Brummitt RK (2006) Am I a bony fish? Taxon 55:268–269

Cracraft J (1983) Species concepts and speciation analysis. Curr Ornithol 1:159–187

de Bustos A, Jouve N (2002) Phylogenetic relationships of the genus *Secale* based on the characterisation of rDNA ITS sequences. Plant Syst Evol 235:147–154

de Queiroz K (2007) Species concepts and species delimitation. Syst Biol 56:879–886. doi:10.1080/10635150701701083

Dewey DR (1984) The genomic system of classification as a guide to intergeneric hybridization with the perennial Triticeae. In: Gustafson JP (ed) Gene manipulation in plant improvement. Plenum Publishing Corporation, New York, NY, pp 209–279
Dorofeev VF, Filatenko AA, Migushova EF et al. (1979) Wheat vol 1. In: Dorofeev VF, Korovina ON (eds) Flora of cultivated plants, vol 1. Kolos, Leningrad, p 346
Dvořák J, Zhang H-B (1992) Reconstruction of the phylogeny of the genus *Triticum* from variation in repeated nucleotide sequences. Theor Appl Genet 84:419–429
Escobar JS, Cenci A, Bolognini J et al. (2010) An integrative test of the dead-end hypothesis of selfing evolution in Triticeae (Poaceae): Selfing evolution in grasses. Evolution 64:2855–2872. doi:10.1111/j.1558-5646.2010.01045.x
Escobar JS, Scornavacca C, Cenci A et al. (2011) Multigenic phylogeny and analysis of tree incongruences in Triticeae (Poaceae). BMC Evol Biol 11:181
Fan X, Liu J, Sha L-N et al. (2014) Evolutionary pattern of rDNA following polyploidy in *Leymus* (Triticeae: Poaceae). Mol Phylogenet Evol 77:296–306. doi:10.1016/j.ympev.2014.04.016
Farooq S, Iqbal N, Shah TM (1990) Promotion of homoeologous chromosome pairing in hybrids of *Triticum aestivum* × *Aegilops variabilis*. Genome 33:825–828
Feldman M, Kislev ME (2007) Domestication of emmer wheat and evolution of free-threshing tetraploid wheat. Isr J Plant Sci 55:207–221. doi:10.1560/IJPS.55.3-4.207
Feldman M, Levy AA (2012) Genome evolution due to allopolyploidization in wheat. Genetics 192:763–774. doi:10.1534/genetics.112.146316
Flaksberger KA (1935) Wheat. In: Wulff EV (ed) Flora of cultivated plants. State Agric Publ Co Mosc St Petersburg, Russia, pp 19–434
Frederiksen S, Petersen G (1998) A taxonomic revision of *Secale* (Triticeae, Poaceae). Nord J Bot 18:399–420
Golovnina KA, Glushkov SA, Blinov AG et al. (2007) Molecular phylogeny of the genus *Triticum* L. Plant Syst Evol 264:195–216. doi:10.1007/s00606-006-0478-x
Golovnina KA, Kondratenko EY, Blinov AG, Goncharov NP (2009) Phylogeny of the A genomes of wild and cultivated wheat species. Russ J Genet 45:1360–1367
Goncharov NP (2002) Comparative genetics of wheats and their related species. Siberian University Press, Novosibirsk
Goncharov NP (2011) Genus *Triticum* L. taxonomy: The present and the future. Plant Syst Evol 295:1–11. doi:10.1007/s00606-011-0480-9
Goncharov NP, Golovnina KA, Kondratenko EY (2009) Taxonomy and molecular phylogeny of natural and artificial wheat species. Breed Sci 59:492–498. doi:10.1270/jsbbs.59.492
Gottschalk W (1973) The genetic control of meiosis. Genetics 74:s99
Hackel E (1887) Gramineae. In: Engler A, Prantl K (eds) Die natürlichen Pflanzenfamilien. Wilhelm Engelmann, Leipzig, pp 1–97
Hammer K, Filatenko AA, Pistrick K (2011) Taxonomic remarks on *Triticum* L. and ×*Triticosecale* Wittm. Genet Resour Crop Evol 58:3–10
Hitchcock AS (1935) Manual of the grasses of the United States. U.S. Department of Agriculture Miscellaneous Publication No. 200. U.S. Government Printing Office, Washington, DC
Hitchcock AS (1951) Manual of the grasses of the United States. Edition 2, revised by A. Chase. U.S. Department of Agriculture Miscellaneous Publication No. 200. U.S. Government Printing Office, Washington, DC
Hsiao C, Chatterton NJ, Asay KH, Jensen KB (1995) Phylogenetic relationships of the monogenomic species of the wheat tribe, Triticeae (Poaceae), inferred from nuclear rDNA (internal transcribed spacer) sequences. Genome 38:211–223. doi:10.1139/g95-026
Jakob SS, Blattner FR (2010) Two extinct diploid progenitors were involved in allopolyploid formation in the *Hordeum murinum* (Poaceae: Triticeae) taxon complex. Mol Phylogenet Evol 55:650–659. doi:10.1016/j.ympev.2009.10.021
Jarvie JK, Barkworth ME (1990) Isozyme similarity in *Thinopyrum* and its relatives (Triticeae: Gramineae). Genome 33:885–891
Kellogg EA (1989) Comments on genomic genera in the Triticeae (Poaceae). Am J Bot 76:796–805

Kellogg EA (1992) Tools for studying the chloroplast genome in the Triticeae (Gramineae): An *Eco*RI map, a diagnostic deletion, and support for *Bromus* as an outgroup. Am J Bot 79:186–197. doi:10.2307/2445107

Kellogg EA, Appels R, Mason-Gamer RJ (1996) When genes tell different stories: The diploid genera of Triticeae (Gramineae). Syst Bot 21:321–347

Kihara H (1930) Genomanalyse bei *Triticum* und *Aegilops*. Cytologia (Tokyo) 1:263–284

Kilian B, Martin W, Salamini F (2010) Genetic diversity, evolution and domestication of wheat and barley in the Fertile Crescent. In: Glaubrecht M (ed) Evolution in action. Springer, Berlin, pp 137–166

Körnicke F (1885) Die Arten und Varietäten des Getreides. In: Körnicke F, Werner H (eds) Handbuch des Getreidebaus, vol 1. Springer, Berlin

Lemmon EM, Lemmon AR (2013) High-throughput genomic data in systematics and phylogenetics. Annu Rev Ecol Evol Syst 44:99–121. doi:10.1146/annurev-ecolsys-110512-135822

Linnaeus CV (1753) Species Plantarum, vol 2: 696. Impensis Laurentii Salvii Stockholm, Stockholm

Löve A (1984) Conspectus of the Triticeae. Feddes Repert 95:425–521

Mac Key J (1966) Species relationship in *Triticum*. Hereditas 2:237–276

Mac Key J (1977) Sec. *Dicoccoides* Flaksb. of wheat, its phylogeny, diversification and subdivision. In: Proceedings of the symposium on extended availability of wheat genetic resources, Bari, pp 5–46

Mac Key J (1989) Genus *Triticum* and its systematics. In: Shumny VK (ed) Vavilov's legacy in modern biology (in Russian). Nauka, Moscow, pp 170–185

Mac Key J (2005) Wheat: Its concept, evolution and taxonomy. In: Royo C et al. (eds) Durum wheat breeding. Current approaches and future strategies, vol. 1. CRC Press, Boca Raton, pp 3–61

Mansfeld R (1951) Das morphologische System des Saatweizens, *Triticum aestivum* L. sl. Theor Appl Genet 21:41–60

Marcussen T, Sandve SR, Heier L et al. (2014) Ancient hybridizations among the ancestral genomes of bread wheat. Science 345:1250092. doi:10.1126/science.1250092

Martis MM, Klemme S, Banaei-Moghaddam AM et al. (2012) Selfish supernumerary chromosome reveals its origin as a mosaic of host genome and organellar sequences. Proc Natl Acad Sci U S A 109:13343–13346. doi:10.1073/pnas.1204237109

Mason-Gamer RJ (2004) Reticulate evolution, introgression, and intertribal gene capture in an allohexaploid grass. Syst Biol 53:25–37. doi:10.1080/10635150490424402

Mason-Gamer RJ (2005) The β-amylase genes of grasses and a phylogenetic analysis of the Triticeae (Poaceae). Am J Bot 92:1045–1058. doi:10.3732/ajb.92.6.1045

Mason-Gamer RJ (2008) Allohexaploidy, introgression, and the complex phylogenetic history of *Elymus repens* (Poaceae). Mol Phylogenet Evol 47:598–611. doi:10.1016/j.ympev.2008.02.008

Mason-Gamer RJ (2013) Phylogeny of a genomically diverse group of *Elymus* (Poaceae) allopolyploids reveals multiple levels of reticulation. PLoS ONE 8, e78449. doi:10.1371/journal.pone.0078449

Mason-Gamer RJ, Kellogg EA (1996) Testing for phylogenetic conflict among molecular data sets in the tribe Triticeae (Gramineae). Syst Biol 45:524–545. doi:10.1093/sysbio/45.4.524

Mason-Gamer RJ, Kellogg EA (2000) Phylogenetic analysis of the Triticeae using the starch synthase gene, and a preliminary analysis of some North American *Elymus* species. In: Jacobs SWL, Everett J (eds) Grasses: Systematics and evolution. CSIRO Publishing, Collingwood, VIC, pp 102–109

Mason-Gamer RJ, Orme NL, Anderson CM (2002) Phylogenetic analysis of North American *Elymus* and the monogenomic Triticeae (Poaceae) using three chloroplast DNA data sets. Genome 45:991–1002. doi:10.1139/g02-065

Matsuoka Y (2011) Evolution of polyploid *Triticum* wheats under cultivation: The role of domestication, natural hybridization and allopolyploid speciation in their diversification. Plant Cell Physiol 52:750–764. doi:10.1093/pcp/pcr018

Mayr E (1942) Systematics and the origin of species, from the viewpoint of a zoologist. Harvard University Press, Cambridge, MA
McIntyre CL (1988) Variation at isozyme loci in Triticeae. Plant Syst Evol 160:123–142
McNeill J, Barrie FR, Buck WR, et al. (2012) International Code of Nomenclature for algae, fungi, and plants (Melbourne Code). Regnum Veg 154
Mochida K, Shinozaki K (2013) Unlocking Triticeae genomics to sustainably feed the future. Plant Cell Physiol 54:1931–1950. doi:10.1093/pcp/pct163
Monte JV, McIntyre CL, Gustafson JP (1993) Analysis of phylogenetic relationships in the Triticeae tribe using RFLPs. Theor Appl Genet 86:649–655
Moore G (2009) Early stages of meiosis in wheat- and the role of *Ph1*. In: Muehlbauer GJ, Feuillet C (eds) Genetics and genomics of Triticeae. Springer, New York, NY, pp 237–252
Nevski SA (1934) Agrostological studies IV. The systematics of the tribe Hordeeae. Flora Sist. Vysshikh Rasteniin 1. Trudy Botanicheskogo Instituta Adakademii Nauk SSSR, pp 9–32
Ogihara Y, Tsunewaki K (1988) Diversity and evolution of chloroplast DNA in *Triticum* and *Aegilops* as revealed by restriction fragment analysis. Theor Appl Genet 76:321–332
Padial JM, Miralles A, De la Riva I, Vences M (2010) The integrative future of taxonomy. Front Zool 7:1–14
Percival J (1921) The wheat plant: A monograph. Duckworth, London
Petersen G (2002) Molecular evolution and phylogenetic application of *DMC1*. Mol Phylogenet Evol 22:43–50. doi:10.1006/mpev.2001.1011
Petersen G, Seberg O (1997) Phylogenetic analysis of the Triticeae (Poaceae) based on *rpo*A sequence data. Mol Phylogenet Evol 7:217–230. doi:10.1006/mpev.1996.0389
Petersen G, Seberg O (2000) Phylogenetic evidence for excision of Stowaway Miniature Inverted-Repeat Transposable Elements in Triticeae (Poaceae). Mol Biol Evol 17:1589–1596
Petersen G, Seberg O, Aagesen L, Frederiksen S (2004) An empirical test of the treatment of indels during optimization alignment based on the phylogeny of the genus *Secale* (Poaceae). Mol Phylogenet Evol 30:733–742
Petersen G, Seberg O, Yde M, Berthelsen K (2006) Phylogenetic relationships of *Triticum* and *Aegilops* and evidence for the origin of the A, B, and D genomes of common wheat (*Triticum aestivum*). Mol Phylogenet Evol 39:70–82. doi:10.1016/j.ympev.2006.01.023
Petersen G, Seberg O, Salomon B (2011) The origin of the H, St, W, and Y genomes in allotetraploid species of *Elymus* L. and *Stenostachys* Turcz. (Poaceae: Triticeae). Plant Syst Evol 291:197–210
Reveal JL (2004) Latest news on vascular plant family nomenclature: Earlier validation of certain tribal names in Poaceae, version 27 Nov 2004. http://www.plantsystematics.org/reveal/pbio/fam/NEWS.html. Accessed 12 Mar 2015
Reveal JL (2011) Indices Nominum Supragenericorum Plantarum Vascularium, version 17 Feb. 2011. http://www.plantsystematics.org /reveal/pbio/fam/famPO-PZ.html. Accessed 12 Mar 2015
Rosen DE (1979) Fishes from the uplands and intermontane basins of Guatemala: Revisionary studies and comparative geography. Bull Am Mus Natl Hist 162:267–376
Schneider J, Döring E, Hilu KW, Röser M (2009) Phylogenetic structure of the grass subfamily Pooideae based on comparison of plastid *mat*K gene-3′ *trn*K exon and nuclear ITS sequences. Taxon 405–424
Seberg O, Frederiksen S (2001) A phylogenetic analysis of the monogenomic Triticeae (Poaceae) based on morphology. Bot J Linn Soc 136:75–97
Seberg O, Petersen G (1998) A critical review of concepts and methods used in classical genome analysis. Bot Rev 64:372–417
Soltis DE, Soltis PS (1999) Polyploidy: Recurrent formation and genome evolution. Trends Ecol Evol 14:348–352. doi:10.1016/S0169-5347(99)01638-9
Soreng RJ, Davis JI, Doyle JJ (1990) A phylogenetic analysis of chloroplast DNA restriction site variation in Poaceae subfam. Pooideae. Plant Syst Evol 172:83–97

Stebbins GL (1956) Taxonomy and the evolution of genera, with special reference to the family Gramineae. Evolution 10:235–245

Tzvelev NN (1976) Zlaki SSSR Grasses of the Soviet Union. Leningr. Engl. Ed. 1983 Transl Sharma BR

van Slageren MW (1994) Wild wheats: A monograph of *Aegilops* L. and *Amblyopyrum* (Jaub. & Spach) Eig (Poaceae). Wageningen Agric. Univ. Pap.

van Slageren MW, Payne T (2013) Concepts and nomenclature of the Farro wheats, with special reference to Emmer, *Triticum turgidum* subsp. *dicoccum* (Poaceae). Kew Bull 68:477–494. doi:10.1007/s12225-013-9459-8

Vavilov N (1925) Studies on the origin of cultivated plants. Bull Appl Bot 26:1–248

Wang RRC, von Bothmer R, Dvorak J, et al. (1994) Genome symbols in the Triticeae (Poaceae). Proceedings second International Triticeae Symposium. Logan, pp 20–24

Watson L, Clifford HT, Dallwitz MJ (1985) The classification of Poaceae: Subfamilies and supertribes. Aust J Bot 33:433–484

Welker CAD, Kellogg EA, Prado J (2014) Andropogoneae versus Sacchareae (Poaceae: Panicoideae): The end of a great controversy. Taxon 63:643–646. doi:10.12705/633.5

Yan C, Sun G (2012) Multiple origins of allopolyploid wheatgrass *Elymus caninus* revealed by *RPB2*, *PepC* and *trn*D/T genes. Mol Phylogenet Evol 64:441–451. doi:10.1016/j.ympev.2012.04.017

Yan C, Sun G, Sun D (2011) Distinct origin of the Y and St genome in *Elymus* species: Evidence from the analysis of a large sample of St genome species using two nuclear genes. PLoS ONE 6, e26853. doi:10.1371/journal.pone.0026853

Yen C, Yang JL (2009) Historical review and prospect of taxonomy of tribe Triticeae Dumortier (Poaceae). Breed Sci 59:513–518

Yen C, Yang J-L, Yen Y (2005) Hitoshi Kihara, Áskell Löve and the modern genetic concept of the genera in the tribe Triticeae (Poaceae). Acta Phytotax Sin 43:82–93

Zohary D, Hopf M, Weiss E (2012) Domestication of plants in the Old World: The origin and spread of domesticated plants in Southwest Asia, Europe, and the Mediterranean Basin. Oxford University Press, Oxford

Chapter 2
Origin and Evolution of Wheat and Related Triticeae Species

Moshe Feldman and Avraham A. Levy

2.1 Introduction

Common and durum wheat are among the first plants that were domesticated during the Agricultural Revolution, about 10,000 years ago. Throughout this period of cultivation these two crops have been of supreme importance as a staple food that facilitated and sustained the development of human civilization. Hence, the origin of domesticated wheat and their evolution under domestication as well as their relationships with their wild relatives have been central questions for botanists, geneticists, agronomists, breeders, ethno-botanists and students of human civilization. This great interest has led to extensive cytogenetic, molecular, and evolutionary studies on the genetic and genomic structure of the various species of the wheat group (the genera *Aegilops, Amblyopyrum,* and *Triticum*) and on the relationships among the various wild relatives as well as between them and the domesticated wheat species.

Given that only a small number of wild genotypes were taken into cultivation, the genetic basis of domesticated wheat in the early stages of cultivation was relatively narrow representing only a fraction of the large variation that exists in their wild progenitors. Yet, during the 10,000 years of wheat cultivation the genetic basis of domesticated wheat has been broadened to some extent due to mutations and sporadic gene flow from their wild progenitors and other related species in Southwest Asia. Moreover, the tendency of traditional farmers in many parts of the world to grow a mixture of genotypes in one field (polymorphic fields) that hybridized and recombined made possible a selection of genotypes that were more desirable to the farmers. Selection pressure was thus, exerted consistently but in different direction

M. Feldman (✉) • A.A. Levy
Department of Plant and Environmental Sciences, The Weizmann Institute of Science, Rehovot, Israel
e-mail: moshe.feldman@weizmann.ac.il; Avi.levy@weizmann.ac.il

M. Molnár-Láng et al. (eds.), *Alien Introgression in Wheat*,
DOI 10.1007/978-3-319-23494-6_2

by different farmers. These efforts resulted in numerous landraces that had a better adaptation to a wider range of climatic and edaphic conditions and to diverse farming regimes. But, modern plant breeding practices have eroded the genetic basis of domesticated wheat, particularly due to the replacement in many countries of the numerous traditional varieties (landraces) by a small number of high-yielding ones (mega varieties). The relatively narrow genetic basis of domesticated wheat reduces their adaptability to abiotic stresses, increases their susceptibility to biotic pressures, and limits very much the ability to improve their performance. This boosted further the interest of wheat geneticists and breeders in the wild relatives of wheat in an attempt to exploit their broad gene pool for the improvement of domesticated wheat. All the species of the wheat group and many of the two subtribes of the Triticeae, the Triticineae and the Hordeineae, can be crossed with domesticated hexaploid and tetraploid wheat and economically important genes can be transferred to the domesticated background. To make most effective the introduction of desirable alien variation into domesticated forms it is necessary to understand the cytogenetic, molecular, genomic, and evolutionary relationships among the wild relatives as well as between them and the domesticated wheats.

2.2 The Triticeae

The wheat group (the genera *Amblyopyrum*, *Aegilops,* and *Triticum*) is classified in the tribe Triticeae Dumort of the grass family Poaceae (Gramineae). The tribe is relatively young; it diverged about 25 million years ago (MYA) from the other Poaceae (Huang et al. 2002b; Gaut 2002). It is economically the most important tribe of the family, giving rise to the domesticated cereals wheat, rye, and barley, and to several important forage grasses. Based on traditional taxonomic sense, the tribe includes a total of 19 genera, 18 described by Clayton and Renvoize (1986), and one, *Amblyopyrum* that was separated from the genus *Aegilops* by Eig (1929b) and van Slageren (1994). The Triticeae includes about 330 species (Clayton and Renvoize 1986), growing in Temperate and Arctic zones, principally in the northern hemisphere. About 250 species are perennials that are distributed mainly in Temperate-Arctic regions while the annuals, including the three major Triticeae crops, wheat, barley, and rye, are mainly distributed in the east Mediterranean and Central Asiatic regions (Table 2.1).

The tribe Triticeae makes a distinct natural group, having a characteristic spike morphology that distinguishes it from other tribes in the Poaceae. It consists of diploid and polyploid species with the basic haploid chromosome number $x=7$. The ancestral Triticeae karyotype have been proposed to have derived from the $n=12$ ancestral grass karyotype through four centromeric ancestral chromosome fusions (leading to functional monocentric neochromosomes), one fission and two telomeric ancestral chromosome fusions (Luo et al. 2009; Salse et al. 2008; Murat et al. 2014). The chromosomes of all the species are large. About 31 % of the species are diploids (or rather palaeopolyploids), 1 % triploids, 45 % tetraploids, 17 % hexaploids, 5 % octoploids, 0.2 % Decaploids, and 0.2 % dodecaploids. *Elymus* displays the larger

Table 2.1 The genera of the Triticeae

Subtribe	Genus	Number of species	Ploidy level	Type of polyploidy	Growth habit	Pollination mode	Distribution	Number of spikelets on node
Hordeineae	*Elymus* L.	~150	2*x*–12*x*	Au-Al	P	C,S	TA,M	G,S
	Hystrix Moench	9	4*x*		P	C	TA	G
	Sitanion Raf.	4	4*x* (?)		P	C	TA	G
	Leymus Hochst	~40	4*x*		P	C	TA	G
	Psathyrostachys Nevski	7	2*x*		P	C	TA	G
	Hordelymus (Jessen) Harz	1	4*x*	Au	P	C	M	G
	Hordeum L.	~40	2*x*–6*x*	Au	P,A	C,S	TA,M	G
	Taeniatherum Nevski	1	2*x*		A	S	M	G
	Crithopsis Jaub.&Spach.	1	2*x*		A	S	M	G
Trtiticineae	*Agropyron* Gaertn.	~15	2*x*–6*x*	Au-Al	P	C	M	S
	Eremopyrum (Ledeb.) Jaub.&Spach.	5	2*x*–4*x*	Al	A	S	M	S
	Heteranthelium Jaub.&Spach.	1	2*x*		A	S	M	S
	Secale L.	3[a]	2*x*		P,A	C	M	S
	Dasypyrum (Coss. & Dur.) Dur.	2	2*x*–4*x*	Au	P,A	C	M	S
	Triticum L.	6[b]	2*x*–6*x*	Al	A	S	M	S
	Amblyopyrum (Jaub.&Spach.) Eig	1[b]	2*x*		A	C	M	S
	Aegilops L.	24[b]	2*x*–6*x*	Al	A	C,S	M	S
	Henrardia C.E. Hubbard	2	2*x*		A	S (?)	M	S

Au autopolyploidy, *Al* allopolyploidy, *P* perennials, *A* annuals, *C* cross pollination, *S* self pollination, *TA* temperate-Arctic, *M* mediterranean, *G* spikelets in group, *S* solitary spikelets

[a]According to Frederiksen and Petersen (1998)

[b]According to van Slageren (1994)

series and highest level of polyploidy, from $2x$ to $12x$ (Table 2.1). The neopolyploids are of two kinds: auto- and allopolyploids. The autopolyploids can be divided further into two types: typical autopolyploids, characterized by multivalent pairing and multisomic inheritance, and bivalent-forming autopolyploids, characterized by exclusive bivalent pairing and presumably disomic inheritance. Several species of the latter group contain a lower DNA content than expected from the additive sum of the diploid parent (Eilam et al. 2009).

The taxonomy of the tribe is complicated by several special factors such as allopolyploidy and ancient and recent inter-specific and inter-generic hybridizations that are largely responsible for the blurred boundaries between species and even between genera. Being a relatively young tribe, it shows an exceptional capacity for inter-generic hybridization involving most of its members, which creates problems both in the theoretical concept of genetic rank, and in the practical construction of keys (Clayton and Renvoize 1986). It also implies a more close-knit reticulate pattern of relationships between the various genera. Most of the species, if not all, can be crossed with wheat, barley, and rye, and thus, their gene pool can serve as an important source of useful traits for the improvement of the domesticated cereals.

Because of their economic importance, species of this tribe have been subjected to intensive cytogenetic, genetic, molecular, and phylogenetic studies. Cytogenetic research has provided an extensive knowledge of the tribe's genomic constitution (Love 1982, 1984; Dewey 1982, 1984). Consequently, the generic makeup of the Triticeae has varied widely. Since there were essentially no genetic barriers between the Triticeae genera, Stebbins (1956) and Bowden (1959) recommended merging all these genera into one genus or, at least, unite *Aegilops* and *Triticum* in one genus, *Triticum*. Morris and Sears (1967) and Kimber and Feldman (1987) included the two genera in one genus, *Triticum*, since tetraploid wheat contains one genome that derived from an *Aegilops* species and hexaploid wheat contains two such genomes. On the other hand, based on the taxonomic philosophy that a system of classification should reflect phylogeny and biological relationship (Love 1982, 1984), the classification of most genera was drastically reorganized, mainly based on the genome constitution of each taxon (Love 1982, 1984; Dewey 1984; Wang et al. 1994). Hence the tribe was divided, on the genome basis, into 38 different mono-generic genera. A problem with this idea stems from the difficulty to reach agreement on what is a similar genomic constitution. Baum et al. (1987) criticized this classification by presenting a number of arguments against genomic genera. Some of these include the observation that the genomic system is in effect a single-character classification and that genomic genera are not recognizable morphological units. Baum et al. (1987) claim that the genomic system makes far too many monotypic genera; a genomic system is unstable, necessitating changes with every new genome combination recognized; and that genomes are not good taxonomical characters anyway. The genome classification, though attractive in theory, is sometimes difficult to translate into practical morphological diagnoses, i.e., there is incongruence between the genomic and morphological data. The net result of this is that the generic classification of the Triticeae is currently in a state of flux, subject to major disagreements whose outcome is still uncertain. In this article, the genera were classified according to Clayton and Renvoize (1986), who based their classification on morphological and eco-geographical characteristics (Table 2.1). Taxonomists who dealt with the classification of *Aegilops* and *Triticum*, e.g., Zhukovsky (1928), Eig (1929a), Hammer (1980), Gupta and Baum (1986), Van Slageren (1994), and others, kept the two genera separated. Cytogeneticists (e.g., Kihara 1954; Kihara et al. 1959) also kept the two genera separated.

The various genera of this tribe are classified in two subtribes; Hordeineae, the barley lineage, (nine genera) and the Triticineae, the wheat lineage, (nine genera) (Clayton and Renvoize 1986; Table 2.1; Fig. 2.1). [The genus *Brachypodium* P. Beauv, diverged approximately 32–39 MYA from the Triticeae (Bossolini et al. 2007; International *Brachypodium* Initiative 2010), is not included in this tribe (Catalán et al. 1995; Hasterok et al. 2004)].

Based on their geographical distributions, Sakamoto (1973, 1991) classified the genera into two major groups: the "Mediterranean" group and the "Arctic-Temperate" group (Table 2.2). This classification was supported by the study of Hsiao et al. (1995) and Fan et al. (2013). The Arctic-Temperate group, distributing in the Arctic-Temperate regions of the world, have evolved into many endemic species in each area. Five of the genera are perennial with two or three spikelets at each rachis node [*Elymus* (several species have solitary spikelets at each rachis node), *Histrix Hordelymus, Psathyrostachys* and *Sitanion*]. The genus *Agropyron* is an exception that has only solitary spikelets. Only *Hordeum* includes perennial and annual species. One noteworthy characteristic of this group is the extensive inter-generic and inter-specific hybridization over the whole distribution area.

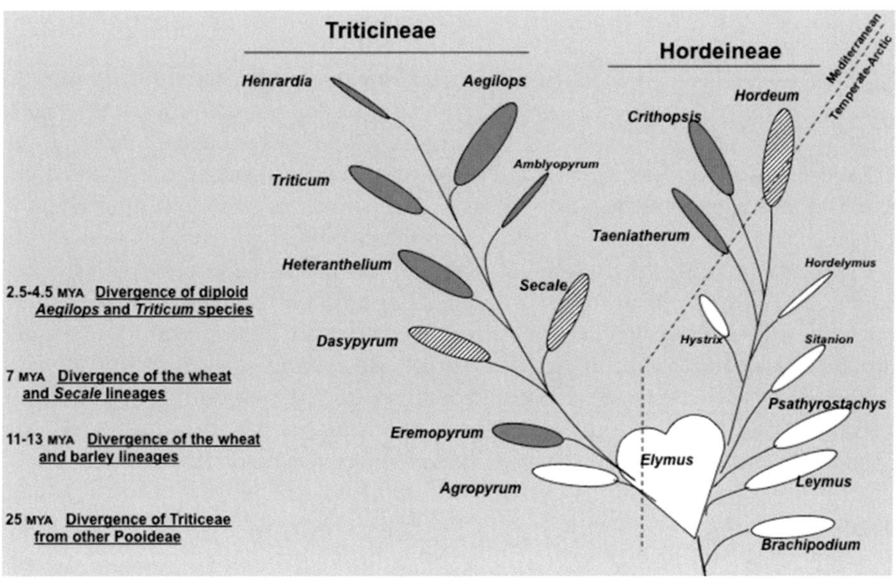

Fig 2.1 Relationships in the Triticeae (modified from Clayton and Renvoize 1986). Perennial genera = *white*; Annual genera = *gray*; Annual and perennial genera = patterned

Table 2.2 Classification of the genera of the Triticeae according to their distribution group and growth habit (after Sakamoto 1991)

	Growth habit			Rachis
Distribution group	Perennial	Perennial + annual	Annual	node with:
Arctic-temperate group (Primitive group)	*Elymus*			Spikelets in groups
	Hystrix			
	Sitanion			
	Leymus			
	Psathyrostachys			
	Hordelymus	*Hordeum*		
			Taeniatherum	
Mediterranean-Central Asiatic group (Advanced group)			*Crithopsis*	
	Elymus			
	Agropyron	*Dasypyrum*	*Heteranthelium*	Solitary spikelets
		Secale	*Eremopyrum*	
			Triticum	
			Amblyopyrum	
			Aegilops	
			Henrardia	

The Mediterranean group, distributed in the Mediterranean-Central Asiatic region, consists mainly of annual species that have solitary spikelet at each rachis node (except in the Mediterranean species of *Hordeum* that have three spikelet at each rachis node, and species of *Crithopsis* and *Taeniatherum* that have two spikelets at each rachis node). Muramatsu (2009) suggested that six genes located on homoeologous group 2 chromosomes of common wheat determine solitary spikelet per rachis node and the 2–3 spikelets per rachis node is a recessive trait. Two small genera, *Dasypyrum* and *Secale*, have both perennial and annual species. Another characteristic of this group is that each genus is morphologically distinct. *Aegilops* is the largest genus (24 species) and *Taeniatherum, Crithopsis, Heteranthelium,* and *Amblyopyrum* are monotypic. Natural inter-generic hybridization usually more restricted in this group in contrast to that in the Arctic-Temperate group, perhaps due to a larger genomic diversification that creates strong inter-generic barriers. Allopolyploidy is more prevalent in this group.

The evolutionary trends in the tribe are in spike structure (erect or nodding, spikelets per node, rachis fragile or tough, pedicels absent or present, rachilla fragile or tough, glume tip unawned or awned, glume back rounded or keeled, caryopsis free or adherent to lemma), longevity (perennialism or annualism), pollination mode (cross or self), polyploidy (autopolyploidy or allopolyploidy), geographical distribution (Temperate-Arctic or Mediterranean), habitat (open and dry or close and humid), and specialization of genetic systems.

Little is known about processes that lead to speciation in the tribe. The progenitors of the Triticeae were probably all diploid (palepolyploid), perennial, and allogamous. From an originally paniculate inflorescence the group developed a spike with three, later mostly one spikelet at each rachis node (Runemark and Heneen 1968; Sakamoto 1973). These authors recognized five steps in the development of the various Triticeae genera: (1) evolvement of diploid (paleopolyploid), cross-pollinated perennial genera in the Temperate-Arctic zones; (2) intensified autopolyploidization and allopolyploidization and geographical spread; (3) evolvement of the diploid (palepolyploid) annual genera in the hot and dry summer area of the east Mediterranean and Southwest Asia regions; (4) evolvement of self-pollinated genera; (5) intensified allopolyploidization of the self-pollinated annuals, especially in the genera *Aegilops*, *Triticum*, and *Eremopyrum*. Since all the primitive genera distribute in the Arctic-Temperate zone, this is probably the center of origin of the tribe while the Mediterranean is the center of variation.

Much of the critical diversification of the tribe, mainly at the diploid level, started during the Oligocene in the middle of the Tertiary (Fan et al. 2013; Table 2.3).

Table 2.3 Geological epochs in the Cenozoic era (65 million years ago to the present) and major climatic, ecological, and Triticeae evolutionary events

Period	Epoch	Million years ago	Major climate and ecological events	Events in the evolution of the Triticeae
Quaternary	Holocene	0.01 to present	Warmer climates; conversion of many grasslands and forests into cultivated areas; increase in human population	Domestication of wheat, barley, and rye; formation of hexaploid (bread) wheat and domesticated forms of tetraploid wheat
	Pleistocene	1.8–0.01	Global cooling; four major ice ages; most temperate zones were covered by glaciers during the cool periods and uncovered during the warmer interglacial periods	Formation of the allotetraploid species of *Triticum* and presumably also the allotetraploid and allohexaploid species of *Aegilops*
	Pliocene	5.3–1.8	Cooler and drier global climates; accumulation of ice at the poles; development seasonal climate (cold and humid winters and hot and dry summers) in the east Mediterranean and Southwest Asia; development of today's landscapes; further spread of grasslands	Evolvement of the diploid species of *Aegilops* and *Triticum*

(continued)

Table 2.3 (continued)

Period	Epoch	Million years ago	Major climate and ecological events	Events in the evolution of the Triticeae
	Miocene	23.8–5.3	Warmer global climates; towards the end, disappearance of the Tethys Sea and the climate cooled off; diversification of temperate ecosystems and new ecological niches opened; expansion of grasslands;	Diversification of grasses; divergence of the Triticinae (wheat lineage) from the Hordeinae (barley lineage)
Tertiary	Oligocene	33.7–23.8	Cold and dry climates; transformation of vegetation to something similar to that of today	Appearance of early Triticeae
	Eocene	54.8–33.7	Warm and humid climates, became cooler towards the end; forests got smaller and grasslands and savannas increased	Further development of grasses
	Paleocene	65.0–54.8	Mild and uniform climate; mammals became the dominant land-living life form; some plant forms (pines, cacti, palms) first appeared	Appearance of early grasses

The two lineages have diverged from one another during the Miocene, about 8–15 MYA (Wolfe et al. 1989; Ramakrishna et al. 2002; Huang et al. 2002a, b; Dvorak and Akhunov 2005; Chalupska et al. 2008; Fan et al. 2013; Middleton et al. 2013, 2014; Marcussen et al. 2014; Gornicki et al. 2014; Table 2.4). Based on analysis of nuclear genes, Marcussen et al. (2014) concluded that the ancestors of the wheat group have diverged from rye during the Miocene, about 7 MYA, while, on the basis of analysis of chloroplast DNA, Middleton et al. (2014) suggested that the divergence from rye was in the Pliocene, 3–4 MYA. The diploid *Triticum* and *Aegilops* species started to diverge from one another during the Pliocene, about 2 to 4 MYA (Huang et al. 2002b; Dvorak and Akhunov 2005; Middleton et al. 2014), and the allopolyploids of this group were formed between 0.01 and 1.0 MYA, during the Pleistocene and Holocene (Huang et al. 2002b; Dvorak and Akhunov 2005; Marcussen et al. 2014; Gornicki et al. 2014; Tables 2.3, 2.4, 2.7, 2.9 and 2.10; Fig. 2.1).

Fan et al. (2013 and reference therein) found a major radiation of the Triticeae during a relatively narrow period of time (6.1–9.2 MYA). It is therefore assumed that the radiation of the Triticeae might have been triggered by the late Miocene climate that was warmer in summer and cooler in winter, simultaneously with the

Table 2.4 Time of beginning divergence of the Triticeae lineages in million years ago

Lineages	Beginning divergence time	Method of study	Reference
Barley and rye-wheat	10–14	Sequencing chloroplast DNA	Wolfe et al. (1989)
	11–15	RFLP of BAC libraries	Ramakrishna et al. (2002)
	11.0	Sequencing of two nuclear genes	Huang et al. (2002a, b)
	10.1	Sequencing of four nuclear genes	Dvorak and Akhunov (2005)
	11.6	Sequencing of nuclear genes	Chalupska et al. (2008)
	8–9	Sequencing of chloroplast genome	Middleton et al. (2014)
	15	Sequencing of three chloroplast genes	Marcussen et al. (2014)
	10.6	Sequencing chloroplast DNA	Gornicki et al. (2014)
Rye and wheat	3–4	Sequencing of chloroplast genome	Middleton et al. (2014)
	7	Sequencing of three chloroplast genes	Marcussen et al. (2014)
Wheat and *Aegilops*	2.5–4.5	Sequencing of two nuclear genes	Huang et al. (2002b)
	2.7 (1.44.1)	Sequencing of four nuclear genes	Dvorak and Akhunov (2005)
	2.1–2.9	Sequencing of chloroplast genome	Middleton et al. (2014)
	6.5	Sequencing of three chloroplast genes	Marcussen et al. (2014)

disappearance of the Tethys Sea and the elevation of the east Mediterranean region; at that geological time temperate ecosystems diversified and new ecological niches opened (Table 2.3). Diversification in the Mediterranean lineage of the Triticeae not only stimulated the formation of many genera but also provided the opportunity for the production of many allopolyploids (Fan et al. 2013).

Recent molecular studies of chloroplast DNA and nuclear genes have contributed to a better understanding of the Triticeae phylogeny. However, there is no consensus concerning the phylogenetic relationships between the various genera of the tribe due to a limited number of studied samples (Kellogg and Appels 1995; Kellogg et al. 1996; Mason-Gamer and Kellogg 1996; Escobar et al. 2011) or a small number of genes that were analyzed (Petersen and Seberg 1997; Hsiao et al. 1995; Kellogg and Appels 1995; Helfgott and Mason-Gamer 2004; Mason-Gamer 2005). Intergeneric hybridizations and introgression events as well as incomplete sorting of ancestral polymorphisms (Mason-Gamer 2005; Kawahara 2009; Escobar et al. 2011) may also contribute to the ambiguity of the phylogenetic relationships.

The eroding genetic basis of domesticated wheat have boosted further the motivation of scientiss to explore the possibility of utilizing the gene pool of the various Triticeae species (Feldman and Sears 1981). This vast genetic resource contains economically important genes that might be exploited for the creation of potentially new variation in domesticated wheat. Many of these species were crossed with each other and consequently, many inter-generic and inter-specific hybrids were produced in this group. These hybrids, most of which were viable, have been used for a variety of purposes including genomic analysis, studies of speciation, phylogeny and evolution, and as the starting point for efforts at the introduction of alien variation into domesticated wheat.

2.3 Difficulties in the Classification of the Wheats

The classification of the species of *Amblyopyrum, Aegilops* and *Triticum* is complicated because of the following reasons: (1) Allopolyploidy—several species contain genomes from two different genera e.g., bread wheat contains one *Triticum* genome and two *Aegilops* genomes; (2) Occurrence of domesticated and wild forms in one biological species; (3) The group is relatively young with possibilities for inter-specific and inter-generic hybridization leading to introgression, formation of hybrid species (e.g., *Ae. sharonensis*) or allopolyploidy. These complications raise the following problems: (1) Definition of the genera (if to keep *Amblyopyrum*, *Aegilops* and *Triticum* as separate genera or to combine them into a single genus); (2) How to regard the wild and domesticated forms of one biological species (taxonomic vs. biological species); (3) How to regard new genera or species of inter-generic or inter-specific synthetic allopolyploids; (4) Nomenclature of synthetic allopolyploids (inter-generic and inter-specific mixoploids).

Following Eig (1929a), Hammer (1980), and Van Slageren (1994), who based their taxonomical classification on the obvious morphological differences between these genera, *Amblyopyrum*, *Aegilops* and *Triticum* are regarded as three separate genera in this review. Nonetheless, it is important to stress the fact that these three genera are very closely related to each other.

The concept of biological species has a cytogenetic and evolutionary (biosystematic) meaning which has a great relevance when genetic resources are considered. On the other hand, taxonomic species are relevant when classification and nomenclature are considered. On this basis, and since this chapter is written from a genetic and evolutionary approaches, the biological species concept is adopted. Consequently, and following Mac Kay (1966) and Van Slageren (1994), the wild and domesticated forms are included as subspecies in one species, even though they are under different selection pressure and consequently, different evolutionary direction.

2.4 Evolution of the Diploid Species

The 13 diploid species of the wheat group (the genera *Aegilops*, *Amblyopyrum*, and *Triticum*) are presented in Table 2.5. These species are differentiated from each other in morphology, having unique and specialized dispersal unit, namely, wedge,

Table 2.5 The diploid species of the wheat group (the genera *Amblyopyrum*, *Aegilops*, and *Triticum*)

Genus	Species[a]	Synonyms	Genome[b]	Genome size[c] (Mean ± SD of 1C DNA in pg)	Region in the group distribution area	Natural groups[d]
Amblyopyrum	*muticum*	*Ae. mutica, Ae. tripsacoides*	T	5.82 ± 0.147	Central	I
Aegilops	*speltoides*	*Ae. speltoides* var. *ligustica* or var. *aucheri*	S	5.81 ± 0123	Central and somewhat southern	II
	searsii	–	S^s	6.65 ± 0.091	Southern	
	bicornis	–	S^b	6.84 ± 0.097	Southern	
	longissima	–	S^l	7.48 ± 0.082	Southern	
	sharonensis	*Ae. longissima* var. *sharonensis*	S^l	7.52 ± 1.000	Southern	
	tauschii	*Ae. squarrosa, T.aegilops*	D	5.17 ± 0.087	Eastern	III
	caudata	*Ae, markgrafii, T, djchasians*	C	4.84 ± 0.089	Central	IV
	comosa	Included *Ae. heldreichii*	M	5.53 ± 0.052	Western	V
	uniaristata		N	5.82 ± 0.105	Western	
	umbellulta		U	5.38 ± 0.073	Central	VI
Triticum	*monococcum*. subsp. aegilopoides	Var *boeoticum*, var. *thaudar*	A^m	6.45 ± 0103	Central	VII
	monococcum. subsp. *monococcum*		A^m	6.48 ± 0.043	Domesticated	
	urartu		A	6.02 ± 0.062	Central	

[a]Species designation after van Slagern (1994)
[b]Genome designations according to Kimber and Tsunewaki (1988)
[c]Genome size from Eilam et al. (2007)
[d]Natural groups based on pairing data in inter-specific and inter-generic hybrids (Kihara 1954)

barrel and umbrella types, and in eco-geography, occupying well-defined ecological niches. All the species have distinct and well-defined genomes, all with $2n=2x=14$ chromosomes but with different genome sizes and pairing patterns in inter-specific and inter-generic hybrids (Table 2.5).

The diploid species are distributed in Southwest and Central Asia. The center of the distribution is in southeast Turkey, i.e., the northern part of the fertile-crescent belt. Six species (*A. muticum*, *T. monococcum* subsp. *aegilopoides*, *T. urartu*, *Ae. speltoides*, *Ae. caudata*, and *Ae. umbellulata*) are distributed in the central part of the group distribution (Table 2.5). Several species of the S genome group (*Ae. bicornis*, *Ae. longissima*, *Ae. sharonensis* and *Ae. searsii*) are found south of the center, the species of the M/N-genome group (*Ae. comosa* and *Ae. uniaristata*) west of the center, and *Ae. tauschii* is in the eastern part of the distribution area. The geographical distribution of the various diploid species indicates that the group has undergone an extensive differentiation in its early stages of development. Southeast Turkey is the geographical center of the group distribution and thus, is presumably the center of origin of the genus.

The species grow in warm countries with a short and moist winter and long, dry hot summer. They do not grow in mountains higher than 1800 m or in the deserts. The species are relatively restricted in their distribution and are specialized in their ecological requirements, usually occupying well-defined habitats with specific edaphic or climatic conditions. Some of the diploids (*Ae. speltoides*, *Ae. tauschii*, *Ae. umbellulata* and the wild subspecies of *T. monococcum*) show wider ecological amplitudes correlated with their weedy tendency.

Natural hybridization between diploid species is a rare phenomenon. In spite of the fact that several of the diploids have massive spatial contact, inter-specific hybrids were reported between only two of these species, *Ae. longissima* and *Ae. sharonensis* (Ankori and Zohary 1962). The two species are closely related, chromosomal pairing in meiosis of the F_1 hybrids is complete (five bivalents and one translocation) and fertility is high, and the only isolation between them is their differential ecological requirements.

The genomes of the diploid species are distinct from each other. On the basis of chromosomal pairing at meiosis of the inter-specific and inter-generic hybrids and on the basis of hybrid fertility, Kihara (1954) defined the genomes of most diploid species, that of *Ae. searsii* was formulated by Feldman et al. (1979). The genomes of *A. muticum*, *Ae. uniaristata*, *Ae. umbellulata*, *T. monococcum* and *T. urartu* were revised by Kimber and Tsunewaki (1988). Based on the genomic divergence, the diploid species of the group are classified in the following seven groups (Table 2.5):

1. T-group (*muticum*).
2. S-group (*speltoides*, *bicornis, longissima*, *sharonensis, and searsii*).
3. A- group (*monococcum* and *urartu*).
4. D-group (*tauschii*).
5. C-group (*caudata*).
6. M-group (*comosa* and *uniaristata*).
7. U-group (*umbellulata*).

Eig (1929a) defined traits that he considered primitive (ancestral) or advanced in this group (Table 2.6). Accordingly, one may consider *A. muticum, Ae. speltoides, T. monococcum and T. urartu,* that exhibit ancestral traits, as the basal species of the group, and *Ae. umbellulata* and *Ae. uniaristata* as the most advanced and therefore, the youngest species. *A. muticum* has morphological traits, e.g., cylindrical awnless spikelets, that differentiate it from the other diploid species of the group (Eig 1929b).

The studies on the karyotypes of the species (Senyaninova-Korchagina 1932; Chennaveeraiah 1960) separated the diploids into two categories: those diploids having only metacentric or submetacentric chromosomes (the T-, A-, S-, and D-genome species) and those with subtelocentric chromosomes (the C-, M-, N-, and U- genome species). Stebbins (1950) considered asymmetry of centromere position more advanced than approximate isobrachial condition since it was found in those diploids with increased specialization of the lemmas, the rachis (with respect to fragility) and awn development.

Dvorak and Zhang (1992), analyzing repeated DNA sequences, concluded that *A. muticum* is close to *Ae. caudata, Ae. comosa, Ae. uniaristata* and *Ae. umbellulata*. Terachi et al. (1984), Murai et al. (1989) and Yamane and Kawahara (2005) suggested that *A. muticum* is close to *Ae. umbellulata* while Terachi and Tsunewaki (1992) suggested that it is close to *Ae. tauschii*. This poses a discrepancy between molecular phylogeny and classification based on morphology. Morphologically *A. muticum* resembles species of *Agropyron* and other perennial relatives (Eig 1929b; Ohta 1991). *Amblyopyrum muticum* and *Ae speltoides* are the only allogamous species in the group. Allogamy is a primitive character since most perennial species in the tribe Triticeae are allogamous. Moreover, *A. muticum* and *Ae. speltoides* are the only species that contain B-chromosomes (Mochizuki 1957; Simchen et al. 1971) and a gene that promotes pairing between homoeologs and suppresses the effect of

Table 2.6 Eig's (1929a) definition of primitive (ancestral) and advanced traits in the genus *Aegilops*

Ancestral traits	Advance traits
Tall plant	Short plants
A small number of tillers	A large number of tillers
Long spike	Short and compact spike
Large number of fertile spikelets per spike	Small number of fertile spikelets per spike
A large number of florets per spikelet	A small number of florets per spikelet
The glumes are shorter than the florets	The glumes are about the same size as the florets
Awns on palea	Awns on the glumes
Glumes of the apical spikelets are awnless	Glumes on the apical spikeletes are with many awns
The spikes disarticulate at maturity into spikelets with the rachis internode that belong to them (wedge type)	The spike falls entire at maturity (umbrella type)
The grain is joined together with the chaff	The grain is free

the *Ph1* gene in hybrids with allopolyploid wheat (Dvorak 1972; Dover and Riley 1972). Indeed, based on chromosome pairing in meiosis of F_1 hybrids between *A. muticum* with B-chromosomes (that suppress pairing between homoeologous chromosomes) and diploid species of *Aegilops*, Ohta (1991) concluded that *A. muticum* is the closest relative of *Ae. speltoides* and as such, it should be placed in section Sitopsis. Sallares and Brown (2004), who analyzed the transcribed spacers of the 18S ribosomal RNA genes, reached a similar conclusion. In-situ hybridization with repeated DNA markers and C-banding patterns suggest that *A. muticum* occupies an isolated position within *Aegilops* and might be closer to the Sitopsis species than to others (Badaeva et al. 1996). These studies support Eig's (1929b) separation of *A. muticum* into a separate genus.

Aegilops speltoides contains two forms, *ligusica* and *aucheri*, that differ markedly in the structure of their seed dispersal unit, *ligusica* has the wedge type and *aucheri* the umbrella type (Eig 1929a). The two types grow sympatrically in mixed populations and are totally inter-fertile (Zohary and Imber 1963). Based on the structure of the dispersal unit, it is reasonable to assume that *aucheri* is a more advance type than *ligustica. Aucheri* type developed independently from the other Sitopsis species or alternatively, derived from hybridization of *Ae. speltoides* (*ligustica* type) with *Ae. longissima* or *Ae. searsii*.

Ae. searsii and *Ae. caudata* have identical chloroplast type (Alnaddaf et al. (2012) and the two species were found to be close to one another (Sliai and Amer 2011). Dvorak and Zhang (1992) and Sasanuma et al. (1996, 2004) found close relationship between *Ae. caudata* and *Ae. umbellulata.* Cytogenetic and phylogenetic studies of the four advanced species of *Aegilops*, *caudata*, *comosa*, *umbellulata* and *uniaristata*, showed that the N genome of *Ae. uniaristata* is one of the most advanced genome in the group and is closer to the U genome of *Ae. umbelluata* than to the genomes of *Ae. caudata* and *Ae. comosa* (Sallares and Brown 2004; Badaeva et al. 1996). PCR fragment polymorphism analyses of chloroplast genomes have placed *Ae. umbellulata* and *Ae. comosa* closer to *Ae. tauschii* than to *T. monococcum* and *Ae. speltoides* (Tsunewaki et al. 1996; Gandhi et al. 2005).

Based on variation in repeated nucleotide sequences, Dvorak and Zhang (1992) constructed a phylogenetic tree of the species of the *Aegilop–Triticum* group. The tree obtained was consistent with many cytotaxonomical data on species relationships in the genus. It clustered the two *Triticum* diploids, *monococcum* and *urartu,* that have been shown cytogenetically to have a common genome (Dvorak 1976; Chapman et al. 1976). Also the species of *Aegilops* section Sitopsis, that contain a similar genome (Kihara 1954; Feldman et al. 1979; Yen and Kimber 1990) were clustered together in the phylogenetic tree. Within section Sitopsis the separation of *Ae. speltoides* from the remaining four species was consistent with the classification by Eig (1929a), who placed on morphological grounds *Ae. speltoides* into subsection Truncata and the other species into subsection Emarginata. [*Ae. searsii* that was described by Feldman and Kislev (1977) was also classified in subsection Emarginata]. Pairing data in meiosis of inter-specific F_1 hybrids between these species (Kihara 1954 and reference therein; Feldman et al. 1979), and studies of starch gel electrophoresis (Brody and Mendlinger 1980), RFLP analysis (Giorgi et al. 2002) and study of telomeric and subtelomeric tandem repeats (Salina et al. 2006), are consistent with this division of section Sitopsis. Brody and Mendlinger (1980) found that *T. monococcum s. lat.* is close to the Sitopsis species; *Ae. comosa*, *Ae. uniari*stata and *Ae. umbellulata* form a second subgroup with *Ae. caudata* most closely related to these species and that *Ae. tauschii* is equally related to all of the species.

Phylogentic relationships of the diploid species were studied on the molecular level in which plasmon [plastome (chloroplast genome) and chondriome (mitochondrial genome)] were analyzed and compared (Tsunewaki 2009; Kawahara 2009). Tsunewaki (2009) reviewed such studies in the various diploid species of the wheat group and concluded that they exhibit a great diversification. A*mblyopyrum muticum* and *Ae. speltoides,* the two outbreeding species, showed especially clear intra-specific chloroplast and mitochondrial differentiation. Earlier studies revealed two types of electromorphs of the rubisco large subunit (the chlotoplast subunit), H- and L-type, in the *Triticum-Aegilops* complex (Chen et al. 1975; Hirai and Tsunewaki 1981). The H-type large subunit was found in the chloroplast of *Ae. speltoides* (and also in that of allopolyploid *Triticum* species) while the L-type large subunit distributed among the chloroplast of all diploid *Aegilops* and *Triticum* species (Hirai and Tsunewaki 1981).

DNA sequencing has had a dramatic effect on the field of molecular phylogenetics. Such studies in the wheat group, based on data from nuclear DNA sequences (genes or repetitious DNA) and chloroplast DNA sequences, show significant inconsistencies possibly due to ancient and recent inter-specific and inter-generic hybridizations and introgressions (Kawahara 2009). Incongruence between chloroplast and nuclear genomic data was often detected (Sasanuma et al. 2004; Kawahara 2009; Li et al. 2014a, b). Monophyletic origin of *Aegilops* and *Triticum* was inferred from some of the studies (e.g., Hsiao et al. 1995; Kellogg and Appels 1995; Kellogg et al. 1996; Huang et al. 2002a, b) whereas a polyphyletic origin of the group was deduced from other studies (Petersen and Seberg 1997, 2000; Seberg and Frederiksen 2001; Sallares and Brown 2004; Mason-Gamer 2005; Petersen et al. 2006). It is probable that inter-generic hybridizations and introgressions form other genera of the Triticeae e.g., *Agropyron*, blurred and obscured the monophyletic origin of the wheat group.

Several recent estimates of the divergence time of the basal genomes of the wheat group indicated that the divergence of the A and the S genomes from an ancestral Triticineae genome established the basal lineages of the wheat group while the D-genome diverged from these two genomes somewhat later (Dvorak and Akhunov 2005; Marcussen et al. 2014; Gornicki et al. 2014). Results of molecular analyses have shown that genomes S, A, and D are much more closely related to each other than to other genomes (Monte et al. 1993; Dvorak and Zhang 1990; Dvorak et al. 1998). Marcussen et al. (2014) proposed that a homoploid ancient hybridization event occurred long before the divergence of the diploid species of *Aegilops* and indicated that the use "D-genome lineage" in their publication is nonsynonymous with the D genome of *Ae. tauschii* but rather, refers to the entire D+S+M clade. They suggested that these ancestral genomes have diverged from a common ancestor and from one another about 6.5 million years ago (MYA) (Table 2.7). Huang et al. (2002b) estimated that theses diploid species diverged from one another much

Table 2.7 Time of beginning divergence of the diploid genomes of the wheat group (species of the genera *Aegilops* and *Triticum*) in million years ago

Genomes	Beginning divergence time	DNA sequences studied	Reference
Genomes A and S	6.5	Several hundreds nuclear genes	Marcussen et al. (2014)
	2.3	Chloroplast DNA	Gornicki et al. (2014)
Genomes A and D	5.50	Several hundreds nuclear genes	Marcussen et al. (2014)
	2.70 (1.4–4.1)	Four nuclear genes	Dvorak and Akhunov (2005)
	2.30	Nuclear genes	Chalupska et al. (2008)
	1.20	Chloroplast DNA	Gornicki et al. (2014)
Genomes S and D	5.5	Several hundreds nuclear genes	Marcussen et al. (2014)
	2.3	Chloroplast DNA	Gornicki et al. (2014)
Genomes A^m and A	0.57	Chloroplast genome	Middleton et al. (2014)
Genomes S^b and S^l	1.4	Six nuclear genes	Marcussen et al. (2014)
Genomes S^l (of *Ae. longissima*) and S^l (of *Ae. sharonensis*)	0.4	Six nuclear genes	Marcussen et al. (2014)
Genomes S and C,M,N,U	2,5	Six nuclear genes	Marcussen et al. (2014)
Genomes C,M and U	1.4	Six nuclear genes	Marcussen et al. (2014)
Genomes C and M	0.7	Six nuclear genes	Marcussen et al. (2014)
Genomes N and C, M,U	2.0	Six nuclear genes	Marcussen et al. (2014)

later, about 2.5–4.5 (MYA) and Dvorak and Akhunov (2005) suggested that the divergence time of these species these was about 2.7 (1.4–4.1) MYA (Table 2.7). In agreement with the conclusion of Dvorak and Akhunov (2005), recent data from sequencing of whole chloroplast genome indicated that the species having the A, B, and D genomes began to diverge between 2.1 and 2.9 MYA (Middleton et al. 2014; Gornicki et al. 2014; Tables 2.4 and 2.7). Thus, it is reasonable to assume that the diploid species of the group evolved at different times, the primitive species about 2.5–3.0 MYA and the advanced ones, namely, *Ae caudata*, *Ae. comosa*, *Ae. uniaristata* and *Ae. umbellulata,* diverged from the basal species later on, about 2.5 MYA (Marcussen et al. 2014). This period corresponds to the geological epoch Pliocene (1.8–5.3 MYA; Table 2.3), that was characterized by development of seasonal climate (cold and humid winters and hot and dry summers) in the east Mediterranean and Southwest Asia, the center of origin of the diploid species of this group. The adaptation to dry habitats with seasonal growth periods led presumably to the development of annual growth habit associated with increased self-fertilization and large grains.

Aegilops speltoides started to diverge from subspecies Emarginata (*Ae. bicornis, Ae. longissima, Ae sharonensis,* and *Ae. searsii*) about 2.5 MYA (Gornicki et al. 2014). Wang et al. (1997) and Mendlinger and Zohary (1995) found that *Ae. sharonensis*,

Ae. longissima and *Ae. bicornis* are closely related, and that *Ae. sharonensis* is equally close to *Ae. bicornis* and *Ae. longissima. Aegilops searsii* is close to *Ae. longissima* (Feldman et al. 1979), and is the only Emaginata species with advanced traits such as free grains, length of glumes close to the length of florets and short culms. *Ae. searsii,* being the advanced species in this group, diverged from the other Emarginata species about 1.0 to 2.0 MYA (Marcussen et al. 2014). *Ae. bicornis* diverged from *Ae. longissima* about 1.4 MYA and *Ae. sharonensis* diverged from *Ae. longissima* and *Ae. bicornis* about 0.4 MYA (Marcussen et al. 2014). *Ae. longissima* has spikes that fall entire at maturity but, later on, during the summer, may disarticulate into spikelets like that of *Ae. tauschii* (the barrel type) indicating its relationships to *Ae. tauschii. Ae. sharonensis*, the close relative of *Ae. longissima*, was recently suggested to be closer to *Ae. tauschii* or its progenitor than to *Ae. speltoides* (Steuenagel et al. 2014a, b). This was proposed from sequencing the genome of two accessions of *Ae. sharonensis* and sequencing the transcriptomes of 16 accessions that cover the natural habitat of *Ae. sharonensis* (Steuenagel et al. 2014a, b). Consequently, they concluded that the *Sitopsis* classification is inconsistent with their genome-wide analysis of *Ae. sharonensis*, provoking the need to reconstruct the current taxonomy of *Aegilops*.

How the diploid species of this group evolved? Speciation at the diploid level might result from accumulation of mutations in coding and in noncoding sequences as well as structural changes that lead to the buildup of genetic barriers between the diverging taxa. Feldman and Strauss (1983) described a genome-restructuring gene in *Ae. longissima* that produced a large number of chromosomal rearrangements in plants homozygous for it. Genome restructuring is an ongoing process in natural plant populations of *Ae. speltoides* (Belyayev 2013). Numerical chromosomal aberrations, spontaneous aneuploidy, B-chromosomes, and repatterning and reduction in the species-specific tandem repeats have been detected in marginal populations of *Ae. speltoides* (Raskina et al. 2004; Belyayev et al. 2010) indicating that chromosomal re-patterning might be one mechanisms of plant genome evolution and speciation (Raskina et al. 2004). Such genetic, epigenetic and structural changes might have promoted the formation of genetic barriers between the diverging species that resulted in sterility of the inter-specific or inter-generic hybrids.

Other species might evolve through inter-specific or even inter-generic hybridization. Judging from the patterns of seed proteins, Waines and Johnson (1972) suggested that *Ae. sharonensis* was originated from hybridization between *Ae. longissima* and *Ae. bicornis.* Recent comparisons of nuclear genes indicated that an ancestral D lineage was derived from hybridization between ancient A and S lineages (Marcussen et al. (2014). Li et al. (2014a, b), reevaluating the origin of *Ae. tauschii* by using recently published data from sequencing of nuclear DNA (Marcussen et al. 2014) and chloroplast DNA (Gornicki et al. 2014) as well as additional data, confirmed the hybrid origin of the D-genome clade but concluded that this clade have had a more complex origin, one that may have involved multiple rounds of hybridizations. Nakamura et al. (2009) studied in a number of species in the group the *PolA1* gene that codes the largest subunit of RNA polymerase I and concluded that the Sitopsis species might have originated by an ancient hybridization between an ancestral *Triticum-Aegilops* species (as female) and an *Hordeum* species (as male) and backcrossing of the resulting hybrid(s) with the ancestor of *Triticum-Aegilops*. Fan et al. (2013) suggested that *Agropyron elongatum* var *bessarabicum* (*Thinopyrum bessarabicum*, *Elymus farctus subsp. bessarabicus*) is close to species of *Aegilops* section Sitopsis and consequently might have hybridized and introgressed with them.

Transposable elements (TEs) have the potential to affect genome structure and function through transposition, ectopic recombination and epigenetic repatterning (Bennetzen 2005; Slotkin and Martienssen 2007; Fedoroff 2012). They may mutate genes, alter gene regulation, and generate new genes allowing response to environmental challenges thus providing fuel for evolution (Kidwell and Lisch 2000). Moreover, TEs have served as building blocks for epigenetic phenomena, both at the level of single genes and across larger chromosomal regions (Slotkin and Martienssen 2007). Since the variation that is induced by TEs depends on their activity that is governed by epigenetic regulation (Slotkin and Martienssen 2007), their activation might be induced by genetic and environmental stresses (Fedoroff 2012). Senerchia et al. (2013) suggested that ancestral TE families followed independent evolutionary trajectories among related species, highlighting the evolution of TE populations as a key factor of genome differentiation. The balance between genome expansion through TE proliferation and contraction through deletion of TE sequences drives variation in genome size and organization (Bennetzen and Kellogg 1997). Hence, the large differences in genome size between the various diploid species of the wheat group (Eilam et al. 2007; Table 2.5) suggest that TE activity has played an important role in the genomic evolution of these species. Indeed, Yaakov et al. (2013) determined the relative copy numbers of 16 TE families in these species and found high variation and genome-specificity of TEs, indicating the apparent involvement of TEs in the evolution of the diploid species. Similarly, Ben-David et al. (2013) determined the copy number of short interspersed nuclear elements (SINEs) of the wheat family Au SINE in these diploid species and found that SINEs may play a prominent role in the genomic evolution of these species through stress-induced activation. Hence, the variation in copy number of TEs among the diploid species of the wheat group may imply that the main genomic differences between these species are the results of differential activity of TEs (Yaakov et al. 2013).

Indeed, TEs, accounting for a very large fraction of the genomes of the diploid species of the wheat group, [80 % of well-annotated TEs, with a majority of LTR retrotransposons (Senerchia et al. 2013)], were found to be one of the main drivers of genome divergence and evolution in this group (Yaakov et al. 2012). Middleton et al. (2013) also found that several TE families differ strongly in their abundance between the diploid species of this group, indicating that TE families can thrive extremely successfully in one species while going virtually extinct in another. Several TE families have undergone either proliferation or a reduction in abundance during species diversification. Using miniature inverted-repeat transposable element (MITE) markers, Yaakov et al. (2012) assessed genetic diversity among several diploid species, namely, *Ae. speltoides*. *Ae. searsii*. *Ae. sharonensis*, *Ae.longissima*, *Ae. tauschii*, and *T. urartu*, and found that about 80 % of the markers showed polymorphic insertions among species and accessions. Charles et al. (2008) estimated from the insertion dates of TEs that the majority of differential proliferation of TEs in the B and A genomes of bread wheat (87 and 83 %, respectively), leading to a rapid sequence divergence, occurred in these genomes already at the diploid level, prior to the allotetraploidization event that brought them together in *Triticum turgidum,* about 0.5 MYA. Finally, rewiring of gene expression in hybrids might dysregulate the silencing of transposons, resulting in activation of transposons, in reduction of the hybrid fitness or viability and thus might contribute to speciation (Levy 2013).

To sum up, the diploid species of the wheat group underwent a reticulated evolution. The divergence of the S and A lineages established the basal branches of the wheat group. The time of this divergence determines the divergence of the group. *Amblyopyrum muticum* diverged presumably earlier since it has many ancestral traits, is allogamous (self-incompatible), and intermediate in morphology between *Aegilops* and *Agropyron* (Eig 1929a, b; Ohta 1991). *Aegilops speltoides* exhibits ancestral traits and is also allogamous. Divergence of D genome from S and A genomes was presumably through inter-generic hybridization. Steuernagel et al. (2014a, b) suggested that *Ae. sharonensis* might be a product of hybridization between *Ae. speltoides* and *Ae. tauscii.* Yet *Ae. longissima* is closer to *Ae. tauschii* than *Ae. sharonensis. Ae. sharonensis* might be the result of hybridization between *Ae. lomgissma* and *Ae. bicornis* (Waines and Johnson 1972). *Ae. bicornis* presumably derived from Ae. *speltoides* via mutations and TEs activities. *Ae. searsii* is the most advanced species in *Emarginata* because it has several advanced morphological traits. *Ae. caudata* diverged from *Ae. searsii* or from hybridization between *Ae. searsii* and Ae. *tauschii. Ae. comosa, Ae. Uniaristata,* and *Ae. umbellulata* derived from *Ae. caudata.*

Because of the fairly recent speciation of the diploid species, most of them have relatively limited morphological and molecular variation, occupy only a few well-defined ecological habitats, and are distributed throughout relatively small geographical areas (Eig 1929a; Zohary and Feldman 1962; Kimber and Feldman 1987; Van Slageren 1994).

2.5 Evolution of the Allopolyploid Species

The discovery of the accurate chromosome numbers of the different *Triticum* species (Sakamura 1918) showed that this genus comprises a polyploid series based on $x=7$, containing diploids, tetraploids, and hexaploid species. Soon after, studies of chromosomal pairing in hybrids between species of different ploidy levels disclosed that the polyploids are allopolyploids, i.e., each polyploid species was formed by inter-specific or inter-generic hybridization followed by chromosome doubling (Kihara 1919, 1924; Percival 1921; Sax 1921, 1927). These studies showed that tetraploid wheat contains one genome, A, that derived from a diploid *Triticum* species and a second genome, B, from another, yet unknown diploid species. Hexaploid wheat contains the two genomes of tetraploid wheat, A and B, and a third genome, D, from another species. The conclusion that the polyploids are allopolyploids was also supported by the fact that only bivalents were formed at first meiotic metaphase of the polyploids (except that of the autoallohexaploid *T. zhukovskyi* that has the genomic constitution AAA^mA^m GG and therefore produces few quadrivalents at the first meiotic metaphase). The diploid-like meiotic pairing pattern, i.e., regular bivalent formation due to exclusive homologous pairing, is characteristic of allopolyploids.

Since the discovery that the polyploid species of wheat comprise an allopolyploid series attempts have been made to identify the diploid donors of the B and D genomes to allopolyploid wheat. In this endeavor studies were extended to the wild relatives of wheat, particularly to the closely related genus *Aegilops*, which also comprises an allopolyploid series with diploid, tetraploid, and hexaploid species (Lilienfeld 1951, and reference therein). The species of these two genera have been subjected to extensive taxonomic, cytogenetic, genetic, biochemical, molecular, and evolutionary study by numerous scientists (see reviews of Kihara 1954; Mac Key 1966; Morris and Sears 1967; Kimber and Sears 1987; Feldman et al. 1995; Feldman 2001; Gupta et al. 2005; Dvorak 2009). One of the most extensively used methods in this search was the cytogenetic approach of genome analysis, developed by Kihara (1919, 1924) and that was based on the concept of genome stability, assuming that the genomes of the allopolyploid species remain similar to those of their diploid parents. Genome analysis by the "analyzer-method" (Lilienfeld 1951) has provided the most consistent recognition of genomic similarities in the wheat group. Genome analysis of wheat and its relatives also provided an insight into the evolutionary past of these species. With this method, Kihara used the diploid species of *Aegilops* and *Amblyopyrum* as analyzers of the genomes in the allopolyploids of the genera *Aegilops* and *Triticum* (Lilienfeld 1951). Based upon all possible cross combinations among the *Aegilops* and *Amblyopyrum* species, nine diploid analyzers were established: one from *Amblyopyrum* (*muticum*), and eight from *Aegilops* (*caudata, umbellulata, comosa, uniaristata, tauschii* (*squarrosa*), *bicornis*, *longissima* (including *sharonensis*) and *speltoides*). All allopolyploid *Aegilops* and *Triticum* species are made up from contributions of the genomes of these diploid analyzers except *A. muticum*, *Ae. bicornis*, *Ae. Sharonensis,* and *T. monococcum* that did not contribute a genome to the allopolyploid species (Kihara 1954). [*Ae. searsii* was discovered later on (Feldman and Kislev 1977) and was not included in Kihara's analysis]. *Ae. bicornis* and *Ae. sharonensis* are geographically isolated from non-Sitopsis diploids but the lack of participation in the allopolyploid formation of *A. muticum*, *Ae. Searsii,* and *T. monococcum* requires explanation).

Due to these extensive studies, the polyploid species of *Triticum* have served as a classical example for evolution through allopolyploidy (Kihara 1924; Sax 1921, 1927; Sears 1948, 1969; Kihara et al. 1959; Morris and Sears 1967; Van Slageren, and reference therein). Von Tschermak and Bleier (1926) were the first to identify a spontaneous chromosome doubling in the cross of wild emmer (*T. turgidum* ssp. *dicoccoides*) with *Ae. geniculata,* thus demonstrating the possibility of species formation via allopolyploidy in the wheat group. Subsequent studies showed that the frequency of unreduced gametes in inter-generic F_1 hybrids of wheat could be in some hybrids as high as 50 % (Kihara and Lilienfeld 1949). Matsuoka et al. (2013) studied the genetic mechanism that causes spontaneous genome doubling in the F_1

hybrid between *T. turgidum,* and *Ae. tauschii.* They identified six QTLs in *Ae. tauschii* that are involved in hybrid genome doubling regulating nonreductional meiosis and its subsequent unreduced gamete production processes. Therefore, one might assume that there is a high potential for the frequent and recurrent formation of allopolyploids in the wheat group. The discovery by Blakeslee (1937) that colchicine can induce chromosome doubling opened new possibilities for the study of wheat evolution through allopolyploidization. It also provided an easy method to synthesize different wheat allopolyploids some of which have similar genomes to natural allopolyploids and others have new genomic combinations (see examples in Ozkan et al. 2001). Synthetic allopolyploids, either induced or occurring spontaneously, offer excellent tools to mimic the evolutionary speciation events that occurred in nature and to test in a controlled manner the new features of the allopolyploid genome compared to those of its parents.

The accumulating cytogenetic and molecular evidence have indicated that the concept of genome stability, that is, the assumption that the genomes of the allopolyploid species remain similar to those of their diploid parents, is not always the case and that one genome in every allopolyploid remains relatively unchanged while the second genome(s) has undergone considerable changes from that of its diploid parent. These genomes were termed modified genomes by Kihara (1954) and other wheat cytogeneticists (Table 2.8). Every allopolyploid species of the genera *Aegilops* and *Triticum* contains a relatively unchanged genome side by side with a modified one whose diploid origin has been intricate to trace (Zohary and Feldman 1962). Nevertheless, genome analysis studies revealed that allotetraploid wheat (genome BBAA) originated from hybridization events involving two diploid progenitors classified in the genera *Aegilops* and *Triticum.* Genome B, which is a modified genome, derived from *Ae. speltoides* or from a closely related species to *Ae. speltoides,* which is extinct or extant (Feldman et al. 1995), and underwent changes at the polyploid level. Genome A, that has been modified relatively little, was derived from *T. urartu* (Chapman et al. 1976; Dvorak 1976). Allohexaploid wheat (genome BBAADD) originated from hybridization between allotetraploid wheat and *Ae. tauschii*, the donor of the D genome (Kihara 1944; McFadden and Sears 1944, 1946).

The wheat group contains 18 allopolyploid species, 12 tetraploids, and 6 hexaploids (Table 2.8). Modern classification for the genus *Triticum* (Van Slageren 1994) recognizes two diploid species, *T. monococcum* L. and *T. urartu* Tum. ex Gand., two tetraploid species, *T. turgidum* L. and *T. timopheevii* (Zhuk.) Zhuk., and two hexaploid species, *T. aestivum* L. and *T. zhukovskyi* Men. & Er. The economically important wheats are *T. aestivum* (bread wheat, comprising 95 % of the global wheat production) and *T. turgidum* (macaroni wheat).

Most possible allopolyploid combinations were produced synthetically indicating that most genomic combinations can be created. The reasons why some combinations are missing in nature can be explained partly by the eco-geographical isolation of the corresponding analyzers and possibly also by the low viability of some combinations that might have hybrid weakness and consequently, could not compete with their parental diploids and establish themselves in nature.

Table 2.8 The allopolyploid species of the wheat group

Ploidy level	Genus	Species[a]	Genome[b]	Genome size[c] (Mean ± SD of 1C DNA in pg)	Expected genomic size[d]	Natural groups[e]
Tetraploids	*Aegilops*	*biuncialis*	UM	10.37 ± 0.037	10.91	The U group
		geniculata (=*ovata*)	MU	10.29 ± 0.008	10.91	
		neglecta (=*triaristata*)	UM	10.64 ± 0.404	10.91	
		columnaris	UM	10.86	10.91	
		triuncialis	UC; CU	9.93 ± 0.041	10.22	
		kotschyi	SU	12.64 ± 0.183	12.86	
		peregrina (=*variabilis*)	SU	12.52 ± 0.181	12.86	
		cylindrica	CD	9.59	10.01	The D group
		crassa	DM	10.86	10.70	
		ventricosa	DN	10.64n	10.99	
	Triticum	*turgidum* subsp. *dicoccoides*	BA	12.91 ± 0.199	11.83	The A group
		turgidum subsp. *durum*	BA	12.84 ± 0.175	11.83	
		timopheevii subsp. *armeniacum*	GA	11.82 ± 0.071	11.83	
		timopheevii subsp. *timopheevii*	GA	11.87 ± 0.630	11.83	
Hexaploids	*Aegilops*	*Recta* (=*triaristata*)	UMN	16.22	16.46	The U group
		vavilovi	DMS	17.13 ± 0.139	18.34	The D group
		crassa	DDM	–	–	
		juvenalis	DMU	–	–	
	Triticum	*aestivum* subsp. *spelta*	BAD	17.72 ± 0.039	18.04	The A group
		aestivum subsp. *aestivum*	BAD	17.67 ± 0.311	18.04	
		zhukovskyi	GAA^{m}	17.74	18.35	

[a]Species designation after van Slageren (1994)
[b]Genome designations according to Kimber and Tsunewaki (1988); underlined designation indicates modified genomes, the first genome is the donor of the cytoplasm.
[c]Genome size from Eilam et al. (2008)
[d]{Expected genomic size from the sum of the DNA amounts of the two parental species (see Eilam et al. 2007 for 1C DNA amount in the diploid species). The calculation of the expected DNA amount in the allopolyploids of Triticum was based on the assumption that genomes G and B were donated by Aegilops speltoides.
[e]Groups according to Zohary and Feldman(1962)

Allopolyploidization, being a radical, rapid mode of speciation, produces a new species in a single step, a novel taxon that is isolated genetically from its two parental species. Besides, newly formed allopolyploid is a hybrid species with two or more different genomes enveloped within the same nucleus. This situation exerts a major genetic stress on the newly formed allopolyploids that must overcome several immediate challenges in order to survive in nature (Levy and Feldman 2002, 2004; Feldman and Levy 2005, 2009, 2011). Overcoming these challenges is achieved through genomic plasticity, namely through the immediate triggering of a variety of cardinal genetic and epigenetic changes that affect genome structure and gene expression. Consequently, a newly formed allopolyploid species must secure an exclusive intra-genomic pairing at meiosis that will lead to increased fertility i.e., to prevent pairing between the homoeologous chromosomes of the different genomes that are still closely related (Sears 1954; Morris and Sears 1967), and orchestrate inter-genomic gene expression and DNA replication, i.e., expression of a number of genes is quickly altered or silenced to enable harmonic inter-genomic coexistence along with reduction in genome size (Levy and Feldman 2002, 2004; Feldman and Levy 2005, 2009, 2011, 2012; Feldman et al. 2013). Meeting these challenges that are achieved through either genetic or epigenetic alterations in the DNA or in chromatin structure is presumably critical for the survival of the newly formed allopolyploids, ensures their increased fitness and successful establishment in nature as competitive entities.

Further divergence of the homoeologous chromosomes, i.e., related chromosomes of the different genomes, is rapidly achieved through elimination of a number of DNA sequences, mainly low-copy and, to some extent, high-copy, noncoding and coding sequence (Feldman et al. 1997; Liu et al. 1998a, b; Ozkan et al. 2001; Shaked et al. 2001; Han et al. 2003, 2005; Salina et al. 2004; Baum and Feldman 2010; Guo and Han 2014). Such rapid elimination was observed also in *Triticale*, a synthetic allopolyploid between wheat and rye (Boyko et al. 1984, 1988; Ma and Gustafson 2005, 2006; Bento et al. 2011). These sequences exist in all the diploid progenitors of the allopolyploid species of the wheat group, whereas in the allopolyploids they exist in only one pair of homologues (Feldman et al. 1997; Ozkan et al. 2001). Evidently, these sequences were eliminated from one of the two genomes in tetraploids and from two of the three genomes in hexaploids. This elimination is rapid and occurs during or soon after the formation of the allopolyploids and was designated as revolutionary changes (Feldman et al. 1997; Ozkan et al. 2001). No further elimination of sequences occurs during the successive generations of the allopolyploids. The elimination of sequences is reproducible as the same sequences were eliminated in newly formed and in natural allopolyploids having the same genomic combinations (Ozkan et al. 2001). It was concluded that instantaneous elimination of DNA sequences in the first generations of the newly formed allopolyploids of *Triticum* and *Aegilops*, is one of the major and immediate responses of the wheat genome to allopolyploidization.

The extent of DNA elimination was estimated by determining the amounts of nuclear DNA in natural allopolyploids and in their diploid progenitors, as well as in newly synthesized allopolyploids and their parental plants (Ozkan et al. 2003;

Eilam et al. 2008, 2010). Natural wheat and related allopolyploids contain 2–10 % less DNA than the sum of their diploid parents, and synthetic allopolyploids exhibit a similar loss, indicating that DNA elimination occurs soon after allopolyploidization (Nishikawa and Furuta 1969; Furuta et al. 1974; Eilam et al. 2008, 2010). Also, the narrow intra-specific variation in DNA content of the allopolyploids shows that the loss of DNA occurred immediately after the allopolyploid formation, and that there was almost no subsequent change in DNA content during the allopolyploid species evolution (Eilam et al. 2008). In *Triticale* Boyko et al. (1984, 1988) and Ma and Gustafson (2005) found that there was a major reduction in DNA content in the course of *Triticale* formation, amounting to about 9 % for the octoploid and 28–30 % for the hexaploid *Triticale*. In this synthetic allopolyploid, the various genomes were not affected equally: the wheat genomic sequences were relatively conserved, whereas the rye genomic sequences underwent a high level of variation and elimination (Ma et al. 2004; Ma and Gustafson 2005, 2006; Bento et al. 2011). Similarly, in hexaploid wheat, genome D underwent a considerable reduction in DNA, while the A and B genomes were not reduced in size (Eilam et al. 2008, 2010). Bento et al. (2011) reanalyzed data concerning genomic analysis of octoploid and hexaploid *Triticale* and found that restructuring depends on parental genomes, ploidy level, and sequence type (repetitive, low copy, and (or) coding) (Bento et al. 2011).

DNA elimination seems not to be random at the intra-chromosomal level as well (Liu et al. 1997). For example, these authors found that the chromosome-specific sequences on chromosome arm 5BL in allohexaploid wheat are not distributed randomly but cluster in terminal (subtelomeric), subterminal and interstitial regions of this arm. Such structures make these regions extremely chromosome-specific — or homologous. Hence, it was tempting to suggest that these chromosome-specific regions, the only regions that determine homology, are equivalents to the classical "pairing-initiation sites" that play a critical role in homology search and initiation of pairing at the beginning of meiosis (Feldman et al. 1997).

Sequence elimination from one pair of homoeologous chromosomes in tetraploids or from two pairs in hexaploids, leaving the sequence in only one homologous pair, renders it a chromosome-specific sequence that can determine chromosome homology. Such differential elimination leads to a cytological diploidization process that strongly augments the physical divergence of the homoeologous chromosomes so that they cannot pair and recombine at meiosis. Thus, cytological diploidization leads to exclusive intra-genomic pairing, i.e., diploid-like meiotic behavior. Computer modeling also shows that homoeologue divergence in association with pairing stringency drives disomic inheritance (Le Comber et al. 2010),

Superimposed on the divergence of the homoeologous chromosomes due to differential sequence elimination, is a genetic system that involves in sustaining the exclusive bivalent pairing in the allopolyploid *Triticum* species. This system consists mainly of the genes *Ph1* that is located on chromosome arm 5BL of common wheat (Okamoto 1957; Sears 1976, and reference therein). From the time of its discovery in the late 50s (Okamoto 1957; Riley and Chapman 1958; Sears and Okamoto 1958), the *Ph1* gene (pairing homoeologous; Wall et al. 1971) of polyploid *Triticum* has had a great impact on wheat cytogenetics and beyond. This dominant gene, located about 1.0 cM from the centromere (Sears 1984), has been assumed to suppress pairing of homoeologues in allopolyploid *Triticum* species while allowing regular pairing of homologous chromosomes (Riley 1960; Sears 1976).

The suppressive effect of *Ph1* on homoeologous pairing in inter-specific and inter-generic *Triticum* hybrids is absolute. In contrast, its effect on homoeologous pairing in bread wheat itself might not be indispensible as plants deficient for this gene exhibit relatively little homoeologous pairing (less than one multivalent per cell resulting from inter-genomic pairing (Sears 1976). Interestingly, and in accord with the above, *Ph*-like gene(s) have not been found in any of the allopolyploid species of the closely related *Aegilops* genus (Sears 1976). Nevertheless, these species also exhibit exclusive bivalent pairing of fully homologous chromosomes presumably due to the sequence-elimination system.

The mechanism controlling the *Ph1* mode of action is still unclear. Six extra doses of the *Ph1*-containing arm of chromosome 5B of hexaploid wheat caused partial asynapsis of homologues at meiosis, concomitantly allowing some pairing of homoeologous chromosomes and inducing a high frequency of interlocking bivalents (Feldman 1966). Similar effects were observed by pre-meiotic treatment with colchicine (Driscoll et al. 1967; Feldman and Avivi 1988). These effects were explained by the assumption that the hexaploid nucleus still maintains some organizational aspects of the individual ancestral genomes, i.e., each genome occupies a separate region in the nucleus, which in turn, is recognized by *Ph1* (Feldman 1993, and references therein). A model was thus proposed whereby *Ph1* exerts its effect at pre-meiotic stages, before the commencement of synapsis, where it affects the pre-meiotic alignment of homologous and homoeologous chromosomes and as a result controls the regularity and pattern of pairing (Feldman 1993). In euploid common wheat, two doses of *Ph1,* while scarcely affecting homologous chromosomes, keep the homoeologues apart, thereby ensuring exclusive homologous pairing in meiosis. In the absence of *Ph1*, the three genomes are mingled together in the nucleus and consequently, the homoeologues can also pair to a small extent with each other. Six doses of *Ph1* or pre-meiotic treatment with colchicine induces separation of chromosomal sets and random distribution of chromosomes in the pre-meiotic nucleus, leading to an increased distance between homologues. This results in partial asynapsis of homologues that are relatively distant, and some pairing of homoeologues that happen to lie close to one another. An interlocking of bivalents can occur with their chromosomal constituents coming to pair from a relative distance and catching other chromosomes between them (Feldman 1993, and references therein). This model is supported by the fact that while extra six doses of chromosome arm 5BL, as well as pre-meiotic treatment with colchicine, caused partial asynapsis of homologues including the inter-chromosomal pairing of isochromosomes, they did not affect the intra-chromosomal pairing of isochromosomes (Feldman and Avivi 1988). This indicates that when chromosomes are close to one another, as the two homologous arms in an isochromosome that are connected to each other by the common centromere, extra dose of 5BL as well as colchicine do not suppress their pairing at meiosis.

Two deletions that include the *Ph1* locus were induced by x-irradiation: *ph1b* (Sears 1977), induced in the bread wheat cultivar Chinese Spring (CS), is a submicroscopic interstitial deletion, about 3 Mb in size (Gill and Gill 1991; Gill et al. 1993), and *ph1c* (Giorgi 1978, 1983), induced in the durum cultivar Cappelli, is a microscopic deletion in the middle of chromosome arm 5BL (Jampates and Dvorak 1986). Mapping data suggested that *Ph1* might encode for a gene that affects cell cycle progression through the regulation of cyclin-dependent kinases (Griffiths et al. 2006; Al-Kaff et al. 2008; Yousafzai et al. 2010), however, direct evidence through complementation studies is still missing. Recently, Bhullar et al. (2014) identified a candidate gene in the *Ph1* region that is different from the one that was proposed previously (Griffiths et al. 2006; Al-Kaff et al. 2008; Yousafzai et al. 2010). Virus-induced silencing of this newly identified gene disrupts the alignment of chromosomes on the metaphase plate in early stages of meiosis suggesting a role of microtubules in its function (Bhullar et al. 2014). This in accord with the finding of Avivi and Feldman (1973, and reference therein) and Vega and Feldman (1998) who found that *Ph1* affects the interaction between the centromeres and microtubules, possibly through a microtubule-associated protein (MAP) whose activity is located near the centromere.

To sum up, cytological diploidization in allopolyploid *Triticum* was engendered by two independent, complementary systems. One is based on the physical divergence of chromosomes, and the second, on the genetic control of pairing. The *Ph*-gene system superimposes itself on and takes advantage of—and thereby reinforces—the above-described system of the physical differentiation of homoeologous chromosomes. In addition, stringent selection for fertility in the allopolyploid *Triticum* species might well favor the development of two systems to effect the suppression of multivalent formation and the promotion of bivalent pairing in nature and, more so, in domesticated material.

The process of cytological diploidization in the allopolyploid species of the wheat group has been critical for their successful establishment in nature. The restriction of pairing to completely homologous chromosomes ensures regular segregation of genetic material, high fertility, genetic stability, and disomic inheritance that prevents the independent segregation of chromosomes of the different genomes. This mode of inheritance leads to permanent maintenance of favorable inter-genomic genetic interactions, thus, enables to fix heterotic interaction between genomes. On the other hand, disomic inheritance sustains the asymmetry in the control of many traits by the different genomes (Feldman et al. 2012). In addition, since cytological diploidization facilitates genetic diploidization, existing genes in double and triple doses can be diverted to new functions through mutations, thereby giving preferentiality to the creation of favorable, new inter-genomic combinations.

How the two or three divergent genomes present in a single nucleus of an allopolyploid were driven to operate in a harmonious manner. From one hand, homoeoalleles in allopolyploid wheat may differ from one another by allelic variation, and in this case, activity of all the duplicated genes may produce desirable inter-genomic interactions and heterotic effects. Inter-genomic gene interactions may be, in some cases, expressed in novel traits that do not exist in their parental diploids. Some of these traits may have great adaptive value. Inter-genomic gene interactions have direct relevance also to wheat cultivation. For example, the baking quality of allohexaploid wheat (bread wheat) is due to the unique properties of its gluten—a product derived from the combined contribution of the three genomes of hexaploid wheat and thus, exists only at the hexaploid level.

On the other hand, increased gene dosage may lead to redundancy or in some cases may have a deleterious effect. Moreover, allopolyploids contain two or three diverse genomes in one nucleus and therefore, there is an immediate need to achieve a harmonious function of the genes of the different genomes. The processes that bring redundant or unbalanced gene systems in allpolyploids toward a diploid-like mode of expression is genetic or functional diploidization. Studies with synthetic allopolyploids as well as genome sequencing data indicate that a broad range of DNA rearrangements have occurred during, soon or after allopolyploidization leading to genetic diploidization (International Wheat Genome Sequencing Consortium, 2014). These rearrangements including deletion, pseudogenization, and subfunctionalization of genes, and transposons activation or deactivation, are extensive and relatively rapid. Feldman and Levy (2005) distinguished between revolutionary changes, occurring during or immediately after allopolyploidization and evolutionary changes that take place throughout the life of the allopolyploid. Revolutionary changes include genetic and epigenetic alterations, that lead to cytological diploidization, improve the harmonious functioning of the divergent genomes, stabilize the nascent allopolyploid and facilitate its establishment as a new species in nature—all of which are species specific. Evolutionary changes comprise mostly genetic changes, promote genetic diversity, flexibility and adaptability—all of which are biotype or population specific. Examples of functional diploidization in allopolyploid wheat involves mainly genes that code for structural or storage proteins, e.g., histones, subunits of tubulins, subunits of glutenins and gliadins, and ribosomal RNA (and possibly also tRNA). In such genes, expression of all homoeoalleles might be redundant and even deleterious, due to over-production and inefficiency. In this case, traits controlled by genes from only or mostly one genome may have a higher adaptive value. It is therefore expected that such gene loci would have been targets for genetic diploidization. In allotetraploid wheat for example, αCENH3 and βCENH3 transcripts are derived more from the A genome than from the B genome (Yuan et al. 2014). The *Hardness (Ha)* locus constitutes another example of genetic diploidization, through gene deletion, in polyploid wheat (Chantret et al. 2005). In addition, recent analysis of the sequences of wheat group 1 chromosomes shows significant deviations from synteny with many of the nonsyntenic genes representing pseudogenes (Wicker et al. 2011). The discovery of premature termination codons in 38 % of expressed genes in 3A double ditelosomic lines in the genetic background of wheat (*Triticum aestivum* 'Chinese Spring') was consistent with ongoing pseudogenization of the wheat genome (Akhunov et al. 2013).

Genetic diploidization might be achieved through epigenetic silencing of one of the homeoalleles via cytosine methylation or activation of silenced genes due to their demethylation. Such changes in gene expression were observed in newly formed wheat allopolyploids (Shaked et al. 2001; Kashkush et al. 2002). Epigenetic changes can result in chromatin modifications or remodeling as well as alter the activity of small RNA molecules. Changes in microRNAs, such as miR168 which targets the *Argonaute1* gene, were shown to occur in newly synthesized wheat

(Kenan-Eichler et al. 2011). A high proportion of microRNAs showed nonadditive expression upon polyploidization, potentially leading to differential expression of important target genes (Li et al. 2014a, b). Their data may provide insights into small RNA–mediated dynamic homoeolog regulation mechanisms that may contribute to heterosis in nascent hexaploid wheat.

New interactions between regulatory factors of the parents, i.e., between the trans factor from one species and the cis or trans factors of the other parental species may account for the observed case of inter-genomic suppression (Galili and Feldman 1984; Tirosh et al. 2009). Genetic suppression or reduction of gene activity can be also caused by DNA elimination (Liu et al. 1998a; Kashkush et al. 2002). Methylation and demethylation of retrotransposons were also observed in wheat allopolyploids (Yaakov and Kashkush 2011a, b) affecting their state of activity (Sabot et al. 2005). Active retrotransposons, that constitute most of the DNA of the allopolyploids of this group but they arc normally transcriptionally silent, may silence or activate neighboring genes (Kashkush et al. 2003). Retrotransposon promoters retain activity under normal conditions and initiate either read-in transcripts of the transposon itself, or read-out transcripts into flanking host sequences (Kashkush et al. 2002; Kashkush et al. 2003; Kashkush and Khasdan 2007). Following allopolyploidization events in wheat, the steady-state level of expression of LTR retrotransposons was massively elevated (Kashkush et al. 2002, 2003). In many cases, these read-out transcripts were associated with the expression of adjacent genes, depending on their orientation: knocking-down or knocking-out the gene product if the read-out transcript was in the antisense orientation relative to the orientation of the gene transcript (such as the *iojap-like* gene); or over-expressing the gene if the read-out transcript was in the sense orientation (such as the *puroindoline-b* gene) (Kashkush et al. 2003). Recent studies on tracking methylation changes around a LTR retrotransposon in the first four generations of a newly formed wheat allopolyploid (Kraitshtein et al. 2010), indicate that this read-out activity is restricted to the first generations of the nascent polyploid species.

Functional diversification of duplicated genes, i.e., differential or partitioning of expression of homoeoalleles in different tissues and in different developmental stages, is also a form of genetic diploidization. Bottley et al. (2006) reported that differential expression of homoeoalleles in different plant tissues is common in hexaploid wheat. The activity of silenced genes could be restored in aneuploid lines, suggesting that no mutation was involved but rather new cis-trans interactions took place or reversible epigenetic alterations. Mochida et al. (2006) also presented evidence for differential expression of homoeoalleles in wheat and suggested that inactivation of homoeoalleles is a nonrandom effect. Similarly, subfunctionalization of all the homoeologues of the *Q*-locus in hexaploid wheat was recently described (Zhang et al. 2011).

Many studies on gene expression compare the expression level in the allopolyploid to those of its parents and/or to the average of its parents, expressed as the mid-parental value. In hexaploid wheat, Pumphrey et al. (2009) found that approximately 16 % of the 825 analyzed genes displayed nonadditive expression in the first generation of synthetic hexaploid wheat. Chague et al. (2010) analyzed 55,052 transcripts in two lines of synthetic allohexaploid wheat and found that 7 % of the genes had nonadditive expression, while Akhunova et al. (2010) found in synthetic allohexaploid wheat that about 19 % of the studied genes showed nonadditive expression. Li et al. (2014a, b) found that nonadditively expressed protein-coding genes were rare but relevant to growth vigor. Moreover, a high proportion of protein-coding genes exhibited parental expression level dominance. Similar studies by He et al. (2003) showed that the expression of a significant fraction of genes (7.7 %) was altered in the synthetic allohexaploid *T. turgidum-Ae. tauschii*, and that *Ae. tauschii* genes were affected much more frequently than those of *T. turgidum*. Interestingly, silencing of the same genes was found also in natural hexaploid wheat, indicating the reversibility of the effect and that the regulation of gene expression is established immediately after allohexaploidization and maintained over generations (He et al. 2003; Chague et al. 2010). In accord with these results, increased small interfering RNA density was observed for transposable element–associated D homoeologs in the progeny of newly formed allohexaploid wheat, which may account for biased repression of D homoeologs (Li et al. 2014a, b). It is of interest to note that several genes, that are silent in the parental species, became active in the newly formed allohexaploid (He et al. 2003). Similarly, cDNA-AFLP gels also revealed several cDNAs that were expressed only in the allopolyploids and not in the diploid progenitors (Shaked et al. 2001; Kashkush et al. 2002).

The rapid processes of cytological and genetic diploidization allow for the development and occurrence of two contrasting and highly important genetic phenomena in allopolyploid wheat that contribute to their evolutionary success: (1) build up and maintenance of enduring inter-genomic favorable genetic combinations (inter-genomic heterosis), and (2) genome asymmetry in the control of a variety of morphological, physiological and molecular traits, i.e., complete or principal control of certain traits by only one of the constituent genomes. However, while the first phenomenon was taken for granted by plant geneticists, genomic asymmetry was only recently documented in allopolyploids of the wheat group (Peng et al. 2003a, b; Fahima et al. 2006; Feldman and Levy 2009; Feldman et al. 2012). The Phenomenon of genome asymmetry is manifested in a clear-cut division of tasks between the constituent genomes of allopolyploid wheat (Levy and Feldman 2004; Feldman and Levy 2009; Feldman et al. 2012). Genome A controls morphological traits while genome B in allotetraploid wheat and genomes B and D in allohexaploid wheat control the reaction to biotic and abiotic stresses and regulate the adaptation to ecological conditions. Similarly, Li et al. (2014a, b) reported that genes for which the total homoeolog expression level in the progeny of newly formed allohexaploid wheat was similar to that in *T. turgidum* potentially participating in development and those with similar expression to that in *Ae. tauschii* involved in adaptation.

Inter-genomic pairing would have lead to disruption of the linkage of the homoeoalleles that contribute to positive inter-genomic interactions and, as well as, lead to segregation of genes that participate in the control of certain traits by a single genome. Inter-genomic recombination may result therefore, in many intermediate phenotypes that may affect, in a negative manner, the functionality, adaptability and stability of the allopolyploids.

Allopolyploid species undergo structural genomic changes also during the lifetime of the taxon (evolutionary changes) that generate a new variation that could not take place in the diploid parental genomes and that occur almost exclusively in an allopolyploid background. Allopolyploids harboring two or more different genomes within each nucleus, may facilitate inter-genomic horizontal transfer of chromosomal segments, transposable elements or genes between the constituent genomes. Inter-genomic invasion of chromatin segments from the B genome into the A genome was demonstrated by FISH analysis (Belyayev et al. 2000). Cytogenetic studies have shown that inter-genomic translocations occur in the allopolyploids of the wheat group (Maestra and Naranjo 1999, and reference therein). Moreover, in contrast to diploids, which are genetically isolated from each other and have undergone divergent evolution, allopolyploids in the wheat group exhibit convergent evolution because they contain genetic material from two or more different diploid genomes and can exchange genes with each other via hybridization and introgression, resulting in the production of new genomic combinations (Zohary and Feldman 1962).

The presence of duplication or triplication of the genetic material in tetraploids and hexaploids, respectively, has relaxed constraints on the function of the multiple genes enabling, in the long run, continued genetic diploidization such as silencing of one of the duplicated or triplicated genes or divergence of one homoeologous locus to a new function. Thus, the accumulation of genetic variation through mutations is tolerated more readily in allopolyploid than in diploid species (Dubcovsky and Dvorak 2007a, b).

Note that evolutionary changes might also occur in an accelerated manner, thanks to the buffering of mutations in the polyploid background (Mac Key 1954, 1958; Sears 1972; Dubcovsky and Dvorak 2007a, b), leading to rapid neo- or subfunctionalization of genes and to a further process of diploidization and of divergence from the diploid progenitor genomes. Akhunov et al. (2013) uncovered a high level of alternative splicing pattern divergence between the duplicated homeologous copies of genes in common wheat. Their results are consistent with the accelerated accumulation of alternative splicing isoforms, nonsynonymous mutations, and gene structure rearrangements in the wheat lineage, likely due to genetic redundancy created by allopolyploidization (Akhunov et al. 2013). Whereas these processes mostly contribute to the degeneration of a duplicated genome and its diploidization, they have the potential to facilitate the origin of new functional variations, which, upon selection in the evolutionary lineage, may play an important role in the origin of novel traits (Akhunov et al. 2013).

According to Stebbins (1950), newly formed allopolyploids are often characterized by limited genetic variation, a phenomenon he referred to as the "polyploidy diversity bottleneck." This bottleneck arises because only a few diploid genotypes were involved in the allopolyploid formation events, because the newly formed allopolyploid is immediately isolated reproductively from its two parental species and because time was not sufficient for the accumulation of mutations. Despite this diversity bottleneck and despite the fact that all *Aegilops* and *Triticum* allopolyploids were formed later than their ancestral diploids [e.g., tetraploid wheat about 300,000–500,000 years ago (Huang et al. 2002a) and allohexaploid wheat was formed only about 10,000 years ago (Feldman et al. 1995; Feldman 2001)], they display a greater genetic variation than their diploid progenitors. It is possible that polyploidy enabled genome plasticity that in turn enabled accelerated evolution to take place as proposed in wheat (Dubcovsky and Dvorak 2007a, b) and as was recently shown in experimental evolution studies in yeast (Selmecki et al. 2015). It might be also that recurrent hybridization with the diploid progenitors occurred during the evolutionary history of allopolyploids as discussed below. The diploid species have undergone divergent evolution and consequently, have diverse genomes that are isolated genetically from one another. They exhibit a relatively limited morphological and molecular variation, specialization in their ear structure, occupation of few well-defined ecological habitats, and distribution throughout relatively small geographical areas (Zhukovsky 1928; Eig 1929a, b; Zohary and Feldman 1962; Kimber and Feldman 1987; Van Slageren 1994). The allopolyploids show wider morphological variation, occupy a greater diversity of ecological habitats, and are distributed over larger geographical area than their diploid progenitors (Zhukovsky 1928; Eig 1929a, b; Zohary and Feldman 1962; Kimber and Feldman 1987; Van Slageren 1994). The distribution areas of most of the tetraploids overlap, completely or partly, with that of their two diploid parents and extend beyond it. They grow well in a very wide array of edaphic and climatic conditions and so do not show the marked ecological specificity of the diploids. Their weedy nature is reflected in their ability to colonize rapidly and efficiently a variety of newly disturbed and secondary habitats. Undoubtedly, the expansion of agriculture and the opening up of many new habitats played a key role in the massive distribution of these allopolyploid species throughout the range of the group (Zohary and Feldman 1962; Kimber and Feldman 1987).

Therefore, students of wheat evolution have been fascinated by this paradox and have dedicated themselves to unraveling the processes and mechanisms that contributed to the buildup of genetic and morphological diversity of the allopolyploids and to their great evolutionary success in term of proliferation and adaptation to new habitats, including under domestication.

In clear contrast to the rarity of inter-specific hybridization at the diploid level, hybridization between allotetraploid species, particularly between those sharing a common genome, is a frequent phenomenon (Zohary and Feldman 1962; Feldman 1965a). Such allotetraploid species tend to grow in mixed stands, and many F_1 hybrids as well as backcrossed progeny were repeatedly found in many localities in Israel, Turkey and Greece (Zohary and Feldman 1962; Feldman 1965a). Actually, a range of morphological intermediates between allotetraploids growing together in one mixed stand is a common phenomenon. Additional evidence for the existence of introgressed genomes in allopolyploid *Aegilops* were obtained from C-banding analysis (Badaeva et al. 2004). An introgression of a DNA sequence from allopolyploid common wheat to the allotetraploid *Aegilops* species, *Ae. peregrina,* was recently described (Weissmann et al. 2005). Hence, hybridization between allotetraploids, particularly between those sharing one common genome, and, to a lesser extent, between allotetraploids and diploids, facilitates a rapid buildup of genetic variability at the tetraploid level. The differential genomes that are isolated from one another at the diploid level genetically and geographically where emphasis is on divergence and specialization are brought together and allowed to recombine at the tetraploid level. Thus, the ability to exchange genetic material through spontaneous inter-specific hybridization promotes further the convergent evolution of these species. These reticulate patterns of evolution of course tend produce an increased range of variation and consequent difficulties in the identification of species. Natural hybrids also occur between the wild species and domesticated allopolyploid wheat. Hybrids have been recorded recurrently in *T. turgidum* between subsp. *durum* and its progenitor, wild emmer (subsp. *dicoccoides*), in Israel (Percival 1921; Feldman 2001; Huang et al. 1999; Dvorak et al. 2006; Luo et al. 2007) and between common wheat and wild emmer (Zohary and Brick 1961).

Considerable evidence was obtained for spontaneous hybridization also between allotetraploids and diploid species in mixed natural populations of tetraploids and diploids (Vardi and Zohary 1967; Vardi 1973; Zohary and Feldman unpublished). The occurrence of hybrid derivatives in such populations, particularly as a result of backcrossing to the allotetraploid parents, indicates the possibility of gene flow from diploid to tetraploid species.

In contrast to the wide distribution of the allotetraploid species, the distribution area of the natural allohexaploid species is, in all cases, smaller than that of their tetraploid and diploid parents. Also their ecological amplitudes are much more restricted than those of the related tetraploids and even the diploid parents. They grow in a smaller range of habitats and often are distributed sporadically. The morphological variation of the allohexaploids is also relatively limited. All these indicate a relatively recent origin of the allohexaploids,

On the basis of plant habitus, spike morphology, and cytogenetic data, Zohary and Feldman (1962) classified the allopolyploid species of *Triticum* and *Aegilops* into three natural clusters (Table 2.8). Genome analysis of the allopolyploids within each cluster showed that they share one unaltered genome (the pivotal genome) and a genome or genomes that is/are modified [the differential genome(s)]. In laboratory hybridization studies of allopolyploids, it was found that the common genome acts as a buffer ensuring some seed fertility following pollination by the female parent, while the chromosomes of the differential genomes, brought together from different parents, may pair to some extent and exchange genes (Feldman 1965b, c). Consequently, the dissimilar genomes of these allopolyploids contain chromosomal segments that originated from two or more diploid genomes. Such genomic constitution reveals different evolutionary rates for each of the two or three genomes of every allopolyploid.

Thus, all seven tetraploids and one hexaploid of the U-genome cluster share a genome homologous to that of diploid *Aegilops umbellulata* (Kihara 1954), all the three tetraploids and three hexaploids of the D-genome cluster share a genome homologous to that of diploid *Ae. tauschii* (Kihara 1954; Kihara et al. 1959), and all two tetraploids and two hexaploids of the A-genome cluster, including all the wild and domesticated forms, share a genome homologous to that of diploid *Triticum urartu* (Dvorak 1976; Chapman et al. 1976). The allopolyploids of each cluster resemble the diploid donor of the shared genome in their basic morphology (stature, leaf shape, and spike and spikelet morphology) and in the structure of the seed dispersal unit. They differ in features of the differential genome(s) that are primarily

Table 2.9 Time of formation of the allopolyploid species of *Triticum* in million years ago

Species	Time of formation	Method of study	Reference
Triticum turgidum subsp. *dicoccoides* (wild emmer)	<0.500	Nucleotide sequences of two nuclear genes	Huang et al. (2002b)
	0.360 (0.190–0.540)	Nucleotide sequences of four nuclear genes	Dvorak and Akhunov (2005)
	<0.800	Nucleotide sequences of several hundreds nuclear genes	Marcussen et al. (2014)
	0.700	Sequencing chloroplast DNA	Gornicki et al. (2014)
Triticum timopheevii subsp. *armeniacum* (wild timopheevii)	0.050–0.300	RFLP analysis of nuclear DNA	Mori et al. (1995)
	0.400	Sequencing Chloroplast DNA	Gornicki et al. (2014)
Triticum aestivum subsp. *aestivum* (common wheat)	0.008	Nucleotide sequences of two genes	Huang et al. (2002b)
	0.008	Nucleotide sequences of four nuclear genes	Dvorak and Akhunov (2005)
	<0.400	Nucleotide sequences of several hundreds nuclear genes	Marcussen et al. (2014)

responsible for the eco-geographical adaptation of the various allopolyploid species in each cluster. Thus, inter-specific hybridization has played a decisive role in the production of a wide range of genetic variation in the allopolyploid species and probably significantly contributed to their evolutionary success.

Huang et al. (2002b), Dvorak and Akhunov (2005), and Marcussen et al. (2014), based on sequencing of nuclear genes, proposed that the wild emmer, *T. turgidum* subsp. *dicoccoides*, was formed about 360,000–700,000 years ago (Table 2.9). Gornicki et al. (2014), based on sequencing of chloroplast genomes, concluded that *Aegilops speltoides* forms a monophyletic clade with the allopolyploid *Triticum* species emmer and *timopheevii*, which originated within the last 0.7 and 0.4 MYA, respectively (Table 2.9). Mori et al. (1995), using RLFP analysis of nuclear genes, assumed that wild *timopheevii*, *T. timopheevii* subsp. *armeniacum*, was formed about 400,000 years ago. The geographic distribution of chloroplast haplotypes of the wild tetraploid wheats and *A. speltoides* illustrates the possible geographic origin of the Emmer lineage in the southern Levant and the *timopheevi* lineage in northern Iraq (Gornicki et al. 2014). This is in accord with the finding of Dvorak (personal communication) that wild emmer was formed in the vicinity of Mt Hermon. *Aegilops speltoides* is the closest relative of the diploid donor of the chloroplast and the nuclear genomes to emmer and *timopheevii* lineages (Gornicki et al. 2014).

Middleton et al. (2014) estimated that the B genome donor to allopolyploid wheat diverged from *Ae. speltoides* approximately 0.98 MYA (Table 2.10). The divergence time of genome A^m of *T. monococcum* and A of *T. urartu* is 0.57 MYA (Middleton et al. 2014), that between the A genome of *T. urartu* and A genome of *T. aestivum* is 0.58–0.82 MYA and that of D genome of *Ae. tauschii* and D of

Table 2.10 Beginning-divergence time of the genomes of diploid species and the homologous genomes of *Triticum aestivum* in million years ago

Genomes	Beginning divergence time	Type of DNA sequences studied	Reference
Genome A of *T. urartu* and genome A of *T. aestivum*	0.58–0.82	Several hundreds nuclear genes	Marcussen et al. (2014)
Genome S of *Ae. speltoides* and genome B of *T .aestivum*	0.98	Chloroplast genome	Middleton et al. (2014)
Genome D of *Ae. tauschii* and genome D of *T .aestivum*	0.28–0.47	Several hundreds nuclear genes	Marcussen et al. (2014)

T. aestivum to 0.28–0.47 MYA (Marcussen et al. 2014) (Table 2.10). They suggested that the age of the time of the allopolyploidization events might be older than previously suggested. Since the hybridization between the donor of the B genome and *T. urartu* leading to the formation of allotetraploid wheat was soon after the divergence of *T. urartu*, i.e., 0.36 to 0.7 MYA (Huang et al. 2002b; Dvorak and Akhunov 2005; Marcussen et al. 2014) and that of *T. aestivum* about 0.01 MYA (Feldman and Levy 1995; Feldman 2001), the earlier divergence time between the genomes of the diploid species and those of the polyploid ones indicate that the intra-specific divergence has occurred at the diploid level, namely, in *Ae. speltoides, T. monococcum/ urartu and Ae. tauschii.* The divergence of the S genome of *Ae. speltoides* and the B genome of polyploid wheat might have been followed by a speciation and if so, the donor of the B genome is either extinct or yet undiscovered extant species.

Whereas the diploid species of the group have evolved in the Pliocene the tetraploids have been produced later on, in the Pleistocene (Table 2.3). The polyploids species of *Aegilops* were produced in the east Mediterranean region and have spread out from this center (Kihara 1954).

2.6 Evolution of Wheat Under Cultivation

During the first several millennia of cultivation genotypes with nonfragile spikes of einkorn and emmer gradually replaced the wild genotypes (Feldman and Kislev 2007). It is assumed that the mutants that founded domesticated einkorn and emmer happened in a small number of wild plants. Similarly, the number of hexaploid plants produced by independent hybridization events between domesticated tetraploid wheat and *Ae. tauschii* was limited. Hence, the primitive domesticated *Triticum* species contained a narrow and very restricted genetic variability. Yet, the initial nonbrittle einkorn (*monococcum*) and emmer plants grew for a long period in mixed populations with wild forms (Tanno and Willcox 2006; Kislev 1984; Feldman and Kislev 2007) and presumably exchanged genes with them. Moreover, even after the complete replacement of wild emmer by domesticated emmer, the latter could continued to introgress with wild genotypes that grew in its vicinity and thereby, broadened its genetic basis (Huang et al. 1999; Dvorak et al. 2006; Luo et al. 2007).

Hexaploid wheat, that is only partially isolated from tetraploid wheat, exchanged genes with wild emmer that grew nearby (Zohary and Brick 1961). Gene flow between different domesticated varieties and species has also played an important role in increasing the genetic variability. For thousands of years farmers have been growing mixtures of different genotypes and even different cytotypes (Zeven 1980). Such cytotype mixtures included representatives of two or even three different species of domesticated wheat, namely, *T. monococcum, T. turgidum* (emmer and naked tetraploid wheat) and *T. aestivum* (spelt and bread wheat), each represented by a number of different genotypes, thus facilitating massive inter- and intra-specific gene flow. These endless hybridizations during the 10,000 years of cultivation have enriched the domesticated gene pool.

Mutations have also played an important role in increasing genetic variability. The genetic structure of the allopolyploid species of wheat, i.e., the existence of four (in tetraploids) or six (in hexaploids) doses of gene loci, reinforced by a diploid-like cytological behavior and predominantly self-pollination, has proven a very successful genetic system facilitating a rapid buildup of genetic diversity. In such allopolyploid species the accumulation of genetic variation through mutation or hybridization is tolerated more readily than in diploid species. Recent data show that allopolyploidization can cause rapid genomic changes and, thereby, accelerate their evolution (Levy and Feldman 2004; Feldman and Levy 2005). Moreover, allopolyploidy facilitates genetic diploidization—the process whereby existing genes in multiple doses can be diverted to new functions. Thus, tetraploid and hexaploid wheat can accumulate a significant amount of genetic variation through mutations. Mutations exerting a lethal or semi-lethal effect at the diploid level, such as *Q, s,* (the *sphaerococcum* gene) *C*, (the *compactum* gene) and *Ph1*, have been successfully utilized at higher levels of ploidy. Induction of mutations may have been accelerated by the activity of transposable elements (Kashkush et al. 2003) and genome-restructuring genes (Feldman and Strauss 1983).

The activation of transposable elements, mainly retrotransposons by various environmental and climatic stresses has an important evolutionary significance. In addition to the generation of genetic variability due to epigenetic changes, e.g., DNA methylation, chromatin acetylation and activity of various small RNA molecules, genome restructuring may lead to the formation of new linkage groups. Rapid chromosomal rearrangements can also contribute to the build up of partial isolating mechanisms between differentiating genotypes in polymorphic fields. Activity of genome-restructuring genes as well as transposable elements during wheat cultivation may explain the wide occurrence of chromosomal rearrangements among domesticated wheat taxa.

Being native to the marginal Mediterranean habitats of the Fertile Crescent, wild emmer, the wild progenitors of domesticated tetraploid wheat, was pre-adapted for domestication. It is predominantly self-pollinated species with large and well-protected grains that assist the safe and rapid reestablishment of the stand. Its large seed size rendered it very attractive to the ancient gatherer. Its annual habit made it also amenable for dry farming, while its self-pollination system could have aided in the fixation of desirable mutants and recombinants resulting from occasional outcrossing events. While in its natural habitats this grass occupy poor, thin, rocky soils, it responded well when transferred to richer habitats.

Domestication imposed a new evolutionary direction whereby traits that had the greatest adaptive value in the cultivated field were preferred. Selection pressures have operated in a different, and sometimes, contradictory manner in cultivation and in the wild. During the 10,000 years of wheat cultivation, the criteria for selection varied from time to time and from place to place. Wheat cultivars have been developed through three main phases of selection: occasional and sometimes nonintentional selection, exerted by the earliest farmers simply by the processes of harvesting and planting; more deliberate selection by traditional farmers in polymorphic fields; and selection as part of scientifically planned modern breeding.

During the first phase, a very significant sequence of changes occurred in the transition from the wild into the domesticated forms. Several wild characters that had no pre-adaptive value for cultivation were selected against. These characters included the wild seed dissemination, seed dormancy, and seed protection by the tightly closed glumes.

Wild wheat has a brittle spike that upon maturity disarticulates into arrowhead-shaped spikelets. While these seed-dispersal units facilitate self-burial in the soil (hence protection during the long, dry summer and successful germination after the first rain), they must have proved a nuisance to the ancient farmer who had to collect most of the spikelets from the ground or cut the culms before the grains matured. No wonder, therefore, that those plants with brittle heads were selected against. Yet, it took about a millennium or more until mutants for nonbrittleness appeared in the cultivated fields and gradually became the dominant crops (Feldman and Kislev 2007).

Wild emmer was domesticated through the loss of spike fragility. This step was probably a gradual process as suggested from both genetic and the archeological evidence. The archeological record shows that in ancient sites where agriculture was practiced, a mixture of fragile and nonfragile types were found, and it took 3–4 thousand years until the nonfragile spikes became prominent in farming units (Kislev 1984 (in emmer wheat); Tanno and Willcox 2006 (in einkorn)). Two major genes, *brittle rachis 2* (*Br-A2*) and *brittle rachis 3* (*Br-A3*) located on the short arms of chromosomes 3A and 3B, respectively, control the brittleness of the rachis in wild emmer (Levy and Feldman 1989a; Watanabe and Ikebata 2000; Watanabe et al. 2002, 2005; Nalam et al. 2006; Millet et al. 2013); in addition, another locus for spike brittleness was mapped to chromosome 2A (Peng et al. 2003a, b; Peleg et al. 2011). Comparative mapping analyses suggest that both *Br-A2* and *Br-A3* are present in homoeologous regions on chromosomes 3A and 3B, respectively. Furthermore, *Br-A2* and *Br-A3* from wheat and *Btr1/Btr2* on chromosome 3H of barley (*Hordeum vulgare* L.) also are homoeologues suggesting that the location of major determinants of the brittle rachis trait in these species has been conserved (Nalam et al. 2006). The loss of fragility gave rise to the first known domesticated wheat, *Triticum turgidum* ssp. *dicoccon*, or domesticated emmer wheat, which is grown to this day, albeit on a small scale (De Vita et al. 2006).

The second requirement for the newly domesticated wheat was uniform and rapid germination. Wild wheat exhibits two types of dormancy: a post-harvest type and a long-range type. The first type prevents too early and untimely germination—an important feature, pre-adapted to agriculture, particularly in view of the fact that seeds were often kept under unsuitable storage conditions. The second type of dormancy ensures in nature a temporal distribution of germination: it is invariably the larger grain of the second floret in each spikelet that germinates in the first year, while the smaller grain of the first floret germinates in the second year. However, since this dormancy is induced by the palea, it was overcome by threshing (unpublished data).

The loss of self-propagation and of the temporal control of germination was followed by the loss of self-protection. The wild forms have tightly-closed glumes resulting in a "hulled" grain after threshing. Several primitive forms of domesticated wheat species, e.g., *T. monococcum* subsp. *monococcum, T. turgidum* subsp. *dicoccon*, and *T. aestivum* subsp. *spelta, macha,* and *vavilovii,* retain this feature. Hence, the appearance of naked kernels was the second most important step in domestication of tetraploid wheat after nonbrittle spikes. How did the free-threshing tetraploid wheat evolve from *dicoccon*? Did it get its naked kernels directly, through mutations in the genes that control glume stiffness (the Q factor and the *Tenacious Glumes* (*Tg*) locus) or, was there another intermediate step? *Triticum turgidum* ssp. *parvicoccum,* a tetraploid wheat "fossil" species (Kislev 1980) was relatively abundant in the archeological record starting already 9000 years BP, but disappeared ~ 2000 years BP, had a compact spike and was free-threshing and thus probably already contained the Q and *tg* mutations prior to durum (Feldman and Kislev 2007). This raises the possibility that *durum* received these mutations from *parvicoccum*, rather than evolving them independently from emmer wheat. Durum may thus derive from hybridization between *parvicoccum* and *dicoccon* receiving the free-threshing trait from *parvicoccum* and the large grains from *dicoccon*. The large grain of durum was probably preferred to the small grains of *parvicoccum* that lead to the prominence of durum as a tetraploid wheat and to the extinction of *parvicoccum.* Similarly, it can be assumed that common wheat, *T. aestivum* subsp. *aestivum*, received these mutations from tetraploid wheat but required additional mutation in the Tg gene of genome D (Kerber and Rowland 1974).

The *Q* gene, located on the long arm of chromosome 5A, is one of the most significant domestication loci as it controls the free-threshing character and several other domestication-related traits such as glume shape and tenacity, rachis fragility, spike length and shape (square-head spike), plant height, and spike emergence time. While the 5A homeoallele has the most significant contribution, other homeoalleles (on 5B and 5D) were also shown to be involved in the domestication traits (Zhang et al. 2011). The *Q* gene is the only one that was so far characterized at the molecular level. Simons et al. (2006) isolated the *Q* gene and verified its identity by analysis of knockout mutants and transformation. The *Q* allele is more abundantly transcribed than *q*, and the product of the two alleles differs for a single amino acid at position 329, isoleucine in the *Q* protein and valine in the *q* protein. The *q* allele is the more primitive, and the mutation that gave rise to *Q* occurred only once leading to the world's domesticated wheat. *Q* is thought to be a major regulatory gene for floral development (Muramatsu 1986). It encodes an *AP2*-like transcription factor that played an important role in the domestication of polyploid wheat. (Chantret et al. 2005; Simons et al. 2006; Zhang et al. 2011).

The higher expression of *Q* than *q* is in accord with the finding of Muramatsu (1963) that extra doses (five or six) of *q* mimic the effect of *Q* in common wheat. In addition to *Q*, Kerber and Rowland (1974) found the *Tg* (tenacious glumes) gene, located on chromosome arm 2DS, confers tough glumes. Chromosome arm 2BS of emmer also contains the *Tg* gene that determines tough glumes (Simonetti et al. 1999), and there is a possibility that also chromosome arm 2AS contains such a gene and, therefore, the free- threshing trait might be determined in tetraploid wheat by at least two complementary genes, *Q* and *tg*. Thus, at least several mutations were required to produce the free- threshing character in ssp. *parvicoccum* (Jantasuriyarat et al. 2004).

The genes determining tenacious glumes and soft glumes are two independent loci that affect glume tenacity and spike threshability (Sood et al. 2009).

In addition to the above-mentioned classical domestication traits that were selected in the process of durum evolution, other domestication traits were selected that were advantageous to the farmer, such as plant erectness versus the prostrate grassy types, increased number of seeds per spikelet, and reduced seed dormancy (Feldman 2001). Several domestication-related QTLs that affected various traits were mapped (Peng et al. 2003a, b; Peleg et al. 2009) but the underlying genes were not identified at the molecular level. It is likely that some other QTLs were selected that are not easily visible to the eye, such as resistance to abiotic and biotic stresses, physiological parameters that contribute to yield (Peleg et al. 2009), increased grain size (Gegas et al. 2010) as well as quality parameters (Levy and Feldman 1989b) and in recent decades, following the green revolution, adaptation to the new cultivation conditions including chemical fertilizers and mechanical harvest.

The main achievements of the first phase were therefore nonbrittleness of the spikes, simultaneous ripening of grains, rapid and synchronous germination, and possibly also larger grains, erect rather than prostrate culms, and free threshing (naked grains).

During the second phase of evolution under cultivation wheat culture spread into new areas—an event that required the adaptation of wheat to new climatic, edaphic and biotic conditions. During the spread to Europe, for example, the plant became taller and its leaves as well as spikes and grains increased in size. Photoperiodic and thermoperiodic responses were modified to achieve an optimum balance between vegetative and reproductive phases: the vegetative period was extended to take advantage of the longer summer days and rainy season, while the need for high temperatures for maturation gradually disappeared. In addition, the grain-filling period became longer and flag-leaf senescence was delayed. These latter adaptations, allowing for larger amounts of carbohydrates to be assimilated and translocated to the developing grains, greatly contributed to higher yields through an increase in grain size and number.

This phase involved a long and continuous selection for various agronomic and technological characters in the polymorphic fields of the traditional farmers. In such fields, numerous genotypes were grown in mixtures as landraces. Most fields even contained a mixture of cytotypes, representing tetraploid and hexaploid wheats, and occasionally diploid wheat as well (Zeven 1980). Hence, the unit of selection was a combination of genotypes rather than a single one.

The economic advantage of growing various genotypes, and even species, as a mixture in one field was yield stability—an economic consideration which was much more important than occasional high yield (Zeven 1980). The traditional farmers preferred a "safe", average yield each year, rather than a high yield for several years that might have been followed by a crop failure. The lack of suitable means for long storage and for large-scale wheat-import rendered a crop failure in any year to have drastic consequences. Hence, yield stability was of utmost importance.

In such polymorphic fields inter-genotypic competition played a decisive role. Plants with horizontal leaves, which shade weeds and competitors, had an advantage over genotypes with erect leaves. Also high tillering and vigorous vegetative growth were traits with a high adaptive value.

The main evolutionary advantage in such fields was the possibility for occasional hybridization between genotypes and species. Because of the self-pollinating system, hybridizations resulted in numerous homozygous recombinants, thereby constantly providing the farmer with new stable genotypes for selection.

Since farmers selected and planted grains that were most desirable for their specific needs, selection pressures were exerted consistently but in different directions in various locales. These efforts resulted in increased plant height, increased tillering, development of canopy with wide horizontal leaves, larger seed size, increased grain number per spikelet, better flour quality, improved seed retention (non shattering), increased competitiveness with other wheat genotypes and weeds, and better adaptation to a wider range or climatic and farming regimes.

Under scientifically planned modern breeding, starting at the end of the nineteenth century, the wheat field has become genetically uniform and no longer conducive to spontaneous gene exchange. On the other hand, large-scale gene migration has been promoted by worldwide-introduction services. Massive scientific screening has aided in revealing desirable genes, and modern methods for manipulating and transferring these genes from one genetic background to another became available. Hybridizations have been confined mainly to intra-specific crosses. Lately, however, some inter-specific and inter-generic crosses are also being performed. Individual genotypes, rather than mixtures, became the unit of selection. Selection has been made mostly for traits that improve wheat performance in dense stands, such as minimum intra-genotypic competition, upright leaves to improve light penetration and prevent shadowing of neighboring plants, low tillering, and a higher number of seminal roots whose development is non-dependent on tillering, enhanced response to fertilizers and agrichemicals, increased resistance to diseases, pests, and lodging, improved harvest index, as well as improved baking and bread-making quality. The general goal was to achieve highest possible yields per area the main limiting factor (globally) is water availability. Plants were selected to address this goal and constraint.

Modern high-yielding cultivars are semi-dwarf (90–120 cm) or dwarf (60–90 cm), which have replaced the conventional taller (120–140 cm) cultivars. These cultivars, containing genes for short stature, the *Rht* genes (Gale and Youssefian 1985), respond well to new agrotechnical practices, particularly to high application rates of fertilizers, without lodging. Thc high performance of these cultivars results from an increase in the number of fertile florets per spikelet and, sometimes, to the length or density of the spike, reduction in shattering, and resistance to lodging as well as to fungal, bacterial and viral diseases and pests. During the course of "the green revolution", the semi-dwarf and dwarf varieties were taken to India, Pakistan, Iran and the Mediterranean basin replacing the numerous land races in every locale. Today, the dwarfing genes have been widely incorporated into most existing cultivars.

The main achievements in breeding for grain quality have been improvements in milling and baking characteristics. Certain modern cultivars are easily milled because the pericarp and seed coat are only loosely attached to the endosperm. Flour yield is particularly higher in varieties with short, almost spherical grains. To date, less progress has been made in improving the nutritional value of the grain, and further efforts are needed, particularly to increase the protein content and remedy deficiencies in amino acids composition.

From the dawn of agriculture, cultivated wheat was under constant selection pressure to increase grain yield; prolificacy became more important than seed efficiency, since successful seedling establishment in the cultivated field was achieved by providing the seedlings with optimal growth conditions. According to Evans (1981), the improvement in grain productivity was achieved through increasing leaf size and flag-leaf area, delay flag-leaf senescence, increasing the size of the vascular system in the spike, advancing flowering time, increasing rate and duration of grain-filling period, increasing rate and duration of assimilates translocation to the grains, increasing grain size and grain number per spikelet, increasing spike number per plant or per unit area. One of the most important changes during cultivation was an increase in the proportion of the dry matter allocated to the harvested grains (harvest index). On the other hand, there were few or no significant changes during wheat cultivation in the photosynthetic capacity per unit leaf area, in growth rate, or in the accumulated dry weight of the crops (biomass).

Compared to domesticated wheat, wild emmer, subsp. *dicoccoides*, is characterized by a later anthesis and earlier grain ripening. This shorter grain filling is apparently the consequence of relatively early and rapid flag-leaf senescence occurring about 2 weeks after anthesis. The degradation of most leaf-proteins at this stage reduces and eventually ceases completely the capacity for carbohydrates assimilation in the leaves, resulting in a higher N/C ratio in the assimilates translocated to the grains. Indeed, all wild emmer lines analyzed had high grain protein percentage (GPP) (Avivi 1977, 1979a, b; Feldman et al. 1990). High GPP may contribute to seedling vigor (Millet and Zaccai 1991). Under the semi-arid conditions that prevail in the natural habitats of wild stands, a small number of medium-sized grains with high protein content suffice to ensure rapid germination and successful establishment of the next stand.

Among the traits that were affected by domestication are the storage proteins, in particular the high molecular weight glutenins whose variability and amounts are higher in wild than in domesticated tetraploid wheat (Levy and Feldman 1988; 1989b; Laido et al. 2013). Recently, a *NAC* genes from emmer wheat that contributes to high protein percent, a trait that affects both the nutritive value and the processing of wheat and was lost during domestication, has been isolated (Uauy et al. 2006)

Joppa and colleagues (Joppa and Cantrell 1990; Joppa et al. 1991; Joppa 1993) found that genes on chromosomes 6B, 2A, 5B, and 6A of wild emmer increased grain protein percentage (GPP) in durum wheat. Genotypes with wild emmer chromosomes 6B had the highest GPP (Cantrell and Joppa 1991). Olmos et al. (2003) mapped the high protein gene of chromosome 6B as a Mendelian locus and named it *Gpc-B1*. This gene was isolated and found to be responsible for earlier senescence and shorter grain-filling period in wild emmer (Uauy et al. 2006). In domesticated free-threshing durum and common wheat, this gene was mutated to a nonfunctional allele resulting in longer grain-filling period (more than 3 weeks longer) and consequently, a higher percentage of starch and lower percentage of proteins and minerals in the grains (Uauy et al. 2006). Since the wild type allele exists in most analyzed domesticated emmer (subsp. dicoccon) lines, it is assumed that the mutation occurred in free-threshing tetraploid wheat (Uauy et al. 2006).

The increased yield in cultivated wheat stems to a large extent from delayed flag-leaf senescence, which prolongs the duration of post-anthesis photosynthesis. This trait can also be achieved by accelerating spike development, which advances anthesis by a few days. In semi-arid regions, where the available amount of water is often a limiting factor during grain development, senescence is delayed for only a short period while in mesic regions senescence and, consequently, maturity in general are delayed for longer periods.

The recent development of new arsenal of genomic tools may facilitate the identification of additional loci that control domestication-related traits in wheat. There has been remarkable progress in the amount of datasets and tools for wheat genomics. Whole genome sequences are also available for the A (Ling et al. 2013) and D (Jia et al. 2013) genomes, yet a good assembly of contigs is still missing. SNP mapping in a broad collection of wheat landraces and modern varieties has indicated the genomic regions that underwent selection (selective sweep) during post-domestication wheat breeding (Cavanagh et al. 2013). Recently, a major advance in *durum* transcriptome analysis was the development of tools for the discrimination of homeologues from the A and B genomes from expression sequence data such as RNA-Seq (Krasileva et al. 2013). Data sets from small RNAs are also becoming available (Kenan-Eichler et al. 2011; Yao and Sun 2012).

A comparison of wild and domesticated wheat may reveal changes that have occurred under domestication. Ben-Abu et al. (2014) focused on the analysis of genomic changes that are correlated with the process of domestication and evolution of modern durum by comparing four genetic groups: wild emmer, domestic emmer, durum landraces and modern durum varieties. Changes in gene expression and copy number variation of genes and transposons were analyzed in these four groups. Genes were clustered based on their pattern of change in expression during durum evolution, e.g. gradual increase, or decrease, or increase at the onset of domestication and plateauing later on. There were not many genes that changed >2 fold in copy number. However, interestingly, the copy number of transposons increased with domestication, possibly reflecting the genomic plasticity that was required for adaptation under cultivation. Extensive changes in gene expression were seen in developing grains. For example, there was an enrichment for certain functions: genes involved in vesicle trafficking in the endosperm showed a gradual increase in expression during durum evolution and genes related to germination and germination inhibition increased in expression in the embryo in the more recent stages of durum evolution (Ben-Abu et al. 2014). Yuan et al. (2014) described differential expression of the CENH3 genes between wild and domesticated tetraploid wheat. *Triticum araraticum ssp. armeniacum*, the wild progenitor of *Triticum timopheevii* subsp. *timopheevii*, had a higher transcript level of αCENH3 while the domesticated subspecies had a lower expression of αCENH3 and increased expression of βCENH3. Similar changes in the CENH3 expression model were found in wild and in domesticated types of *T. turgidum*; the wild subspecies of *T. turgidum* exhibited higher expression level of αCENH3 whereas in the domesticated subspecies the differences in expression between αCENH3 and βCENH3 were not so obvious. In contrast to the markedly higher expression level of αCENH3 in wild tetraploids, expression of βCENH3 was enhanced to a level near that of αCENH3 in domesticated tetraploids (Yuan et al. 2014). These genome-wide analyses are providing insight into the molecular basis of plant domestication, in particular for the non-obvious cellular functions that were selected during evolution under cultivation.

2.7 Concluding Remarks

Evolution in the Triticeae is a complex network of general processes and singular events which have occurred over long periods of time and which are still happening in a dynamic environment. Perhaps the most conspicuous feature of the evolution of the group is the divergence of several forms at the diploid level and their convergence at the polyploid one. This reticulate pattern of evolution provides the basis for many of the characteristics of individual diploid species and the often-unclear demarcations between closely or partially related allopolyploid species.

Although the diploid species of the wheat group are presumably descended from a common ancestor, they have diverged considerably from one another in a relatively short time (2.0 to 4.5 MYA). Cytogenetic studies have shown that almost every species has a distinct genome. The homoeologous chromosomes of the different genomes exhibit varying degrees of reduced affinity for one another, and consequently, they do not pair regularly at meiosis in the inter-specific hybrids. As a result, almost all of the inter-specific hybrids are completely or almost completely sterile and thus, the species are genetically isolated from one another. So, the diploid level is characterized by chromosomal and genetic divergence that is expressed in the considerable eco-geographical and morphological specialization that is evident in the well-defined dispersal unit marking each diploid species.

The allopolyploids in the wheat group have experienced a multifaceted process of evolution, in nature, through allopolyploidization and natural hybridization between polyploids and, under domesticaton, in the case of the *Triticum* species. Early cytogenetic studies showed that the polyploids of the wheat group are allopoyploids. Morphologically and cytogenetically the allopolyploid species are clustered in three natural groups: one sharing the U genome of *Ae.umbellulata*, one the D genome of *Ae. tauschii*, and one the A genome of *T. urartu*. The A-genome cluster includes all the domesticated forms of wheat. The species of each group share one common genome and differ in their other genome in tetraploids and their two other genomes in hexaploids. In basic morphology and particularly in the structure of the seed dispersal unit, the allopolyploids of each group resemble the diploid donor of the common genome and differ in features of the other genomes. This genomic structure accounts for the relatively high rate of successful gene flow between the allopolyploids that belong to one cluster resulting in the production of new recombinant genomes. Hence, genomes that are isolated at the diploid level are able to exchange genes and recombine at the polyploid one.

The above reviewed studies indicate that in the wheat group the process of allopolyploidization generates a genetic shock that induces a burst of genetic and epigenetic, non-Mendelian, alterations, (referred to as revolutionary changes), some of which could not occur at the diploid level. Some of these changes presumably improve the ability of the newly formed allopolyploids to survive in nature and to compete with their parental species. Transposable elements seem to play a significant role in the various responses to allopolyploidization due to their abundance and also due to their tendency to be dysregulated as a result of genomic shocks.

Other changes occur sporadically over a long time period during the life of the allopolyploid species (evolutionary changes). From a population point of view, the chances of a new individual, such as a nascent hybrid/allopolyploid, to establish itself as a new species is almost null, unless it has some increased fitness over its parents. This must happen within a few generations otherwise the new species will rapidly be extinct. The revolutionary changes described here may contribute to the establishment of the new species. Instantaneous elimination of sequences from one genome in the newly formed allopolyploids increases the divergence of the homoeologous chromosomes and thus, leads to exclusive intra-genomic pairing that improves fertility. Mechanisms such as loss of deleterious genes (e.g. genetic incompatibilities), or positive dosage effects or new inter-genomic heterotic interactions may all rapidly increase the fitness of the nascent species. The evolutionary changes, on the other hand, contribute to the build up of genetic variability and thus, increase adaptability, fitness, competitiveness, and colonizing ability.

Altogether, the reported revolutionary and evolutionary genomic changes, emphasize the dynamic plasticity of the wheat allopolyploid genome with regards to both structure and function. Presumably these changes have improved the adaptability of the newly formed allopolyploids and facilitated their rapid colonization of new ecological niches. No wonder, therefore, that domesticated allopolyploid wheats exhibit a wider range of genetic flexibility than diploid wheats and could adapt themselves to a great variety of environments.

The evolutionary relationships of the *Aegilops*, *Amblyopyrum* and *Triticum* species are not confined to the generic boundaries of the wheat group. It is obvious from the ability of chromosomes of alien species generally to substitute for the chromosomes of only one homoeologous group that homoeologous relationships exist between the chromosomes of the wheat group and other genera in the Triticeae.

The wheat group can therefore be viewed as part of a greater continuum of genetic relationships extending to many other grasses. This abundant gene pool containing many economically important traits can be utilized for the improvement of domesticated wheat.

References

Akhunov ED, Sehgal S, Liang H, Wang S, Akhunova AR, Kaur G, Li W, Forrest KL, See D, Šimková H, Ma Y, Hayden MJ, Luo M, Faris JD, Doležel J, Gill BS (2013) Comparative analysis of syntenic genes in grass genomes reveals accelerated rates of gene structure and coding sequence evolution in polyploid wheat. Plant Physiol 161:252–265

Akhunova AR, Matniyazov RT, Liang H, Akhunov ED (2010) Homoeolog-specific transcriptional bias in allopolyploid wheat. BMC Genomics 11:505

Al-Kaff N, Knight E, Bertin I, Foote T, Hart N, Griffiths S, Moore G (2008) Detailed dissection of the chromosomal region containing the *Ph1* locus in wheat *Triticum aestivum*: with deletion mutants and expression profiling. Ann Bot 101:863–872

Alnaddaf LM, Moualla MY, Haider N (2012) the genetic relationships among *Aegilops* L. and *Triticum* L. species. Asian J Agric Sci 4:352–367

Ankori H, Zohary D (1962) Natural hybridization between *Aegilops sharonensis* and *Ae. longissima*: a morphological and cytological study. Cytologia (Tokyo) 27:314–324

Avivi L (1977) High grain protein content in wild wheat. Can J Gent Cytol 19:569–570

Avivi L (1979a) Utilization of *Triticum dicoccoides* for the improvement of grain protein quantity and quality in cultivated wheats. Monogr Genet Agric 4:27–38

Avivi L (1979b) High grain protein content in wild tetraploid wheat Triticum dicoccoides Körn. Proceedings of 5th International Wheat Genetics Symposium. New Delhi, India, Vol. 1, pp. 372–380

Avivi L, Feldman M (1973) Mechanism of non-random chromosome placement in common wheat. In: Sears ER, Sears LMS (eds.), Proceedings of 4th International Wheat Genetics Symposium, Columbia, Missouri, pp. 627–633

Badaeva ED, Friebe B, Gill BS (1996) Genome differentiation in *Aegilops*. Distribution of highly repetitive DNA sequences on chromosomes of diploid species. Genome 39:293–306

Badaeva ED, Amosova AV, Samatadze TE, Zoshchuk SA, Shostak NG, Chikida NN, Zelenin AV, Raupp WJ, Friebe B, Gill BS (2004) Genome differentiation in Aegilops. 4. Evolution of the U-genome cluster, Plant Syst. Evol. 246: 45–76

Baum BR, Feldman M (2010) Gene loss of paralogous 5S DNA unit classes in newly formed allopolyploids of *Aegilops* and *Triticum*. Genome 53:430–438

Baum BR, Estes JR, Gupta PK (1987) Assessment of the genomic system of classification in the Triticeae. Am J Bot 74:1388–1395

Belyayev A (2013) Chromosome evolution in marginal populations of *Aegilops speltoides*: causes and consequences. Ann Bot 111:531–538

Belyayev A, Raskina O, Korol A, Nevo E (2000) Coevolution of A and B genomes in allotetraploid *Triticum dicoccoides*. Genome 43:1021–1026

Belyayev A, Kalendar R, Brodsky L, Nevo E, Schulman AH, Raskina O (2010) Transposable elements in a marginal plant population: temporal fluctuations provide new insights into genome evolution of wild diploid wheat. Mobile DNA 1: 6. doi: http://dx.doi.or/10.1186/1759-8753-1-6

Ben-Abu Y, Tzfadia O, Maoz Y, Kachanovsky D, Melamed-Bessudo C, Feldman M, Levy AA (2014) Durum wheat evolution—a genomic analysis In: Porceddu E (ed) Proceedings of Symposium on Genetics and Breeding of durum. Rome, 2014, pp.29–44

Ben-David S, Beery Y, Kashkush K (2013) Genome-wide analysis of short interspersed nuclear elements SINES revealed high sequence conservation, gene association and retrotranspositional activity in wheat. Plant J 76:201–210

Bennetzen JL (2005) Transposable elements, gene creation and genome rearrangement in flowering plants. Curr Opin Genet Dev 15:621–627

Bennetzen JL, Kellogg EA (1997) Do plants have a one-way ticket to genomic obesity? Plant Cell 9:1509–1514

Bento M, Gustafson JP, Viegas W, Silva M (2011) Size matters in Triticeae polyploids: larger genomes have higher remodeling. Genome 54:175–183

Bhullar R, Nagarajan R, Bennypaul H, Sidhu GK, Sidhu G, Rustgi S, Von Wettstein D, Gill KS (2014) Silencing of a metaphase I-specific gene results in a phenotype similar to that of the pairing homeologous 1 (Ph1) gene mutations. Proc Natl Acad Sci 111:14187–14192. doi:10.1073/pnas.1416241111K, www.pnas.org/cgi

Blakeslee AF (1937) Redoublement du nombre de chromosomes chez les plantes par traitement chimique. Compt Rend Acad Sci Paris 205:476–479

Bossolini E, Wicker T, Knobel PA, Keller B (2007) Comparison of orthologous loci from small grass genome *Brachypodium* and rice: implications for wheat genomics and grass genome annotation. Plant J 49:704–717. doi:10.1111/j.1365-313

Bottley A, Xia GM, Koebner RMD (2006) Homoeologous gene silencing in hexaploid wheat. Plant J 47:897–906

Bowden MW (1959) The taxonomy and nomenclature of the wheats, barleys and ryes and their wild relatives. Can J Bot 37:657–684

Boyko EV, Badaev NS, Maximov NG, Zelenin AV (1984) Does DNA content change in the course of triticale breeding. Cereal Res Commun 12:99–100

Boyko EV, Badaev NS, Maximov NG, Zelenin AV (1988) Regularities of genome formation and organization in cereals. I DNA quantitative changes in the process of allopolyploidization. Genetika 24:89–97

Brody T, Mendlinger S (1980) Species relationships and genetic variation in the diploid wheats (*Triticum*, *Aegilops*) as revealed by starch gel electrophoresis. Plant Syst Evol 136:247–258

Cantrell RG, Joppa LR (1991) Genetic analysis of quantitative traits in wild emmer (*Triticum turgidum L.* var. *dicoccoides*). Crop Sci 31:645–649

Catalán P, Shi Y, Amstrong L, Draper J, Stace CA (1995) Molecular phylogeny of the grass genus *Brachypodium* P. Beauv based on RFLP and RAPD analysis. Bot J Linnean Soc 117:263–280

Cavanagh CR, Chao S, Wang S, Huang BE, Stephen S, Kiani S, Forrest K, Saintenac C, Brown-Guedira GL, Akhunova A, See D, Bai G, Pumphrey M, Tomar L, Wong D, Kong S, Reynolds M, da Silva ML, Bockelman H, Talbert L, Anderson JA, Dreisigacker S, Baenziger S, Carter A, Korzun V, Morrell PL, Dubcovsky J, Morell MK, Sorrells ME, Hayden MJ, Akhunov E (2013) Genome-wide comparative diversity uncovers multiple targets of selection for improvement in hexaploid wheat landraces and cultivars. Proc Natl Acad Sci U S A 110:8057–8062

Chague V, Just J, Mestiri I, Balzergue S, Tanguy AM, Huneau C, Huteau V, Belcram H, Coriton O, Jahier J, Chalhoub B (2010) Genome-wide gene expression changes in genetically stable synthetic and natural wheat allohexaploids. New Phytol 187:1181–1194

Chalupska D, Lee HY, Faris JD, Evrard A, Chalhoub B, Haselkorn R, Gornicki P (2008) ACC homoeoloci and the evolution of wheat genomes. Proc Natl Acad Sci U S A 105:9691–9696

Chantret N, Salse J, Sabot F, Rahman S, Bellec A, Laubin B, Dubois I, Dossat C, Sourdille P, Joudrier P, Gautier MF, Cattolico L, Beckert M, Aubourg S, Weissenbach J, Caboche M, Bernard M, Leroy P, Chalhoub B (2005) Molecular basis of evolutionary events that shaped the hardness locus in diploid and polyploid wheat species (*Triticum* and *Aegilops*). Plant Cell 17:1033–1045

Chapman V, Miller TE, Riley R (1976) Equivalence of the A genome of bread wheat and that of *Triticum urattu*. Genet Res 27:69–76

Charles M, Belcram H, Just J, Huneau C, Viollet A, Couloux A, Segurens B, Carter M, Huteau V, Coriton O, Appels R, Samain S, Chalhoub B (2008) Dynamics and differential proliferation of transposable elements during the evolution of the B and A genomes of wheat. Genetics 180:1071–1086

Chen K, Gray JC, Wildman SG (1975) Fraction 1 protein and the origin of polyploid wheats. Science 190:1304–1306

Chennaveeraiah MS (1960) Karyomorphologic and cytotaxonomic studies in *Aegilops*. Acta Horti Gotoburgensis 23:85–178
Clayton SD, Renvoize SA (1986) Genera Graminum, Grasses of the World, Distributed for Royal Botanic Gardens, Kew bulletin. Additional series; 13, Kew, London, pp.1-389
De Vita P, Riefolo C, Codianni P, Cattivelli L, Fares C (2006) Agronomic and qualitative traits of T-turgidum ssp dicoccum genotypes cultivated in Italy. Euphytica 150:195–205
Dewey DR (1982) Genomic and phylogenetic relationships among North American perennial Triticeae grasses. In: Estes JE (ed) Grasses and Grasslands. University of Oklahoma Press, Norman, pp 51–80
Dewey DR (1984) The genomic system of classification as a guide to inter-generic hybridization with the perennial Triticeae. In: Gustafson JP (ed) Gene Manipulation in Plant Improvement. Plenum Press, New York, NY, pp 209–279
Dover GA, Riley R (1972) Variation at two loci affecting homeologous meiotic chromosome pairing in *Triticum aestivum*×*Aegilops mutica* hybrids. Nat New Biol 235:1–6
Driscoll CJ, Darvey NL, Barber HN (1967) Effect of colchicine on meiosis of hexaploid wheat. Nature 216:687–688
Dubcovsky J, Dvorak J (2007) Genome plasticity a key factor in the success of polyploid wheat under domestication. Science 316:1862–1866. doi:10.1126/science.1143986
Dvorak J (1972) Genetic variability in *Aegilops speltoides* affecting homoeologous pairing in wheat. Can J Genet Cytol 14:371–380
Dvorak J (1976) The relationship between the genome of *Triticum urattu* and the A and B genomes of *Triticum aestivum.* Can J Genet Cytol 18:371–377
Dvorak J (2009) Triticeae genome structure and evolution. In: Feuiller C, Muehlbauer GJ (eds) Genetics and genomics of the Triticeae, plant genetics and genomics: crops and models 7. Springer, Berlin, pp 685–711
Dvorak J, Akhunov ED (2005) Tempos of gene locus deletions and duplications and their relationship to recombination rate during diploid and polyploid evolution in the *Aegilops-Triticum* alliance. Genetics 171:323–332. doi:10.1534/genetics.105.041632
Dvorak J, Zhang H-B (1990) Variation in repeated nucleotide sequences sheds light on the phylogeny of the wheat B and G genomes. Proc Natl Acad Sci U S A 87:9640–9644
Dvorak J, Zhang H-B (1992) Application of molecular tools for study of the phylogeny of diploid and polyploid taxa in Triticeae. Hereditas 116:37–42
Dvorak J, Luo M-C, Yang Z-l (1998) Restriction fragment length polymorphism and divergence in the genomic regions of high and low recombination in self-fertilizing and cross-fertilizing *Aegilops* species. Genetics 148:423–434
Dvorak J, Akhunov ED, Akhunov AR, Deal KR, Luo M-C (2006) Molecular characterization of a diagnostic DNA marker for domesticated tetraploid wheat provides evidence for gene flow feom wild tetraploid wheat to hexaploid wheat. Mol Biol Evol 23:1386–1396
Eig A (1929a) Monographisch-Kritische Ubersicht der Gattung Aegilops. Reprium nov Spec Regni veg 55:1–288
Eig A (1929b) *Amblyopyrum* Eig. A new genus separated from the genus *Aegilops*. PZE Inst Agri Natural History Agric Res 2:199–204
Eilam T, Anikster Y, Millet E, Manisterski J, Sagi-Assif O, Feldman M (2007) Nuclear DNA amount in diploid Triticeae species. Genome 50:1029–1037
Eilam T, Anikster Y, Millet E, Manisterski J, Feldman M (2008) Nuclear DNA amount and genome downsizing in natural and synthetic allopolyploids of the genera Aegilops and Triticum. Genome 51(8):616–627
Eilam T, Anikster Y, Millet E, Manisterski J, Feldman M (2009) Genome size in natural and synthetic autopolyploids and in a natural segmental allopolyploid of several Triticeae species. Genome 52:275–285
Eilam T, Anikster Y, Millet E, Manisterski J, Sagi-Assif O, Feldman M (2010) Genome size in diploids, allopolyploids, and autopolyploids of Mediterranean Triticeae. J Botany 2010:12. doi:10.1155/2010/34138, Article ID 341380

Escobar JS, Scornavacca C, Cenci A, Guilhaumon C, Santoni S et al (2011) Multigenic phylogeny and analysis of tree incongruences in Triticeae (Poaceae). BMC Evol Biol 11:181–198. doi:10.1186/1471-2148-11-181
Evans LT, (1981) Yield improvement in wheat: empirical or analytical? In: L.T. Evans, W.J. Peacock (eds.), Wheat Science - Today and Tomorrow, Cambridge University Press, Cambridge, 1981, pp. 203–222.
Fahima T, Cheng JP, Peng JH, Nevo E, Korol A (2006) Asymmetry distribution of disease resistance genes and domestication synrome QTLs in tetraploid wheat genome. 8th International Congress of Plant Molecular Biology, Adelaide, Australia
Fan X, Sha LN, Yu SB, Wu DD, Chen XH, Zhuo XF, Zhang HQ, Kang HY, Wang Y, Zheng YL, Zhou YH (2013) Phylogenetic reconstruction and diversification of the Triticeae (Poaceae) based on single-copy nuclear Acc1 and Pgk1 gene data. Biochem Syst Ecol 50:346–360
Fedoroff NV (2012) Transposable elements, epigenetics, and genome evolution. Science 338:758–767
Feldman M (1965a) Further evidence for natural hybridization between tetraploid species of *Aegilops* section Pleionathera. Evolution 19:162–174
Feldman M (1965b) Fertility of interspecific F1 hybrids and hybrid derivatives involving tetraploid species of *Aegilops* section Pleionathera. Evolution 19:556–562
Feldman M (1965c) Chromosome pairing between differential genomes in hybrids of tetraploid *Aegilops* species. Evolution 19:563–568
Feldman M (1966) The effect of chromosomes 5B, 5D and 5A on chromosomal pairing in *Triticum aestivum*. Proc Natl Acad Sci U S A 55:1447–1453
Feldman M (1993) Cytogenetic activity and mode of action of the pairing homoeologous *(Phl)* gene of wheat. Crop Sci 33:894–897
Feldman M (2001) The origin of cultivated wheat. In: Bonjean A, Angus W (eds) The wheat book. Lavoisier Tech and Doc, Paris, pp 1–56
Feldman M, Avivi L (1988) Genetic control of bivalent pairing in common wheat: the mode of *Ph1* action. The Third Kew Chromosome Conference, pp. 269–279
Feldman M, Kislev M (1977) *Aegilops searsii,* a new species of section *Sitopsis (Platystachys).* Isr J Bot 26:190–201
Feldman M, Kislev M (2007) Domestication of emmer wheat and evolution of free-threshing tetraploid wheat. Isr J Plant Sci 55:207–221
Feldman M, Levy AA (2005) Allopolyploidy—a shaping force in the evolution of wheat genomes. Cytogenet Genome Res 109:250–258
Feldman M, Levy AA (2009) Genome evolution in allopolyploid wheat—a revolutionary reprogramming followed by gradual changes. J Genet Genomics 36:511–518
Feldman M, Levy AA (2011) Instantaneous genetic and epigenetic alterations in the wheat genome caused by allopolyploidization. In: Gissis SB, Jablonka E (eds) Transformations of Lamarckism, From Subtle Fluids to Molecular Biology. The MIT press, Cambridge, MA, pp 261–270
Feldman M, Levy AA (2012) Genome evolution due to allopolyploidization in wheat. Genetics 192:763–774
Feldman M, Sears ER (1981) The wild gene resources of wheat. Sci Am 244:102–112
Feldman M, Strauss I (1983) A genome-restructuring gene in *Aegilops longissima.* In: Sakamoto S (ed.), Proceedings of 6th International Wheat Genet. Symp., Kyoto, Japan, pp 309–314
Feldman M, Strauss I, Vardi A (1979) Chromosome pairing and fertility of F_1 hybrids of *Aegilops longissima* and *Ae. searsii.* Can J Genet Cytol 21:261–272
Feldman M, Avivi L, Levy AA, Zaccai M, Avivi Y, Millet E (1990) High protein wheat. In: Bajaj YPS (ed) Biotechnology in agriculture and forestry, vol 6, Crops II. Springer, Berlin, Heidelberg, pp 593–614
Feldman M, Lupton FGH, Miller TE (1995) Wheats. In: Smartt J, Simmonds NW (eds) Evolution of crop plants, 2nd edn. Longman Scientific, London, pp 184–192
Feldman M, Liu B, Segal G, Abbo S, Levy AA, Vega JM (1997) Rapid elimination of low-copy DNA sequences in polyploid wheat: a possible mechanism for differentiation of homoeologous chromosomes. Genetics 147:1381–1387

Feldman M, Levy AA, Fahima T, Korol A (2012) Genomic asymmetry in allopolyploid plants—wheat as a model. J Exp Bot 63:5045–5059. doi:10.1093/jxb/ers192

Feldman M, Levy AA, Chalhoub B, Kashkush K (2013) Genome Plasticity in Polyploid Wheat. In: Soltis PS, Soltis D (eds.), Polyploidy and Genome Evolution, doi: 10.1007/978-3-642-31442-1-7, Springer-Verlag Berlin Heidelberg

Frederiksen S, Petersen G (1998) A taxonomic revision of Secale (Triticeae, Poaceae). - Nord J Bot 18:399–420

Furuta Y, Nishikawa K, Tanino T (1974) Stability in DNA content of AB genome component of common wheat during the past seven thousand years. Jpn J Genet 49:179–187. doi:10.1266/jjg.49.179

Gale MD, Youssefian S (1985) Dwarfing genes in wheat. In: Russell GE (ed) Progress in plant breeding. Butterworths, London, pp 1–35

Galili G, Feldman M (1984) Inter-genomic suppression of endosperm- protein genes in common wheat. Can J Genet Cytol 26:651–656

Gandhi HT, Vales MI, Watson CJW, Mallory-Smith CA, Mori N et al (2005) Chloroplast and nuclear microsatellite analysis of *Aegilops cylindrica*. Theor Appl Genet 111:561–572. doi:10.1007/s00122-005-2047-z

Gaut BS (2002) Evolutionary dynamics of grass genomes. New Phytol 154:15–28

Gegas VC, Nazari A, Griffiths S, Simmonds J, Fish L, Orford S, Sayers L, Doonan JH, Snape JW (2010) A genetic framework for grain size and shape variation in wheat. Plant Cell 22:1046–1056

Gill KS, Gill BS (1991) A DNA fragment mapped within the submicroscopic deletion of *Ph1*, a chromosome pairing regulator gene in polyploid wheat. Genetics 129:257–260

Gill KS, Gill BS, Endo TR, Mukai Y (1993) Fine physical mapping of *Ph1*, a chromosome pairing regulator gene in polyploid wheat. Genetics 134:1231–1236

Giorgi B (1978) A homoeologous pairing mutant isolated in *Triticum durum* cv. Cappelli. Mutat Breed Newsl 11:4–5

Giorgi B (1983) Origin, behaviour and utilization of a *ph1* mutant of durum wheat, *Triticum turgidum* (L.) var. *durum*. In: Sakamoto S (ed.), Proceedings of 6th International. Wheat Gen. Symp. Kyoto, Japan, 1033–1040

Giorgi D, D'Ovidio R, Tanzarella OA, Porceddu E (2002) RFLP analysis of *Aegilops* species belonging to the *Sitopsis* section. Genet Res Crop Evol 49:145–151

Gornicki P, Zhu H, Wang J, Challa GS, Zhang Z, Gill BS, Li W (2014) The chloroplast view of the evolution of polyploid wheat. New Phytol 204:704–714. doi:10.1111/nph.12931

Griffiths S, Sharp R, Foote TN, Bertin I, Wanous M, Reader S, Colas I, Moore G (2006) Molecular characterization of Ph1 as a major chromosome pairing locus in polyploid wheat. Nature 439:749–752

Guo X, Han F (2014) Asymmetric epigenetic modification and elimination of rDNA sequences by polyploidization in wheat. Plant Cell 26:4311–4327

Gupta PK, Baum BR (1986) Nomenclature and related taxonomic issues in wheats, triticales and some of their wild relatives. Taxon 35:144–149

Gupta PK, Kulwal PL, Rustgi S (2005) Wheat cytogenetics in the genomics era and its relevance to breeding. Cytogenet Genome Res 109:315–327

Hammer K (1980) Vorarbeiten zur monographischen Darstellung von Wildpflanzensortimenten: *Aegilops* L. Kulturpflanze 28:33–180

Han FP, Fedak G, Ouellet T, Liu B (2003) Rapid genomic changes in interspecific and intergeneric hybrids and allopolyploids of Triticeae. Genome 46:716–723

Han FP, Fedak G, Guo WL, Liu B (2005) Rapid and repeatable elimination of a parental genome specific DNA repeat (pGcIR-1a) in newly synthesized wheat allopolyploids. Genetics 170:1239–1245

Hasterok R, Draper J, Jenkins G (2004) Laying the cytotaxonomic foundations of a new model grass, *Brachypodium distachyon* (L.) Beauv. Chromosom Res 12:397–403

He P, Friebe BR, Gill BS, Zhou JM (2003) Allopolyploidy alters gene expression in the highly stable hexaploid wheat. Plant Mol Biol 52:401–414

Helfgott M, Mason-Gamer RJ (2004) The evolution of North American *Elymus* (Triticeae, Poaceae) allotetraploids: evidence from phosphoenolpyruvate carboxylase gene sequences. Syst Bot 29:850–861

Hirai A, Tsunewaki K (1981) Genetic diversity of the cytoplasm in *Triticum* and *Aegilops*. VIII. Fraction I protein of 39 cytoplasms. Genetics 99:487–493

Hsiao C, Chatterton NJ, Asay KH, Jensen KB (1995) Phylogenetic relationships of the monogenomic species of the wheat tribe, Triticeae (Poaceae), inferred from nuclear rDNA (internal transcribed spacer) sequences. Genome 38:211–223

Huang L, Millet E, Rong JK, Wendel JF, Anikster Y, Feldman M (1999) Restriction fragment length polymorphism in wild and cultivated tetraploid wheat. Isr J Plant Sci 47:213–224

Huang S, Sirikhachornkit A, Faris JD, Su X, Gill BS et al (2002a) Phylogenetic analysis of the acetyl-CoA carboxylase and 3-phosphoglycerate kinase loci in wheat and other grasses. Plant Mol Biol 48:805–820

Huang S, Sirikhachornkit A, Su X, Faris JD, Gill BS et al (2002b) Genes encoding plastid acetyl-CoA carboxylase and 3-phosphoglycerate kinase of the *Triticum/Aegilops* complex and the evolutionary history of polyploid wheat. Proc Natl Acad Sci U S A 99:8133–8138. doi:10.1073/pnas.072223799

International Brachypodium Initiative (2010) Genome sequencing and analysis of the model grass *Brachypodium distachyon*. Nature 463:763–768. doi:10.1038/nature08747

International Wheat Genome Sequencing Consortium (2014) A chromosome-based draft sequence of the hexaploid bread wheat genome. Science 345, doi: 10.1126/science.1251788

Jampates R, Dvorak J (1986) Location of the *Ph1* locus in the metaphase chromosome map and the linkage map of the 5Bq arm of wheat. Can J Genet Cytol 28:511–519

Jantasuriyarat C, Vales MI, Watson CJW, Riera-Lizarazu O (2004) Identification and mapping of genetic loci affecting the free-threshing habit and spike compactness in wheat (*Triticum aestivum* L.). Theor Appl Genet 108:261–273

Jia J, Zhao S, Kong X, Li Y, Zhao G et al (2013) *Aegilops tauschii* draft genome sequence reveals a gene repertoire for wheat adaptation. Nature 496:91–95

Joppa LR (1993) Chromosome engineering in tetraploid wheat. Crop Sci 33:908–913

Joppa LR, Cantrell RG (1990) Chromosomal location of genes for grain protein content of wild tetraploid wheat. Crop Sci 30:1059–1064

Joppa LR, Hareland GA, Cantrell RG (1991) Quality characteristics of the Langdon *durum-dicoccoides* chromosome substitution lines. Crop Sci 31:1513–1517

Kashkush K, Khasdan V (2007) Large-scale survey of cytosine methylation of retrotransposons, and the impact of readout transcription from LTRs on expression of adjacent rice genes. Genetics 177:1975–1978

Kashkush K, Feldman M, Levy AA (2002) Gene loss, silencing and activation in a newly synthesized wheat allotetraploid. Genetics 160:1651–1659

Kashkush K, Feldman M, Levy AA (2003) Transcriptional activation of retrotransposons alters the expression of adjacent genes in wheat. Nat Genet 33:102–106

Kawahara T (2009) Molecular phylogeny among *Triticum-Aegilops* species and of the tribe *Triticeae*. Breed Sci 59:499–504

Kellogg EA, Appels R (1995) Intraspecific and interspecific variation in 5S RNA genes are decoupled in diploid wheat relatives. Genetics 140:325–343

Kellogg EA, Appels R, Mason-Gamer RJ (1996) When genes tell different stories: the diploid genera of Triticeae (Gramineae). Syst Bot 21:321–347

Kenan-Eichler M, Leshkowitz D, Tal L, Noor E, Melamed-Bessudo C, Feldman M, Levy AA (2011) Wheat hybridization and polyploidization results in deregulation of small RNAs. Genetics 188:263–272

Kerber ER, Rowland GG (1974) Origin of the free threshing character in hexaploid wheat. Can J Genet Cytol 16:145–154. doi:10.1139/g74-014

Kidwell MG, Lisch DR (2000) Transposable elements and host genome evolution. Trends Ecol Evol 15:95–99

Kihara H (1919) Über cytologische studien bei einigen getreidearten. I. Species-bastarde des weizens und weizenroggen-bastarde. Bot Mag 33:17–38

Kihara H (1924) Cytologische und genetische studien bei wichtigen getreidearten mit besonderer rücksicht ouf das verhalten der chromosomen und die sterilitat in den bastarden. Mem Cell Sci, Kyoto Imp Univ B1:1–200

Kihara H (1944) Discovery of the DD-analyser, one of the ancestors of *Triticum vulgare*. Agric Hortic 19:13–14

Kihara H (1954) Considerations on the evolution and distribution of *Aegilops* species based on the analyzer-method. Cytologia 19:336–357

Kihara H, Lilienfeld F (1949) A new synthesized 6x-wheat. In: Larsson GBAR (ed.), Proceedings of Eighth International Congress of Genetics, Stockholm, Sweden, 1949. Hereditas (Suppl), pp 307–319

Kihara H, Yamashita H, Tanaka M (1959) Genomes of 6x species of *Aegilops*. Wheat Inf Ser 8:3–5

Kimber G, Feldman M (1987) Wild wheats: an introduction. Special Report 353. College of Agriculture, Columbia, Missouri, USA, pp 1–142

Kimber G, Sears ER (1987) Evolution in the genus Triticum and the origin of cultivated wheat. In: Heyne EG (ed.), Wheat and Wheat Improvement (2nd edition), Amer Soc Agr, Inc., Madison, pp 154–164

Kimber G, Tsunewaki K (1988) Genome symbols and plasma types in the wheat group. In: Miller TE, Koebner RMD (eds.), Proceedings of 7th International Wheat Genet Symp, Cambridge, pp 1209–1210

Kislev ME (1980) *Triticum parvicoccum* sp. nov, the oldest naked wheat. Isr J Bot 28:95–107

Kislev ME (1984) Emergence of wheat agriculture. Paléorient 10(2):61–70

Kraitshtein Z, Yaakov B, Khasdan V, Kashkush K (2010) Genetic and epigenetic dynamics of a retrotransposon after allopolyploidization of wheat. Genetics 186:801–812

Krasileva KV, Buffalo V, Bailey P, Pearce S, Ayling S, Tabbita F, Soria M, Wang S, Consortium I, Akhunov E, Uauy C, Dubcovsky J (2013) Separating homeologs by phasing in the tetraploid wheat transcriptome. Genome Biol 14:R66

Laido G, Mangini G, Taranto F, Gadaleta A, Blanco A, Cattivelli L, Marone D, Mastrangelo AM, Papa R, De Vita P (2013) Genetic Diversity and Population Structure of Tetraploid Wheats (Triticum turgidum L.) Estimated by SSR, DArT and Pedigree Data. Plos One 8

Le Comber SC, Ainouche ML, Kovarik A, Leitch AR (2010) Making a functional diploid: from polysomic to disomic inheritance. New Phytol 186:113–122

Levy AA (2013) Transposons in plant speciation. In: Fedoroff NV (ed) Plant transposons and genome dynamics in evolution. John Wiley and Sons Inc., Hoboken, NJ, pp 164–179

Levy AA, Feldman M (1988) Ecogeographical distribution of HMW glutenin alleles in populations of the wild tetraploid wheat Triticum turgidum var. dicoccoides. Theor Appl Genet 75:651–658

Levy AA, Feldman M (1989a) Genetics of morphological traits in wild wheat, Triticum turgidum var. dicoccoides. Euphytica 40:275–281

Levy AA, Feldman M (1989b) Location of genes for high grain protein percentage and other quantitative traits in wild wheat *Triticum turgidum* var. *dicoccoides*. Euphytica 41:113–122

Levy AA, Feldman M (2002) The impact of polyploidy on grass genome evolution. Plant Physiol 130:1587–1593

Levy AA, Feldman M (2004) Genetic and epigenetic reprogramming of the wheat genome upon allopolyploidization. Biol J Linn Soc 82:607–613

Li L, Liu B, Olsen K, Wendel J (2014a) A reevaluation of the homoploid hybrid origin of *Aegilops tauschii*, the donor of the wheat D-subgenome. New Phytol. doi:10.1111/nph.13294

Li A, Liu D, Wu J, Zhao X, Hao M, Geng S, Yan J, Jiang X, Zhang L, Wu J, Yin L, Zhang R, Wu L, Zheng Y, Mao L (2014b) mRNA and small RNA transcriptomes reveal insights into dynamic homoeolog regulation of allopolyploid heterosis in nascent hexaploid wheat. Plant Cell 26:1878–1900

Lilienfeld FA (1951) H. Kihara: Genome-analysis in Triticum and Aegilops. X. Concluding review. Cytologia 16:101–123

Ling HQ, Zhao S, Liu D, Wang J, Sun H et al (2013) Draft genome of the wheat A-genome progenitor *Triticum urartu*. Nature 496:87–90. doi:10.1038/nature11997

Liu B, Segal G, Vega JM, Feldman M, Abbo S (1997) Isolation and characterization of chromosome-specific DNA sequences from a chromosome arm genomic library of common wheat. Plant J 11:959–965

Liu B, Vega JM, Feldman M (1998a) Rapid genomic changes in newly synthesized amphiploids of *Triticum* and *Aegilops*. II. Changes in low-copy coding DNA sequences. Genome 41:535–542

Liu B, Vega JM, Segal G, Abbo S, Rodova H, Feldman M (1998b) Rapid genomic changes in newly synthesized amphiploids of *Triticum* and *Aegilops*. I. Changes in low-copy noncoding DNA sequences. Genome 41:272–277

Love A (1982) Generic evolution of the wheatgrasses. Biol Zentralbl 101:199–212

Love A (1984) Conspectus of the Triticeae. Feddes Repert 95:425–521

Luo M-C, Yang Z-L, You FM, Kawahara T, Waines JG, Dvorak J (2007) The structure of wild and domesticated emmer wheat populations, gene flow between them, and the site of emmer domestication. Theor Appl Genet 114:947–959. doi:10.1007/s00122-006-0474-0

Luo MC, Deal KR, Akhunov ED, Akhunova AR, Andeson OD et al (2009) Genome comparisons reveal a dominant mechanism of chromosome number reduction in grasses and accelerated genome evolution in Triticeae. Proc Natl Acad Sci U S A 106:15780–15785

Ma XF, Gustafson JP (2005) Genome evolution of allopolyploids: a process of cytological and genetic diploidization. Cytogenet Genome Res 109:236–249

Ma XF, Gustafson JP (2006) Timing and rate of genome variation in triticale following allopolyploidization. Genome 49:950–958

Ma XF, Fang P, Gustafson JP (2004) Polyploidization-induced genome variation in triticale. Genome 47:839–848

Mac Key J (1954) Mutation breeding in polyploidy cereals, Acta Agriculturae Scandinavica, Abstract content, vol. 4, issue 1

Mac Key J (1958) Mutagenic response in Triticum at different levels of ploidy, In: Jenkins BC (ed) Proc. 1st Inter Wheat Genetics Symp, Winnipeg, Manitoba, Canada pp. 88–111

Mac Key J (1966) Species relationship in *Triticum*. Proc 2nd Intern Wheat Genet Symp, Lund, Hereditas (Suppl) 2: 237–276

Maestra B, Naranjo T (1999) Structural chromosome differentiation between *Triticum timopheevii* and *T. turgidum* and *T. aestivum*. Theor Appl Genet 98:744–750

Marcussen T, Sandve SR, Heier L, Spannagl M, Pfeifer M, The International Wheat Genome Sequencing Consortium, Jakobsen KS, Wulff BBH, Steuernagel B, Klaus FX, Mayer KFX, Olsen, OA (2014) Ancient hybridizations among the ancestral genomes of bread wheat. Science 345(6194), doi: 10.1126/science.1250092

Mason-Gamer RJ (2005) The ß-amylase genes of grasses and a phylogenetic analysis of the Triticeae (Poaceae). Am J Bot 92:1045–1058

Mason-Gamer RJ, Kellogg EA (1996) Testing for phylogenetic conflict among molecular data sets in the tribe Triticeae (Gramineae). Syst Biol 45:524–545

Matsuoka Y, Nasuda S, Ashida Y, Nitta M, Tsujimoto H, Takumi S, Kawahara T (2013) Genetic basis for spontaneous hybrid genome doubling during allopolyploid speciation of common wheat shown by natural variation analyses of the paternal species. PLoS One 8, e68310. doi:10.1371/journal.pone.0068310

McFadden ES, Sears ER (1944) The artificial synthesis of *Triticum spelta*. Records Genet Soc Am 13:26–27

McFadden ES, Sears ER (1946) The origin of *Triticum spelta* and its free-threshing hexaploid relatives. J Heredity 37:107–116

Mendlinger S, Zohary D (1995) The extent and structure of genetic variation in species of the Sitopsis group of *Aegilops*. Heredity 74:616–627

Middleton C, Stein N, Keller B, Kilian B, Wicker T (2013) Comparative analysis of genome composition in Triticeae reveals strong variation in transposable element dynamics and nucleotide diversity. Plant J 73:347–356. doi:10.1111/tpj.12048

Middleton CP, Senerchia N, Stein N, Akhunov ED, Keller B, Wicker T, Kilian B (2014) Sequencing of chloroplast genomes from wheat, barley, rye and their relatives provides a detailed insight into the evolution of the Triticeae tribe. PLoS One 9:e85761. doi:10.1371/journal.pone.0085761

Millet E, Zaccai M (1991) Effects of genotypically- and environmentally-induced differences in seed protein content on seedling vigor in wheat. J Genet Breed 45:45–50

Millet E, Rong JK, Qualset CO, McGuire PE, Bernard M, Sourdille P, Feldman M (2013) Production of chromosome-arm substitution lines of wild emmer in common wheat. Euphytica 190:1–17

Mochida K, Kawaura K, Shimosaka E, Kawakami N, Shin-I T, Kohara Y, Yamazaki Y, Ogihara Y (2006) Tissue expression map of a large number of expressed sequence tags and its application to in silico screening of stress response genes in common wheat. Mol Genet Genomics 276(3):304–312

Mochizuki A (1957) B chromosomes in *Aegilops mutica* Boiss. Wheat Inf Serv 5:9–11

Monte JV, McIintyre CL, Gustafson JP (1993) Analysis of phylogenetic relationships in the Triticeae tribe using RFLPs. Theor Appl Genet 86:649–655

Mori M, Liu Y-G, Tsunewaki K (1995) Wheat phylogeny determined by RFLP analysis of nuclear DNA. 2. Wild tetraploid wheats. Theor Appl Genet 90:129–134

Morris R, Sears ER (1967) The cytogenetics of wheat and its relatives. In: Quisenberry KS, Reitz LP (eds.), Wheat and wheat improvement. The American Society of Agronomy, Madison, USA, pp 19–87

Murai K, Naiyu X, Tsunewaki K (1989) Studies on the origin of crop species by restriction endonuclease analysis of organellar DNA. III. Chloroplast DNA variations and intersprcific relationships in the genus Swcale. Jpn J Genet 64:36–47

Muramatsu M (1963) Dosage effect of the spelta gene q of hexaploid wheat. Genetics 48:469–482

Muramatsu M (1986) The *vulgare* super gene, *Q*: its universality in *durum* wheat and its phenotypic effects in tetraploid and hexaploid wheats. Can J Genet Cytol 28:30–41

Muramatsu M (2009) A presumed genetic system determining the number of spikelets per rachis nodein the tribe Triticeae. Breed Sci 59:617–620

Murat F, Pont C, Salse J (2014) Paleogenomics in Triticeae for translational research. Curr Plant Biol 1:34–39

Nakamura I, Rai B, Takahashi H, Kato K, Sato Y-I, Komatsuda T (2009) *Aegilops* section Sitopsis species contains the introgressive *PolA1* gene with a closer relationship to that of *Hordeum* than *Triticum-Aegilops* species. Breed Sci 59:602–610

Nalam VJ, Vales MI, Watson CJW, Kianian SF, Riera-Lizarazu O (2006) Map-based analysis of genes affecting the brittle rachis character in tetraploid wheat (*Triticum turgidum* L.). Theor Appl Gen 112:373–381

Nishikawa K, Furuta Y (1969) DNA content per nucleus in relation to phylogeny of wheat and its relatives. Jpn J Genet 44:23–29. doi:10.1266/jjg.44.23

Ohta S (1991) Phylogenetic relationship of *Aegilops mutica* Boiss with the diploid species of congeneric Aegilops-Triticum complex, based on the new method of genome analysis using its B-chromosomes. Mem Coll Agric Kyoto Univ 137:1–116

Okamoto M (1957) Asynaptic effect of chromosome V. Wheat Inf Serv 5:6

Olmos S, Distelfeld A, Chicaiza O, Schlatter AR, Fahima T, Echenique V, Dubcovsky J (2003) Precise mapping of a locus affecting grain protein content in durum wheat. Theor Appl Gent 107:1243–1251

Ozkan H, Levy AA, Feldman M (2001) Allopolyploidy—induced rapid genome evolution in the wheat (*Aegilops-Triticum*) group. Plant Cell 13:1735–1747

Ozkan H, Tuna M, Arumuganathan K (2003) Nonadditive changes in genome size during allopolyploidization in the wheat (*Aegilops-Triticum*) group. J Hered 94:260–264

Peleg Z, Fahima T, Krugman T, Abbo S, Yakir D, Korol AB, Saranga Y (2009) Genomic dissection of drought resistance in durum wheat x wild emmer wheat recombinant inbreed line population. Plant Cell Environ 32:758–779

Peleg Z, Fahima T, Korol AB, Abbo S, Saranga Y (2011) Genetic analysis of wheat domestication and evolution under domestication. J Exp Bot 62:5051–5061

Peng I, Ronin Y, Fahima T, Röder MS, Li Y, Nevo E, Korol A (2003a) Genomic distribution of domestication QTLs in wild emmer wheat, *Triticum dicoccoides*. In Proceedings of 10th International Wheat Genetics Symposium, Paestum, Italy, pp. 34–37

Peng JH, Ronin Y, Fahima T, Roder MS, Li YC, Nevo E, Korol A (2003b) Domestication quantitative trait loci in *Triticum dicoccoides*, the progenitor of wheat. Proc Natl Acad Sci U S A 100(5): 2489–2494

Percival J (1921) The wheat plant. EP. Dutton and Company, New York, NY, pp 1–463

Petersen G, Seberg O (1997) Phylogenetic analysis of the Triticeae (Poaceae) based on *rpo*A sequence data. Mol Phylogenet Evol 7:217–230

Petersen G, Seberg O (2000) Phylogenetic evidence for the excision of *Stowaway* miniature inverted-repeat transposable elements in Triticeae (Poaceae). Mol Biol Evol 17:1589–1596

Petersen G, Seberg O, Yde M, Berthelsen K (2006) Phylogenetic relationships of *Triticum* and *Aegilops* and evidence for the origin of the A, B, and D genomes of common wheat (*Triticum aestivum*). Mol Phylogenet Evol 39:70–82. doi:10.1016/j.ympev.2006.01.023

Pumphrey M, Bai J, Laudencia-Chingcuanco D, Anderson O, Gill BS (2009) Nonadditive expression of homoeologous genes is established upon polyploidization in hexaploid wheat. Genetics 181:1147–1157

Ramakrishna W, Dubcovsky J, Park Y-J, Busso C, Emberton J, SanMiguel P, Bennetzen JL (2002) Different types and rates of genome evolution detected by comparative sequence analysis of orthologous segments from four cereal genomes. Genetics 162:1389–1400

Raskina O, Belyayev A, Nevo E (2004) Activity of the En/Spm-like transposons in meiosis as a base for chromosome repatterning in a small, isolated, peripheral population of *Aegilops speltoides* Tausch. Chromosom Res 12:153–161

Riley R (1960) The diploidization of polyploid wheat. Heredity 15:407–429

Riley R, Chapman V (1958) Genetic control of the cytologically diploid behavior of hexaploid wheat. Nature 182:713–715

Runemark H, Heneen WK (1968) *Elymus* and *Agropyron*, a problem of generic delimitation. Bot Notiser 121:51–79

Sabot F, Guyot R, Wicker T, Chantret N, Laubin B, Chalhoub B, Leroy P, Sourdille P, Bernard M (2005) Updating of transposable element annotations from large wheat genomic sequences reveals diverse activities and gene associations. Mol Genet Gen 274:119–130

Sakamoto S (1973) Patterns of phylogenetic differentiation in the tribe Triticeae. Seiken Zihô 24:11–31

Sakamoto S (1991) The cytogenetic evolution of Triticeae grasses. In: Tsuchiya T, Gupta PK (eds) Chromosome engineering in plants: genetics, breeding. Evolution. Elsevier Science Publications B.V, Amsterdam, The Netherland, pp 469–482

Sakamura T (1918) Kurze mitteilung über die chromosomenzahalen und die verwandtschaftsverhältnisse der Triticum Arten. Bot Mag 32:151–154

Salina EA, Numerova OM, Ozkan H, Feldman M (2004) Alterations in subtelomeric tandem repeats during early stages of allopolyploidy in wheat. Genome 47:860–867

Salina EA, Lim KY, Badaeva ED, Shcherban AB, Adonina IG, Amosova AV, Samatadze TE, Vatolina TY, Zoshchuk SA, Leitch AR (2006) Phylogenetic reconstruction of *Aegilops* section *Sitopsis* and the evolution of tandem repeats in the diploids and derived wheat polyploids. Genome 49:1023–1035. doi:10.1139/g06-050

Sallares R, Brown TA (2004) Phylogenetic analysis of complete 5′ external transcribed spacers of the 18S ribosomal RNA genes of diploid *Aegilops* and related species (Triticeae, Poaceae). Genet Resour Crop Evol 51:701–712

Salse J, Bolot S, Throude M, Jouffe V, Piegu B, Quraishi UM, Calcagno T, Cooke R, Delseny M, Feuillet C (2008) Identification and characterization of conserved duplications between rice and wheat provide new insight into grass genome evolution. Plant Cell 20: 11–24

Sasanuma T, Miyashita NT, Tsunewaki K (1996) Wheat phylogeny determined by RFLP analysis of nuclear DNA. 3. Intra- and interspecific variations of five Aegilops Sitopsis species. Theor Appl Genet 92:928–934

Sasanuma T, Chabane K, Endo TR, Valkoun J (2004) Characterization of genetic variation in and phylogenetic relationships among diploid *Aegilops* species by AFLP: incongruity of chloroplast and nuclear data. Theor Appl Genet 108:612–618
Sax K (1921) Chromosome relationships in wheat. Science 54:413–415
Sax K (1927) Chromosome behavior in *Triticum* hybrids, Verhandlungen des V Int. Kongresses für Vererbungswissenchaft (Berlin) 2:1267–1284
Sears ER (1948) The Cytology and Genetics of the Wheats and their Relatives. Adv Genet 2:239–270
Sears ER (1954) The aneuploids of common wheat. Missouri Agric Exp Stn Res Bull 572:1–58
Sears ER (1969) Wheat cytogenetics. Ann Rev Genet 3:451–468
Sears ER (1972) The nature of mutation in hexaploid wheat , Symp Biol Hung ,PP 73–82
Sears ER (1976) Genetic control of chromosome pairing in wheat. Ann Rev Genet 10:31–51
Sears ER (1977) An induced mutant with homoeologous chromosome pairing. Can J Genet Cytol 19:585–593
Sears ER (1984) Mutations in wheat that raise the level of meiotic chromosome pairing. In: Gustavson JP (ed) Gene manipulation in plant improvement. Stadler Sympoium, Columbia, MO, pp 295–300
Sears ER, Okamoto M (1958) Intergenomic chromosome relationships in hexaploid wheat. Proc X Int Congr Genet 2:258–259
Seberg O, Frederiksen S (2001) A phylogenetic analysis of the monogenomic Triticeae (Poaceae) based on morphology. Bot J Linn Soc 136:75–97
Selmecki AM, Maruvka YE, Richmond PA, Guillet M, Shoresh N, Sorenson AL, De S, Kishony R, Michor F, Dowell R, Pellman D (2015) Polyploidy can drive rapid adaptation in yeast. doi: 10.1038/nature14187
Senerchia N, Wicker T, Felber F, Parisod C (2013) Evolutionary dynamics of retrotransposons assessed by high throughput sequencing in wild relatives of wheat. Genome Biol Evol 5:1010–1020
Senyaninova-Korchagina MV (1932) Karyosystematic analysis of the genus Aegilopsl. Bull Appl Bot Genet Pl Breed, Ser II, pp. 1–90
Shaked H, Kashkush K, Ozkan H, Feldman M, Levy AA (2001) Sequence elimination and cytosine methylation are rapid and reproducible responses of the genome to wide hybridization and allopolyploidy in wheat. Plant Cell 13:1749–1759
Simchen G, Zarchi Y, Hillel J (1971) Supernumerary chromosomes in second outbreeding species of wheat group. Chromosoma 33:63–69. doi:10.1007/bf00326384
Simonetti MC, Bellomo MP, Laghetti G, Perrino P, Simeone R, Blanco A (1999) Quantitative trait loci influencing free-threshing habit in tetraploid wheats. Genet Res Crop Evol 46:267–271
Simons KJ, Fellers JP, Trick HN, Zhang Z, Tai YS, Gill BS, Faris JD (2006) Molecular characterization of the major wheat domestication gene Q. Genetics 172:547–555
Sliai AM, Amer SAM (2011) Contribution of chloroplast DNA in the biodiversity of some *Aegilops* species. Afr J Biotechnol 10:2212–2215
Slotkin RK, Martienssen R (2007) Transposable elements and the epigenetic regulation of the genome. Nat Rev Genet 8:272–285. doi:10.1038/nrg2072
Sood S, Kuraparthy V, Bai G, Gill BS (2009) The major threshability genes soft glume (sog) and tenacious glume (Tg), of diploid and polyploid wheat, trace their origin to independent mutations at non-orthologous loci. Theor Appl Genet 119:341–351
Stebbins GL (1950) Variation and evolution in plants. Columbia University Press, New York, NY
Stebbins GL (1956) Taxonomy and evolution of genera, with special reference to the family Gramineae. Evolution 10:235–245
Steuernagel B, Moscou MJ, Hernández-Pinzón I, Champouret N, Wright J, Marcussen T, Wulff BBH (2014a) Whole genome sequence of Aegilops sharonensis reveals inconsistency in the Sitopsis section taxonomy. Plant and Animal Genome XXII Conference, January 10–15, 2014, San Diego, California, USA
Steuernagel B, Marcussen T, Moscou MJ, Champouret N, Hernández-Pinzón I, Wright J, IWGSC, Ward E, Sandve SR, Wulff BBH (2014b) Multiple hybridization events shaped the genome of a wild diploid wheat. Genome Biology (submitted)

Tanno K, Willcox G (2006) How fast was wild wheat domesticated? Science 311:1886

Terachi T, Tsunewaki K (1992) The molecular basis of genetic diversity among cytoplasms of *Triticum* and *Aegilops*. VIII. Mitochondrial RFLP analyses using cloned genes as probes. Mol Biol Evol 9:911–931

Terachi TY, Ogihara Y, Tsunewaki K (1984) The molecular basis of genetic diversity among cytoplasms of Triticum and Aegilops. III. Chloroplast genomes of the M and modified M genome-carrying species. Genetics 108:681–695

Tirosh I, Reikhav S, Levy AA, Barkai N (2009) A yeast hybrid provides insight into the evolution of gene expression regulation. Science 324:659–662

Tsunewaki K (2009) Plasmon analysis in the *Triticum-Aegilops* complex. Breed Sci 59:455–470

Tsunewaki K, Wang GZ, Matsuoka Y (1996) Plasmon analysis of *Triticum* (wheat) and *Aegilops*. 1. Production of alloplasmic common wheats and their fertilities. Genes Genet Syst 71:293–311. doi:10.1266/ggs.71.293

Uauy C, Distelfeld A, Fahima T, Blechl A, Dubcovsky J (2006) A NAC Gene regulating senescence improves grain protein, zinc, and iron content in wheat. Science 314:1298–1301

Van Slageren MW (1994) Wild wheats: a monograph of Aegilops L. and Amblyopyrum (Jaub. and Spach) Eig (Poaceae). Agricultural University, Wageningen, The Netherlands

Vardi A (1973) Introgression between different ploidy levels in the wheat group. In: Sears ER, Sears LMS (eds.), Proceedings of the 4th International Wheat Genetic Symposium, Columbia, Mo, pp. 131–141

Vardi A, Zohary D (1967) Introgression in wheat via triploid hybrids. Heredity 22:541–560

Vega JM, Feldman M (1998) Effect of the pairing gene *Ph1* on centromere misdivision in common wheat. Genetics 148:1285–1294

Von Tschermak E, Bleier H (1926) Über fruchtbare Aegilops-weizenbastarde, der deutsch. Bot Ges 44:110–132

Waines JG, Johnson BL (1972) Genetic differences between *Aegilops longissima*, *A. sharonensis*, and *A. bicornis*. Can J Genet Cytol 14:411–415

Wall AM, Riley R, Gale MD (1971) The position of a locus on chromosome 5B of *Triticum aestivum* affecting homoeologous meiotic pairing. Genet Res 18:329–339

Wang RR-C, Von Bothmer R, Dvorak J, Fedak G, Linde-Laursen I, Muramatsu M (1994) Genome symbols in the Triticeae (Poaceae). In Wang RR-C, Jensen KB, Jaussi C (eds.), Proceedings 2nd International Triticeae Symp, June 20-24, Logan, UT, 1994, pp. 29–34

Wang G-Z, Miyashita NT, Tsunewaki K (1997) Plasmon analyses of *Triticum* (wheat) and *Aegilops*: PCR-single strand conformational polymorphism (PCR-SSCP) analyses of organellar DNAs. Proc Natl Acad Sci U S A 94:14570–14577

Watanabe N, Ikebata N (2000) The effects of homoeologous group 3 chromosomes on grain colour dependent seed dormancy and brittle rachis in tetraploid wheat. Euphytica 115:215–220

Watanabe N, Sugiyama K, Yamagishi Y, Sakata Y (2002) Comparative telosomic mapping of homoeologous genes for brittle rachis in tetraploid and hexaploid wheats. Hereditas 137:180–185

Watanabe N, Takesada N, Fujii Y, Martinek P (2005) Comparative Mapping of Genes for Brittle Rachis in *Triticum* and *Aegilops*. Czech J Genet Plant Breed 41:39–44

Weissman S, Feldman M, Gressel, J (2005) Sequence evidence for sporadic inter-generic DNA introgression from wheat into wild Aegilops species. Mol. Biol. Evol. 22: 2055–2062

Wicker T, Mayer KFX, Gundlach H, Martis M, Steuernagel B, Scholz U, Simkova H, Kubalakova M, Choulet F, Taudien S, Platzer M, Feuillet C, Fahima T, Budak H, Doležel J, Keller B, Stein N (2011) Frequent gene movement and pseudogene evolution is common to the large and complex genomes of wheat, barley, and their relatives. Plant Cell 23:1706–1718

Wolfe KH, Gouy M, Yang Y-W, Sharp PM, Li W-H (1989) Date of the monocot-dicot divergence estimated from chloroplast DNA sequence data. Proc Natl Acad Sci U S A 86:6201–6205

Yaakov B, Kashkush K (2011a) Massive alterations of the methylation patterns around DNA transposons in the first four generations of a newly formed wheat allohexaploid. Genome 54:42–49

Yaakov B, Kashkush K (2011b) Methylation, transcription, and rearrangements of transposable elements in synthetic allopolyploids. Int J Plant Genomics. doi:10.1155/2011/569826

Yaakov B, Ceylan E, Domb K, Kashkush K (2012) Marker utility of miniature inverted-repeat transposable elements for wheat biodiversity and evolution. Theor Appl Genet 124:1365–1373

Yaakov B, Meyer K, Ben-David S, Kashkush K (2013) Copy number variation of transposable elements in *Triticum– Aegilops* genus suggests evolutionary and revolutionary dynamics following allopolyploidization. Plant Cell Rep, doi: 10.1007/ s00299-013-1472-8

Yamane K, Kawahara T (2005) Intra- and interspecific phylogenetic relationships among diploid *Triticum-Aegilops* species (Poaceae) based on base-pair substitutions, indels, and microsatellites in chloroplast noncoding sequences. Am J Bot 92:1887–1898

Yao Y, Sun Q (2012) Exploration of small non coding RNAs in wheat (Triticum aestivum L.). Plant Mol Biol 80:67–73

Yen Y, Kimber G (1990) Genomic relationships of *Triticum searsii* to other S-genome diploid *Triticum* species. Genome 33:369–374

Yousafzai FK, Al-Kaff N, Moore M (2010) Structural and functional relationship between the Ph1 locus protein 5B2 in wheat and CDK2 in mammals. Funct Integr Genomics 10:157–166

Yuan J, Xiang Guo X, Hu J, Lv Z Han F (2014) Characterization of two *CENH3* genes and their roles in wheat evolution. Article first published online: 31 DEC 2014, doi: 10.1111/nph.13235

Zeven AC (1980) Polyploidy and domestication: the origin and survival of polyploids in cytotype mixtures. In: Lewis WH (ed) Polyploidy—biological relevance. Plenum Press, New York, NY, pp 385–407

Zhang ZC, Belcram H, Gornicki P, Charles M, Just J, Huneau C, Magdelenat G, Couloux A, Samain S, Gill BS, Rasmussen JB, Barbe V, Faris JD, Chalhoub B (2011) Duplication and partitioning in evolution and function of homoeologous Q loci governing domestication characters in polyploid wheat. Proc Natl Acad Sci 108:18737–18742

Zhukovsky PM (1928) A critical systematical survey of the species of the genus *Aegilops* L. Bull Appl Bot Genet and Plant Breed 18:417–609

Zohary D, Brick Z (1961) *Triticum dicoccoides* in Israel: notes on its distribution, ecology and natural hybridization. Wheat Informat Service 1961(13):6–8

Zohary D, Feldman M (1962) Hybridization between amphiploids and the evolution of polyploids in the wheat (*Aegilops-Triticum*) group. Evolution 16:44–61

Zohary D, Imber D (1963) Genetic dimorphism in fruit types in *Aegilops speltoides*. Heredity 18:223–231

Chapter 3
Wheat Breeding: Current Status and Bottlenecks

Zoltán Bedő and László Láng

3.1 Aims and Importance of Wheat Breeding

Wheat is traditionally one of the main food sources for mankind, providing 19 % of the food calories and 20 % of the protein consumed by the world population (Braun et al. 2010). In addition to its favourable nutritional properties and to the fact that it can be cultivated under a wide range of conditions, its grain can be easily stored and transported, which explains why both food reserves and the food trade are based largely on wheat.

Over the last 50 years, the sowing area of wheat has fluctuated over a narrow range, being between 204 and 239 million hectares. It is grown on the largest scale between latitudes of 30° and 60°N and 27° and 40°S, primarily in South and East Asia, North America and Eastern Europe. The countries producing wheat on the largest areas are the EU, China, India, Russia, the USA, Australia and Kazakhstan. It can be grown in such a wide variety of environments partly because several different wheat species are cultivated, and partly because most of these species have winter, facultative and spring types.

Common wheat (*Triticum aestivum* L.) is grown on the vast majority of wheat-growing areas. The winter type has the highest yield potential, as it is able to develop a deep root system during its long vegetation period, which helps the plants to survive dry periods and also contributes to efficient nutrient uptake. The vegetative/generative transition requires several weeks of cold weather (2–5 °C), i.e., the plants have a well-defined vernalisation requirement. Wheat can thus be grown on areas with a continental climate, where the winter is cold enough for vernalisation, but not cold enough to kill the plants.

Z. Bedő (✉) • L. Láng
Agricultural Institute, Centre for Agricultural Research,
Hungarian Academy of Sciences, 2462 Martonvásár, Hungary
e-mail: bedo.zoltan@agrar.mta.hu

M. Molnár-Láng et al. (eds.), *Alien Introgression in Wheat*,
DOI 10.1007/978-3-319-23494-6_3

Spring wheat is grown in regions with a very cold winter, or in Mediterranean and subtropical zones. Spring wheat may also have a short vernalisation requirement. On areas with a mild winter and relatively high temperatures (6–10 °C), vernalisation is protracted, so cultivars with a vernalisation requirement and those that are photoperiod-sensitive can be sown in autumn without any danger of the plants shooting before spring arrives, in which case ears in the booting stage could be damaged by the cold or become sterile. Spring wheat also develops a deep root system when sown in autumn, so it has better yield potential and yield stability than when sown in the spring.

Durum wheat (*Triticum durum* Desf.) has better drought and heat tolerance than common wheat (*Triticum aestivum* L.), so it is more suitable for cultivation in a hot climate than common wheat. The wider cultivation of the winter type is inhibited by the fact that its winter hardiness is poorer than that of common wheat. Durum wheat is cultivated on approx. 6 % of the total wheat-growing area, i.e., on 12–14 million ha (USDA 2010). Countries with the largest areas sown to durum are the EU, Canada, the USA, Turkey and North African countries with a Mediterranean climate.

Spelt wheat (*Triticum spelta* L.) is the cereal grown in cool, hilly regions of Europe, but it is also grown as an alternative crop in the USA and Australia. Its total sowing area is only a fraction of the durum area.

Wheat yields have tripled over the last 50 years, as demonstrated in Fig. 3.1. The yield increase can be attributed chiefly to the rise in average yields per hectare,

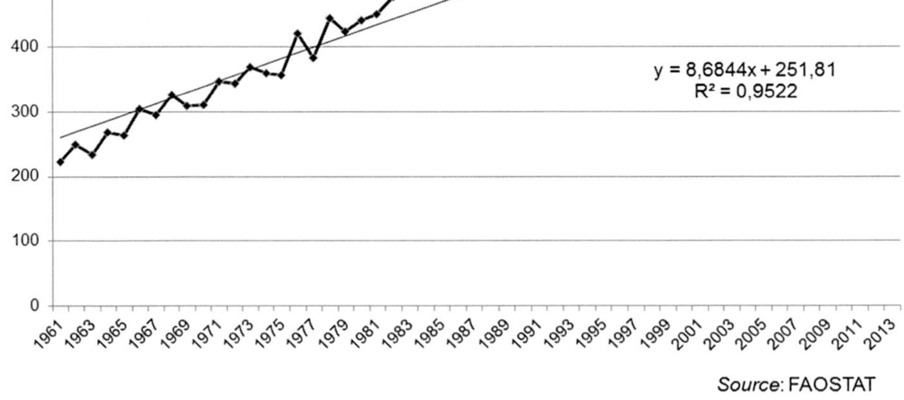

Fig. 3.1 World wheat production

amounting to 40 kg/year, while the sowing area has only changed slightly over the same period. Yield increases are always the joint result of developments in crop production methods and the appearance of new cultivars. The two factors are interdependent and the importance of each may vary from time to time and from place to place, exhibiting a clear fluctuation. After a time the old cultivars are unable to exploit further changes in production conditions, especially an increase in the intensity of crop production, so they become an obstacle to development. By developing new types of cultivars, a further increase in production can be achieved and other factors, such as nutrient replacement, plant protection and other agronomic measures become vital for the full exploitation of the yield potential of the new cultivars. If agricultural production is to be really efficient, there must be a harmonious balance between the two factors, cultivar and technology.

The development of new cultivars adapted to new management practices or growing conditions has in turn allowed management practices to be further improved, within the limits set by the given environmental conditions. The three decades starting in 1960 were a period of intensive mechanisation and of rapid developments in the production technology. The rate of mineral fertiliser applied per hectare rose from 50 kg/ha to over 200 kg/ha (FAOSTAT), and there was a similar increase in the application of fungicides and insecticides.

The increasing use of fertilisers and the greater level of mechanisation required the cultivation of new plant varieties that could be harvested without lodging and which responded to larger quantities of inputs with higher yields. The new wheat cultivars developed by breeders to suit the climatic conditions in specific ecological regions had greater productivity than the cultivars previously sown, with stable yields and grain quality meeting the requirements of processors and consumers. The potential area on which a particular cultivar can be grown is relatively restricted in the case of winter cereal species, but considerably wider for spring types. Due to the cultivar × environment interaction, regional breeding centres or regional experimental networks were established, which continually supply growers with new cultivars.

Increasing or even simply maintaining the yield level requires the regular introduction of new cultivars. Farmers are constantly seeking cultivars with greater yield potential, because the increasing costs of land, input materials and labour mean that crop production will only remain profitable if yield averages continually increase. Maintaining food safety acts in the same direction, as a 60 % increase in wheat production will be needed by 2050 to meet the demands of a growing population with a changing diet.

Though it varied in each period and region, the annual genetic gain for grain yield in developed countries typically ranged from 0.5 to 1 %/year in the second half of the last century (Peltonen-Sainio and Peltonen 1994; Khalil et al. 1995). Genetic gain is responsible for approximately half the yield increment (Kuhr et al. 1985; MacKey 1993), while the other half can be attributed to developments in the technology. A retardation in the rate of genetic gain has been observed in several regions in recent years.

A regular change of cultivar is essential to counteract the appearance of new, virulent fungal races. According to a survey carried out in 26 countries, resistance was effective for 1–15 years, depending on the cultivar, the area on which the cultivar was sown, the country and the type of rust (Kilpatrick 1975). If a few cultivars occupy a large sowing area, there is an increasing danger that mutant pathogen strains will become established and break down the resistance of the cultivars (Brennan and Byerlee 1991).

The technology used for wheat production undergoes constant changes, which may exhibit opposing trends in different growing regions. In many areas fertiliser doses are on the increase, while in many countries in the European Union the opposite tendency can be observed due to environment protection concerns. The plant protection technology may also change, either in the interests of maintaining production levels or due to the banning of certain chemicals. Whatever changes may occur in wheat production practice, however, there will always be a need for new cultivars that can be optimally grown in the changed environment.

It is not unusual for the demand for new cultivars to originate from consumers or the processing industry. People now expect to buy better quality products, so the quality standards required by the industry are rising. New wheat-based products from the baking, confectionery, pasta, brewing, canning, etc. industries require raw materials with diverse composition and technological quality, and this can only be achieved by breeding for specific end-uses. Breeding aims are also influenced by the expansion of the wheat trade and changes in export markets.

The targeted increase in yields does not necessarily require an increase in yield potential, since even the potential latent in existing cultivars cannot be exploited under farm conditions (Anderson 2010; Patrignani et al. 2014). In general it can be said that the lower the extent to which the growing environment (location, weather, technology) satisfies the conditions optimum for wheat, the more the actual yield will depend on the adaptability of the cultivars rather than on their potential yielding ability.

Numerous examinations have been carried out to determine what proportion of the yield increase can be attributed to plant breeding, i.e., to the development of new cultivars. The annual genetic gain demonstrated for the grain yield of wheat varies from 0.16 to 2.3 % (Table 3.1). The estimates depend greatly on the length of the period examined, the initial starting level, the size of the growing area and how favourable the environmental and economic conditions are for breeding research and production development.

Table 3.1 Genetic gain/year through wheat breeding in different regions

Genetic gain %	Country	Period	Author(s)
0.16	India	1901–1980	Sinha and Aggarwal (1980)
0.48	Finland	1939–1991	Peltonen-Sainio and Peltonen (1994)
0.9	Mexico	1950–1982	Waddington et al. (1986)
0.74	USA	1958–1980	Schmidt (1984)
1.3	Oklahoma, USA	1969–1993	Khalil et al. (1995)
2.3	Hungary	1930–1986	Balla et al. (1986)

There were several periods in the history of wheat breeding when substantial changes in the growth habit or genetic background of the plants led to a steep rise in productivity. Landraces and the first cultivars bred from them had long stems and poor lodging resistance, making them unsuitable for intensive production and mechanical harvesting. In the first half of the last century, the Italian breeder Nazareno Strampelli succeeded in transferring the *Rht8* dwarfing gene from the Japanese cultivar Akakomugi into numerous new cultivars, thus creating the first group of intensive wheats, which had an enormous influence not only on wheat production, but also on further breeding efforts in many parts of the world. A few decades later Norman Borlaug developed semi-dwarf wheat cultivars well adapted to intensive production. Cultivars carrying the dwarfing genes *Rht1* or *Rht2* had a plant height only half or two-thirds of that of traditional cultivars and were so successful in field production that by 1963 95 % of the wheat-growing area in Mexico was sown to semi-dwarf wheats. When cultivated using a satisfactory technology, these new cultivars resulted in a spectacular rise in yield averages. Large quantities of seed from dwarf Mexican cultivars (Lerma Rojo, Sonora 64, etc.) were sent to famine-struck India and Pakistan, and by 1970 wheat production had almost doubled in both countries, putting an end to the famine.

This process, dubbed as the 'Green Revolution', also took place in countries growing winter wheat. *Rht1* and *Rht2* genes responsible for semi-dwarf plant height are widespread in the English, French and other West European wheat cultivars giving them better lodging resistance and higher yield (Worland and Snape 2001). In winter wheat-growing regions of Eastern Europe and Middle East the semi-dwarf cultivar Bezostaya 1 with dwarfing gene Rht8 had the most far-reaching effect.

The positive effect of dwarfing genes was manifested not only in the fact that the better lodging resistance resulting from reduced plant height enabled them to be grown at a higher nutrient level, but also in the substantial change that occurred in the distribution of the assimilates between the vegetative and generative organs. A major reduction in plant height may, however, also have a negative effect on the biological yield. According to Richards (1992) each 10 cm decrease in straw length below a plant height of 100 cm leads to a 4.4 % drop in the aboveground biomass. This could be observed in India, where the introduction of dwarf cultivars resistant to lodging was accompanied by a decline in the biomass (Sinha and Aggarwal 1980).

Most experiments designed to compare old and new wheat cultivars either detected no genetic gain for the biological yield (Siddique et al. 1989; Slafer and Andrade 1989; Bodega and Andrade 1996) or only a very slight change (Balla et al. 1986; Perry and Antuono 1989; Nedel 1994). Due to the negative correlation between biomass and lodging resistance, breeders automatically selected for greater harvest index (HI) (Riggs et al. 1981). While in the case of non-intensive wheats around two-thirds of the aboveground dry matter was made up of stems and leaves, the harvest index of modern cultivars approached or even exceeded 50 % (Peltonen-Sainio and Peltonen 1994; Hay 1995). The main reason for the rise in HI is that modern cultivars form a larger number of seeds per unit area (Perry and Antuono

1989; Calderini et al. 1995; Bodega and Andrade 1996), usually without a reduction in size (Ledent and Stoy 1988; Hay 1995).

A further increase in harvest index is difficult to achieve without a deterioration in adaptability, so in the future higher yields are likely to be the result of an increase in biomass or an improvement in water and nutrient use efficiency (White and Wilson 2006; Parry and Hawkesford 2010).

3.2 Traditional Wheat Breeding Methods

The breeding of a wheat cultivar can be divided into several stages. The first step is to develop populations with very broad genetic variability. After crossing, the diverse properties of the parents recombine in the progeny, producing genetically variable populations from which stable, homozygous lines can be selected over a period of years. The third stage of breeding is testing. After several years of experiments in an increasing number of environments it is possible to identify lines that have significantly better productivity than earlier cultivars, combined with a satisfactory level of yield stability and with quality that meets the demands of users.

The most important method applied to develop populations with wide genetic variability is the crossing of parental cultivars with diverse genetic backgrounds. This may be a single cross (A×B), a top cross [(A×B) × C], a double cross [(A×B) × (C×D)] or an even more complicated multiple cross. The choice of crossing scheme depends greatly on the properties of the parental cultivars.

If the aim is to introduce into the breeding material a new trait that is only to be found in initial stocks which are not adapted to local environmental and technological conditions, it is advisable to use a crossing scheme where the donor genotype contributes to a smaller extent to the development of the progeny generation. A special crossing method designed for this purpose is the backcross [(A×B) × A] × A, where the aim is to improve a recipient or recurrent parent (an elite cultivar) by introducing one or a few favourable traits (e.g., resistance, chemical quality, etc.) from a donor (B) carrying alien genes. After a few backcrosses, followed by continuous selection for the desired trait, the result is practically the same as the initial recipient genotype (A), except for the improvement in resistance or some other trait. When applying this method it is important to remember that a new genotype developed in this way is essentially a derived version of the recurrent cultivar, so if the initial material is protected by plant breeder's rights, this protection extends to the modified/improved form of the cultivar.

The success of crossing programmes is greatly influenced by the properties of the initial breeding material. A survey that involved breeders from 52 countries demonstrated that in the majority of cases breeders use their own advanced lines (39.7 %) or registered cultivars (17.4 %) for crossing, while 11 % of the crossing partners are foreign lines and 17.4 % registered cultivars. Landraces are used in 3–4.5 % of cases. A considerable proportion of the parental partners (18.6 %) are made up of genotypes included in international nurseries organised by CIMMYT

(Rejesus et al. 1996). Each type of parent material appears to be used more in the pursuit of certain goals than others. Wild relatives and landraces are more likely to be used for disease resistance and quality than for yields, while advanced lines and released cultivars are more frequently used in breeding for yield potential than for other breeding goals.

Starting from the F_2 generation, the parental traits are combined and become fixed in the progeny. The smaller the number of genes determining the expression of a trait, i.e., the simpler the inheritance, the earlier the generation in which the progeny plants become homozygous and the sooner selection can be commenced. The most important plant traits, however, such as yield, quality, biotic resistance, etc., are polygenic, so selection for these traits is only effective in later generations. Depending on when plant selection is started, two methods of breeding can be distinguished, bulk selection and pedigree selection.

When the bulk method is applied, the populations arising from a cross are sown and harvested for several years without human selection. The essence of this method is that while the level of homozygosity rises from year to year due to self-fertilisation, natural selection leads to the increasing frequency of favourable genes. The bulk method is a cheap way of handling segregating populations in cases where the environmental conditions make efficient natural selection possible for a trait critical for cultivation. The method is ideal when breeding for tolerance of unfavourable ecological factors, e.g., in locations with a very cold winter. In such areas, plants not adapted to the environment, i.e., with poor winter hardiness, will die, leading to the greater frequency of plants with the desired traits. Natural selection may have a similar success rate in the case of disease epidemics, where the survival and reproduction rates of resistant plants will be greater than that of susceptible plants.

The bulk method can be supplemented by mechanical mass selection, which relies only partially on the selection effect of the natural environment. In addition the population is artificially reduced to increase the ratio of plants with favourable traits. Selection may be direct or indirect. Direct selection can be performed for enhancing grain size (Derera and Bhatt 1972) and adaptability by selecting for grain size. An efficient way of reducing plant height is to cut the population back to the desired height (Romero and Fey 1966). In a special case of mass selection for earliness and drought tolerance, the plant stand is desiccated prior to ripening, after which selection is made for large grains (Blum et al. 1983). After several years of self-fertilisation, pure line selection is begun in the F_5–F_6 generation of the bulks, followed by the testing of selected lines with satisfactory homogeneity.

In the pedigree method, plant selection is commenced in early generations (F_2 or F_3) and is repeated several times until a satisfactory level of homozygosity is reached. If plant selection is not begun in the F_2 population, a manageable population size is generally obtained by manual mass selection. In this case the selected ears are threshed together and the grains thus obtained are sown for further selection.

In order for every possible recombination in a segregating population to occur at least once, a much larger population size would be needed than can be achieved in practice. Even if a desirable genotype occurred in a population with low frequency, it is doubtful whether it would be noticed on the basis of phenotype. Due to the inevitable limits to population size and to practical considerations regarding the possibility of selection, most breeders use small population size and a large number of combinations, rather than the other way round. In practice, plant selection usually means ear selection, because in most cases it is impossible to separate the plant from the population, while ears are easy to handle and contain enough seeds to allow the sowing of ear progeny rows in the following year. Plant selection is more frequently applied when plant selection follows bulk selection. This has the advantage that the whole yield of a near-homozygous plant allows a greater volume of testing and more rapid multiplication than is possible after ear selection.

In the course of traditional breeding, visual selection is carried out on the basis of phenotype in populations consisting of a few thousand plants. In this stage, selection for monogenic/oligogenic traits is efficient for traits such as earliness, plant height or disease resistance. Plants with favourable traits can be labelled during the vegetation period and given priority during the final selection process. Selection can best be made from populations with great variability and where the plants can be fairly easily identified, i.e., the stand is neither too dense nor too thin. In a dense stand it is difficult to decide whether an ear in a low position is the main ear of a short plant or the side ear of a tall plant, while in a thin stand the growth habit of the plants is not a reliable prediction of how they would behave in the dense stands typical of farm production.

The third stage of breeding is testing. Yield level, yield stability and end-use quality can only be identified after several years of experiments in a range of environments.

3.3 Alien Gene Introgression into Wheat

Crosses between wheat and related wild or cultivated species have been carried out ever since breeding was begun. The first sterile interspecific wheat × rye hybrid was reported by Wilson in 1875, after which Rimpau developed similar hybrids in 1891 (cit. Lelley and Rajháthy 1955). Similar research was initiated worldwide, but for a time the results were not utilised in practical plant breeding. The method used for transferring genes from related species to wheat depends greatly on the evolutionary distance between the species involved.

Species belonging to the primary gene pool of common wheat share homologous genomes. Gene transfer from these species can be achieved by direct hybridisation, homologous recombination, backcrossing and selection (Friebe et al. 1996). The secondary gene pool of common wheat includes polyploid *Triticum*/*Aegilops*spec ies that have at least one homologous genome in common with *T.aestivum*. Gene transfer from these species is possible by homologous recombination if the target gene is located on a homologous chromosome. Species belonging to the tertiary gene pool are more distantly related. Their chromosomes are not homologous to those of wheat. Other strategies need to be employed, because gene transfer from these species cannot be achieved by homologous recombination (Friebe et al. 1996; Molnár-Láng et al. 2014).

One of the major obstacles to interspecific breeding is the incompatibility between species, as the male and female gametes from different species are unable to unite to form a zygote. There may be many reasons for this, for instance: the inability of pollen grains to germinate on the receptive stigma; the failure of pollen tubes to grow successfully in the style; non-attraction of the pollen tube to the ovary; or the inability of the male gamete to fuse with the egg cell. In many cases embryo or endosperm abortion occurs after successful fertilisation. To avoid this problem, in vitro techniques such as ovule culture or embryo rescue are applied (Brown and Caligari 2008). It often happens that the F_1 plants from interspecific crosses are sterile. Many methods have been introduced to overcome incompatibility between the species and achieve successful hybridisation. These include doubling the ploidy level, protoplast fusion, etc. Colchicine treatment is usually applied to F_1 hybrids to induce chromosome doubling in order to produce fertile amphidiploids (triticale, synthetic hexaploid wheat, etc.).

The use of a genome homozygous for the crossability alleles (*kr1kr1kr2kr2*) may contribute to higher seed set when wheat is crossed with rye, barley, etc. The recessive crossability allele *kr1* was transferred from the spring wheat cultivar Chinese Spring (CS) into the winter wheat cultivar Martonvásár 9 (Mv9) by backcrossing Mv9 × CS hybrids with Mv9. As a result of five backcrosses with Mv9 and two selfings after each backcross, the selected progenies had over 50 % seed set with rye when tested on a large number of individual plants (Molnár-Láng et al. 1996).

In the case of wild relative × wheat crosses the pairing homoeologous gene, *Ph1*, suppresses the pairing and recombination of wheat and alien chromosomes, so no alien genetic transfer can occur. However, in plants nullisomic for the *Ph1* gene, and in *ph1b* mutant stocks having a large deletion at the *Ph1* locus, homoeologous wheat and alien chromosomes can pair and recombine. An original *ph1b* mutant stock in the Chinese Spring background, which has poor agronomic characteristics, was transferred into an adapted Kansas winter wheat, which was released as KS12WGGRC55. This new germplasm will accelerate the evaluation and utilization of wheat alien recombinants in cultivar improvement (Friebe et al. 2012).

3.4 New Approaches to Widen Genetic Variation in Wheat Breeding

Wheat improvement in the twentieth century was carried out using traditional breeding methods. However, genetic, pathological, physiological and chemical knowledge has significantly contributed to a better understanding of this crop. Selection for better adaptability has become particularly important in the light of climate change during the last two decades, especially with the increasing frequency of extreme weather events (Veisz et al. 1996). Considerable fluctuation is observed not only between regions, but also between years. Price volatility in recent years has also been influenced by unstable yields. On a world scale, wheat is attacked by 200–250 pathogens and pests, and may cause significant economic losses in commercial wheat production. Under weather conditions favourable to the pathogens, yield losses in susceptible cultivars may be as great as 40–100 %.

It has become evident that the development of wheat genotypes for future agriculture and further crop improvement will require new genetic variation and the selection of wheat cultivars that meet new challenges. Efficient wheat breeding programmes will require breeding efforts, including

- New strategies in genebank research to exploit the genetic variation existing in wild relatives.
- The utilisation of the genetic variation in wild relatives to develop new germplasm in pre-breeding programmes.
- The introgression of new germplasm into the elite wheat pool.

3.4.1 Genebank Research

The role played in plant breeding by the wild and cultivated relatives of wheat and by the old cultivars and populations stored in gene banks has greatly increased recently. The efficiency of research programmes in plant gene banks depends on the accuracy and precision of evaluation techniques. The evaluation of large volumes of germplasm using only traditional tools such as geographic origin, pedigree information, botanical and agronomic descriptions is too time-consuming.

The establishment of a cost-effective core collection to represent the genetic variability of the large germplasm collections traditionally available to breeders is of vital interest. Core collections are useful for the association mapping of disease resistance, seed quality and domestication-related traits. A good example is the core collection of 372 accessions based on passport and simple sequence repeat (SSR) marker data selected by Balfourier et al. (2007) in the Clermont-Ferrand Genetic Resources Center to explore the diversity present in wheat accessions. Breseghello and Sorrells (2006) also underlined the usefulness of association mapping when breeders analyse traits with low heritability, such as yield components.

The use of wild relatives in traditional breeding is less efficient when valuable genes are linked with numerous unfavourable agronomic traits, in addition to which these genotypes also have low yield potential. The transfer of useful genes is also complicated by linkage drag, crossing barriers, the absence of pairing between homoeologous chromosomes, etc. The efficiency of gene bank research can be improved through the joint application of new methods of genotyping and phenotyping, which enable plant breeders to

- Screen collections for genes important for breeding programmes.
- Identify unique alleles and characterise genetic resources at the gene level.
- Dissect the populations of wild relatives to provide insights into the allelic content of germplasm potentially useful for breeding.

3.4.2 New Germplasm Development in Pre-breeding

Allard (1988) considered the events taking place in wheat breeding in the twentieth century as a two-way process, involving on the one hand the erosion of the plant germplasm due to the spread of modern cultivars consisting of homogeneous populations and the disappearance of landraces with heterogeneous populations, and on the other hand the constant accumulation of alleles providing broad adaptability and increased productivity in modern plant cultivars as the result of breeding cycles. One consequence of commercial breeding is that differences between alleles are diminishing in modern elite cultivars, so the development of new germplasm in pre-breeding programmes has become important in wheat breeding. Fundamentally, five types of genetic resources are available to breeders for the development of new germplasm:

- Adapted cultivars and lines
- Exotic germplasm
- Old landraces and cultivar populations
- Wild and cultivated relatives
- Mutant genotypes

Each type of parent material appears to be used more in the pursuit of certain goals than others. However, the chance of broadening genetic variation is lowest when utilising adapted cultivars and lines. The strategy to "cross the best with the best" resulted in a narrowing genetic variation of new cultivars and a stagnation in the yield improvement during the last 10–20 years in most European wheat-growing regions. Compared to adapted parents, the use of wild and cultivated relatives as parents in traditional breeding is time-consuming in many cases. A good example of this is wheat varieties carrying the 1RS.1BL ryetranslocation, which required 33 years from the original cross to the registration of the first cultivar (Rabinovich 1998).

The increasing need for new genetic variation via alien gene introgression has enhanced the importance of pre-breeding activities. Separate pre-breeding projects parallel to the cultivar development were established in many large breeding programs recently to exploit the genetic variation of wild and cultivated relatives as well as old landraces and cultivar populations. The main goal is to improve the efficiency of the introduction from alien genetic resources into adapted germplasm with new breeding tools. Pre-breeding offers a background of germplasm research for commercial plant cultivar breeding and also represents an excellent opportunity for cooperation between the public and private sectors.

3.4.3 New Breeding Methods

Several new methods have been elaborated over the last 30 years to improve the efficiency of wheat breeding, to widen the genetic variation and to shorten the selection time. One method which has been routinely introduced in wheat programmes is the doubled haploid technology used to produce homozygous progenies from the F_1 generation in a single step. This technology is an excellent tool not only for cultivar development, but also for pre-breeding and for the establishment of mapping populations (Laurie and Bennett 1988). Another technology employed in larger wheat research programmes is the use of in vitro cultures for somaclonal selection.

The use of molecular markers for the development of new germplasm and to improve the efficiency of pre-breeding is a great step forward in cereal breeding. It enables breeders to accelerate the introgression and backcrossing of genes into diverse genetic backgrounds, while allowing genes with similar phenotypic effects to be pyramided. Marker-assisted backcross breeding (MABC) is a useful tool to accelerate the introgression of genes into an adapted background. The simulation of recombination during meiosis proves that breeders can recover the recurrent parent using molecular markers more efficiently than by traditional backcrossing (Frisch et al. 2000). MABC provides efficient positive foreground selection for the donor trait, positive background selection for the recurrent parental genome and negative background selection against undesirable donor parent alleles. Marker-assisted selection (MAS) is a valuable technology for pyramiding genes for resistance to diseases. The agronomic traits of BC_5 and BC_6 lines are very similar to those of the recurrent parents. To exploit the advantages of new methods, pyramided gene combinations have been developed (Fig. 3.2), and a doubled haploid programme has been set up in order to stabilize them in the frame of BIOEXPLOIT EU FP6 project to improve durable leaf rust resistance of wheat (Vida et al. 2009).

In the framework of the HEALTHGRAIN FP6 EU project, a backcross programme was carried out to introduce high amylose genes into current commercial European cultivars using the triple SGP mutant lines developed by Lafiandra and his team (Sestili et al. 2010) at the University of Tuscia, Italy. After three BC generations it was possible to select genotypes with amylose contents similar to that of the mutant lines. The projected use of MABC to introgress transgenes into elite germplasm will permit the rapid deployment of agronomic traits.

Selection for mutations induced using radiation or chemical agents was formerly based on the detection of morphological traits or by phenotyping quality traits, an approach not easily applied to a polyploid species such as wheat, as most mutations

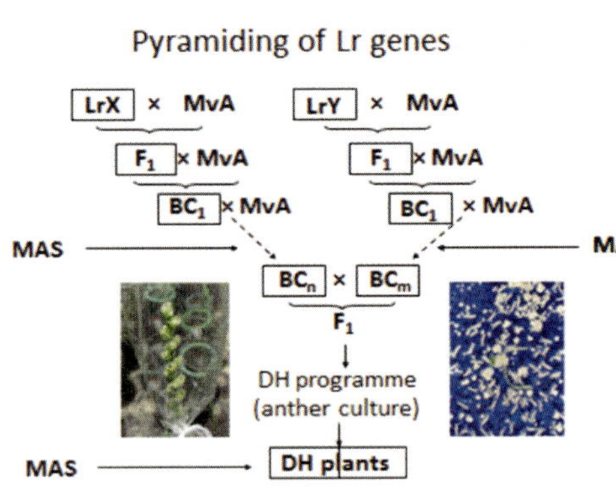

Fig. 3.2 Pyramiding of *Lr* genes using marker assisted backcross and doubled haploid technology (Vida et al. 2009)

are recessive and complemented by functional copies in the additional genomes. This was one reason why breeders only used mutation methods when they had exhausted every other possibility to increase genetic variation. Recently the TILLING (Targeted Induced Local Lesions in Genomes) approach has contributed significantly to the efficiency of mutant discovery. It can be applied in both diploid and polyploid species, allowing the identification of DNA polymorphism in large numbers of genotypes, based on DNA sequence differences. Selection for mutant candidate genes can therefore be made at the molecular level, after which the genes can be combined to develop new genetic resources for breeding.

TILLING is ideal for common wheat, as this technique allows the detection of recessive mutations in polyploid species (Comai et al. 2004). Slade et al. (2005) identified 246 alleles of the waxy starch genes in a mutagenised wheat population. M_4 seeds from the cultivar Cadenza were used to identify mutations through a combination of SDS-PAGE, to analyse granule-bound starch proteins, and TILLING (Sestili et al. 2010). Work has focused on two groups of synthase genes, those encoding starch synthase II (*Sgp-1*) and those corresponding to the waxy proteins. The identification of lines with null alleles for starch biosynthesis genes is a realistic approach for modifying the wheat starch structure, using traditional breeding to transfer the mutations into commercial varieties.

Nowadays transgenesis is a potentially important breeding technology to introduce genes for various agronomical traits from wild and cultivated relatives into wheat. Transgenic wheat production is a two-step process, involving the integration of foreign DNA and the regeneration of fertile plants, followed by the selection of a genetically modified cultivar. The first results in transgenic plant development were obtained using model plants and some important agricultural crops like maize and rice, while the transformation protocol for small-grain cereals was developed later. The first fertile transgenic wheat was developed by Vasil et al. (1992) using a biolistic method. *Agrobacterium*-mediated transformation protocols were developed a decade later for all the major cereal species (Lazzeri and Jones 2009). Well-established transformation protocols are available for wheat and barley with increasingly efficient *Agrobacterium*-mediated transformation techniques, better integration patterns and improved co-transformation. The increasing number of traits tested in field trials is proof of the rapid development of the transgenic breeding technology. Several transgenic wheat and barley transgenic genotypes with agronomically important traits have been submitted for registration (Dunwell 2008).

One promising new technology is cisgenesis, which differs from transgenesis in that DNA fragments from wheat or cross-compatible species are incorporated into the genome, so cisgenic plants do not contain foreign or modified genetic material. Cisgenic plants are produced by the same transformation techniques as transgenic plants. As wheat has many wild and cultivated relatives, the prospects for developing precision breeding via the cisgenesis technology are excellent. This technology would be a modern alternative to traditional breeding and would offer an ideal way of incorporating agronomically important genes from wild and cultivated wheat relatives into common wheat. Cisgenesis has great potential to overcome many problems of traditional breeding, such as linkage drag, crossing barriers, the introduction of many deleterious genes linked with one useful gene, etc. The first results published by Lusser et al. (2011) were achieved for potato and apple.

The prospects and opportunities opened up by developments in crop genomics must be translated into outcomes that improve agricultural productivity and sustainability in the future. Genomic selection is a promising new tool for estimating breeding value, which will contribute to the improvement of the quantitative characters that are controlled by many loci with small effects by means of genome-wide marker coverage (Heffner et al. 2009).

3.4.4 Alien Gene Introgressions in Wheat Germplasm

In the last century, the introgression of the genes or gene complexes responsible for the favourable properties of rye into wheat proved to be one of the most effective ways of enhancing the adaptability and grain yield of wheat. It has long been the desire of wheat breeders to combine the high level of adaptability to extensive conditions, winter hardiness, earliness and tillering ability of rye with the good quality of wheat. Many breeders in various countries have attempted to develop wheat × rye hybrids, primarily with the aim of transferring the disease resistance and winter hardiness of rye into wheat. The year 1917 played an important role in the theoretical work of the Russian scientist Vavilov (Szunics, personal communication), because the hot dry weather caused rye and wheat to flower at the same time, leading to a considerable extent of cross-pollination and natural hybridisation and allowing Vavilov to collect large quantities of wheat × rye hybrid seed. The 1RS.1BL translocations developed by Riebesel and Katterman in Germany in the 1920s and 1930s were of great significance. In 1964 Tsunewaki crossed two octoploid triticales to develop the Salmon wheat cultivar, which, according to Zeller (1973), contains a translocation homologous to that in Salzmunder 14–14 and Zorba. The 1RS.1AL translocation was introduced into the American cultivar Amigo via the Argentine diploid rye, Insave F.A.

The 1RS.1BL wheat-rye translocations, the 1B(R) substitution and the 1RS.1AL translocations contributed to improvements in the yield potential, adaptability, disease resistance and insect resistance of wheat. Since then the occurrence of the 1RS.1BL translocation has been reported in more than 1000 wheat cultivars (Zeller 1972; Mettin et al. 1973; Schlegel and Korzun 1997; Lukaszewski 1990; Rabinovich 1998). A gene complex that includes the resistance genes *Sr31* for stem rust (*Puccinia graminis*), *Lr26* for leaf rust (*P. recondita*), *Yr9* for yellow rust (*P. striiformis*), *Pm8* for powdery mildew (*Erisiphe graminis*) and Gb for leaf aphids (*Schizapis graminum*) is now to be found in a large proportion of the cultivars currently grown (Heun and Fischbeck 1987; Sebesta et al. 1995). Although the gene complex has now lost its resistance to powdery mildew and leaf rust, it still protects wheat from many stem rust races, with the exception of Ug99. Some 11–12 % of the increase in the biological yield of wheat can be attributed to the 1RS.1BL rye translocation

(Carver and Rayburn 1994). According to Rajaram (2001), the main characters associated with this translocation and, consequently, with better adaptation to marginal environments, are drought tolerance, higher biomass (Villareal et al. 1995) and better phosphorus extraction (Manske et al. 1996).

Good use of the 1RS.1BL translocation was made in Krasnodar, Russia, by Lukyanenko (1973), who developed the winter wheat cultivars Kavkaz, Avrora, Bezostaya 2, Skorospelka 35 and Predgornaya 2 based on the Neuzucht germplasm, which contains a rye chromosome segment. These had direct, practical significance, as they spread widely, many of them being grown in various Eastern European countries. The cultivars Kavkaz and Avrora, for instance, spread rapidly in Hungary in the early 1970s, though they later became susceptible to powdery mildew.

What is perhaps of greater importance, however, is that these cultivars proved to be excellent crossing partners for the development of new genotypes. They had numerous valuable traits, the most outstanding of which was their disease resistance. Lukyanenko's cultivars contain the *Pm8*, *Lr26*, *Sr31* and *Yr9* resistance genes originating from rye. Kavkaz was intensively used in the CIMMYT programme and its favourable traits were transferred via the cultivar Veery into many successful genotypes. In Hungary, Kavkaz and Avrora were successfully used as crossing partners by breeders in Martonvásár and Szeged due to their high yield potential, good adaptability and stem rust resistance, so by the 1990s the ratio of cultivars carrying the 1RS.1BL translocation in crop production was around 40–50 % (Fig. 3.3).

According to Kőszegi et al. (2000), 35 (53 %) of the 66 Hungarian-bred wheat cultivars registered in Hungary between 1978 and 1999 carried the 1RS.1BL translocation. This translocation was introduced into Hungarian breeding stock with the

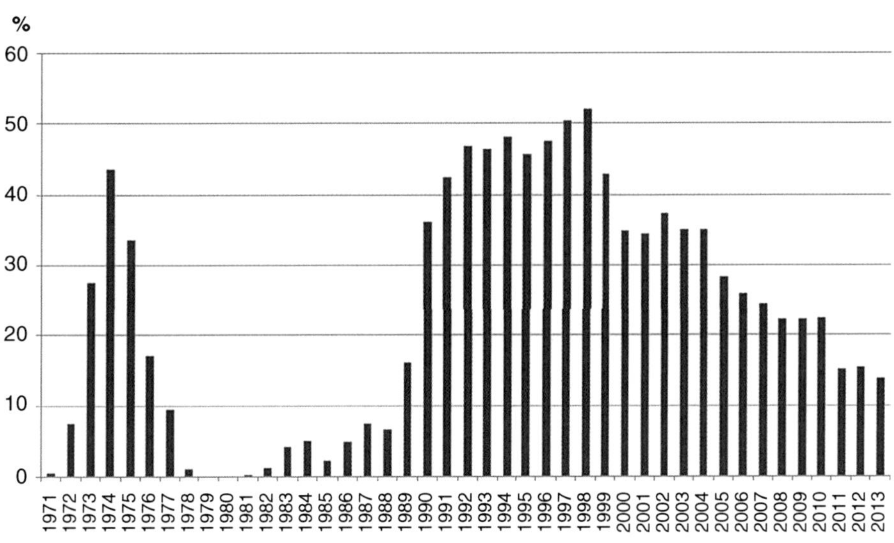

Fig. 3.3 Ratio of the growing area sown to wheat cultivars carrying the 1BL.1RS ryetranslocation Hungary, 1971–2013

cultivars Avrora, Kavkaz, Skorospelka 35 and Bezostaya 2. Bedő et al. (1993) demonstrated the presence of the 1RS.1BL translocation in the cultivars Martonvásári (Mv) 14, Mv 15, Mv 16, Mv 17, Mv 18, Mv 19, Mv 20, Mv 21, Mv 22 and Mv 23. On the basis of their pedigrees the translocational chromosome originated from four different sources: in Mv 14, Mv 15, Mv 16 and Mv 20 from Kavkaz, bred in Krasnodar, in Mv 18 from Solaris, bred in Solary, Slovakia, in Mv 21 and Mv 22 from Posavka 2, bred in Novi Sad, Yugoslavia and in Mv 19 from GT 5239-2, bred in General Toshevo, Bulgaria. Kavkaz is among the ancestors of Solaris, and Skorospelka 35 among those of Posavka 2, while Avrora is the source of the translocation in the case of GT 5239-2.

Wheat-rye chromosome substitutions and translocations have had a great influence on quality improvements. Although the gene complex originating from rye carries numerous disease resistance genes, these are combined with an unfavourable effect on the technological quality, partly due to the presence of gamma secalin instead of omega and gamma gliadins, and partly to the lack of the LMW glutenin subunits found on wheat chromosome 1BS (Dhaliwal et al. 1988). The substitution of the seed storage proteins of wheat by those of rye (*Secale cereale*) in the 1RS.1BL translocation line resulted in inferior wheat quality (Lukaszewski 1993). The undesirable technological traits include small loaf volume and sticky crumb (Zeller 1973), together with poor rheological traits. Breeders have made considerable efforts in recent years to eliminate unfavourable technological quality by incorporating the translocated fragment into chromosomes of homoeologous group 1 or by altering the present genetic background. Graybosch et al. (1990) drew attention to the fact that a change in the genetic background could considerably modify the unfavourable traits arising when the flour is used for bread-making. Similar conclusions were drawn by Javornik et al. (1991) when analysing the cultivars Jugoslavia, Balkan and Zvezda, which have good bread-making quality and contain the 1RS.1BL translocation. Results obtained in Martonvásár, Hungary, also suggest that a favourable genetic background is capable of reducing the deleterious effect of the 1RS.1BL translocation, but it cannot be completely eliminated and a negative influence on quality must generally be expected (Bedő et al. 1993). Knackstedt et al. (1994) reported a new wheat-rye translocation on the long arm of chromosome 2B, carrying a resistance gene against Hessian fly, which did not influence the wheat storage proteins or the technological quality.

A method was elaborated for the efficient introduction of new allelic variation into 1RS chromosome arm in wheat cultivars with 1RS.1BL translocation by Molnár-Láng et al. (2010). A wheat genotype containing both the recessive crossability alleles (*kr1kr1kr2kr2*) allowing high crossability between 6*x* wheat and diploid rye, and the 1BL.1RS wheat/rye translocation chromosomeswas developed. This wheat genotype was used as a recipient partner in wheat × rye crosses. Chromosome pairing between the 1BL.1RS translocation and the 1R chromosome of the rye cultivar was detected in meiosis of the wheat × rye hybrids using sequential GISH and FISH, thus recombination occurred between the 1RS arms (Molnár-Láng et al. 2014).

From among the *Thinopyrum*species*Thinopyrum ponticum* [(Podp.) Barkw and DR Dewey (syn *Agropyron elongatum*)] is carrying many potentially favourable traits for wheat improvement. A number of disease resistance genes, including *Lr19*, *Lr24* and *Lr29* (Knott 1968; Sears 1973, 1977), *Sr25* and *Sr26* (McIntosh et al. 1977; Jin et al. 2007), Cmc2 (Whelan et al. 1986) and Qfhs.pur-7EL (Shen and Ohm 2007), as well as genes controlling salt tolerance (Chen et al. 2004), yield and biomass (Reynolds et al. 2001; Monneveux et al. 2003), were transferred from *Th. ponticum* to wheat. The dwarf line 31505-1 was obtained after backcrossing 31505 to Lumai 5 (Chen et al. 2012). In Russia, Tzitzin selected winter wheat cultivars from *Agropyron* sp. × *T.aestivum* crosses, which were grown in commercial production (Zhukovsky 1957). Interspecific crosses were used to incorporate a number of disease resistance and storage protein genes into the common wheat germplasm in the Odessa breeding programme (Litvinenko et al. 2001).

The *Lr19* leaf rust resistance gene was transferred to common wheat from *Thinopyrum ponticum* by Sharma and Knott (1966). *Lr19* is an important gene not only due to the rust resistance conferred by this gene, but also because of the yield increases produced in different backgrounds when alien chromatin carrying *Lr19* is introgressed in wheat. CIMMYT scientist Ricardo Rodriguez, a pre-breeder, successfully transferred the gene into a Bluebird 2 background in the mid 1980s. Later the Mexican National Agricultural Research Service (INI-FAP) released a wheat cultivar carrying *Lr19* named Oasis 89, which was higher yielding than its recurrent parent Yecora 70, even in the absence of rust. This observation led to further studies and the development of more lines carrying *Lr19* (Rajaram 2001). According to Reynolds et al. (2001) *Lr19*was associated with increases in yield (average 13 %), final biomass (10 %) and grain number (15 %) in all the backgrounds studied. *Lr19* was also associated with the increased partitioning of biomass to spike growth at anthesis (13 %), a higher grain number per spike, and higher values of radiation use efficiency (RUE) and flag-leaf photosynthetic rate during grain filling. According to Miralles et al. (2007) the *Lr19* gene promoted the partitioning of assimilates to the reproductive organs and nitrogen partitioning to the spike, resulting in an increased number of fertile florets per spike and grains per unit area, without affecting the number of spikes per unit area or crop development.

Unfortunately, the *Lr19*translocation also carries a gene(s) for yellow endosperm pigmentation that renders the resistance useless in countries where this trait is regarded as undesirable because of quality aspects. Several modifications were made to construct a shorter form of the translocation, lacking the yellow pigment genes (*Lr19-149*). Marais et al. (2001) carried out a yield trial with near isogenic lines of both the original and shortened translocations, which suggested that *Lr19* may cause a small reduction in kernel size and an increase in loaf volume, effects which are not associated with *Lr19-149*.

Linkage drag is a great challenge for wheat breeders carrying out alien gene introgression. A resistance gene against eyespot, caused by *Tapesia* spp. (formerly *Pseudocercosporella herpotrichoides*), was introduced from *Aegilops ventricosa* with a closely linked gene depressing yield performance. After a conscious selection programme this unfavourable linkage was broken, and UK breeders developed cultivars with the *Pch1* gene, which exhibited no yield depression, such as the cultivar Lynx (Worland and Snape 2001).

One of the most dangerous threats to world wheat production is the stem rust (*Puccinia graminis* f. sp. *tritici* Pers.) race Ug99, which originated in Africa. Resistance gene *Sr22* is present on a chromosomal translocation derived from *Triticum boeoticum* Boiss. Only a limited number of cultivars were selected due to the poor agronomic performance of lines carrying this resistance gene. Linkage analysis of simple sequence repeat (SSR) markers was performed on chromosome 7A to identify loci closely linked to *Sr22*. Resistant lines were identified, having less than 6 % of the chromosome arm derived from *T. boeoticum*. These lines may provide a more agronomically desirable source of *Sr22* that can be readily deployed to develop cultivars resistant to Ug99 (Olson et al. 2010).

Several other genes for resistance to rust diseases have been introduced from *Aegilops* and *Thinopyrum* species to common wheat. According to the catalogue published by McIntosh et al. (2008, 2010), the following leaf and yellow rust resistance genes were isolated from the wild relatives of wheat: *Lr9—Ae. umbellulata*; *Lr24* and *Lr29—Thinopyrum ponticum*; *Lr37* and *Yr17—Ae. ventricosa*; *Lr38—Th. intermedium*; *Lr28, Lr35, Lr36, Lr51* and *Lr66—Ae. speltoides*; *Lr21*, *Lr22a*, *Lr32*, *Lr39*, *Lr40*, *Lr41* and *Yr28—Ae. tauschii*; *Lr57* and *Yr40—Ae. geniculata*; *Lr58—Ae. triuncialis*; *Lr53*, *Yr15*, *Yr35* and *Yr36—Ae. longissima* and *T. dicoccoides* (Amandeep et al. 2012).

Distant hybridisation has been given particular attention in regions where natural populations of wild species have great genetic variability, and where they were used in agricultural production before the introduction of modern cultivation practices due to their agronomically valuable, locally adapted traits, which made them popular with local farmers. The main advantages were their adaptability and disease resistance traits, which have been utilized by breeders in order to transfer disease resistance genes with simple inheritance into wheat. The cultivars Plainsman V, Plainsman IV, Encore and Frontiersman, which contain genes from *Aegilops ovata*, were selected specifically for increased protein content and technological quality (Sharma and Gill 1983). In addition to their favourable disease resistance, *Triticum dicoccoides, Triticum monococcum*, etc. are also valuable sources for broadening the genetic base of *T.aestivum* in terms of increased protein content and protein quality traits. Addition lines containing the 1E chromosome from *Thinopyrum elongatum* improved seed storage protein quality (Garg et al. 2009). Similar results were obtained with *Thinopyrum intermedium* when Cao et al. (2007) examined substitution for chromosome 1D of wheat. The V genome of *Dasypyrum villosum* (L.) Candargy (Dv), also known as *Haynaldia villosa* (L.) Schur, an annual wild diploid relative of common wheat, contains genes for elevated seed protein content and also has a positive effect on gluten strength (Shewry et al. 1987, 1991). Zhang et al. (2014) suggested that the *Glu-V1* and *Gli-V1/Glu-V3* loci, located in the chromosome 1 V of *Dasypyrum villosum* (L.), were proved to have positive effects on grain quality parameters, like protein content, Zeleny sedimentation value and the rheological characteristics of wheat flour dough.

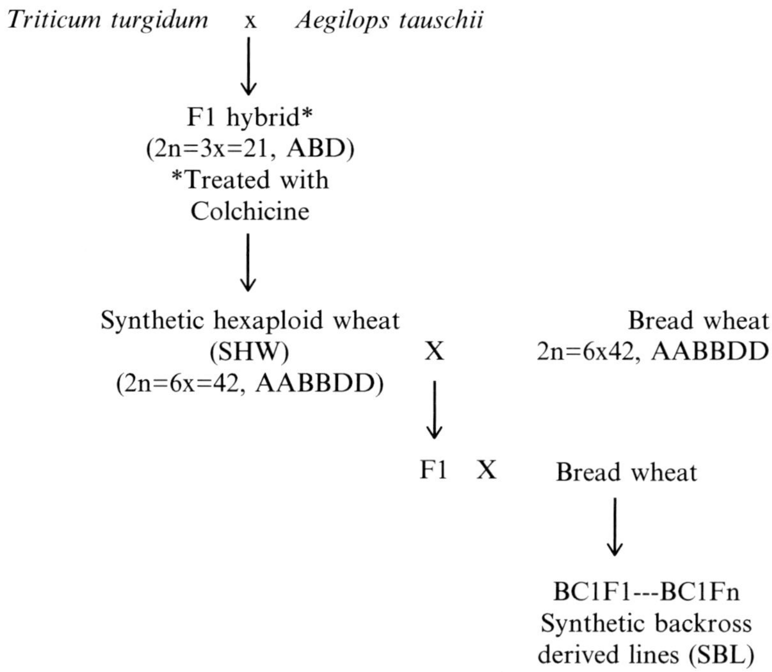

Fig. 3.4 Production scheme of synthetic backcross derived lines (after Ogbonnaya 2011)

The artificial synthesis of hexaploid wheat is an excellent tool to widen genetic variability and to select genotypes with new agronomical properties (Fig. 3.4). The first attempts to develop synthetic hexaploid wheats through the artificial hybridisation of durum wheat [*Triticum turgidum ssp. durum* (Desf.) Husn.] with *Aegilops tauschii* Coss. were published by McFadden and Sears (1944). In the early 1990s, synthetic hexaploid wheats were backcrossed to many CIMMYT and global elite breeding lines. The advanced backcross lines are popular with breeders worldwide due to their excellent levels of resistance to biotic and abiotic stresses, and in some cases, high grain quality. The successful incorporation of genetic diversity from the wild relatives of wheat has created wheats containing more variation (Rajaram 2001).

Synthetic-derived lines were compared with their recurrent parent in field experiments conducted in the mid-1990s near Ciudad Obregon, Sonora, Mexico. More than 80 % of the synthetic-derived lines were significantly superior to their recurrent parent for kernel weight. Eight lines had significantly higher grain yield compared with their recurrent parent. The grain yields of superior lines were up to 11 % higher than those of their recurrent parents. A strong association between grains m^{-2}, biomass, spikes m^{-2}, grain and biomass production rates and grain yield was observed in all the populations (del Blanco et al. 2001).

Synthetic hexaploid lines and their derivatives were intensively used in breeding programmes for stress resistance. They provide new genetic variability for adaptation to drought, high temperature, salinity, waterlogging and soil micronutrient imbalances (Trethowan and Mujeeb-Kazi 2008). Concerning their resistance to biotic stresses, these lines carry resistance to all three rust diseases, caused by *Puccinia triticina, P. graminis* f.sp. *tritici* and *P. striiformis* f.sp. *tritici*. Their resistance reactions varied from 1 % for *Septoria nodorum* glume blotch to 82 % for *Septoria tritici* blotch. In addition, numerous authors reported the succeesful use of synthetic-derived backcross lines in environments with constraints such as drought, and white-grained lines resistant to pre-harvest sprouting were selected (Ogbonnaya 2011).

According to studies carried out by Warburton et al. (2006) using SSR molecular markers, there has been a very significant increase in the genetic diversity of recently developed synthetic hexaploid wheat derivatives. The inherent diversity of the new synthetic derivatives is comparable to that of the landraces; however, they also exhibit improved yield, disease resistance, abiotic stress tolerance and better end-use quality.

The aim to use alien gene introgression to increase genetic diversity goes back to a long tradition, it is almost as old as the conscious selection. When a trait of interest was not available in the same species plant breeders screened related species to identify the desired character. Common wheat is an excellent crop in this respect as it has several relatives within the same genus to produce interspecific hybrids and intergeneric hybrids between different genera to develop new genotypes with useful traits. Plant breeders put a lot of emphasis to select new genetic resources derived from alien gene introgression. Many decades of tremendous work characterized these activities to overcome "bottlenecks" to introduce genes into the cultivated crops and new varieties, useful for the whole mankind. Efficient alien gene introgression programs require new molecular methods and genomic tools to develop new germplasm in pre-breeding programmes, and the introgression of new germplasm into the elite wheat pool. It is evident using alien genes in crop breeding that both the food security and safety became better in the World agriculture.

References

Allard RW (1988) Genetic changes associated with the evolution of adaptedness in cultivated plants and their wild progenitors. J Heredity 79:225–238

Amandeep KR, Satinder K, Dhaliwal HS, Singh K, Chunneja P (2012) Introgression of a leaf rust resistance gene from *Aegilops caudata* to bread wheat. J Genet 91:155–161

Anderson WK (2010) Closing the gap between actual and potential yield of rainfed wheat: the impacts of environment, management and cultivar. Field Crops Res 116:14–22

Balfourier F, Roussel V, Strelchenko P, Exbrayat-Winson F, Sourdille P, Boutet G, Koenig J, Ravel C, Mitrofanova O, Beckert M, Charmet G (2007) A worldwide bread wheat core collection arrayed in a 384-well plate. Theor Appl Genet 114:1265–1275

Balla L, Bedő Z, Láng L, Szunics L (1986) Genetic advance in wheat breeding and its contribution to yield gains. Acta Agron Hung 35:219–225

Bedő Z, Balla L, Szunics L, Láng L, Kramarikné-Kissimon J (1993) A martonvásári 1B/1R transzlokációt hordozó búzafajták agronómiai tulajdonságai. (Agronomic properties of Martonvásár wheat varieties bearing the 1B/1R ranslocation). Növénytermelés 42:391–398

Blum A, Poiarkova H, Golan G, Mayer J (1983) Chemical desiccation of wheat plants as a simulator of post-anthesis stress. I. Effect on translocation and kernel growth. Field Crops Res 6:51–58

Bodega JL, Andrade FH (1996) The effect of genetic improvement and hybridization on grain and biomass yield of bread wheat. Cereal Res Comm 24:171–177

Braun HJ, Atlin G, Payne T (2010) Multi-location testing as a tool to identify plant response to global climate change. In: Reynolds MP (ed) Climate change and crop production. CABI, London, pp 115–138

Brennan JP, Byerlee D (1991) The rate of crop varietal replacement on farms. Measures and empirical results for wheat. Plant Varieties Seeds 4:99–106

Breseghello F, Sorrells M (2006) Association analysis as a strategy for improvement of quantitative traits in plants. Crop Sci 46:1323–1330

Brown J, Caligari P (2008) An introduction to plant breeding. Blackwell Publishing, Oxford, p 209

Calderini DF, Dreccer MF, Slafer GA (1995) Genetic improvement in wheat yield and associated traits. Re-examination of previous results and latest trends. Plant Breed 114:108–112

Cao S, Xu H, Li Z, Wang X, Wang D, Zhang A, Jia X, Zhang X (2007) Identification and characterization of a novel *Ag.intermedium* HMW-GS gene from *T. aestivum* to *Ag. intermedium* addition lines TAI-I series. J Cereal Sci 45:293–301

Carver BF, Rayburn AL (1994) Comparison of related wheat stocks possessing 1B or 1RS.1BL chromosomes: agronomic performance. Crop Sci 34:1505–1510

Chen SY, Xia GM, Quan TY, Xiang FN, Jin Y, Chen HM (2004) Introgression of salt-tolerance from somatic hybrids between common wheat and *Thinopyrum ponticum*. Plant Sci 167:773–779

Chen G, Zheng Q, Bao Y, Liu S, Wang H, Li X (2012) Molecular cytogenetic identification of a novel dwarf wheat line with introgressed *Thinopyrum ponticum* chromatin. J Biosci 37:149–155

Comai L, Young K, Till BJ, Reynolds SH, Green EA, Codomo CA, Enns LC, Johnson JE, Burtner C, Odden AR, Henikoff S (2004) Efficient discovery of DANN polymorphism in natural population by ecotilling. Plant J 34:778–786

del Blanco IA, Rajaram S, Kronstad WE (2001) Agronomic potential of synthetic hexaploid wheat-derived populations. Crop Sci 41:670–676

Derera NF, Bhatt GM (1972) Effectiveness of mechanical mass selection in wheat (*Triticum aestivum* L.). Aust J Agric Res 23:761–768

Dhaliwal AS, Mares DJ, Marshall DR, Skeritt JH (1988) Protein composition and pentosan content in relation to dough stickiness of 1B/1R translocation wheats. Cereal Chem 62:143–149

Dunwell JM (2008) Transgenic wheat, barley, and oats: future prospects. In: Jones HD Shewry PR (eds) Transgenic wheat, barley and oats: production and characterization protocols, Springer Protocols, Methods in Molecular Biology. Humana Press, New York, pp 333–346

FAOSTAT, FAO Database. Food and Agriculture Organization of the United Nations. Available at: http://faostat.fao.org

USDA Foreign Aricultural Service (2010) Commodity intelligence Report, December 22, 2010

Friebe B, Jiang J, Raupp WJ, McIntosh RA, Gill BS (1996) Characterization of wheat-alien translocations conferring resistance to diseases and pests: current status. Euphytica 91:59–87. doi:10.1007/BF00035277

Friebe B, Qi LL, Liu C, Liu WX, Gill BS (2012) Registration of a hard red winter wheat genetic stock homozygous for *ph1b* for facilitating alien introgression for crop improvement. J Planet Reg 6:121–123

Frisch M, Bohn M, Melchinger AE (2000) PLABSIM: Software for simulation of marker-assisted backcrossing. J Hered 91:86–87

Garg M, Tanaka H, Ishikawa N, Takata K, Yanaka M, Tsujimoto H (2009) *Agropyron elongatum* HMW-glutenins have a potential to improve wheat end-product quality through targeted chromosome introgression. J Cereal Sci 50:358–363

Graybosch RA, Peterson CJ, Hansen LE, Mattern PJ (1990) Relationship between protein solubility characteristics, 1BL.1RS, high molecular weight glutenin composition, and end-use quality in winter wheat germplasm. Cereal Chem 67:342–349
Hay RKM (1995) Harvest index: a review of its use in plant breeding and crop physiology. Ann Appl Biol 126:197–216
Heffner EL, Sorrells ME, Jannink J-L (2009) Genomic selection for crop improvement. Crop Sci 49:1–12
Heun M, Fischbeck G (1987) Identification of wheat powdery mildew resistance genes by analysing host-pathogen interactions. Plant Breed 98:124–129
Javornik B, Sinkovic T, Vapa L, Koebner RMD, Rogers WJ (1991) A comparison of methods for identifying and surveying the presence of 1BL.1RS translocations in bread wheat. Euphytica 54:45–53
Jin Y, Singh RP, Ward RW, Wanyera R, Kinyua MG, Njau P, Fetch JT, Pretorius ZA, Yahyaoui A (2007) Characterization of seedling infection types and adult plant infection responses of monogenic Sr gene lines to race TTKS of *Puccinia graminis f.* sp. tritici. Planet Dis 91:1096–1099
Khalil IH, Carver BF, Smith EL (1995) Genetic gains in two selection phases of a wheat-breeding programme. Plant Breed 114:117–120
Kilpatrick RA (1975) New wheat cultivars and longevity of rust resistance, 1971-75. No. ARS-NE-64, 20pp
Knackstedt MA, Sears RG, Rogers DE, Lookhart GL (1994) Effects of T2BS-2RL wheat-rye translocation on breadmaking quality in wheat. Crop Sci 34:1066–1070
Knott DR (1968) Translocations involving *Triticum* chromosomes and *Agropyron* chromosomes carrying rust resistance. Can J Genet Cytol 10:695–696
Kőszegi B, Linc G, Juhász A, Láng L, Molnár-Láng M (2000) Occurence of the 1RS/1BL wheat-rye translocation in Hungarian wheat varieties. Acta Agron Hung 48:227–236
Kuhr SL, Johnson VA, Peterson CJ, Mattern PJ (1985) Trends in winter wheat performance as measured in international trials. Crop Sci 6:1045–1049
Laurie DA, Bennett MD (1988) The production of haploid wheat plants from wheat x maize crosses. Theor Appl Genet 73:393–397
Lazzeri PA, Jones HD (2009) Transgenic wheat, barley and oats: production and characterization. In: Jones HD Shewry PR (eds.) Transgenic wheat, barley and oats: production and characterization protocols, Springer Protocols, Methods in Molecular Biology, Humana Press, New York, 478. pp 3–23
Ledent JF, Stoy V (1988) Yield of winter wheat, a comparison of genotype from 1910 to 1976. Cer Res Comm 16:151–156
Lelley J, Rajháthy T (1955) A búza és nemesítése. Akadémiai Kiadó, Budapest, p 544
Litvinenko M, Lyfenko S, Poperelya F, Babajants L, Palamatchuk A (2001) Ukrainien wheat pool. In: Bonjean AP, Angus WJ (eds) The world wheat book: a history of wheat breeding. Lavoisier Publishing, Paris, pp 351–375
Lukaszewski AJ (1990) Frequency of 1RS.1AL and 1RS.1BL translocations in United States wheats. Crop Sci 30:1151–1153
Lukaszewski A (1993) Recontstruction in wheat of complete chromosome-1b and chromosome-1r from the 1rs.1bl translocation of Kavkaz origin. Genome 36:821–824
Lukyanenko PP (1973) Szelekcija ozimüh psenyic Avrora i Kavkaz. Izbrannüje trudii. Kolos, Moskva, pp 338–343
Lusser M, Parisi C, Plan D and Rodríguez-Cerezo E (2011) New plant breeding techniques: state-of-the-art and prospects for commercial development. JRC Scientific and Technical Reports. JRC European Commission, EUR 24760, pp 24–25
MacKey J (1993) Demonstration of genetic gain from Swedish cereal breeding. Sveriges Utsadesforenings Tidskrift 103:33–43
Manske GGB, Behl RK, Vlek PLG (1996) The effect of total root length and arbuscular mycorrhizal fungi on the nutrient and water efficiency in summer wheat. In: Crop Productivity and

Sustainability—Shaping the Future: Abstracts of the 2nd International Crop Science Congress, New Delhi, India, Vol I, p 105
Marais GF, Marais AS, Groenewald JZ (2001) Evaluation and reduction of Lr19-149, a recombined form of the *Lr19* translocation of wheat. Euphytica 121:289–295
McFadden ES, Sears ER (1944) The artificial synthesis of *Triticum spelta*. Rec Genet Soc Am 13:26–27
McIntosh RA, Dyck PL, Green GJ (1977) Inheritance of leaf rust and stem rust resistances in wheat cultivars Agent and Agatha. Aust J Agric Res 28:37–45
McIntosh RA, Yamazaki Y, Dubcovsky J, Rogers WJ, Morris CF, Somers DJ, Appels R, Devos KM (2008) Catalogue of gene symbols for wheat. In: Appels R, Eastwood R, Lagudah E, Langridge P, Mackay M, McIntyre L, Sharp P (eds) Proceedings of the 11th international wheat genetics symposium Sydney, Australia. Sydney University Press, Australia
McIntosh RA, Yamazaki Y, Rogers WJ, Morris CF, Devos KM (2010) Catalogue of gene symbols for wheat (http://www.shigen.nig.ac.jp/wheat/komugi/genes/download.jsp)
Mettin D, Blüthner WD, Schlegel G (1973) Additional evidence on spontaneous 1BL.1RS wheat rye substitutions. Proceedings of 4th International Wheat Genetics Symposium, Columbia, USA, 179–184
Miralles DJ, Resnicoff E, Carretero R (2007) Yield improvement associated with *Lr19* translocation in wheat. In: Spiertz JHJ JHJ, Struik PC, van Laar HH (eds) Scale and complexity in plant systems research: gene-plant-crop relations. Springer, New York, pp 171–178
Molnár-Láng M, Linc G, Sutka J (1996) Transfer of the recessive crossability allele *kr1* from Chinese Spring into the winter wheat variety Martonvásári 9. Euphytica 90:301–305
Molnár-Láng M, Cseh A, Szakács É, Molnár I (2010) Development of a wheat genotype combining the recessive crossability alleles *kr1kr1kr2kr2* and the 1BL.1RS translocation, for the rapid enrichment of 1RS with new allelic variation. Theor Appl Genet 120(8):1535–1545. doi:10.1007/S00122-010-1274-0
Molnár-Láng M, Molnár I, Szakács E, Linc G, Bedő Z (2014) Production and molecular cytogenetic identification of wheat-alien hybrids and introgression lines. In: Tuberosa R, Graner A, Frison E (eds) Genomics of plant genetic resources. Springer, Springer Science + Business Media, Dordrecht, pp 255–283. doi:10.1007/978-94-007-7572-5
Monneveux P, Reynolds MP, Aguilar JG, Singh RP (2003) Effect of the *7DL.7Ag* translocation from *Lophopyrum elongatum* on wheat yield and related morphological traits under different environments. Plant Breed 122:379–384
Nedel JL (1994) Genetic improvement in grain yield of wheat cultivars released between 1940 and 1992. Assuit J Agric Sci 29:1565–1570
Ogbonnaya FC (2011) Development, management and utilization of synthetic hexaploids in wheat improvement. In: Bonjean AP, Angus WJ, van Ginkel M (eds) The world wheat book: a history of wheat breeding, vol 2. Lavoisier, Paris, pp 823–849
Olson EL, Brown-Guedira G, Marshall D, Stack E, Bowden RL, Jin Y, Rouse M, Pumphrey MO (2010) Development of wheat lines having a small introgressed segment carrying stem rust resistance gene *Sr22*. Crop Sci 50:1823–1830
Parry MAJ, Hawkesford MJ (2010) Food security: increasing yield and improving resource use efficiency. Proc Nutr Soc 69:1–9
Patrignani A, Lollato PL, Ochsner TE, Godsey CB, Edward JT (2014) Yield gap and production gap of rainfed winter wheat in the southern Great Plains. Agron J 106:1329–1339
Peltonen-Sainio P, Peltonen J (1994) Progress since the 1930s in breeding for yield, its components, and quality traits of spring wheat in Finland. Plant Breed 113:177–186
Perry MW, Antuono MFD (1989) Yield improvement and characteristics of some Aust. spring wheat cultivars between 1860 and 1982. Aust J Agric Res 40:457–472
Rabinovich SV (1998) Importance of wheat-rye translocations for breeding modern cultivars of *Triticum aestivum L.* Euphytica 100:323–340
Rajaram S (2001) Prospects and promise of wheat breeding in the 21*st* century. Euphytica 119:3–15

Rejesus RM, van Ginkel M, Smale M (1996) Wheat breeders' perspectives on genetic diversity and gemplasm use: findings from an international survey. Wheat Special Report. No.40, Mexico, D.F.: CIMMYT pp 129-147

Reynolds MP, Calderini DF, Condon AG, Rajaram S (2001) Physiological basis of yield gains in wheat associated with the *Lr19* translocation from *Agropyron elongatum*. Euphytica 119:137–141

Richards RA (1992) The effect of dwarfing genes in spring wheat in dry environments I. agronomic characteristics. Aust J Agric Res 43:517–527

Riggs TJ, Hanson PR, Start ND, Milles DM, Morgan CL, Ford MA (1981) Comparison of spring barley varieties grown in England and Wales between 1880 and 1980. J Agric Sci 97:599–610

Romero CA, Frey KJ (1966) Mass selection for plant height in oat populations. Crop Sci 6:283–287

Schlegel R, Korzun V (1997) About the origin of 1RS.1BL wheat-rye chromosome translocations from Germany. Plant Breed 116:537–540

Schmidt JW (1984) Genetic contributions to yield gains in wheat. In: Fehr WR (ed) Genetic contributions to yield gains of five major crop plants. CSSA Special Publication Number 7, Madison, pp 89–101

Sears ER (1973) Agropyron-wheat transfers induced by homoeologous pairing. In: Sears ER, Sears LMS (eds) Proc 4th Int Wheat Genet Symp Columbia, MD University of Missouri pp 191–199

Sears ER (1977) Analysis of wheat-Agropyron recombinant chromosomes. In: Proc 8th Eucarpia Congr, Madrid Spain pp 63–72

Sebesta EE, Wood EA, Porter DR, Webster JA, Smith EL (1995) Registration of 'Amigo' wehat germplasm resistance to greenbug. Crop Sci 35:293

Sestili F, Botticella E, Bedő Z, Phillips A, Lafiandra D (2010) Production of novel allelic variation for genes involved in starch biosynthesis through mutagenesis. Mol Breed 25:145–154

Sharma HC, Gill BS (1983) Current status of wide hybridization in wheat. Euphytica 32:17–31

Sharma D, Knott DR (1966) The transfer of leaf rust resistance from *Agropyron* to *Triticum* by irradiation. Can J Genet Cytol 8:137–143

Shen X, Ohm H (2007) Molecular mapping of *Thinopyrum* derived *Fusarium* head blight resistance in common wheat. Mol Breed 20:131–140

Shewry PR, Parmar S, Pappin DJC (1987) Characterization and genetic control of the prolamins of *Haynaldia villosa*: relationship to cultivated species of the *Triticeae* (rye, wheat and barley). Biochem Genet 25:309–325

Shewry PR, Sabelli PA, Parmar S, Lafiandra D (1991) λ-type prolamins are encoded by genes on chromosomes 4Ha and 6Ha of *Haynaldia villosa* Schur (syn. *Dasypyrum villosum* L.). Biochem Genet 29:207–211

Siddique KHM, Belford RK, Perry MW, Tennant D (1989) Growth, development and light interception of old and modern wheat cultivars in a Mediterranean-type environment. Aust J Agric Res 40:473–487

Sinha SK, Aggarwal PK (1980) Physiological basis of achieving the productivity potential of wheat in India. Indian J Genet Plant Breed 40:375–384

Slade AJ, Fuerstenberg SI, Loeffler D, Steine MN, Facciotti D (2005) A reverse genetic, nontransgenic approach to wheat crop improvement by TILLING. Nat Biotech 23:75–81

Slafer GA, Andrade FH (1989) Genetic improvement in bread wheat (*T. aestivum*) yield in Argentina. Field Crops Res 21:289–296

Trethowan RM, Mujeeb-Kazi A (2008) Novel germplasm resources for improving environmental stress tolerance of hexaploid wheat. Crop Sci 48:1255–1265

Tsunewaki K (1964) Genetic studies of a 6*x* derivative from an 8*x triticale*. Can J Genet Cytol 6:1–11

Vasil V, Castillo AM, Fromm ME, Vasil IK (1992) Herbicide resistant fertile transgenic wheat plants obtained by microprojectile bombardment of regenerable embryogenic callus. Bio/Technology 10:667–674

Veisz O, Harnos N, Szunics L, Tischner T (1996) Overwintering of winter cereals in Hungary in the case of global warming. Euphytica 92:249–253

Vida GY, Gál M, Uhrin A, Veisz O, Syed NH, Flavell AJ, Wang Z, Bedő Z (2009) Molecular markers for the identification of resistance genes and marker-assisted selection in breeding wheat for leaf rust resistance. Euphytica 170:67–76

Villareal RL, del Toro E, Mujeeb-Kazi A, Rajaram S (1995) The 1BL/1RS chromosome translocation effect on yield characterization in a *Triticum aestivum* L. cross. Plant Breed 114:497–500

Waddington SR, Ransom JK, Osmanzai M, Saunders DA (1986) Improvement int he yield potential of bread wheat adopted to Northwest Mexico. Crop Sci 26:698–703

Warburton ML, Crossa J, Franco J, Kazi M, Trethowan R, Rajaram S, Pfeiffer W, Zhang P, Dreisigacker S, van Ginkel M (2006) Bringing wild relatives back into the family: recovering genetic diversity in CIMMYT improved wheat germplasm. Euphytica 149:289–301

Whelan EDP, Conner RL, Thomas JB, Kuzyk AD (1986) Transmission of a wheat alien translocation with resistance to the wheat curl mite in common wheat, *Triticum aestivum* L. Can J Genet Cytol 28:294–297

White EM, Wilson FEA (2006) Responses of grain yield, biomass and harvest index and their rates of genetic progress to nitrogen availability in ten winter wheat varieties. Irish J Agric Food Res 45:85–101

Worland T, Snape JW (2001) Genetic basis of worldwide wheat varietal improvement. In: Bonjean A, Angus JW (eds) The world wheat book, a history of wheat breeding. Lavoisier Publishing, Paris, pp 59–94

Zeller FJ (1972) Cytologischer Nachweis einer Chromosomensubstitution in der Weizenstamm Salzmünde 14/44 (*T. aestivum* L.). Z Pflanzenzühtg 67:90–94

Zeller FJ (1973) 1B/1R wheat-rye chromosome substitutions and translocations. Proceedings 4th International Wheat Genet Symposium Columbia, MO pp 209–222

Zhang R, Zhang M, Wang X, Chen P (2014) Introduction of chromosome segment carrying the seed storage protein genes from chromosome 1V of *Dasypyrum villosum* showed positive effect on bread-making quality of common wheat. Theor Appl Genet 127:523–533

Zhukovsky PM (1957) Wheat in the USSR. State Publishing House of Agricultural Literature, Moscow, Leningrad, p 632

Chapter 4
The Crossability of Wheat with Rye and Other Related Species

Márta Molnár-Láng

4.1 Genetic Control of the Crossability of Wheat with Rye

Varieties of common wheat, *Triticum aestivum* L. ($2n=6\times=42$), differ in the ease with which they hybridize with rye, *Secale cereale* L. ($2n=14$). Following emasculation and hand pollination, more than 50 % of the pollinated florets usually set seeds in readily crossable varieties. By contrast less than 5 % of florets set seeds in poorly crossable forms (Riley and Chapman 1967).

The first detailed analysis of crossability was carried out by Lein (1943), using the readily crossable wheat variety Chinese 466 and the poorly crossable varieties Marquis and Peragis. Ready crossability is recessive and, by pollinating F_2 plants from intervarietal crosses with rye, Lein showed that allelic differences at two loci were responsible for the contrasting parental behavior. On this basis it was suggested that Chinese 466 was genotypically *kr1kr1kr2kr2*, while Marquis and Peragis were *Kr1Kr1Kr2Kr2*. Moreover, the variety Blausamtiger Kolben, which had an intermediate level of crossability and which was also included in the analysis, was thought to be genotypically *Kr1Kr1kr2kr2*. Lein concluded from these results that the presence of *Kr1* resulted in a more marked reduction in crossability than the presence of *Kr2*.

According to Lein's classification (1943) the relationship between genotype (*kr* allele composition) and phenotype (percentage of crossability with rye) is the following:

Genotype	Percentage crossability with rye
Kr1Kr1Kr2Kr2	<5
Kr1Kr1kr2kr2	10–30

M. Molnár-Láng (✉)
Agricultural Institute, Centre for Agricultural Research, Hungarian Academy of Sciences, 19, Martonvásár 2462, Hungary
e-mail: molnar.marta@agrar.mta.hu

M. Molnár-Láng et al. (eds.), *Alien Introgression in Wheat*,
DOI 10.1007/978-3-319-23494-6_4

Genotype	Percentage crossability with rye
kr1kr1Kr2Kr2	30–50
kr1kr1kr2kr2	>50

The use of intervarietal chromosome substitution lines showed that the poor crossability of the wheat variety Hope with rye (*S. cereale*) was determined by chromosomes 5A and 5B (Riley and Chapman 1967). Marquis, one of the parents of the wheat cultivar Hope, had a *Kr1Kr1Kr2Kr2* genotype according to Lein (1943). The genes *Kr1* and *Kr2* are located on chromosomes 5B and 5A, respectively (Riley and Chapman 1967). Crossability is actively inhibited by the dominant *Kr1* and *Kr2* alleles of Hope and is apparently not enhanced by the recessive *kr1* and *kr2* alleles of the readily crossable variety Chinese Spring (CS). The *Kr1* and *Kr2* alleles are not completely dominant, but crossability is greatly reduced in heterozygotes. Neither of the substitution lines CS/Hope 5A and CS/Hope 5B was as infertile in crosses with rye, as the variety Hope. Consequently, the effects of the inhibitors of crossability, *Kr1* and *Kr2*, are either complementary or additive (Riley and Chapman 1967). Sasaki and Wada (1966) studied wheat-rye crossability using intervarietal disomic substitution lines, in which pairs of chromosomes from the poorly crossable cv. Cheyenne were substituted in turn into the readily crossable cv. Chinese Spring. The substitution of chromosome 5B had a marked effect on crossability, with an average reduction of 10.5 %. Several other chromosomes showed weaker, variable effects, those of chromosomes 4A, 1D and 7D being the most consistent over the 2 years and over the four lines of rye. Riley and Chapman (1967), who were presumably unaware of study made by Sasaki and Wada, used the same method.

The *Kr1* locus was mapped using a telocentric chromosome consisting of the long arm of chromosome 5B (Lange and Riley 1973). Mapping was carried out by analyzing F_2 and testcross progenies, and it was concluded that the *Kr1* locus is located on the long arm of 5B. The location of *Kr1* was confirmed by Sitch et al. (1985) using the telocentric mapping technique and was found to be on the long arm of chromosome 5B distal to the centromere with a mean recombination frequency of 44.8 ± 3.28 %. *Kr2* was shown by linkage with the major gene markers *Vrn1*, controlling vernalization requirement, and q, controlling ear morphology to be located on the long arm of chromosome 5A. *Kr2* is closely linked to *Vrn1*, with a mean recombination frequency of 38.1 ± 10.60 %. The similar locations of *Kr1* and *Kr2* on homoeologous chromosomes suggest that these two loci are homoeoallelic. The differences found in the recombination frequencies of the *Kr1* allele on the 5B long arm in the work of Lange and Riley (1973) and Sitch et al. (1985) could have arisen from the use of different wheat varieties, having different 5B chromosomes or different alleles that modified crossability in the given genetic background. It may also be that crossability was greatly influenced by the environment in the various experiments.

Crosses between wheat group 5 chromosome substitution lines with three different cultivar backgrounds indicated that there may be multiple alleles for reduced crossability with rye and *H. bulbosum* on both the 5A and 5B chromosomes (Falk and Kasha 1983). No reduction in seed set was observed for any of the 5D substitution lines. The *Kr1* locus on chromosome 5B was found to have a more pronounced effect on both rye and *H. bulbosum* crossability than the *Kr2* locus on chromosome 5A, and the effects of the two loci were shown to be cumulative. Tests with tetrasomic and nullitetrasomic lines of Chinese Spring indicated that the kr allele is "null" or inactive in promoting crossability, while the *Kr* allele is active in reducing crossability with both rye and *H. bulbosum*. Thus, extra doses of the *kr* allele do not increase rye or *H. bulbosum* crossability in the presence of the corresponding *Kr* allele. The crossability of the various *Kr*/*kr* combinations with rye exhibited a significant correlation with Lein's (1943) rye crossability data. The linearity of the crossability response with respect to the dose of *Kr* alleles suggested an additive genetic system, rather than the complete dominance of the *Kr* alleles. Genotypes which carried heterozygous *Kr*/*kr* alleles gave better seed set with rye than those with homozygous dominant *Kr* alleles. Krolow (1970) suggested that a *Kr3* gene was located on the D genome, but the effect of this gene could not be confirmed by other authors. Substitution lines involving the cultivars Hope, Atlas 66, and Cheyenne reduced the crossability with rye to various extents, so it was assumed that these cultivars may have different alleles for reduced crossability at each of the known *Kr* loci on chromosomes 5A and 5B, as compared to the allele in Chinese Spring 5B (Falk and Kasha 1983).

Snape et al. (1979) found a strong positive correlation between the crossability of wheat varieties with *Hordeum bulbosum* and with rye. Crossability with *H. bulbosum* was investigated using chromosome substitution lines of the non-crossable variety, Hope, into the crossable variety Chinese Spring. Two chromosomes, namely 5A and 5B of Hope, were found to cause a marked reduction in the crossability of Chinese Spring with *Hordeum bulbosum*. It was concluded that the crossability of wheat with both species was governed by the same genetic system. Fifty-six wheat (*Triticumaestivum* L.) genotypes representing wide genetic diversity were screened for crossability with rye by Falk and Kasha (1981). The 56 wheats produced seed set values ranging from 0.0 to 95.4 %. The distribution of percentage crossability was similar to that published by Lange and Wojciechowska (1976) for 177 cultivars. A high proportion had no or very low seed set, while the others were distributed fairly evenly across the remaining range. Twenty-eight of the wheat lines, representing a range of crossability with rye, were then pollinated with tetraploid *Hordeum bulbosum* L. (Falk and Kasha 1981). The seed set ranged from 0.0 to 32.2 %. Wheats that set seeds with *H. bulbosum* generally had medium-to-high crossability with rye. A high correlation was obtained for the crossability of rye and *H. bulbosum* with wheat, indicating that the main genetic system (*Kr1* and *Kr2*) governing crossability with rye is also the main system operating in crosses with *H. bulbosum*. Ten F_1 hybrids were produced by crossing the highly crossable cv. Chinese Spring with nine other wheat cultivars and the highly crossable cv. Zaragosa 75 with the medium crossable cv. Songlen. Crossability with rye was generally intermediate between the parents. Leighty and Sando (1928), Lein (1943), Marais and Pienaar (1977) and Tanner and Falk (1981) also reported earlier that hybrids between Chinese Spring and wheats with low rye crossability exhibited intermediate levels of crossability with rye. It was confirmed that plants with the heterozygous *Kr1kr1 Kr2kr2* allele combination had better seed set with rye than homozygous *Kr1Kr1 Kr2Kr2* genotypes, suggesting that the dominance of the *Kr* alleles is not complete.

The wheat cultivars Chinese Spring and Hope were crossed with the barley (*Hordeum vulgare* L.) cultivar Betzes by Fedak and Jui (1982) to establish the level of crossability. The complete set of Chinese Spring/Hope chromosome substitution lines was used to identify chromosomes carrying genes responsible for the crossability of Chinese Spring with Betzes. No progeny were obtained from the Hope×Betzes cross indicating that this combination was incompatible. Hybrids were obtained from the Betzes×Chinese Spring combination at rates of 48.9 % and 0.78 % of pollinated florets for seeds and seedlings, respectively. No progeny were obtained from substitution lines 5A, 5B, and 5D, even though large numbers of florets were pollinated, indicating that the chromosomes of Chinese Spring homoeologous group 5 are the major chromosomes responsible for permitting crossability with Betzes barley. Other chromosomes of Hope had minor effects on crossability with barley. A number of wheats varying in wheat-rye crossability were selected as female parents for interspecific and intergeneric crossess by Thomas et al. (1980). A number of species more or less closely related to common wheat were selected as male parents (*Secale cereale*, *S. montanum*, *Aegilops ovata*, *Ae. variabilis*, *Triticum timopheevii*, *T. turgidum*, *T. monococcum*, *Agropyron junceum*, *A. elongatum*, *Elymus giganteus*, *E. arenarius*). The wheat genes best known for reducing seed set in crosses with cultivated rye also prevent the pollen of other related species from setting seed on common wheat. This correlation between interspecific crossabilities is general in common wheat and is not restricted to Chinese Spring, Hope and their joint derivatives. Average seed set was lower when more distantly related taxa were used as pollen parents. Thus, low wheat-rye crossability in common wheat was not a serious barrier to hybridization with emmer (*T. turgidum* var. *dicoccum*), but even high wheat-rye crossability did not guarantee high rates of seed set in crosses with *Elymus* species (Thomas et al. 1980). Genes for low interspecific crossability in wheat act against a wide spectrum of potential pollinators. The use of common wheats with high wheat-rye crossability in interspecific hybridization will increase the recovery of hybrids in many wide crosses.

When diploid, tetraploid and hexaploid *Triticum* species were pollinated with rye by Kiss and Rajháthy (1956), the highest seed set was achieved for the hexaploid species (*Triticumaestivum*, *T. spelta*), and the lowest seed set for diploid species (*T. monococcum*, *T. boeticum*). *T. durum* and *T. carthlicum* gave higher seed set with rye than *T. timopheevii*, *T. turgidum*, or *T. dicoccum* (Kiss 1966). A wheat line with high crossability with rye, selected from the Hungarian wheat cultivar Bánkúti 1201, gave more than 40 % crossability with rye in the field over several years (Kiss and Rédei 1952; Kiss 1953; Kiss and Rajháthy 1956). In a study on how genetic variation in tetraploid wheat and rye influenced crossability, Halloran (1981) found that the crossability of the tetraploid wheat *T. turgidum* (L.) Thell. ssp. *turgidum* varied over a very wide (0.0–77.8 %) range. The lack or very low level of crossability between certain rye types and a particular tetraploid wheat suggest the occurrence of a mutation significantly inhibiting crossability at the tetraploid level in the evolution of wheat. This conclusion contradicted the earlier findings of Riley and Chapman (1967), who suggested "that the first hexaploid wheats were probably able to hybridize fully with rye." The crossability of tetraploid *T. timopheevii* with rye was also studied by Mujeeb-Kazi (1981).

Various rye species were studied as pollinators in wheat-rye crosses by Kiss (1968). The highest average seed set on a large number of wheat genotypes was achieved with *Secale vavilovi* (18.5 %) and the lowest with *Secale fragile* (0.4 %), while the seed set with *S. montanum*, *S. cereale* and *S. africanum* was intermediate over the course of several years. Halloran (1981) also studied the effect of rye genotypes (various *Secale* species and *S. cereale* cultivars) on crossability with tetraploid wheats. The best seed set was achieved with a *Secale vavilovi* line, as reported by Kiss (1968). Various rye (*S. cereale*) cultivars resulted in different levels of seed set. The species *Secale segetale*, *S. anatolicum*, and *S. montanum* were also included in the experiment. Variation in the crossability level was found to be due to genetic variation, in both tetraploid wheat and rye (Halloran 1981). Zeven and van Heemert (1970) concluded that there was a strong correlation between the crossability of wheat with rye (*S. cereale*) and weed rye (*S. segetale*). The same crossability barrier acted in wheat×rye and in wheat×weed rye crosses. Five inbred lines of rye (*S. cereale*) and an open-pollinated rye cultivar were used by Scoles (1983) to pollinate wheat cultivars with differing crossability. No hybrid seed was produced using a cultivar with low crossability, while a highly crossable cultivar gave an average seed set of 65 %. Significant differences were detected between the inbreds in terms of seed set and the weight of F_1 seed. The significant effect of the rye genotype on crossability in crosses between wheat and rye was also observed by Tanner and Falk (1981) and Oettler (1982).

4.2 Mechanism of Pollen Tube Growth Inhibition in Non-crossable Wheat Varieties (*Kr* Alleles) When Pollinated with Rye

Zeven and van Heemert (1970) pollinated ears of 28 wheat varieties having different crossability using the pollen of weed rye. No crossing barrier was observed in the stigma, style or ovary wall, because pollen tubes of weed rye were seen in these tissues irrespective of the wheat variety used as female parent. The pollen grains germinated within 6 min after pollination and the tubes reached the region of the micropyle after about 40 min. Similar observations were made by Lange and Wojciechowska (1976), who studied the action of the crossability genes of wheat by crossing the readily crossable wheat cv. Chinese Spring, the poorly crossable Hope and the disomic substitution CS/Hope 5B with rye. A comparison of crossability and actual fertilization showed that the poor crossability with rye of both cv. Hope and the CS/Hope 5B substitution line resulted from the absence of fertilization. Pollen germination did not depend on whether wheat was pollinated with wheat or rye pollen, or whether the wheat was readily or poorly crossable with rye. Studies of pollen grain germination and pollen tube growth showed that the dominant alleles of the crossability genes manifested themselves through the retardation and eventual inhibition of pollen tube growth at the style base and in the ovary wall. In Hope the growth of all pollen tubes was inhibited, whereas in CS/Hope 5B fertilization was achieved in rare cases. The recessive alleles of the crossability genes do not seem to have any influence on the growth of rye pollen tubes in wheat pistils.

Fertilization and early seed development were studied by Wojciechowska and Lange (1977) in the common wheat variety Chinese Spring after pollination with rye and selfing, and in the common wheat variety Hope after selfing. Fertilization with rye gave rise to a tetraploid embryo ($2n=28$) and a septaploid endosperm ($2n=49$) in hexaploid maternal tissue. In all three combinations the first pollen tube reached the mycropyle about 40 min after pollination. When pollinated with rye the migration of the sperm nuclei to the egg cell and the polar nuclei was delayed by about an hour. The first mitosis in the zygote of selfed Chinese Spring and Chinese Spring × rye occurred at about the same time, as did the first mitosis of the polar nuclei. The subsequent development of both embryo and endosperm was slightly faster for Chinese Spring × rye than for selfed Chinese Spring. The development of the embryo and endosperm was monitored for 72 h after pollination.

D'Souza (1978) studied differences in the germination percentage of rye pollen on the stigmas and the rate of pollen tube growth of within the styles of wheat varieties known to have good and poor crossability with rye. For this purpose, wheats known to be readily crossable with rye, e.g., Chinese Spring, *T. durum*, and *T. carthlicum*, were compared with varieties which have little or no crossability with rye, e.g., Carsten VIII and *T. monococcum*. In wheat × rye crosses rye pollen germinated as well on wheat stigmas as on rye. There was no retardation in the growth of the pollen tubes, which penetrated the styles and reached the base of the styles regardless of the crossability of the wheat varieties with rye. Irrespective of the ploidy level and the crossability with rye, the pollen germinated on the stigma and the pollen tubes reached the base within an hour of pollination. The failure of seed set in poorly crossable genotypes when crossed with rye is thus not due to the differential behaviour of the style, but probably to physiological differences within the ovules. Jalani and Moss (1980) investigated the manifestation of *Kr* genes on pollen germination and pollen tube growth in a wide range of hexaploid wheat varieties crossed with rye, with the aim of elucidating and identifying the site(s) of action of the crossability genes. The results indicated that the crossability genes have little effect on pollen germination or on the time taken for the pollen tubes to reach the micropyle. These results were in agreement with those of Zeven and van Heemert (1970) and Lange and Wojciechowska (1976). The number of pollen tubes reaching the micropyle is, however, affected by the *Kr* genes, as this number was greater in highly crossable genotypes than in poorly crossable ones. There was a strong correlation between the mean number of pollen tubes at the micropyle and seed set, which also reflects crossability (Jalani and Moss 1980). The inhibitation or retardation of pollen tubes occurred mainly between the style base and the top of the embryo sac, especially in the case of poorly crossable genotypes where both *Kr* genes were present (Jalani and Moss 1981). Cameron and Reger (1991) showed that a soluble, dialyzed lysate extracted from the ovaries of Hope and CS/Hope 5B inhibited rye pollen tube elongation significantly more than a similar lysate from Chinese Spring ovaries.

Zeven (1987) published a list showing the percentage crossability with rye for some 1400 varieties and lines of bread wheat, which included 76 wheat varieties having a crossability with rye of 40 % or higher. These data could be used to select wheat varieties for further interspecific crossing programs. There was one Hungarian wheat cultivar, Bánkúti 1201, among the varieties with high crossability with rye. The other two Hungarian wheat varieties in the list, Dioszegi 200 and Fleischmann 481, both have low crossability with rye. These are old varieties, which are no longer cultivated in Hungary. The crossability of the old Hungarian wheat varieties Bánkuti 1201, Fleischmann 481, Székács 1055 and Alcsuti 21 with rye was tested by Kiss and Rédei (1953) who analysed both seed set and the germination ability of the hybrid seeds. Bánkuti 1201 always gave better seed set than F 481. Twenty-two lines with good compatibility with rye were selected from the Bánkuti 1201 variety, the best of which (Bánkuti 1201-4) gave more than 40 % seed set with rye over several years (Kiss and Rajháthy 1956). A line with good crossability with rye was also selected from the wheat cultivar Thatcher. More than 40 hexaploid wheat cultivars were tested for crossability with rye by Kiss (1968) and the best seed set was achieved with wheat cv. Bánkuti 1201. Belea (1992) also reported that a number of biotypes selected from Bánkuti 1201 (B-1201-6, B 1201-10) gave higher seed set with *Triticum monococcum* than the cultivar average. This character was inherited.

4.3 Geographic Distribution of *kr* Alleles

Falk and Kasha (1981) found 15 wheats with high crossability with rye among the 56 genotypes tested. Among the 15 wheats exhibiting more than 40 % crossability with rye, 14 can be traced directly or indirectly to cv. Chinese Spring or to an Asiatic origin. Songlen, Mendos and 79-72S can be traced back to Chinese Spring through CI 12632, and Zaragosa 75 through Mengavi. CI 12632 and CI 12633 were derived from a cross between Chinese Spring and *Triticum timopheevii*. Salmon reverted to wheat from a triticale which had Chinese Spring in the wheat parentage. Norin 29 originated from Japan and Peking 10 from China. The other lines tested, with the exception of Chancellor, can be traced back to crosses involving Chinese Spring. Riley and Chapman (1967) hypothesized that Central Asia was a wheat origin area where rye was not generally grown and would not have been an admixture in cultivated wheats. Therefore, there would have been no selection pressure for mutations of *kr* to *Kr* in wheat to prevent pollination by rye. While wheats from this area may exhibit crossability, not all wheats of Chinese origin are highly crossable with rye, as indicated by crosses involving China 3 and Pau 45 (Falk and Kasha 1981). Zeven (1987) investigated the geographic distribution of *kr* alleles on crossability data for 1400 wheat genotypes and concluded that *kr* alleles occur with high frequency in bread wheat landraces from China, Japan, East Siberia, and Iran. However not all varieties from these countries have high crossability, as indicated by the Chinese varieties listed among the 1400 cultivars in this publication.

West European wheat cultivars generally have poor crossability with rye (Stefanowska and Cauderon 1983). The hexaploid wheat cultivar Roazon, which is agronomically similar to intensive West European common wheats was derived from interspecific hybridization involving three different species and retained the *Aegilops ventricosa* cytoplasm. It was shown that Roazon, unlike West European wheats, had good crossability with rye cv. Petkus (Stefanowska and Cauderon 1983). According to these authors Roazon has a *kr1kr1Kr2Kr2* genotype.

Chinese wheat (*Triticumaestivum* L.) landraces originating in the Sichuan Basin were tested for crossability with rye by Luo et al. (1992). Sichuan province, which lies in the southwest of China, consists of the Sichuan Basin and the adjacent mountain ranges. Only spring wheats are grown in this region, which is where Chinese Spring originates from. Four groups could be distinguished among the 177 genotypes. The 68 landraces in the first group showed low crossability with rye, with seed set percentages less than 5 %. The 59 landraces in the second group could be crossed with rye, having crossability percentages greater than or equal to 5 %, but significantly lower than that of Chinese Spring. The 34 landraces in group III had similar crossability to Chinese Spring. The 16 landraces in group IV crossed very easily with rye, and had crossability percentages significantly higher than that of Chinese Spring. Most landraces with high crossability occurred in the Qinling and Dabashan mountain ranges in the north of Sichuan and in the valleys of the rivers Minjiang, Fujiang and Jialinjiang in the Sichuan Basin. Seventy-two wheat landraces originating from Shaanxi province and 46 landraces from Henan province in China were pollinated with rye to observe their crossability with rye by Luo et al. (1993). Seven of the 72 landraces from Shaanxi expressed a much higher crossability level than Chinese Spring. As to the geographical distribution landraces with high crossability percentages were mainly distributed in the valley of the River Weihe in central Shaanxi, in the Hanzhong Plain and in the hilly area along the Qinling and Dabashan mountains. Seven of the 46 samples from Henan province were very easy to cross with rye. Landraces with crossability similar to or significantly higher than Chinese Spring mainly occurred in the west and southwest parts of Henan province, showing a geographically continuous distribution with those in Shaanxi. The studies of Luo et al. (1992, 1993) revealed that highly crossable landraces occur in the Sichuan, Shaanxi and Henan provinces of China, and have crossability similar to or even higher than that of Chinese Spring. These landraces carry recessive *kr4* genes (Zheng et al. 1992; Luo et al. 1993). The crossability of landraces from India and of improved semi-dwarf, high-yielding Indian wheat cultivars adapted locally was investigated by Sarvjeet Singh and Sethi (1991). Sixty-two bread wheat accessions (19 landraces, 43 others) representing local wheat germplasm from the Himalayan ranges were pollinated with rye. Two of the 14 landraces from Himachal Pradesh were found to be free of crossability inhibitors, therefore having the genotype *kr1kr1kr2kr2*, as they had more than 50 % crossability with rye. The other 48 cultivars had low crossability with rye.

The crossability of Hungarian wheat varieties was studied in the field in Martonvásár in two different years by pollinating them with rye (Molnár-Láng and Sutka 1989). Russian, Yugoslav, Rumanian, Czech, Italian, and Austrian wheat varieties used as crossing partners in Hungarian wheat breeding programmes were also included in the study. The hybrid nature of the seeds was checked by chromosome counting. A total of 33 genotypes were pollinated with rye. The seed set values of the Chinese Spring, Roazon and Songlen wheat varieties used as controls agreed with the numbers expected according to relationship between genotype and the percentage crossability reported by Lein's (1943). Among the Martonvásár wheat varieties the seed set values of Mv9 and Mv13 were 15.4 % and 10 %, respectively, putting them in the 10–30 % group with the Yugoslav variety NS 3000 (10.7 %) and the Rumanian F-29 (12.5 %). These data suggest that the recessive allele pair *kr2kr2* may have been introduced into Hungarian varieties via Soviet and Yugoslav genotypes.

4.4 Transfer of Crossability Alleles into European Wheat Varieties

A highly crossable winter wheat genotype, Martonvásári 9 kr1 (Mv9 kr1), was developed in Martonvásár by Molnár-Láng et al. (1996). The crossability of the initial wheat cultivar Martonvásári 9 (Mv9) was studied by pollinating with rye in seven different years together with the control wheat cv. Chinese Spring (CS). The maximum seed set of Mv9 with weed rye was 15.4 % in 1 year, but in the following years it never exceeded 5 % when pollinated with rye. The Mv 9 variety possessed dominant *Kr1Kr1* alleles and was heterogeneous at the *kr2* locus, so that some individual plants carried recessive *kr2* alleles. Mv 9 was crossed with CS in 1985 and the F_1 hybrid was backcrossed with Mv9 in 1986. Plants possessing the recessive *kr* alleles were selected from the (Mv9 × CS) Mv9 BC_1 generation according to the seed set achieved when pollinated with rye. The partial dominance of the *kr* alleles made it possible to differentiate between plants heterozygous at the *Kr1* locus and the dominant homozygous *Kr1Kr1* plants. Progenies in two consecutive selfed generations were tested by pollination with rye to select homozygous recessive *kr1kr1kr2kr2* plants and to check the result of selection after each backcross (Fig. 4.1).

As a result of three backcrosses with Mv9 and two selfings after each backcross the selected progenies had 61.6 % seed set with rye when tested on 60 individual plants. These data confirmed that after the third backcross the selected Mv9 kr1 line carried recessive crossability alleles *kr1* and *kr2*, but the genotype was 93.75 % Mv9 (Molnár-Láng et al. 1996). After the sixth backcross and two selfings the CS

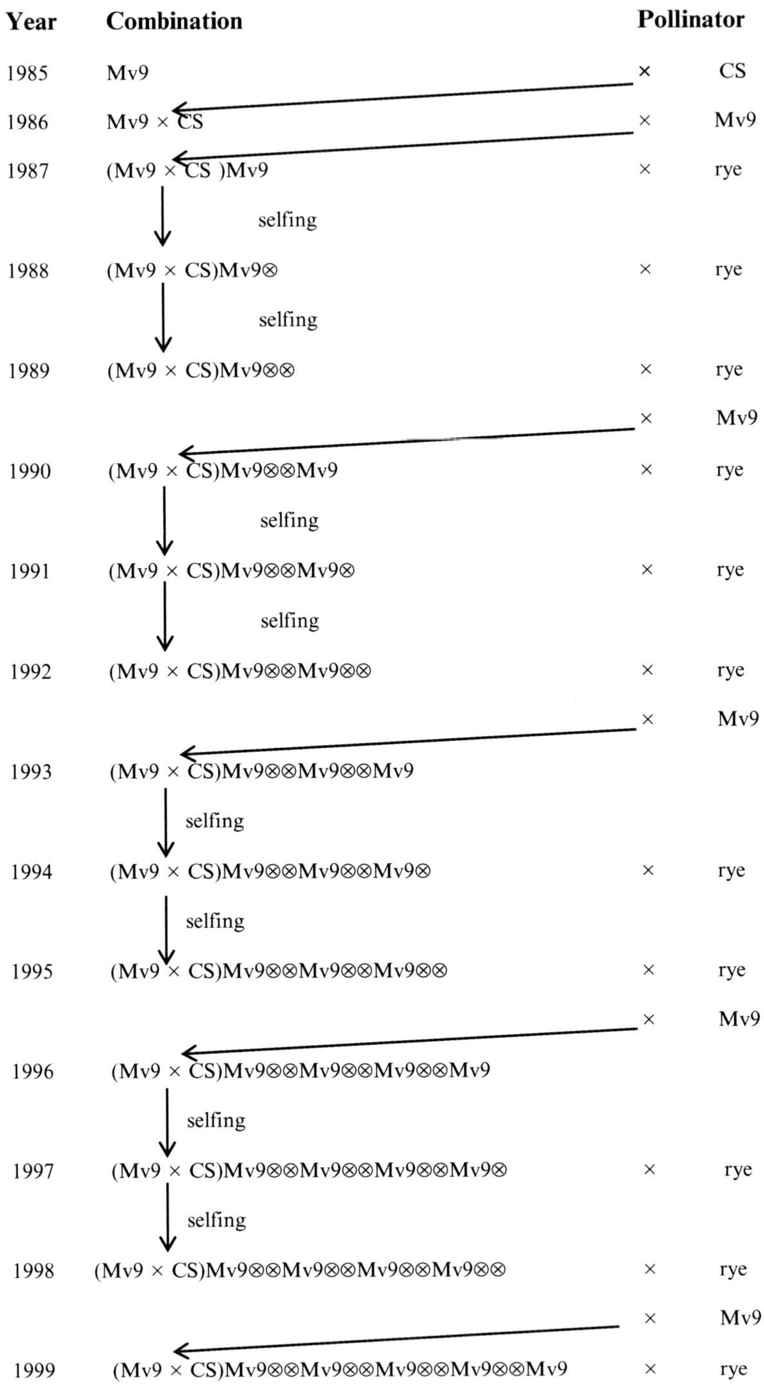

Fig. 4.1 Scheme for developing a Mv9kr1 line from a Mv9×CS hybrid via backcrossing with Mv9. Progenies were selected based on the seed set obtained when pollinated with rye. After each backcross two self-pollinated generations were tested for crossability with rye in order to select plants which carry crossability allele *kr1* in the homozygous state

Table 4.1 Crossability of the Martonvásári 9 kr1 (Mv9 kr1), Mv Magdaléna kr1 and Mv Béres kr1 wheat lines and the initial wheat cultivars Mv9, Mv Magdaléna and Mv Béres when pollinated with rye (*Secale cereale*) cv. Ryefood

Wheat genotype	Year	Pollinator	Pollinated florets/ spike	Seed set %	Deviation
Mv9 kr1/BC_5[a]	2013	Rye	340/9	40.3	14.02
	2014	Rye	546/16	53.1	17.85
Mv9 kr1/BC_6[b]	2013	Rye	520/18	47.0	14.55
	2014	Rye	428/14	60.2	13.27
Mv9	2013	Rye	84/4	3.1	6.25
	2014	Rye	134/5	0	–
Mv Magdaléna kr1[c]	2013	Rye	274/9	65.5	12.77
	2014	Rye	368/9	54.9	14.02
Mv Magdaléna	2013	Rye	88/3	2.15	1.86
	2014	Rye	182/5	6.8	4.33
Mv Béres kr1[d]	2013	Rye	272/9	53.4	19.3
	2014	Rye	168/5	60.9	17.39
Mv Béres	2013	Rye	94/3	0	-
	2014	Rye	184/5	2.1	3.4

The crossability allele *kr1* was transferred from Chinese Spring into the wheat cultivar Martonvásári 9 (Molnár-Láng et al. 1996) and then further transferred from Mv9 kr1 into the modern Martonvásár wheat cultivars Mv Magdaléna and Mv Béres, which carry the 1RS.1BL translocation (Molnár-Láng et al. 2010). The emasculated wheat spikes were pollinated with rye in the field in Martonvásár in May 2013 and 2014

[a]Mv9kr1/BC_5 Wheat cultivar Mv9 was crossed with CS, and the hybrid was backcrossed with Mv9 five times. After each backcross two consecutive self-pollinated generations were tested for crossability with rye in order to select lines carrying crossability allele *kr1* in the homozygous state

[b]Mv9kr1/BC_6 The Mv9×CS hybrid was backcrossed with Mv9 six times. After each backcross two consecutive self-pollinated generations were tested for crossability in order to select lines which carrying crossability allele *kr1* in homozygous state

[c]Mv Magdaléna kr1 Wheat cultivar Mv Magdaléna was crossed with Mv9kr1, and plants with high crossability with rye were selected from the F_2 generation (Molnár-Láng et al. 2010). Mv Magdaléna kr1 is the sixth self-pollinated generation of the selected F_2 plants

[d]Mv Béres kr1. Wheat cultivar Mv Béres was crossed with Mv9kr1, and plants with high crossability with rye were selected from the F_2 generation (Molnár-Láng et al. 2010). Mv Béres kr1 is the sixth self-pollinated generation of the selected F_2 plants

segment carrying the crossability allele was further reduced in the Mv9 background, so the genotype was 99.22 % Mv9, but had very high crossability with rye (Table 4.1).

This crossable winter wheat line (Mv9 kr1) is well adapted to Central European environmental conditions, and has much better agronomic traits than Chinese Spring (Fig. 4.2). The Mv9 kr1 line is now used as maternal parent in wheat-alien hybridization experiments in Martonvásár (Molnár-Láng et al. 2002). This has the advantage that the alien genes can be transferred directly into a winter wheat line with strong stems, good yielding ability, good quality, and good winter hardiness,

Fig. 4.2 Spike morphology of wheat cultivars Mv9 (**a**), and Chinese Spring (**b**), and wheat lines Mv9 kr1/BC_5 (**c**) and Mv9 kr1/BC_6 (**d**) having high crossability with rye. Crossability allele *kr1* was transferred from CS into Mv9 from a Mv9 × CS hybrid via backcrossing with Mv9. Plants with recessive *kr1* alleles were selected on the basis of the seed set with rye. The development of the Mv9kr1 line is demonstrated in Fig. 4.1

instead of into CS, which has many unfavorable features from the agronomic point of view.

When the Mv9 kr1 line was pollinated with the old Hungarian rye cultivar Lovászpatonai (Molnár-Láng et al. 2002), the mean crossability percentage was fairly high (68.4 %) The chromosome number distribution was examined in mitotic chromosome spreads of the octoploid triticale obtained via colchicine treatment on the initial hybrid and was found to range from 51 to 56. All the rye chromosomes were identified in mitotic chromosome spreads of octoploid triticale with the help of C-banding and were detected using genomic in situ hybridization (GISH) (Nagy et al. 1998). High seed set (37.4 %) was achieved when the Mv9 kr1 wheat line was pollinated with *Aegilops biuncialis* (Logojan and Molnár-Láng 2000). This was higher than that reported earlier by Mustafaev and Piralov (1980) and Özgen (1983). Similarly good seed set was achieved when the Mv9 kr1 line was crossed in Martonvásár with other *Aegilops* species (*Ae. geniculata*, *Ae. triuncialis*, *Ae. cylindrica*, *Ae. tauschii*, unpublished data), with *Thinopyrum* species (unpublished data) and *T. timopheevii* (Farshadfar et al. 1994; Mikó et al. 2014) and *T. monococcum* (Megyeri 2014).

The crossability alleles were further transferred from the Mv9 kr1 genotype into two modern Martonvásár wheat cultivars, Mv Magdaléna and Mv Béres (Molnár-Láng et al. 2010). The main objective of this work was to develop a wheat genotype containing both the recessive crossability alleles (*kr1kr1kr2kr2*), allowing high crossability between 6× wheat and diploid rye, and the 1BL.1RS wheat/rye translocation chromosome. This wheat genotype could be used asa recipient partner in wheat-rye crosses for the efficient introduction of new allelic variation into the 1RS arm in translocation wheats. After the wheat cultivars Mv Magdaléna and Mv Béres, in which the 1BL.1RS translocation involves the 1RS chromosome arm from Petkus, were crossed with line Mv9 kr1, 117 F_2 plants were analysed for crossability, 10 of which had more than 50 % seed set with rye and thus presumably carried the *kr1kr1kr2kr2* alleles. Four of these contained the 1BL.1RS translocation in the disomic condition, as detected by genomic in situ hybridization (GISH). The average seed set with rye on selected F_3 progenies from the Mv Béres × Mv9 kr1 and Mv Magdalena × Mv kr1 crosses was 65.5 % and 64.7 %, respectively (Fig. 4.3). On some plants the seed set with rye was as high as 87.5 or 83.3 %. When the wheat × rye F_1 hybrids produced between these lines and the rye cultivar Kriszta were analyzed

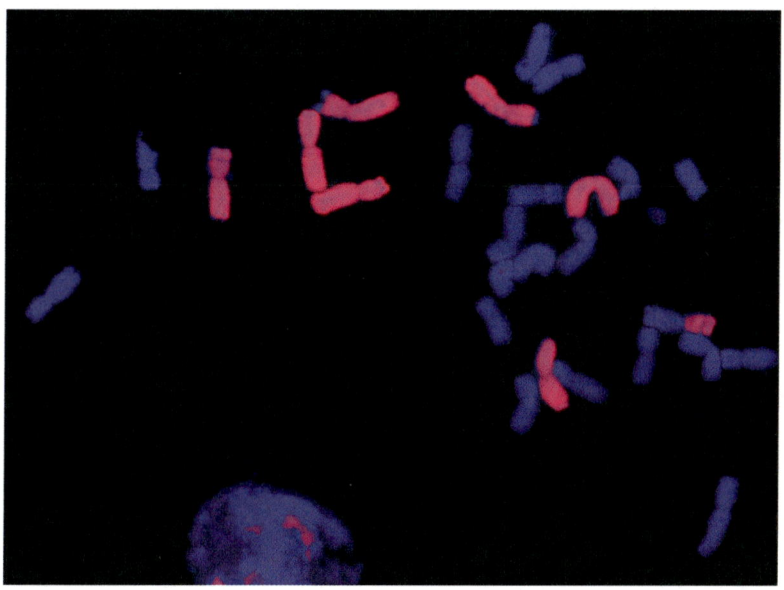

Fig. 4.3 In situ hybridization on mitotic chromosomes of a wheat-rye F_1 hybrid. Genomic in situ hybridization (GISH) on somatic chromosomes of the wheat × rye hybrid produced between the Mv Béres kr1 line and the rye cv. Kriszta. Total rye genomic DNA was labeled with digoxigenin and detected with anti-dig-rhodamine. Seven rye chromosomes and the 1RS arm in the 1BL.1RS translocated chromosome are *magenta colour*. Twenty wheat chromosomes and the 1BL arm of the 1BL.1RS chromosome are *blue* as a result of counterstaining with DAPI

in meiosis using GISH, 1BL.1RS/1R chromosome pairing was detected in 62.4 % of the pollen mother cells (Molnár-Láng et al. 2010). The use of fluorescent in situ hybridization (FISH) with repetitive DNA probes allowed the 1R and 1BL.1RS chromosomes to be identified. The use of rye SSR markers RMS13 and SCM9 demonstrated the presence of the 1RS arm from Kriszta besides that of Petkus in the F_1 hybrids. In four of the 22 BC_1 progenies analyzed only Kriszta-specific bands were observed with these markers, though the presence of the 1BL.1RS translocation was detected using GISH. It can be concluded that recombination occurred between the Petkus and Kriszta 1RS chromosome arms in the translocated chromosome in these plants (Molnár-Láng et al. 2010).

4.5 Genetic Mapping of the Crossability Alleles

The substitution of Chinese Spring chromosome 5B into the variety Hobbit sib was initiated by Snape et al. (1987) to transfer crossability alleles into UK wheat varieties. This task was complicated, as Hobbit sib, like other PBI wheats, did not have a normal 5B, 7B karyotype, but contained the 5BL–7BL and 5BS–7BS translocations. Following five backcrosses to the Hobbit sib monosomic 5BL-7BL line a disomic substitution was extracted. The substitution line Hobbit sib (Chinese Spring 5BL, 7BL) was crossed to Hobbit sib and disomic single chromosome recombinant lines were developed. Three lines were identified which had the 5BL–7BL, 5BS–7BS karyotype but were crossable with *H. bulbosum* and were non-necrotic (Snape et al. 1987). Later, a total of 71 homozygous disomic recombinant substitution lines were generated from these crosses, exhibiting recombination between Hobbit sib and Chinese Spring 5BL (Bertin et al. 2009). Initially the *Kr1* locus was mapped to a 13 cm region between the microsatellite markers *Xgwm213* and *Xgwm371* (Bertin et al. 2009). These markers were used to screen the *Ph1* mutant line, *ph1b*, which revealed that *Kr1* mapped within the *ph1b* deletion region. Three wheat BACs and the *cdc2-2* gene, located within the *Ph1b* deletion region, were used to develop additional markers for mapping. A series of recombinant substitution lines were developed in which the genome of the normally non-crossable wheat variety Hobbit sib carried a recombinant 5BL chromosome arm containing segments from the crossable variety Chinese Spring (Bertin et al. 2009). The crossability of the recombinant lines was studied by pollinating with rye in the greenhouse over four seasons. The crossability locus *Kr1* was fine mapped using molecular markers with the help of 29 lines exhibiting recombinations between the *Xw5145* and *DR740708* markers on the 5BL long arm. A second region responsible for crossability was detected on the 5BL arm (Bertin et al. 2009).

An attempt was made to improve the crossability of the semi-dwarf wheat cultivar Courtot with rye so that it could be used in the development of primary triticale (Gay and Bernard 1994). The Courtot chromosomes 5A, 5B and 5D were replaced by their counterparts from the highly crossable Japanese cultivars Norin 29 and Fukuhokomugi using the monosomic back-crossing technique. Throughout the backcrosses the substituted chromosome was maintained in the monosomic condition. The backcross process was continued up to BC_4 for chromosomes 5A and 5B, but only up to BC_1 for chromosome 5D. After four backcrosses, monosomic plants were selfed, and plants with $2n = 42$ were selected from among the progeny. The most potent compatibility factor was detected on chromosome 5B (*kr₁*); this allele was sufficient to achieve a seed set of about 50 %, depending on environmental factors. Conversely, the effect of the other two chromosomes, whatever the donor parent, was hardly detectable; if they have any crossability alleles, these have a low level of expression, being inhibited by the *Kr1* allele (Gay and Bernard 1994). The recovery of the Courtot genetic background during subsequent backcross generations did not alter the degree of crossability estimated in BC_1.

An intervarietal molecular-marker map was used by Tixier et al. (1998) for the detection of genomic regions influencing crossability between wheat and rye. The mapping population consisted of doubled haploid (DH) lines and was produced at Clermont-Ferrand by anther culture from Courtot × Chinese Spring F_1 hybrids. The two parents of the DH population, Courtot and Chinese Spring, differed greatly, with a 95 % success rate for crosses with Chinese Spring, and only 10 % with Courtot. Testing for the presence of QTLs in the whole genome was carried out using 110 DH lines, while 187 DH lines were employed to explore the effect of homoeologous group 5 chromosomes. Analysis of deviance and logistic marker-regression methods were performed on data from doubled haploid lines. A major quantitative trait locus (QTL) involved in crossability and associated with the marker *Xfba367-5B* was detected on the short arm of chromosome 5B (Tixier et al. 1998). An additional locus, *Xwg583-5B*, was indicated on the long arm of chromosome 5B. This minor QTL might correspond to *Kr1*, which was presumed to be the major gene controlling crossability. Another locus *Xtam51-7A* on chromosome 7A, was significantly associated with this trait. The three-marker model explained 65 % of the difference in crossability between the two parents. The results achieved by Tixier et al. (1998) were not consistent, however with those of Lange and Riley (1973) and Sitch et al. (1985), who mapped the *Kr1* locus on the long arm of chromosome 5B. Tixier et al. (1998) found a gene suppressing crossability, denoted *Skr*, on chromosome arm 5BS. The QTL associated with the locus *Xfba367-5B* on the long arm of 5B had only a low value. The results could have been influenced by the large phenotypic variance and by the different genotypes used by the various authors.

The *Skr* locus, on the short arm of 5B, was identified as a major QTL (Tixier et al. 1998; Lamoureux et al. 2002) inhibiting the crossability of wheat and rye. Alfares et al. (2009) used a recombinant SSD population originating from a cross between the poorly crossable cultivar Courtot (Ct) and the crossable line MP98 to characterize the major dominant effect of *SKr* and to map the gene at the distal end of the chromosome near the 5B homoeologous *GSP* locus. In Courtot *SKr* had a stronger effect than *Kr1* on crossability of wheat to rye. It is possible that Courtot carries a weak *Kr1* allele. The results confirm that there is allelic variability among the different crossability genes at the *Kr1* locus. Colinearity with barley and rice was used to saturate the *SKr* region with new markers and establish orthologous relationships with a 54 kb region on rice chromosome 12. A total of five markers were mapped within a genetic interval of 0.3 cm and 400 kb BAC contigs were established on both sides of the gene to lay the foundation for the map-based cloning of *SKr*. Two SSR markers completely linked to *SKr* were used to evaluate a collection of crossable wheat progenies originating from a primary triticale breeding programme (Alfares et al. 2009). The marker *cfb306*, which is closely linked to *SKr*, was found to be an efficient marker for the introduction of crossability into wheat germplasm through marker-assisted selection. In work performed by Mishina et al. (2009) QTL mapping was carried out using the F_7 population derived from a cross between Chinese Spring (CS) and a CS/Cheyenne 5B chromosome substitution line. In this population, a major QTL region controlling crossability with rye was detected on a locus closely linked to the SSR marker Xgwm443 on the short arm of chromosome 5B, which was assumed to be the *Skr* locus.

Acknowledgements This work was supported by the Hungarian National Scientific Research Fund (OTKA K 104382) and by the "Wheat and Barley Legacy for Breeding Improvement"—WHEALBI EU FP7 project. The technical assistance of Edina Türkösi is gratefully acknowledged. Thanks are due to Barbara Hooper for revising the manuscript linguistically.

References

Alfares W, Bouguennec A, Balfourier F, Gay G, Bergés H, Vautrin S, Sourdille P, Bernard M, Feuillet C (2009) Fine mapping and marker development for the crossability gene *SKr* on chromosome 5BS of hexaploid wheat (*Triticum aestivum* L.). Genetics 183:469–481. doi:10.1534/genetics.109.107706

Belea A (1992) Interspecific and intergeneric crosses in cultivated plants. Akadémiai Kiadó, Budapest, 255 p

Bertin I, Fish L, Foote TN, Knight E, Snape J, Moore G (2009) Development of consistently crossable wheat genotypes for alien wheat gene transfer through fine-mapping of the *Kr1* locus. Theor Appl Genet 119:1371–1381

Cameron RG, Reger BJ (1991) Hybridization barriers between wheat and rye: in vitro pollen assays and electrophoretic survey. Euphytica 52:147–153

D'Souza L (1978) Behaviour of rye pollen on styles of wheat varieties having good and poor crossability with rye. Z Pflanzenzüchtg 80:129–133

Falk DE, Kasha KJ (1981) Comparison of the crossability of rye (*Secale cereale*) and *Hordeum bulbosum* onto wheat (*Triticum aestivum*). Can J Genet Cytol 23:81–88

Falk DE, Kasha KJ (1983) Genetic studies of the crossability of hexaploid wheat with rye and *Hordeum bulbosum*. Theor Appl Genet 64:303–307

Farshadfar M, Molnár-Láng M, Sutka J (1994) The crossability of different wheat (*Triticum aestivum* L.) genotypes with *Triticum timopheevii* Zhuk. under two types of conditions. Cereal Res Commun 22:15–20

Fedak G, Jui PY (1982) Chromosomes of Chinese Spring wheat carrying genes for crossability with Betzes barley. Can J Genet Cytol 24:227–233

Gay G, Bernard M (1994) Production of intervarietal substitution lines with improved interspecific crossability in the wheat cv Courtot. Agronomie 14:27–32

Halloran GM (1981) Tetraploid wheat crossability with rye (*Secale*). Genetica 55:191–194

Jalani BS, Moss JP (1980) The site of action of the crossability genes (*Kr1*, *Kr2*) between *Triticum* and *Secale*. I. Pollen germination, pollen tube growth, and number of pollen tubes. Euphytica 29:571–579

Jalani BS, Moss JP (1981) The site of action of crossability genes (*Kr1*, *Kr2*) between *Triticum* and *Secale*. II. Proportion of pistils containing pollen tubes and effects of alternate pollination on seed set. Euphytica 30:105–112

Kiss Á (1953) Genetic investigations of wheat-rye hybrids and Triticale No1 of Martonvásár. Acta Agron Hung 4:239–278

Kiss Á (1966) Kreuzungsversuche mit Triticale. Der Züchter 36:249–255

Kiss Á (1968) Triticale, a new crop for sandy soils (Triticale, a homok új gabonája – In Hungarian). Mezőgazdasági Kiadó, Budapest, 179 p

Kiss Á, Rajháthy T (1956) Untersuchungen über die Kreuzbarkeit innerhalb des Subtribus Triticinae (Investigations on crossability within the subtribe Triticinae). Züchter 26:127–136

Kiss Á, Rédei G (1952) Experiments to produce wheat-rye hybrids (Triticale). (Kísérletek búza-rozs hibridek (Triticale) előállítására). Növénytermelés 1:67–84

Kiss Á, Rédei G (1953) Experiments to produce rye-wheat (Triticale). Acta Agron Hung 3:257–276

Krolow KD (1970) Untersuchungen über die Kreuzbarkeit zwischen Weizen und Roggen. Z Pflanzenzüchtg 64:44–72

Lamoureux D, Boeuf C, Regad F, Garsmeur O, Charmet G, Sourdille P, Lagoda P, Bernard M (2002) Comparative mapping of the wheat 5B short chromosome arm distal region with rice, relative to a crossability locus. Theor Appl Genet 105:759–765

Lange W, Riley R (1973) The position on chromosome 5B of wheat of the locus determining crossability with rye. Genet Res Camb 22:143–153

Lange W, Wojciechowska B (1976) The crossing of common wheat (*Triticum aestivum* L.) with cultivated rye (*Secale cereale* L.) I. Crossability, pollen grain germination and pollen tube growth. Euphytica 25:609–620

Leighty CE, Sando WJ (1928) Natural and artificial hybrids of a Chinese wheat and rye. J Hered 19:23–27

Lein A (1943) Die genetische Grundlage der Kreuzbarkeit zwischen Weizen und Roggen. Z Indukt Abstamm und Vererb Lehre 81:28–61

Logojan A, Molnár-Láng M (2000) Production of *Triticum aestivum – Aegilops biuncialis* chromosome additions. Cereal Res Commun 28:221–228

Luo MC, Yen C, Yang JL (1992) Crossability percentages of bread wheat landraces from Sichuan Province, China with rye. Euphytica 61:1–7

Luo MC, Yen C, Yang JL (1993) Crossability percentages of bread wheat landraces from Shaanxi and Henan provinces, China with rye. Euphytica 67:1–8

Marais GF, Pienaar RV (1977) Hybridisation between wheat and rye. I. Variations in the frequency of kernel set. Agroplantae 9:135–141

Megyeri M (2014) Molecular cytogenetic analysis of *Triticum monococcum* gene bank accessions and their utilization in wheat improvement. PhD Thesis. Szent István University, Gödöllő, Hungary, p 116

Mikó P, Megyeri M, Molnár-Láng M (2014) Novel utilization of wild relatives in bread wheat prebreeding. In: Kőszegi I (ed) Advances in plant breeding and biotechnology techniques – book of abstracts 27–29 April, 2014. Pannonian Plant Biotechnology Association, Mosonmagyaróvár, pp 59–61

Mishina K, Sato H, Manickavelu A, Sassa H, Koba T (2009) Molecular mapping of *SKr* for crossability in common wheat. Breed Sci 59:679–684

Molnár-Láng M, Sutka J (1989) Crossability of wheat varieties grown in Hungary with rye (Magyarországon termesztett búzafajták keresztezhetősége rozzsal). Növénytermelés 38:479–484

Molnár-Láng M, Linc G, Sutka J (1996) Transfer of the recessive crossability allele *kr1* from Chinese Spring into the winter wheat variety Martonvásári 9. Euphytica 90:301–305

Molnár-Láng M, Linc G, Nagy DE, Schneider A, Molnár I (2002) Molecular cytogenetic analysis of wheat-alien hybrids and derivatives. Acta Agron Hung 50:303–311

Molnár-Láng M, Cseh A, Szakács É, Molnár I (2010) Development of a wheat genotype combining the recessive crossability alleles *kr1kr1kr2kr2* and the 1BL.1RS translocation, for the rapid enrichment of 1RS with new allelic variation. Theor Appl Genet 120(8):1535–1545. doi:10.1007/S00122-010-1274-0

Mujeeb-Kazi A (1981) *Triticum timopheevii*×*Secale cereale* crossability. J Hered 72:227–228

Mustafaev ID, Piralov GR (1980) Evolutionary and genetical aspects of the interrelations between wheat and *Ae. biuncialis* Vis. Genetika 16:1440–1446

Nagy DE, Linc G, Molnár-Láng M (1998) Molecular cytogenetic analysis of a new wheat-rye amphiploid with C-banding and genomic *in situ* hybridization (GISH). (Új búza-rozs amfiploid molekuláris citogenetikai elemzése C-sávozással és genomikus *in situ* hibridizációval (GISH)). Növénytermelés 47:253–260

Oettler G (1982) Effect of parental genotype on crossability and response to colchicine treatment in wheat-rye hybrids. Z Pflanzenzüchtg 88:322–330

Özgen M (1983) Hybrid seed set in wheat×*Aegilops* crosses. Wheat Info Serv 56:9–11

Riley R, Chapman V (1967) The inheritance in wheat of crossability with rye. Genet Res Camb 9:259–267

Sasaki M, Wada O (1966) Chromosomal location of genes for crossability of wheat with rye using chromosome substitution lines. (In Japanese with author's translation). Jpn J Breed 16: 178–179

Scoles GJ (1983) The effect of rye genotype on wheat-rye crossability and on the development of F_1 seed. Can J Genet Cytol 25:668–669

Singh S, Sethi GS (1991) Crossability of some bread wheat landraces and improved cultivars from western Himalayas with rye. Euphytica 53:137–141

Sitch LA, Snape JW, Firman SJ (1985) Intrachromosomal mapping of crossability genes in wheat (*Triticum aestivum*). Theor Appl Genet 70:309–314

Snape JW, Chapman V, Moss J, Blanchard CE, Miller TW (1979) The crossabilities of wheat varieties with *Hordeum bulbosum*. Heredity 42:291–298

Snape JW, Parker BB, Leckie D (1987) Progress reports. Intervarietal transfer of crossability genes. EWAC Newsletter. In: Sutka J, Worland AJ (eds) Proceedings of the EWAC Conference, 1987 Martonvásár. Inst Plant Sci Res, Cambridge Lab, Agric Res Inst Hung Acad Sci, Martonvásár, pp 17–20

Stefanowska G, Cauderon Y (1983) L'aptitude au croisement avec le seigle (*Secale cereale* L.) d'une lignee hexaploide de *Triticum* (cv. "Roazon" résultant d'hybridation interspécifique (*Gramineae* Juss.). Agronomie 3:355–358

Tanner DG, Falk DE (1981) The interaction of genetically controlled crossability in wheat and rye. Can J Gen Cytol 23:27–32

Thomas JB, Kaltsikes PJ, Anderson RG (1980) Relation between wheat-rye crossability and seed set of common wheat after pollination with other species in the *Hordeae*. Euphytica 30: 121–127

Tixier MH, Sourdille P, Charmet G, Gay G, Jaby C, Cadalen T, Bernard S, Nicolas P, Bernard M (1998) Detection of QTLs for crossability in wheat using a double-haploid population. Theor Appl Genet 97:1076–1082

Wojciechowska B, Lange W (1977) The crossing of common wheat (*Triticum aestivum* L.) with cultivated rye (*Secale cereale* L.). II. Fertilization and early post-fertilization developments. Euphytica 26:287–297

Zeven AC (1987) Crossability percentages of some 1400 bread wheat varieties and lines with rye. Euphytica 36:299–319

Zeven AC, van Heemert C (1970) Germination of pollen of weed rye (*Secale segetale* L.) on wheat (*Triticum aestivum* L.) stigmas and the growth of the pollen tubes. Euphytica 19:175–179

Zheng Y, Luo M, Yen C, Yang J (1992) Chromosome location of a new crossability gene in common wheat. Wheat Info Serv 75:36–40

Chapter 5
Gametocidal Genes

Takashi R. Endo

5.1 Introduction

Long before the practice of genetic manipulation, many chromosomes and genes have been introgressed into cultivated plants from related wild and crop plants. Generally alien chromosomes introduced into crops by interbreeding are stably maintained by substituting for homoeologous chromosomes in the hosts. Otherwise, alien chromosomes would be eventually eliminated from the descendants of initial hybrids by occasional nonsegregational events because they are surplus, namely dispensable chromatin to the hosts. Even in an exceptional case of successful alien introgression, namely the substitution of rye chromosome 1R or 1B/1R translocation in bread wheat (Schlegel and Korzun 1997), the introgressed rye chromosome would be lost in the descendants of the inter-varietal crosses without cytological check or phenotypic selection.

However, there are certain chromosomes and genes that stay in host plants in a selfish manner once they are introgressed through interspecific crossing. Such genes or chromosomes are called "pollen killer" (Cameron and Moav 1957; Loegering and Sears 1963) or "gamete eliminator" (Rick 1966; Sano 1990) or gametocidal chromosomes (Endo 1990, 2007); hereafter in this chapter, the author collectively call such genetic factors as gametocidal (Gc) chromosomes/genes and use the term "Gc system" to refer to the mode of action of the Gc gene. The term "gametocidal"

The original version of this chapter was revised. An erratum to this chapter can be found at DOI 10.1007/978-3-319-23494-6_14

T.R. Endo (✉)
Centre of Plant Structural and Functional Genomics, Institute of Experimental Botany AS CR, Slechtitelu 31, Olomouc, 783 71 Holice, Czech Republic

Department of Plant Life Science, Faculty of Agriculture, Ryukoku University, 1-5 Yokotani, Seta Oe-cho, Otsu, Shiga 520-2194, Japan
e-mail: endo.takashi.2e@gmail.com

M. Molnár-Láng et al. (eds.), *Alien Introgression in Wheat*,
DOI 10.1007/978-3-319-23494-6_5

was first adopted in a paper by Maan (1975), in which the behavior of an alien chromosome exclusively transmitted in common wheat was described with the following sentence: "There was thus an apparent gametocidal action of the sporophyte having an *Aegilops* chromosome on the gametes lacking this chromosome."

Gc chromosomes, which are dispensable to the host, ensure the predominance of gametes containing them over gametes lacking them, presumably by killing or damaging the latter gametes. This situation occurs when Gc chromosomes are in hetero- or hemizygous condition, causing preferential transmission and persistent existence of the Gc chromosomes in hosts (Fig. 5.1). Superficially similar cases of preferential transmission of introgressed chromosomes were reported in some cytoplasmic substitution lines of wheat, in which the abnormal behavior of the alien chromosome can be attributed to the interaction between nuclear and cytoplasmic genes (Tsuji and Murata 1976; Nakata et al. 1993). Specific chromosomes derived from the cytoplasm donors are indispensable to the viability of zygotes of the wheat lines carrying the alien cytoplasm; therefore, the alien chromosomes persist in the wheat lines. These chromosomes do not have Gc genes but so-called fertility-restorer genes. The Gc gene might occur by mutation, but it would easily be regarded as a normal allele of a mutated gametic lethal gene. In rice, hybrid sterility genes were identified as Gc genes in linkage analysis of the backcross progeny of intra- and interspecific hybrids. In this chapter two well-documented cases of the Gc system in wheat and rice are described.

5.2 Gametocidal Chromosomes in Wheat

5.2.1 Preferential Transmission of Alien Chromosomes

Many interspecific hybrids were made between wheat and its related wild species, and the hybrids were repeatedly backcrossed to wheat in order to produce alien chromosome addition and substitution lines (Jiang et al. 1994), and also to produce alien cytoplasm substitution lines (Tsunewaki et al. 1996). Even though hybrids are

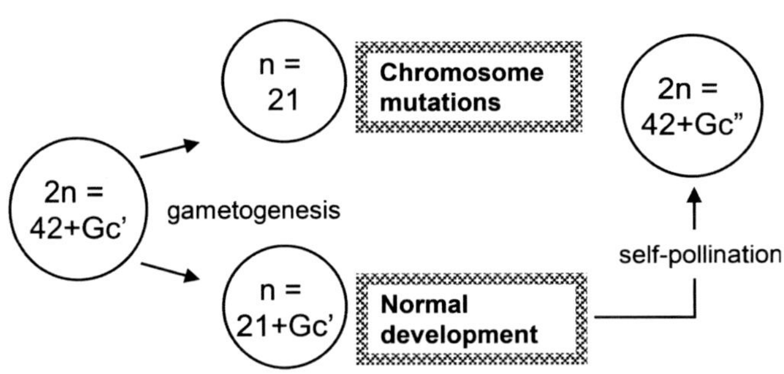

Fig. 5.1 Schematic diagram of Gc action in common wheat carrying a single Gc chromosome. 42 and 21 represent the wheat chromosome numbers in zygotes and in gametes, respectively. +Gc′ and +Gc″ stand for the addition of one dose and two doses of gametocidal chromosome, respectively

highly sterile, in backcross progeny normal fertility, especially female fertility, is restored, as the genome constitution of the progeny gets close to that of the recurrent parent. However, there are some cases where certain alien chromosomes tenaciously persist in a selfish way in the backcross progeny. Endo and Tsunewaki (1975) tried to substitute the nuclear genome of natural and synthetic strains of *Aegilops triuncialis* for that of common wheat by repeated backcrossing. After five backcrosses, however, the backcross progeny still carried an extra chromosome from the cytoplasm donors, without improving their male and female fertility. They did not know whether the sterility was caused by the interaction between the *Aegilops* chromosome and the cytoplasmic genome, or by the *Aegilops* chromosome by itself. Maan (1975) tried to substitute the nucleus of common wheat into the cytoplasm of *Ae. longissima* and *Ae. sharonensis*, and obtained partially sterile plants having one *Aegilops* chromosome. Crossing such plants as pollen parents to euploid common wheat, he again obtained partially sterile offspring carrying the *Aegilops* chromosome, and found the selfed progeny to be exclusively disomic for the *Aegilops* chromosome. Thus the sporophyte having the *Aegilops* chromosome apparently exerted a Gc action on gametes not containing it (Fig. 5.1). Miller et al. (1982) backcrossed the F_1 hybrid between common wheat and *Ae. sharonensis* in an attempt to produce a set of addition lines of *Ae. sharonensis* into common wheat. However, they found all addition lines to contain one and the same alien chromosome, typically corresponding to a Gc chromosome. Finch et al. (1984) showed that such a Gc chromosome (called "cuckoo" chromosome) ensures its transmission by causing chromosome breaks in meiospores lacking it. Similarly, a chromosome derived from decaploid *Thinopyrum ponticum* (formerly *Agropyron elongatum*), named $7el_2$ for its homoeology to wheat group 7 chromosomes, was found to carry a Gc gene(s), which, in the wheat background, induced its preferential transmission through female gametes and abortion of those lacking it (Scoles and Kibirge-Sebunya 1983).

5.2.2 *Diversity of Gc Chromosomes in the Genus* Aegilops

After the first findings of Gc chromosomes in wheat, many more Gc chromosomes were found in various species of the genus *Aegilops* possessing different genome (genomes C, S and M) and belonging to different homoeologous groups (groups 2, 3, 4, and 6) (for review see Endo 1990, 2007; Tsujimoto 2005). Their identity in terms of the Gc action was investigated in double monosomic plants for different Gc chromosomes (Endo 1982, 1985). When a plant carried both chromosome 3C and 2S or both 3C and 4S, only gametes with both Gc chromosomes were functional, and the plant had severely reduced fertility. On the other hand, from a plant carrying both 2S and 4S, gametes carrying 4S were functional, regardless of the presence or absence of 2S, with the plant having a similar fertility to that of the 4S monosomic addition plants (Fig. 5.2). The former case shows that the Gc action of 3C is

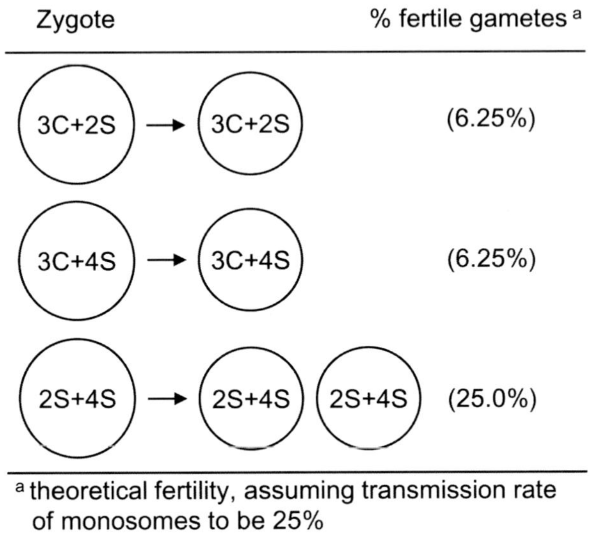

Fig. 5.2 Schematic diagram of Gc action in common wheat carrying two different types of Gc chromosomes 3C, 2S, and 4S. The diagram is based on the data published by Endo (1982). The actual seed set percentages from hand pollination with euploid wheat pollen are 6.5 % for 3C+2C, 5.4 % for 3C+4S, and 15.0 % for 2S+4S (see text for details)

independent from that of 2S and 4S, and therefore both Gc chromosomes should coexist in functional gametes. The latter case, instead, suggests either that 4S epistatically suppresses the Gc action of 2S or that 4S has two Gc genes, one of which is the same as that located on 2S.

5.2.3 Modification of Gc Action

Depending on the host cultivar into which a Gc chromosome is introduced, the Gc action varies. Chromosome 2C, for instance, has a complete Gc action and is therefore exclusively transmitted to the progeny in the common wheat cultivar Jones Fife, whereas its Gc action becomes incomplete in the common wheat cultivar Chinese Spring, in whose background chromosome 2C can be lost in part of the progeny (Endo 1988). Chromosome 3C has a severe Gc action in Chinese Spring and some other common wheat cultivars, but it displays almost no Gc action in Norin 26, which possesses the *Igc1* Gc-inhibitor gene on chromosome 3B (Tsujimoto and Tsunewaki 1985). In both cases of incomplete Gc action, semi-lethal chromosomal mutations occur in gametes lacking the Gc chromosome, and structurally rearranged chromosomes are transmitted to the progeny.

The Gc gene in sporophytes seems to have a dual function, i.e. to induce chromosomal mutations in gametes that lack it and to suppress such mutations in gametes that include it. This was demonstrated by a knockout mutation for the *Gc2* gene on *Ae. sharonensis* chromosome 4S, which renders the former function ineffective, while having no influence on the latter function (Friebe et al. 2003).

5.2.4 Use of Gc Gene in the Production of Deletion and Dissection Lines

Thanks to its hexaploid nature, common wheat can tolerate aneuploidy and chromosomal structural changes to a considerable extent. Using mostly chromosome 2C, Endo and Gill (1996) produced about 350 homozygous deletion lines of Chinese Spring wheat that contain deletions of various size in specific chromosomes. These lines are useful in cytologically mapping (deletion mapping) of genes and especially DNA markers to the missing chromosomal regions (Werner et al. 1992; Qi et al. 2004). Most of the Chinese Spring deletion lines, together with the Gc chromosomes, are available from NBRP-wheat website (http://www.shigen.nig.ac.jp/wheat/komugi/strains/aboutNbrpLgku.jsp).

The Gc system can be usefully applied to the induction of structural changes in alien chromosomes introduced into common wheat, particularly in the case of alien chromosomes from species distantly related to wheat, showing little tendency to undergo homoeologous recombination with wheat chromosomes even under genetically permissive conditions (see Chap. 6). As an example, the *Ae. cylindrica* Gc chromosome 2C was introduced into all barley disomic addition lines into Chinese Spring wheat, except for 1H (Shi and Endo 1997). Chromosomal rearrangements were induced by the 2C gametocidal system for each barley chromosome, including 2H (Joshi et al. 2011), 3H (Sakai et al. 2009), 4H (Sakata et al. 2010), 5H (Ashida et al. 2007), 6H (Ishihara et al. 2014), and 7H (Schubert et al. 1998; Serizawa et al. 2001; Masoudi-Nejad et al. 2005; Nasuda et al. 2005). The Gc system was similarly proved to be effective in inducing structural rearrangements in rye chromosome 1R introduced into common wheat (Endo et al. 1994; Masoudi-Nejad et al. 2002; Gyawali et al. 2009, 2010; Li et al. 2013). Genomic in situ hybridization (GISH) represents a useful tool to identify Gc-induced structural changes of alien chromo-

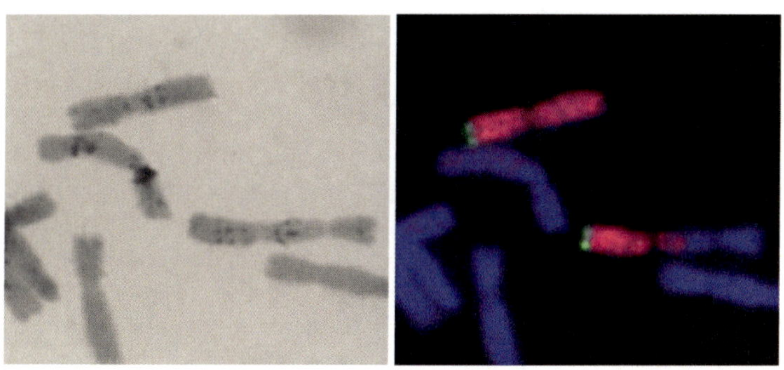

Fig. 5.3 C-banding (*left*) and fluorescence in situ hybridization (*right*) images showing normal barley chromosome 2H and a translocation between 2H and a wheat chromosome induced by the gametocidal system. Green fluorescence represents the barley subtelomeric sequences HvT01 and pink fluorescence shows barley chromatin. This translocation is seemingly part of a dicentric chromosome

somes in common wheat (Fig. 5.3). Since both terminal deletions and wheat-alien translocations enable cytological mapping of alien chromosomes, the present author have been developing many common wheat lines carrying deletions and translocations of alien chromosomes, collectively named "dissection lines." Comparative studies of cytological and genetic maps conducted in the above studies revealed that crossing-over is generally more frequent in the distal region than in the proximal region for all the wheat, barley, and rye chromosomes that were studied.

5.3 Gametocidal Genes in Rice

5.3.1 Hybrid Sterility by Allelic Interaction at a Single Locus and Selective Gamete Abortion

There are two species of cultivated rice, *Oryza sativa* L. ($2n=24$) and *O. glaberrima* Steud. ($2n=24$), which originated in Asia and West Africa, respectively. *O. sativa* has two subspecies ssp. *japonica* and ssp. *indica*. Hybrids between the two species or between the two subspecies normally form 12 bivalents at meiosis, but it is well known that sterility, on the male or both male and female side, is prevalent in these hybrids.

Among many gene loci responsible for female sterility in *indica-japonica* hybrids, S_5 is a major one (Ikehashi and Wan 1996). There are three alleles at the S_5 locus, an indica allele, S_5^i, in *indica* varieties, a japonica allele, S_5^j, in *japonica* varieties, and a neutral allele, S_5^n, in some varieties of a third, *javanica* subspecies or wide compatibility varieties. In the hybrid with genotype S_5^i/S_5^j, gametes carrying the S_5^j allele are aborted, while no gamete abortion occurs in the hybrid with genotypes S_5^i/S_5^n and S_5^j/S_5^n. Therefore, the S_5^n allele has been incorporated into various rice cultivars to obtain fertile hybrids in hybrid rice breeding (Ikehashi 2009). The S_5^i allele acts like a Gc gene (cf. Fig. 5.1) and the S_5^n allele is the equivalent of the abovementioned inhibitor gene *Igc1* which knocks out the Gc gene of 4S (see Sect. 5.2.3).

Sano et al. (1979) repeatedly (eight times) backcrossed the male sterile but partially female fertile hybrid between *O. sativa* and *O. glaberrima* to each of the parents to obtain semi-sterile isogenic lines having the genetic background of the *sativa* and *glaberrima* parents. Self-pollination of these lines produced fully fertile progeny plants, and backcrossing these lines to the parents produced semi-sterile progeny plants. They applied a model, described as "one locus sporo-gametophytic interaction," to the sterility in the interspecific hybrid, assuming that the *sativa* and *glaberrima* parents have two sterility genes $S_1^aS_1^a$ S_2S_2 and S_1S_1 $S_2^aS_2^a$, respectively, and that if an S_1 or S_2 gene is present in the maternal tissue, gametes with S_1^a or S_2^a deteriorate (Fig. 5.4). This explanation conforms to the gametocidal system in wheat involving two different Gc chromosomes, when S_1 and S_2 are assumed to correspond to, e.g., 2C and 3C, respectively (cf. Fig. 5.2).

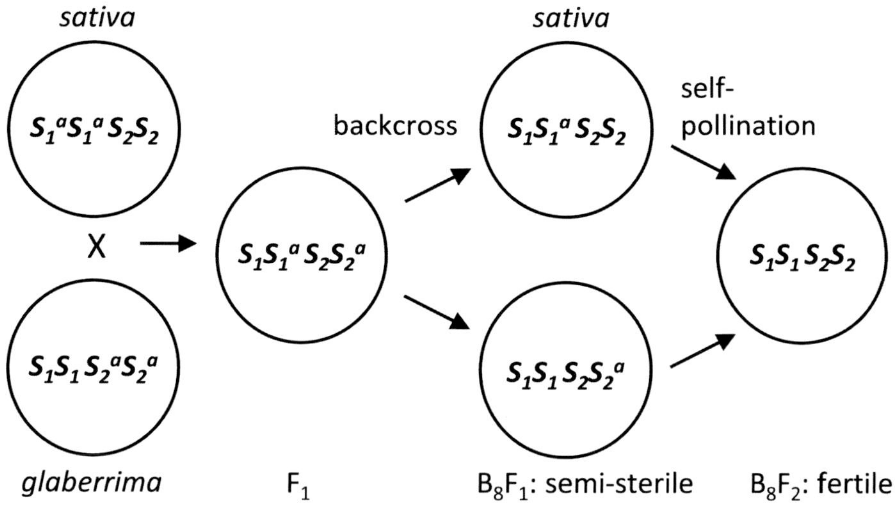

Fig. 5.4 Schematic diagram of hybrid sterility in rice, based on the description of hybrid sterility between *O. sativa* and *O. glaberrima* reported by Sano et al. (1979). See text for details

5.3.2 Epistatic Control of Hybrid Sterility Genes

Sano (1990) found that the intensity of hybrid sterility, namely Gc action, caused by the S_1 gene varied depending on *O. sativa* cultivars into which S_1 was introduced, and that the Gc action decreased as the S_1-containing chromosomal segment from *O. glaberrima* became smaller as a result of recombination during backcrossing. He inferred that S_1 and some modifiers linked to S_1 form a gene complex showing profound hybrid sterility.

Kubo et al. (2008) used rice near-isogenic lines carrying IR24 (ssp. *indica*) genomic segments in the genetic background of Asaminori (ssp. *japonica*) and found two new loci, *S24* and *S35*, causing male semi-sterility in this inter-subspecific combination. The *S24* locus has the IR24 allele (*S24*-ir) and the Asaminori allele (*S24*-as), and *S24*-ir acts as a pollen killer: the heterozygotes (*S24*-ir/*S24*-as) produce two types of microgametophytes after meiosis, with pollen carrying *S24*-as selectively aborted, and, as a consequence, exclusive transmission of pollen with the *S24*-ir allele to the offspring. On the other hand, the *S35* locus, also with two alleles, *S35*-ir and *S35*-as, shows the male semi-sterility phenotype only in the progeny carrying the *S24*-ir allele, i.e., the *S24*-ir/*S24*-ir *S35*-ir/*S35*-as and *S24*-ir/*S24*-as *S35-ir*/*S35*-as genotypes. This suggests an epistatic interaction occurring between the two hybrid sterility genes in rice, similar to what hypothesized in wheat between the Gc genes on *Aegilops* chromosomes 2S and 4S (cf. Fig. 5.2). Kubo et al. (2011) found that the hybrid male sterility caused by *S24* is also epistatically controlled by the *EFS* gene, which has two alleles, the dominant *indica* allele (*EFS-i*) and the recessive *japonica* allele (*efs-j*). The *EFS-i* allele in sporophytes counteracts the pollen sterility caused by *S24* heterozygosity, but the *efs-j* allele does not. Although *S24* and *EFS* are located on different chromosomes, *EFS* seems to act like the S_5^n allele against the S_5^i in *indica-japonica* hybrids.

The mutational origin of hybrid sterility genes in rice was demonstrated in irradiation experiments (Wan and Ikehashi 1996a). Variety Miyukimochi, which is an irradiated mutant from Toyonishiki, has two hybrid sterility genes, $S5^j$ and $S7^j$, while Toyonishiki carries a neutral allele, $S7^n$; therefore $S7^n$ must have been mutated into $S7^j$ by irradiation. A second case is that of the experimental line 02428 that has the wide compatibility allele $S5^n$. This line is derived from a progeny population of a hybrid whose parents both have $S5^j$; therefore, $S5^j$ must have mutated into $S5^n$ in both parents as a result of irradiation. The second case is similar to the knocked-out Gc gene of chromosome 4S in wheat (Sect. 5.2.2), but the first case, i.e., the creation of a Gc gene, is not known in wheat.

5.3.3 Evolutionary Implication of the Gametocidal System

Hybrid sterility prevents the movement of genes from one population to the other within a species, which keeps both populations distinct and eventually leads to speciation. Suppose two Gc genes of different type, which do not compensate for each other, are in different populations; hybrids between the two populations will suffer from sterility due to the gametocidal action. In case the two different Gc genes are on nonhomologous chromosomes, one-fourth of the gametes produced by the hybrid become fertile because the nonhomologous chromosomes segregate at random in meiosis (cf. Fig. 5.2). On the other hand, if the Gc genes are on homologous chromosomes, all gametes of the hybrid become sterile, because the homologous chromosomes pair and segregate from each other in meiosis I, and therefore no gametes will possess both Gc genes. Thus, sexual isolation would be established in a species between two populations that easily cross-fertilized.

The formation of new Gc genes or the alteration of existing Gc genes by mutation is most probable as reported in rice (Wan and Ikehashi 1996a) and wheat (Friebe et al. 2003). There are various hybrid sterility gene loci in *O. glaberrima* (Sano 1990) and *O. sativa* (Wan and Ikehashi 1996b). Gc chromosomes of various homoeologous groups have been introduced into wheat from different *Aegilops* species, including those from the S genome of *Ae. sharonensis*, which involve homoeologous groups 2 and 4, and those from the C genome of *Ae. cylindrica* and *Ae. triuncialis*, which are in homoeologous groups 2 and 3, respectively (Endo 1990, 2007). All the mentioned Gc chromosomes/genes have presumably been involved in the sexual isolation and speciation of wheat and rice. A suggestive example of sexual isolation within a species is seen in hybrids between allopatric accessions of *Ae. caudata*, which have normal meiotic chromosome pairing, but produce completely sterile pollen (Ohta 1992). This sterility might be explained as the result of the occurrence of two different alleles at the Gc loci on homologous chromosomes of the allopatric accessions. If so, since the Gc alleles segregate during meiosis I into separate daughter cells, none of the microgametophytes produced by the hybrids will receive both Gc alleles and will be thus able to develop into fertile pollen.

The presence of incomplete Gc action suggests that the Gc system is involved in the karyotype evolution of the genus *Aegilops*. Incomplete Gc action induces chromosomal rearrangements in hybrids heterozygous for a Gc gene, and gametes with rearranged chromosomes will survive and take part in self-fertilization. The karyotype of the selfed progeny will stabilize when the Gc gene will become homozygous, and some well-balanced karyotypes might be established in separate populations. Although not well investigated yet, chromosomal mutations also occur in the zygotes of hybrids between 4S addition line and euploid common wheat, only when chromosome 4S is transmitted through pollen (Tsujimoto 2005). Thus, the Gc system might have induced karyotype changes in gametes and zygotes of interspecific hybrids that were formed during polyploid evolution of wheat and *Aegilops* species.

The abovementioned Gc genes or chromosomes are only those showing pronounced Gc action. Considering the omnipresence of such highly penetrant Gc genes in plants, there must be more Gc genes with low penetrance throughout almost all plant species, and, altogether, they must have played a major role in the evolution of plants, in terms of karyotype diversification and speciation by sexual isolation. Probably the same is true for animals. This sort of selfishness of Gc genes might be the nature of living organisms as is more prevalent in human society.

References

Ashida T, Nasuda S, Sato K, Endo TR (2007) Dissection of barley chromosome 5H in common wheat. Genes Genet Syst 82:123–133

Cameron DR, Moav R (1957) Inheritance in *Nicotiana tabacum* XXVII. Pollen killer, an alien genetic locus inducing abortion of microspores not carrying it. Genetics 42:326–335

Endo TR (1982) Gametocidal chromosomes of three *Aegilops* species in common wheat. Can J Genet Cytol 24:201–206

Endo TR (1985) Two types of gametocidal chromosome of *Aegilops sharonensis* and *Ae. longissima*. Jpn J Genet 60:125–135

Endo TR (1988) Induction of chromosomal structural changes by a chromosome of *Aegilops cylindrica* L. in common wheat. J Hered 79:366–370

Endo TR (1990) Gametocidal chromosomes and their induction of chromosome mutations in wheat. Jpn J Genet 65:135–152

Endo TR (2007) The gametocidal chromosome as a tool for chromosome manipulation in wheat. Chromosome Res 15:67–75

Endo TR, Gill BS (1996) The deletion stocks in common wheat. J Hered 87:295–307

Endo TR, Tsunewaki K (1975) Sterility of common wheat with *Aegilops triuncialis* cytoplasm. J Hered 66:13–18

Endo TR, Yamamoto M, Mukai Y (1994) Structural changes of rye chromosome 1R induced by a gametocidal chromosome. Jpn J Genet 69:13–19

Finch RA, Miller TE, Bennett MD (1984) 'Cuckoo' *Aegilops* addition chromosome in wheat ensures its transmission by causing chromosome breaks in meiospores lacking it. Chromosoma 90:84–88

Friebe B, Zhang P, Nasuda S, Gill BS (2003) Characterization of a knock-out mutation at the Gc2 locus in wheat. Chromosoma 111:509–517

Gyawali YP, Nasuda S, Endo TR (2009) Cytological dissection and molecular characterization of chromosome 1R derived from 'Burgas 2' common wheat. Genes Genet Syst 84:407–416

Gyawali YP, Nasuda S, Endo TR (2010) A cytological map of the short arm of rye chromosome 1R constructed with 1R dissection stocks of common wheat and PCR-based markers. Cytogenet Genome Res 129:224–233

Ikehashi H (2009) Why are there Indica type and Japonica type in rice? – history of the studies and a view for origin of two types. Rice Sci 16:1–13

Ikehashi H, Wan J (1996) Differentiation of alleles at seven loci for hybrid sterility in cultivated rice (*Oryzsa sativa* L.). In: Khush GS (ed) Rice genetics III. International Rice Research Institute, Manila, Philippines, pp 404–408

Ishihara A, Mizuno N, Islam AKMR, Doležel J, Endo TR, Nasuda S (2014) Dissection of barley chromosomes 1H and 6H by the gametocidal system. Genes Genet Syst 89:203–214

Jiang J, Friebe B, Gill BS (1994) Recent advances in alien gene transfer in wheat. Euphytica 73:199–212

Joshi GP, Nasuda S, Endo TR (2011) Dissection and cytological mapping of barley chromosome 2H in the genetic background of common wheat. Genes Genet Syst 86:231–248

Kubo T, Yamagata Y, Eguchi M, Yoshimura A (2008) A novel epistatic interaction at two loci causing hybrid male sterility in an inter-subspecific cross of rice (*Oryza sativa* L.). Genes Genet Syst 83:443–453

Kubo T, Yoshimura A, Kurata N (2011) Hybrid Male sterility in rice is due to epistatic interactions with a pollen killer locus. Genetics 189:1083–1092

Li J, Nasuda S, Endo TR (2013) Dissection of rye chromosomes by the gametocidal system. Genes Genet Syst 88:321–327

Loegering WQ, Sears ER (1963) Distorted inheritance of stem-rust inheritance of Timstein wheat caused by a pollen-killing gene. Can J Genet Cytol 5:67–72

Maan SS (1975) Exclusive preferential transmission of an alien chromosome in common wheat. Crop Sci 15:287–292

Masoudi-Nejad A, Nasuda S, McIntosh RA, Endo TR (2002) Transfer of rye chromosome segments to wheat by a gametocidal system. Chromosome Res 10:349–357

Masoudi-Nejad A, Nasuda S, Bihoreau M-T, Endo TR (2005) An alternative to radiation hybrid mapping for large-scale genome analysis in barley. Mol Genet Genomics 274:589–594

Miller TE, Hutchinson J, Chapman V (1982) Investigation of a preferentially transmitted *Aegilops sharonensis* chromosome in wheat. Theor Appl Genet 61:27–33

Nakata N, Tomita M, Yasumuro Y (1993) Origin of midget chromosome present in the rye cytoplasmic Chinese Spring. Wheat Info Serv 76:54–55

Nasuda S, Kikkawa Y, Ashida T, Islam AKMR, Sato K, Endo TR (2005) Chromosomal assignment and deletion mapping of barley EST markers. Genes Genet Syst 80:357–366

Ohta S (1992) Intraspecific hybrid sterility in *Aegilops caudata* L. Hereditas 116:247–251

Qi LL, Echalier B, Chao S, Lazo GR, Butler GE, Anderson OD et al (2004) A chromosome bin map of 16,000 expressed sequence tag loci and distribution of genes among the three genomes of polyploid wheat. Genetics 168:701–712

Rick CM (1966) Abortion of male and female gametes in the tomato determined by allelic interaction. Genetics 53:85–96

Sakai K, Nasuda S, Sato K, Endo TR (2009) Dissection of barley chromosome 3H in common wheat and comparison of 3H physical and genetic maps. Genes Genet Syst 84:25–34

Sakata M, Nasuda S, Endo TR (2010) Dissection of barley chromosome 4H in common wheat by the gametocidal system and cytological mapping of chromosome 4H with EST markers. Genes Genet Syst 85:19–29

Sano Y (1990) The genic nature of gamete eliminator in rice. Genetics 125:185–191

Sano Y, Chu YE, Oka HI (1979) Genetic studies of speciation in cultivated rice, 1. Genic analysis for the F_1 sterility between *O. sativa* L. and *O. glaberrima* Steud. Jpn J Genet 54:121–132

Schlegel R, Korzun V (1997) About the origin of 1RS.1BL wheat-rye chromosome translocations from Germany. Plant Breed 116:537–540

Schubert I, Shi F, Jorg Fuchs J, Endo TR (1998) An efficient screening for terminal deletions and translocations of barley chromosomes added to common wheat. Plant J 14:489–495

Scoles GJ, Kibirge-Sebunya IN (1983) Preferential abortion of gametes in wheat induced by an *Agropyron* chromosome. Can J Genet Cytol 25:1–6

Serizawa N, Nasuda S, Shi F, Endo TR, Prodanovic S, Schubert I, Künzel G (2001) Deletion-based physical mapping of barley chromosome 7H. Theor Appl Genet 103:827–834

Shi F, Endo TR (1997) Production of wheat-barley disomic addition lines possessing an *Aegilops cylindrica* gametocidal chromosome. Genes Genet Syst 72:243–248

Tsuji S, Murata M (1976) Specific interactions between the D genome and the three alien cytoplasms in wheat. II. Seed inviability induced by the alien cytoplasms. Jpn J Genet 51:327–336

Tsujimoto H (2005) Gametocidal genes in wheat as the inducer of chromosome breakage. Wheat Info Serv 100:33–48

Tsujimoto H, Tsunewaki K (1985) Gametocidal genes in wheat and its relatives. II. Suppressor of the chromosome 3C gametocidal gene of *Aegilops triuncialis*. Can J Genet Cytol 27:178–185

Tsunewaki K, Wang G-Z, Matsuoka Y (1996) Plasmon analysis of *Triticum* (wheat) and *Aegilops*. 1. Production of alloplasmic common wheats and their fertilities. Genes Genet Syst 71:293–311

Wan J, Ikehashi H (1996a) Evidence for mutational origin of hybrid sterility genes in rice (*Oryza sativa* L.). Breed Sci 46:169–172

Wan J, Ikehashi H (1996b) List of hybrid sterility gene loci (HSGLi) in cultivated rice (Oryza sativa L.). Rice Genet Newsl 13:110–114

Werner JE, Endo TR, Gill BS (1992) Toward a cytogenetically based physical map of the wheat genome. Proc Natl Acad Sci U S A 89:11307–11311

Chapter 6
The Mode and Regulation of Chromosome Pairing in Wheat–Alien Hybrids (*Ph* Genes, an Updated View)

Tomás Naranjo and Elena Benavente

Abbreviations

Cdk	Cyclin-dependent kinase
C-Ph1	Candidate *Ph1*
DMC1	Disrupted Meiotic cDNA 1
DSBs	Double-strand DNA breaks
EMS	Ethil Methanesulfonate
FISH	Fluorescence in situ hybridization
GISH	Genome in situ hybridization
IWSGC	The International Wheat Genome Sequencing Consortium
KL	The Chinese Kaixian-luohanmai landrace
ltp	Low temperature pairing
MI	Metaphase I
MLH1	MutL homologue 1
MSH	MutS homologues
Ph	Pairing homoeologous
Ph^I	Inhibitor of *Ph*
PrBn	Pairing regulator in *B. napus*
SC	Synaptonemal complex
SNP	Single-nucleotide polymorphism
WM1	Wheat Meiosis 1

T. Naranjo (✉)
Departamento de Genética, Facultad de Biología, Universidad Complutense,
28040 Madrid, Spain
e-mail: toranjo@bio.ucm.es

E. Benavente
Departamento de Biotecnología-Biología Vegetal, Escuela Técnica Superior de Ingenieros Agrónomos, Universidad Politécnica de Madrid, 28040 Madrid, Spain
e-mail: e.benavente@upm.es

M. Molnár-Láng et al. (eds.), *Alien Introgression in Wheat*,
DOI 10.1007/978-3-319-23494-6_6

6.1 Introduction

Efficient production of balanced gametes in diploid organisms depends on the regular bivalent formation at the prophase of the first meiotic division. Homologous chromosomes find each other and interact at early meiosis to become physically connected by chiasmata and form bivalents. Clustering of telomeres in the meiotic bouquet at the leptotene–zygotene transition facilitates the approximation of terminal regions of homologous chromosomes and their recognition (Naranjo and Corredor 2008). Homologous interactions are initiated after the production of a programmed set of double strand DNA breaks (DSBs) catalyzed by the topoisomerase II related enzyme SPO11 at the initiation of meiosis (Keeney et al. 1997). DSBs are immediately immersed in a repairing process, which is controlled to ensure that each homologous pair receives at least the obligate crossover necessary for accurate segregation. Concomitant with the bouquet organization, a linear protein axis is formed along each homolog. Some of the proteins that make up the axis, for example, ASY1 in *Arabidopsis*, play a key role by influencing DSB repair through inter-homologous recombination (Sánchez-Morán et al. 2007). Then, homologous chromosomes become aligned, at least in their distal region, and undergo synapsis through the formation of a proteinaceous structure, the synaptonemal complexes (SC) that maintains the homologous chromosomes intimately juxtaposed along their length. Repair of DSBs starts with the invasion of the double strand DNA in any of the chromatids of the homologous chromosome. A crossover and a noncrossover (non-reciprocal exchange) represent the two possible outcomes in the pathway that the homologous recombination machinery follows to repair one DSB. The majority of DSBs are destined to become noncrossover products; the few that are crossovers show a distal location in many plant species (Higgins et al. 2014) and create chiasmata, which maintain homologues bound after the SC dissolution at diplotene. Chiasmata enable the homologues to correctly orientate on the meiotic spindle and segregate at anaphase I. Four haploid gametes are produced after the second meiotic division.

Aneuploids and polyploids, with more than two potential partners for each chromosome, generate deviations of regular bivalent pairing at metaphase I (MI). Polyploidy is widespread in plants and is considered to be a central feature of plant diversification (Grandont et al. 2013). Most polyploid plant species, including important crops like wheat, are allopolyploids that arose after hybridization between related diploid progenitors. The polyploid condition confers some advantages such as heterosis or gene redundancy but implies disadvantages such as the propensity to produce aneuploid meiotic products that reduce fertility (Comai 2005). Many polyploid species have evolved genetic regulatory systems that ensure a diploid-like behavior with efficient disjunction of homologous chromosomes at the first meiotic division (Jenczewski and Alix 2004). The best-studied example is common bread wheat, *Triticum aestivum*, most likely because of the outstanding importance of this species as a crop, but also because its meiotic pairing regulatory system is effective both in wheat and interspecific hybrids, which has implications in the transfer to wheat of alien genes controlling important agronomical traits. In this chapter we review the advances produced in the identification of genes that control meiotic pairing in wheat and in hybrids of wheat with related species, with special reference to the homoeologous pairing suppressor genes. Genes in other species with effect on the control of homoeologous pairing are also concerned. In the strict sense, the term chromosome pairing refers to the alignment of homologous chromosomes prior to synapsis. However, the terms homologous pairing and homoeologous pairing, which appear repetitively along the text, have been employed in the wide sense that they had at the time of the discovery of the pairing regulating genes, i.e., the association of homologous or homoeologous chromosomes at metaphase I.

6.2 Genes That Control Chromosome Pairing in Wheat

Common bread wheat is an allohexaploid species ($2n = 6x = 42$) with three genomes, A, B, and D, from three related diploid species, that arose after two hybridization events. The first hybridization event involved *T. urartu* and *Aegilops speltoides*, the donors of the A and B genomes, respectively, and originated the emmer tetraploid wheat (*T. turgidum*), and the second hybridization involved *T. turgidum* and *Ae. tauschii*, the donor of the D genome (Petersen et al. 2006). Tetraploid wheat is considered to have arisen 500,000 years ago while hexaploid wheat appeared within modern agriculture, 8000 years ago (Huang et al. 2002). In spite of the genetic synteny between homoeologous chromosomes, bread wheat forms 21 bivalents at diakinesis and metaphase I of meiosis. Obviously, in addition to genes that control meiosis in diploid species, polyploid wheats have developed genetic systems that restrict to homologues the formation of bivalent associations at metaphase I. The large size (17 gigabase) and the repetitive nature of the wheat genome have hindered and delayed the generation of a genome reference sequence (The International Wheat Genome Sequencing Consortium (IWGSC) 2014). Despite the lack of the molecular genomic tools necessary for gene identification and isolation, the number and diversity of aneuploid stocks available in wheat have permitted to identify and locate on chromosomes some of the genes involved in the control of meiotic events.

Two types of genes, promoters and suppressors, with effect on meiotic chromosome pairing in wheat itself or in hybrids of wheat and related species, were identified (Sears 1976). The *Ph1* (*Pairing homoeologous 1*) locus, located on the long arm of chromosome 5B (5BL), is the most effective and acts by suppressing homoeologous pairing (Okamoto 1957; Sears and Okamoto 1958; Riley and Chapman 1958, 1964; Riley and Kempana 1963). *Ph2*, another suppressor located on the short arm of chromosome 3D (3DS), shows an intermediate effect (Mello-Sampayo 1971a). Besides *Ph1* and *Ph2*, a third suppressor, even less effective than *Ph2*, was located on the short arm of chromosome 3A (Driscoll 1972; Mello-Sampayo and Canas 1973). Deficiency for both 3AS and 3DS results in a level of pairing similar to that caused by the deficiency for chromosome 5B, or about twice as high as the pairing produced by the lack of 3DS (Mello-Sampayo and Canas 1973). Because of the redundant functional activity of these two minor suppressors and their location on homoeologous arms (3AS and 3DS) they were proposed to be homoeologous loci. The level of homoeologous pairing in the double mutant *Ph1⁻/Ph2⁻* is not quite different from that of the single mutant *Ph1⁻* (Ceoloni and Donini 1993). This suggests no additive cooperation between both pairing suppressor genes. The absence of *Ph1* seems to induce the maximum possible level of homoeologous pairing in wheat. Two additional minor suppressors with a similar effect to that of 3AS were reported on chromosome 4D (Driscoll 1973) and chromosome arm 2DL (Ceoloni et al. 1986).

Genes that promote pairing are located on group 2, 3, and 5 chromosomes. All three homoeologous arms 2AS, 2BS and 2DS carry a pairing promoter gene with a minor effect (Sears 1954; Ceoloni et al. 1986). A relatively strong promoter gene, whose presence is necessary for normal synapsis and chiasma formation, is located on 3BL (Sears 1954; Kempana and Riley 1962). Its effect is more intense than the effect of promoters on group 2 chromosomes. The arms 3AL and 3DL carry also pairing promoters somewhat weaker than 3BL (Mello-Sampayo 1971a; Driscoll 1972; Mello-Sampayo and Canas 1973).

The 5BS arm carries a gene that promotes pairing (Riley and Chapman 1967; Feldman and Mello-Sampayo 1967). However, the intensity of its promoter activity is less than the suppressor effect of *Ph1*. This pairing promoter appears to reside in the distal third of 5BS as homozygosity for a deletion of this segment results in chiasma frequency similar to that found in wheat lacking the entire 5BS arm (Kota and Dvořák 1986). Chromosomes 5D and 5A carry also pairing promoters since chiasma frequency decreases in plants nulli-5D, while four doses of chromosome 5A compensate for the lack of 5D and restore the number of chiasmata produced in euploid plants (Feldman 1966). Chromosome 5D has been shown to carry a gene or genes, called *ltp* (low temperature pairing) that stabilize the number of chiasmata formed at high and low temperatures. Chiasma frequency in nullisomic-5D between 19 and 29 °C resembles that in euploids. Above and below these temperatures chiasma frequency is sharply reduced in 5D-deficient plants but little changed in euploids (Riley 1966a; Riley et al. 1966). Reduced chiasma frequency at low and high temperatures is accompanied by failure of synapsis during zygotene (Bayliss and Riley 1972a; Morais et al. 1992). The temperature sensitive stage was located in the premeiotic interphase prior to DNA synthesis (Bayliss and Riley 1972b). A weak stabilizing effect on chiasma formation at low temperature is derived from the presence of four doses of chromosome 5A in plants nulli 5D-tetra 5A (Riley et al. 1966). By contrast, tetraploid wheat, which has only the A and B genomes, shows no reaction to temperature changes (Riley et al. 1966). This suggests that the homoeoallele present in *T. turgidum*, most likely on chromosome 5A, is of greater effect than that in *T. aestivum*.

6.3 Meiotic Phenotype of *Ph* Mutants

The first attempt to isolate meiotic mutants in hexaploid wheat was carried out by treatment of plants ditelocentric 5BL with ethyl methanesulfonate (EMS) in order to produce mutations in the *Ph1* locus. The occurrence of mutations was checked in the meiosis of hybrids of the EMS-treated plants with rye. A mutant called 10/13 was later fixed in homozygosis both in the 5BL ditelocentric and in euploid condition. Although the mutant genotype formed regularly 21 bivalents, its hybrids with rye showed a higher level of homoeologous pairing than those of the wild-type genotype (Wall et al. 1971a, b). This mutation was associated to the 5BL arm and segregated independently of the centromere (Wall et al. 1971a, b). Another mutant consisting in a deletion of the *Ph1* locus was obtained after normal pollen irradiation with X-rays and pollination of plants monosomic-5B (Sears 1977). Wheat plants homozygous for this mutation showed multivalents at metaphase I denoting the occurrence of homoeologous pairing. This mutant was called *ph1b* since it was isolated later than the 10/13 (*ph1a*) mutation.

In the search for the *ph1b* mutation, Sears isolated a second X-ray induced mutation, termed *ph2a*, which conditions an intermediate level of pairing in hybrids with *Ae. kotschyi*. This mutation is located on chromosome arm 3DS and was suggested to be a deficiency that includes the *Ph2* locus (Sears 1982). Complementation tests demonstrated that mutations *ph1b* and 10/13 are not allelic and segregate independently, while *ph2a* and 10/13 are allelic. Then the 10/13 mutation was redesignated as *ph2b*. The *ph1b* and *ph2b* mutations map 0.9 cM from the centromere of 5B and 2.3 cM from the centromere of 3D, respectively (Sears 1984).

A *Ph*⁻ (*ph1c*) mutant was also isolated in cv. Capelli of *T. turgidum* after irradiation of seeds with neutrons (Giorgi 1978, 1983). This mutation consists of an intercalary deletion in the middle of 5BL, which includes the *Ph1* locus, and originated together with a tandem duplication involving the same region (Dvořák et al. 1984). Deletion and duplication are assumed to be produced by the same event, an unequal interchange between sister chromatids or homologous chromosomes. Intergeneric hybrids and haploids of mutant *ph1c* show a high level of chromosome pairing (Giorgi and Cuozzo 1980; Giorgi and Barbera 1981a, b; Jauhar et al. 1999; Cifuentes et al. 2006; Cifuentes and Benavente 2009a).

Meiotic phenotypes of wild type and mutants of tetraploid and hexaploid wheat have been described at prophase I and metaphase I. Ultrastructural analysis of SC formation in fully traced spread nuclei at zygotene and pachytene denoted the formation of multivalent associations involving homoeologous chromosomes in wild type and *Ph* mutants of both hexaploid and tetraploid wheat (Holm 1986; Holm and Wang 1988; Martínez et al. 2001a, b). On average 28 % (12/42) of chromosomes of hexaploid wheat, and 39 % (11/28) of tetraploid wheat, are involved in multivalent SCs at mid zygotene both in wild type and *Ph1* mutants. In the *ph2b* mutant of hexaploid wheat, only 22 % of the chromosomes are involved in multivalent SCs. While synapsis progresses to become completed in most pachytene nuclei, multivalent SCs are corrected and transformed into bivalents except in the case of *ph1b* and *ph1c*, in which half of them persists until metaphase I. Pairing correction concerns also the resolution of interlocking produced as a result of synapsis. In plants of hexaploid wheat lacking chromosome 5B or with four of six doses of chromosome arm 5BL, synapsis is arrested before completion, most likely due to the absence of the 5BS arm. The frequency of interlocked bivalent decreases with the level of synapsis achieved in the different genotypes, which suggests that completion of synapsis is required for the interlocking correction (Holm and Wang 1988).

Meiotic phenotypes at metaphase I (Table 6.1) are relatively similar in the wild type and *ph2b* genotypes of hexaploid wheat. None of them show multivalents although the number of ring bivalents is slightly lower in the *ph2b* mutant. However, the *ph1b* mutant shows multivalent configurations (Fig. 6.1) that identify the absence of the *Ph1* function. The same happens in the *ph1c* mutant of tetraploid wheat. As a result of homoeologous recombination, the *ph1b* mutant line accumulates extensive chromosome rearrangements and eventually becomes infertile (Sánchez-Morán et al. 2001). Gross genome changes produced in the absence of *Ph1* emphasize the key role of the *Ph1* locus in maintaining the genome integrity and ensuring the fertility of this important crop.

In hybrids of hexaploid wheat with related species, all three genotypes show homoeologous chromosome associations at metaphase I with low, intermediate, and high pairing levels in the wild type, *ph2b*, and *ph1b* hybrid genotypes, respectively (Fig. 6.2). The average number of the different types of configuration produced in hybrids of wheat with, rye, *Ae longissima*, *Ae. sharonensis*, and *Ae speltoides* are shown in Table 6.2. MI pairing frequencies are markedly lower in wheat–rye hybrids than in wheat–*Aegilops* hybrids mainly due to their different level of interspecific, wheat–rye or wheat–*Aegilops*, homoeologous pairing. This variation is related to the evolutionary history of the Triticeae: rye diverged 7,000,000 years ago in the timeline of wheat evolution while the clade including the diploid species of the *Triticum–Aegilops* complex radiated 2.5–4.5 million years ago (Huang et al. 2002). Thus, *Aegilops* chromosomes are more closely related to wheat than rye chromosomes. Identification of individual chromosomes demonstrated that homoeologous wheat–wheat associations are unevenly distributed between genomes. Associations of the A–D type are the most frequent among wheat–wheat associations in all

Table 6.1 Meiotic phenotype at metaphase I of wild type and *ph* mutants of hexaploid and tetraploid wheat

Genotype	I	IIrod	IIring	Multiv	Bonds/cell	No. of cells
Wild type (6×)[a]	0.02	1.48	19.50	0	40.49	90
ph2b[a]	0.48	2.95	17.78	0	38.57	120
ph1b[a]	2.76	4.76	14.05	0.77	34.22	120
Wild type (4×)[b]	0.04	0.34	13.64	0	27.62	50
ph1c[b]	0.94	3.69	9.46	0.19	23.16	100

I, univalents; IIrod, rod bivalents; IIring, ring bivalents; Multiv, multivalents
[a]Data from Martínez et al. (2001a)
[b]Data from Martínez et al. (2001b)

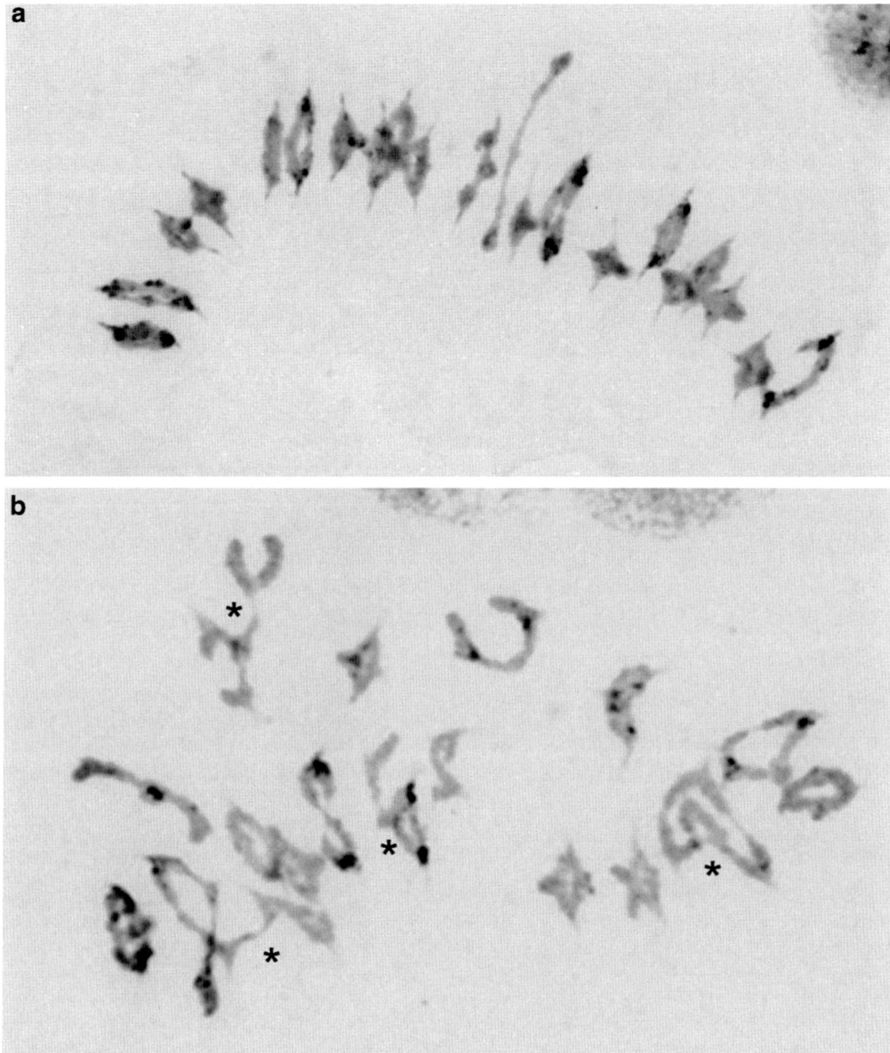

Fig. 6.1 Metaphase I pairing in Giemsa stained cells of hexaploid wheat. (**a**) Wild type wheat with 21 bivalents. (**b**) *ph1b* mutant wheat with four multivalents (*asterisks*) and 13 bivalents

hybrids studied. B–D associations occur somewhat more often than A–B associations in hybrids with rye, *Ae. longisima*, *Ae sharonensis*, or *Ae. speltoides* (Naranjo 1992; Naranjo and Maestra 1995; Maestra and Naranjo 1997, 1998) but not in hybrids with the allotetraploid *Ae. geniculata* (Cifuentes and Benavente 2009b). A recent genome assembly analysis of bread wheat and five diploid related species supports that both at the base-pair level as well as in gene content, the A and B genome lineages are more similar to the D genome lineage than they are to each other. The D genome is suggested to be originated from the A and B genomes

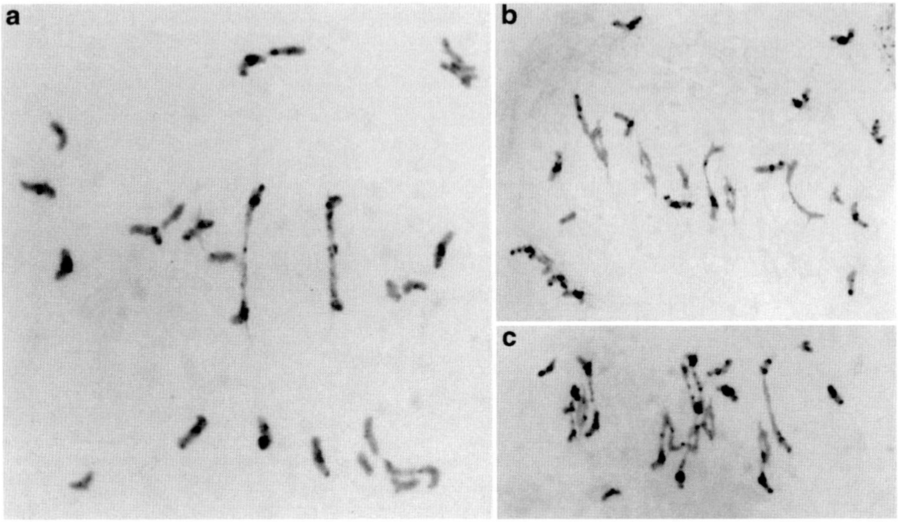

Fig. 6.2 Metaphase I pairing in Giemsa stained cells of wheat×*Ae. longissima* hybrids. (**a**) Low pairing level in the wild type genotype. (**b**) Intermediate pairing level in the *ph2b* mutant. (**c**) High pairing level in the *ph1b* mutant

Table 6.2 Average number of chromosome configurations at metaphase I in hybrids of wild type, *ph2b* and *ph1b* mutant wheat (ABD) with rye (R), *Ae. longissima* (S^l), *Ae. sharonensis* (S^{sh}), and *Ae. speltoides* (S)

Hybrid	I	IIrod	IIring	Multiv	Bonds/cell	No. of cells
ABDR	26.31	0.80	0.03	0.01	0.88	500
*ph2b*ABDR	19.23	3.04	0.57	0.51	5.26	500
*ph1b*ABDR	11.76	2.33	2.36	2.16	12.35	500
$ABDS^l$	24.55	1.59	0.06	0.05	1.81	300
ph2b$ABDS^l$	14.93	5.08	0.58	0.55	7.44	300
ph1b$ABDS^l$	3.48	4.04	2.99	2.86	18.28	300
$ABDS^{sh}$	25.21	1.18	0.03	0.03	1.29	250
ph2b$ABDS^{sh}$	10.16	5.58	1.42	1.13	11.17	300
ph1b$ABDS^{sh}$	4.37	3.74	3.79	2.39	17.93	300
ABDS	3.97	4.09	3.11	2.61	17.79	100
*ph2b*ABDS	3.25	3.41	3.28	3.02	19.41	100
*ph1b*ABDS	2.53	3.36	4.29	2.68	20.08	100

Data for the three ABDR hybrid combinations correspond to cells studied for other purposes by Naranjo et al. (1987, 1988). For $ABDS^l$ hybrids see Naranjo and Maestra (1995), Maestra and Naranjo (1997) for $ABDS^{sh}$ hybrids and Maestra and Naranjo (1998) for ABDS hybrids

through homoploid hybridization (Marcussen et al. 2014). However, the preferential A–D pairing consistently observed in the hybrids denotes a much lower differentiation between these two genomes than between A and B or B and D, at least in the highly recombinational distal chromosome region.

6.4 Unravelling the Structure of Loci *Ph1* and *Ph2*

The *ph1b* deletion covers an intercalary region, in which, the *Ph1* locus was confined to a submicroscopic region of less than 3 Mb flanked by the breakpoints of deletions *ph1c* and 5BL-1 (Gill et al. 1993). The 5BL-1 deletion belongs to the deletion stocks of wheat, a set of truncated chromosome lines generated using the gametocidal genes of *Ae. cylindrica*, *Ae. triuncialis*, or *Ae. speltoides* (Endo and Gill 1996).

A set of five lines carrying specific deletions generated by fast-neutron irradiated wheat were used to physically map the *Ph1* region, which corresponds to a subregion of 400 Kb of the ricegenome (Roberts et al. 1999). Markers of this rice region, and the equivalent region in *Brachypodium sylvaticum*, were used to saturate and sequence a 2.5-Mb region of the chromosome of wheat containing the *Ph1* locus and homoeologous regions of chromosomes 5A and 5D. The 5BL region includes 36 genes among which the *Ph1* locus was assigned to a structure consisting of a segment of subtelomeric heterochromatin from 3AL that inserted into a cluster of seven cyclin-dependent kinase (*Cdk*)-like genes, with a similar sequence to mammalian *Cdk2* (Griffiths et al. 2006; Al-Kaff et al. 2008). Two other clusters with five and two copies of *Cdk*-like genes are present in chromosomes 5A and 5D, respectively. The *Cdk* locus on 5B is dominant to loci on 5A and 5D in determining the overall kinase activity. Among the seven *Cdk*-like genes of chromosome 5B, 5B2 was considered to be the wheat equivalent of the gene encoding the human CDK2 kinase (Yousafzai et al. 2010). The level of phosphorylation measured in specific sites of histone H1 was twofold higher in the absence than in the presence of *Ph1* (Greer et al. 2012). This result and the fact that mammalian CDK2 is required for proper homologous pairing and recombination (Ortega et al. 2003; Viera et al. 2009) were considered to support the multi *Cdk*-like gene structure proposed for *Ph1* (Greer et al. 2012). Homoeologous recombination can be induced in hybrids of wild-type wheat with rye by treatment prior to meiosis with okadaic acid, a potent drug that activates CDKs causing a premature chromosome condensation. Then such an effect was assumed to be a phenocopy of *Ph1* (Knight et al. 2010).

However, Bhullar et al. (2014) concludes that *Cdk2-4* is present in chromosome deletion lines that show the meiotic mutant phenotype, and therefore, it is not a good candidate for the *Ph1* locus. Seven markers localized in the *Ph1* region were used to identify a 450-Kb segment of rice chromosome 9 including genes involved in chromatin reorganization, microtubule attachment, acetyltransferases, methyltransferases, DNA binding, and meiosis/anther specific proteins (Sidhu et al. 2008). Among 91genes present in this region, 26 were selected for further characterization by gene silencing. Transient or stable silencing of one of these genes, called *C-Ph1* (candidate *Ph1*), located on chromosome 5B, results the same meiotic phenotype as the *ph1b* mutant. Homoeologous copies of *C-Ph1* present in chromosomes 5A and 5D show a dramatically different structure and expression pattern, and their inactivation does not modify the wild-type meiotic phenotype (Bhullar et al. 2014). On the other hand, silencing of the *Cdk2-4* gene does not change the wild-type MI phenotype. However, multivalents are formed and centromere clumping is observed at prophase

I when the *C-Ph1* orthologue is silenced in *Arabidopsis*, which supports *C-Ph1* as an excellent candidate for the *Ph1* locus (Bhullar et al. 2014). This further suggests that okadaic acid might phenocopy a suppressor gene different than *Ph1*.

The *Ph2* locus has not yet been identified although some candidate genes have been proposed. Based on the conserved regions of the bacterial mismatch repair gene *MutS* and its homologues (*MSH*) in yeast and human, Dong et al. (2002) identified three *MSH7* homoeoloci located on chromosome arms 3AS, 3BS, and 3DS with highest similarity to maize and *Arabidopsis MSH7*. The copy on 3DS is included in the region deleted in the *ph2a* mutant, and is a good candidate for *Ph2*. *MSH7* genes of wheat are expressed in tissues associated with a high level of mitotic and meiotic activity, with maximum expression in the reproductive organs of young flower spikes. A reduction of the *MSH7* expression is observed in the wheat deletion mutant *ph2a* while the *ph2b* mutant shows a similar transcription level than the wild type. This suggests that *ph2b* is a point mutation that has affected the activity of the protein rather than the expression of the gene (Dong et al. 2002). In fact, two SNPs (single-nucleotide polymorphisms) leading to amino acid changes in the *ph2b* MSH7 protein were reported, although it remains to be established whether such changes affect the protein function (Lloyd et al. 2007). The reduction of fertility in plants of barley transformed with a *MSH7* RNAi knockdown construct is consistent with some role of *MSH7* on meiotic recombination in cereals (Lloyd et al. 2007). The mismatch repair system plays indeed an important role in reducing meiotic recombination between divergent sequences in *Arabidopsis* (Li et al. 2006), which has been specifically associated to *MSH7* in tomato (Tam et al. 2011).

Based on syntenic relationships between rice and wheat a length of 80 Mb was estimated for the region deleted in *ph2a* (Sutton et al. 2003). A 173-Kb DNA sequence localized within this region contains the *Wheat Meiosis 1* (*WM1*) gene family, which is composed of seven members. These genes are predominantly expressed in floral tissues at early meiosis and encode mainly membrane-anchored leucine-rich repeat-like receptor proteins with possible roles on disease response and floral development (Whitford et al. 2007). In addition to 3DS, the *WM1* gene cluster is present on barley 3HS but missing from the A and B genomes of hexaploid wheat. There is yet, no evidence supporting that the *WM1* gene family or any individual gene family member represents the *Ph2* gene.

6.5 The Mode of Action of *Ph1*

From the discovery of *Ph1* a number of hypotheses to explain its mode of action have been advanced. Some rely on assumptions concerning the presynaptic organization of chromosomes. Feldman (1966) studied meiosis in hexaploid wheat carrying four (diisosomic 5BL) or six doses (triisosomic 5BL) of *Ph1*, respectively. Extradoses of *Ph1* caused partial asynapsis of homologues, pairing between homoeologous, and interlocking of bivalents. Similar effects were observed by premeiotic treatment with colchicine (Driscoll et al. 1967; Feldman and Avivi 1988). It was therefore assumed that the hexaploid nucleus still maintains some organizational aspects of the individual ancestral genomes, i.e., each genome occupies a separate region in the nucleus, which in turn, is recognized by *Ph1*. A model was thus proposed whereby *Ph1* affects the premeiotic alignment of homologous and homoeologous chromosomes (Feldman 1993). In the absence of *Ph1*, the three genomes are mingled together in the nucleus and, consequently, both homologues and homoeologous compete in pairing. In euhexaploid wheat, two doses of *Ph1* keep the homoeologous apart, thereby leading to exclusive homologous pairing in meiosis. Six doses of *Ph1* or premeiotic treatment with colchicine induce random distribution in the premeiotic nuclei leading to separation of homologues, which results in overall reduction of synapsis, some level of synapsis between homoeologous and interlocking of bivalents. Based upon the observation of complex structures formed by multiple centromeres at leptotene, a later hypothesis proposed that *Ph1* exert its effect through the control of presynaptic centromere association (Martínez-Pérez et al. 1999, 2001, 2003). In wild type wheat, homologous centromeres coalescing in such structures become associated, and then the entire chromosomes collocate prior to synapse, whereas the recognition of homologous centromeres is not operative in *Ph1* mutants.

However, these models are contradictory with the synaptic pattern observed in wild-type wheat, which failed to reveal a large scale alignment of homologues in the leptotene–zygotene transition stage (Holm 1986), as well as with the absence of presynaptic association of two homologous chromosomes of rye, or their centromeres, in wheat–rye addition lines both in the presence and the absence of *Ph1* (Maestra et al. 2002; Corredor et al. 2007). Furthermore, two homoeologous arms physically connected to the same centromere, in a homoeoisochromosome, are bound at metaphase I in the absence of *Ph1* but not in its presence (Dvořák and Lukaszewski 2000) while two homologous arms can recombine even if they are linked to homoeologous centromeres (Corredor et al. 2007). Premeiotic colchicine treatment does not affect premeiotic centromere association but inhibits bouquet formation, i.e., the meiotic structure that brings together the majority of homologous pairs that normally occupy separated territories in premeiotic nuclei (Corredor and Naranjo 2007). Disturbed organization of the bouquet is the main reason of the failure of both synapsis (Corredor and Naranjo 2007) and resolution of interlockings (Feldman and Avivi 1988).

The meiotic phenotype of *Ph1* mutants indicates that *Ph1* exerts its action on two main steps of bivalent formation in polyploid wheat: first, on the suppression of recombination between homoeologous chromosomes in favor of that between homologues and, second, on the correction of SC multivalents formed during zygotene, which are transformed into bivalents.

In chromosomes of hexaploid wheat composed of homologous and homoeologous segments, recombination is absent from homoeologous segments while occurs normally in juxtaposed homologous segments in the presence of *Ph1*. However, extensive homoeologous recombination takes place in the absence of *Ph1*. This happens irrespective of the length of the homoeologous segment, its location on the centromere–telomere axis, and homology of both the telomeric and centromeric regions (Dubcovsky et al. 1995; Luo et al. 1996). Thus, *Ph1* confers to the DSBs repairing mechanism the ability of discriminating between homologous and homoeologous chromosomes thus restricting the progression through the crossover pathway to homologous interactions.

How this *Ph1* role is accomplished remains an unresolved question. The presence of *Cdk2*-like genes in wheat deletion lines showing the *Ph1* mutant meiotic phenotype (Bhullar et al. 2014) questions that the *Cdk2*-like gene cluster structure has any functional role on the *Ph1* activity (Greer et al. 2012). A *Ph1* gene encoding a product capable of detecting some degree of divergence (homoeology) among DNA sequences involved in the recombinational machinery, and excluding them from the crossover pathway, seems now feasible in the light of the discovery of the *C-Ph1* gene. The fact that mutations in genes, such as *Asy1*, mimic the meiotic phenotype of the *ph1b* mutant (Boden et al. 2009) might seem contradictory with this model. [*Asy1* encodes a protein of the SC lateral element, which promotes DMC1-mediated interhomologue recombination in *Arabidopsis* (Sánchez-Morán et al. 2007)]. However, there is a substantial difference between *Asy1* and the *C-Ph1* orthologue in *Arabidopsis*. The *asy1* mutant of *Arabidopsis* shows asynapsis and a reduced number of chiasmata between homologous chromosomes, while silencing of the *C-Ph1* orthologue results in the formation of multivalents and therefore, in nonhomologous interactions (Bhullar et al. 2014). Furthermore, the clumping of centromeres observed in the silenced plants of *Arabidopsis* denotes a different arrangement of chromatin. This may be in agreement with some conformational changes of chromatin produced in the absence of *Ph1* (Upadhya and Swaminathan 1967; Mikhailova et al. 1998; Maestra et al. 2002; Greer et al. 2012).

The involvement of *Ph1* in resolving SC multivalents into bivalents has also been explored. Many polyploid species form SC multivalent configurations at zygotene. In most cases, multiple associations decrease in number during prophase I through late zygotene to pachytene (Grandont et al. 2013). Irrespective of the nature, homologous or homoeologous, of the chromosomes involved, the initiation of recombination is thought to be necessary for synapsis. Accordingly, early recombination nodules associate with the SC in wheat synaptic multivalents (Hobolth 1981). The transformation of each multivalent in two or more bivalents, in wild-type wheat, implies that the early homoeologous recombination events abort and are redirected into intersister or noncrossover pathways at the stage of late zygote and pachytene. This is accompanied by disassembly of homoeologous SC stretches, which are resynapsed completing the homologous pattern (Holm 1986; Holm and Wang 1988; Martínez et al. 2001a, b). Separation of initially synapsed homoeologous segments may be facilitated by chromosome movements leading to disorganization of the bouquet as occurs in heterozygotes for an inversion involving most of the 1RL chromosome arm. The subtelomeric region of the normal chromosome, which associates with the subtelomeric region of the inverted arm at the bouquet stage, synapses with the proximal region of the inverted arm once the bouquet disorganization starts (Valenzuela et al. 2012).

The homologous synapsis that follows disassembly of SC between homoeologous is expected to be concluded either by extension of SC stretches initially formed between homologues, or by additional interactions in sites apart from the already synapsed region, or both. A number of ring bivalents at metaphase I similar to the number of bivalents at pachytene in wild-type hexaploid and tetraploid wheats (Martínez et al. 2001a, b) indicates that each bivalent present at pachytene forms at least one chiasma in each chromosome arm. Because chiasmata are distally located in wheat, some recombinational events processed in the crossover pathway should occur between homologous sites interacting after the synaptic multivalent correction. This supports that the recombination machinery may be loaded at different times during the prophase I.

The loading of the recombination machinery has been studied in hexaploid wheat and wheat–rye hybrids by immunolocalization of MLH1, a DNA mismatch protein considered to mark interfering crossovers at diplotene (Martín et al. 2014). Bread wheat yields a number of MLH1 foci which fits the number of chiasmata estimated at metapahase I, and is not affected by the presence or absence of *Ph1*. Wheat–rye hybrids show also a similar number of MLH1 sites both in the presence and absence of *Ph1*. However, these numbers are much higher than those expected from the number of MI chiasmata. Almost 70 % of MLH1 sites formed in the *ph1b* wheat × rye hybrids, and 95 % in the wild type hybrids, are processed to produce noncrossover products. Martín et al. (2014) suggest that these MLH1 sites, in which the crossover pathway is aborted, are disassembled later in the hybrids than when chromosomes have a homologous partner, and that this behavior is conditioned by *Ph1*.

In the absence of *Ph1*, recombinational events involving homoeologous chromosomes are not impeded or aborted at zygotene but some of them progress through the crossover pathway to culminate in the formation of homoeologous chiasmata. Such a DSBs repairing option would facilitate the stability of the homoeologous SC stretches and, therefore, the persistence of multivalent through the remaining prophase I stages and metaphase I.

6.6 Additional Genetic Factors Regulating Wheat–Alien Chromosome Pairing

6.6.1 Wheat Genotype

The diploid-like meiotic behavior of wheat is the result of a complex interaction between two main loci, *Ph1* and *Ph2*, and other minor genes that either promote or suppress homoeologous chromosome pairing (Sears 1976). The occurrence of natural allelic variation for genes responsible of the regular meiotic behavior in wheat is supported by numerous studies that report a wheat genotype-dependent variation in the level of homoeologous pairing in hybrids with related species. Driscoll and

Quinn (1970) and Farooq et al. (1990a) compared the number of MI chromosome association in hybrids between *A. variabilis* ($2n=4\times=28$, UUSS) and a series of bread wheat varieties. Both analyses evidenced significant differences in the extent of homoeologous pairing between hybrids and concluded on the existence of wheat varieties with a relatively higher promoter pairing effect than the canonical variety Chinese Spring. Similar results have been reported by Farooq et al. (1996) in hybrids with *Ae. geniculata* (syn. *Ae. ovata*; $2n=4\times=28$, UUMM). Interestingly, the level of synapsis in prophase-I nuclei and the frequency of chiasma at metaphase I are higher in haploids of wheat varieties Thatcher and Chris than in haploids of Chinese Spring (Martinez et al. 2005).

Up to date, the *phKL* gene of the Chinese landrace Kaixian-luohanmai (KL) is the best well-characterized *ph*-like gene described in a wheat variety other than Chinese Spring. This gene, assigned to chromosome 6A, shows an effect similar to that of inactivity for the *Ph2* locus in hybrid combinations with *Ae. variabilis*and rye (Liu et al. 1997; cited by Liu et al. 2003). Liu and coworkers developed a KL-derived line, which lacked 3DS, and demonstrated the additive effect of *phKL* and the absence of *Ph2* on promoting homoeologous pairing in wheat–alien combinations (Liu et al. 2003). A further comparative analysis of the effects of the *phKL* gene and the *ph1b*, *ph2a*, and *ph2b* mutations on homoeologous pairing in ABDUS and ABDR hybrids, concluded that *phKL* has an intermediate effect between the mutations of the *Ph1* and *Ph2* genes (Xiang et al. 2005; cited by Hao et al. 2011). However, *phKL* seems to have an homoeologous pairing promoting effect even stronger than *ph1b* in hybrids with the less related species *Psathyrostachys huashanica* ($2n=2\times=14$, NsNs) (Kang et al. 2008). By using GISH to discriminate wheat from rye partners, Hao et al. (2011) showed a similar amount of wheat–rye MI pairing in ABDR hybrids derived from *phKL* and *ph1b* wheat lines, the higher overall level of pairing in the *ph1b* hybrids being attributed to a higher amount of wheat–wheat chromosome pairing. While its molecular characterization is in progress (Dr. DC Liu, personal communication), all suggests that the lower the relatedness between homoeologous genomes, the higher the promoting effect of the *phKL* gene.

Allelic variation for the control of homoeologous pairing has also been demonstrated in crop and wild tetraploid wheats. Dhaliwal (1977) found that some lines of emmer and timopheevii wheats compensated for the absence of chromosome 5B while Ozkan and Feldman (2001) reported frequencies of chiasmata per cell ranging from 0.56 to 6.32 in hybrids between a line of *Ae. peregrina* and several genotypes of *T. turgidum* ssp. *dicoccoides* or *T. timopheevii*. Substitution of chromosome 5B of *T. aestivum* cv. Chinese Spring by 5B of *T. turgidum* ssp. *dicoccoides* or 5G of *T. timopheevii* yielded also some homoeologous pairing promotion in the interspecific hybrids, which suggested the presence of weak *Ph1* alleles in these species. On the other hand, a GISH analysis of MI association in durum wheat × *Ae. geniculata* hybrids (ABUM) supports the presence of a *ph* allele with a lower promoting effect than *ph1c* in the cultivated variety Ardente (Cifuentes et al. 2006).

6.6.2 Ph-*Like Genes in Wheat Related Species*

6.6.2.1 *Ae. speltoides*

Many studies have reported the ability of certain genotypes of the outcrossing species *Ae. speltoides* to induce wheat–alien pairing even in the presence of *Ph1* (e.g., Maestra and Naranjo 1998; see Table 6.2). Since promoter genes of *Ae. speltoides* behave as epistatic to *Ph* genes of wheat, these genes have been generically designated as Ph^I (inhibitors of *Ph*) (Chen et al. 1994).

The first evidence on the existence of Ph^I genes in *Ae. speltoides* was communicated soon after the identification of a meiotic control activity on wheat chromosome arm 5BL (Riley et al. 1961). Crossing with *Ae. speltoides* was indeed the critical intermediate step of the backcrossing program that resulted in the obtention of 'Compair', the first wheat line incorporating a useful alien trait by genetic induction of homoeologous recombination (Riley et al. 1968). Allelic variation for *Ae. speltoides* genes interacting with the wheat diploidizing meiotic control system was also reported (Dvořák 1972).

Promoter genes were successfully transferred from a high-pairing accession of *Ae. speltoides* to hexaploid wheat Chinese Spring by Chen et al. (1994). These Ph^I wheat stocks have demonstrated to be effective in inactivating the *Ph* system in varied wheat–alien hybrid combinations (Aghaee-Sarbarzeh et al. 2000).

Chen and Dvořák (1984) proposed that the genetic system in *Ae. speltoides* is controlled by two major genes, with different effectiveness and additive effects, in addition to some minor genes. This was supported by the segregation of low, intermediate, and high chromosome pairing phenotypes in the progenies of Ph^I wheat lines differing in their wheat–alien homoeologous pairing promoting effect (Chen et al. 1994). The presence of a structurally modified chromosome 4D in Ph^I wheat lines as well as its altered pairing behavior led also to suggest that a high-pairing gene(s) was introgressed as a 4S segment transferred on 4D (Chen et al. 1994).

By using a QTL mapping approach, Dvŏrák and coworkers have detected two *Ae. speltoides* genes with major suppressing effect on *Ph1* activity (Dvořák et al. 2006). These genes, designated as *Su1-Ph1* and *Su2-Ph1*, locate on the long arm of chromosomes 3S and 7S, respectively. *Su1-Ph1*, with a greater promoting effect on homoeologous chromosome pairing, is epistatic to *Su2-Ph1*. These authors mapped also a QTL with a minor effect on the short arm of chromosome 5S, a finding supported by the observations of multivalents in wheat lines where 5B is substituted by 5S (Friebe et al. 2011). Based upon the failure to map any of the major *Ph1* suppressors on chromosome 4S, Dvŏrák and coworkers concluded that neither *Su1-Ph1* nor *Su2-Ph1* were the genes operating in Ph^I lines. This is consistent with the absence of *Ae. speltoides* species-specific amplification products for SSR markers located on the long arm of group-3 and -7 chromosomes in three CS-Ph^I lines reported by Li et al. (2011).

The molecular tagging of Ph^I is being approached. Meanwhile, the Ph^I lines have been successfully employed for the incorporation into wheat chromosomes of alien genes carrying agronomically important traits, such as rust resistance from diverse *Aegilops* species (Aghaee-Sarbarzeh et al. 2002; Chhuneja et al. 2008), or salt tolerance from the less related species *Thinopyrum junceum* (Wang et al. 2003). The great practical advantage of *Ph*^{I} strains, over *Ph1*$^{-}$ or *ph1b* mutant stocks, for obtaining alien introgression into wheat by meiotic recombination, is that *Ph*^{I} has a dominant effect capable of inducing homoeologous pairing in the wheat–alien F1 hybrid itself as well as in the following backcross generations (Marais et al. 2010; Li et al. 2011).

6.6.2.2 *Ph*-Like Genes in Other Wheat Relatives

Inhibition by alien species genes of the wheat *Ph* system has mostly been characterized in *Ae. speltoides*. However, many studies demonstrate that genes affecting the wheat diploidizing control system exist in several other species of *Aegilops*, *Secale*, *Agropyron*, or *Elymus*.

The presence of homoeologous pairing promoting genes in the diploids *Ae. longissima* and *Ae. mutica* was evidenced as early as reported for *Ae. speltoides* (Riley 1966b; Mello-Sampayo 1971b). Mello-Sampayo (1971b) observed that the amount of MI pairing in wheat×*Ae. longissima* hybrids was significantly higher than in hybrids with *Ae. sharonensis*, though lower than in hybrids with *Ae. speltoides*. The segregation for various levels of homoeologous pairing found in hybrids of distinct *Ae. mutica* accessions with Chinese Spring led to suggest allelic variation at two *Ae. mutica* loci for the high and low pairing phenotypes (Dover and Riley 1972a). The expression of those *ph*-like genes was only evidenced when at least one chromosome 5B was present but seemed to be unaffected by the *Ph1* dosage or by the presence of *Ae. mutica* B chromosomes (Dover and Riley 1972a, b).

A promoting effect on wheat×alien homoeologous pairing was also reported in some lines of *Ae. variabilis* (syn. *Ae. peregrina*) (Farooq et al. 1990a, b; Fernández-Calvín and Orellana 1991) and *Ae. geniculata* (syn. *Ae. ovata*) (McGuire and Dvořák 1982; Farooq et al. 1996). Between-lines variation for *Ae. variabilis ph*-like genes concerns not only their presence or absence but also their interaction with the *Ph1* gene. To this regard, the effect of some promoting genotypes is only detectable in *Ph1* hybrids (Fernández-Calvín and Orellana 1991), whereas the pairing promoter genes present in other genotypes additively interact with the *ph1b* mutation (Farooq et al. 1990b). Paring regulators influencing the meiotic behavior of wheat–alien hybrids are suspected in other allotetraploid *Aegilops* species (McGuire and Dvořák 1982; see also Jenczewski and Alix 2004).

Lelley (1976a) conducted MI pairing analyses of hybrids between the seven wheat–rye cv. 'Imperial' addition lines and the *Secale* species *S. cereale* and *S. montanum*. This study evidenced the presence of alleles suppressing the wheat *Ph* control on chromosome 3R.

Within the genus *Agropyron*, genes interfering the homoeologous pairing restricting system of wheat have been identified in diploid and autotetraploid lines of *A. elongatum* (E genome) and *A. cristatum* (P genome) (Dvořák 1987; Charpentier et al. 1988; Chen et al. 1992; Jauhar 1992). The availability of addition and substitution lines of the E and P genome chromosomes in hexaploid wheat has provided further information on their chromosomal location. Either Dvořák (1987) and Charpentier et al. (1988) allocated strong promoters to chromosomes 3E and 5E of *A. elongatum*, which suggests some homoeoallelic relationships to the *Ph* loci of hexaploid wheat (Sears 1976). Additional promoters and suppressors seem to exist in other E chromosomes. The meiotic behavior of hybrids between Chinese Spring and addition lines of the P genome chromosomes, or chromosome arms, into wheat has demonstrated that the *Ph1* suppressor system of *A. cristatum* is polygenic, with no major gene (Jubault et al. 2006). However, this *Ph1* switching off system, though effective in promoting synapsis among wheat A, B, and D homoeologous, is unable to promote synapsis between wheat and the less related P genome chromosomes (Yang et al. 2010).

The existence of *Ph*-suppressors in *Elymus* species has been deduced from the meiotic analysis of hybrids between durum or common wheat and *Elytricum fertile*, an amphiploid of wheat and *Elymus sibiricus* ($2n=4\times=28$, SSHH) (Motsny and Simonenko 1996; Simonenko et al. 1998). As described for *A. cristatum*, this homoeologous pairing promoter system seems too weak to induce pairing between wheat and the distant related S/H genome homoeologous (Motsny and Simonenko 1996).

All studies mentioned above are based on the level of meiotic chromosome pairing observed in interspecific hybrids or wheat and related species. However, the existence of alien genotypes that interact with the meiotic control system of wheat has also been inferred from analyses of karyotype or genome stability in specially designed newly derived amphiploids. For instance, Zhang et al. (2013) have analyzed, by means of a FISH/GISH approach, the chromosomal constitution of artificial allotetraploids derived from crosses between diploid *Triticum*/*Aegilops* species related to the progenitors of tetraploid and hexaploid wheats. The combinations *Ae. sharonensis–T. monococcum* and *Ae. longissima–T. urartu* (comparable to tetraploid wheat, AABB) showed extremely low incidence of chromosomal rearrangements compared to *T. urartu–Ae. tauschii* and *Ae. bicornis–Ae. tauschii* amphiploids (AADD- and BBDD-like allotetraploids, respectively). Such a difference was proposed to be due to the presence of *Ph*-suppressors in the parental accessions of *Ae. sharonensis* and *Ae. longissima*. This interpretation is in close agreement with the absence of extensive genomic changes in a wheat–*Ae. speltoides* amphiploid, and derived lines, produced from an *Ae. speltoides* accession which does not induce homoeologous pairing in hybrids with wheat (Kumar et al. 2010).

6.6.3 B-Chromosomes

B-chromosomes coexist with the basic chromosome complement (A chromosomes) in *Ae. mutica*, *Ae. speltoides*, and *S. cereale*. The first report on the influence of B-chromosomes on the wheat meiotic control system was authored by Mochizuki

(1964) who proposed that B-chromosomes of *Ae. mutica*suppress homoeologous pairing in *T. aestivum*×*Ae. mutica*hybrids lacking chromosome 5B. Vardi and Dover (1972) reported a temperature-dependent significant decreasing effect of the B-chromosomes of *Ae. mutica* on the level of meiotic pairing in wheat×*Ae. mutica* hybrids, as well as some spindle irregularities linked to the presence of *Ae. speltoides* and *Ae. mutica* Bs in their respective hybrids with wheat.

Jones (1991) and Jenkins and Jones (2004) reviewed the effect of rye B-chromosomes in wheat×rye hybrids. Some studies failed to detect any significant influence of two rye B-chromosomes on meiotic pairing in wheat–rye hybrids regardless chromosome 5B is present or absent (Roothan and Sybenga 1976; Lelley 1976b) However, other studies report a homoeologous pairing suppressor effect of the rye Bs in a wheat–rye *ph* mutant background (Cuadrado et al. 1988) or irrespective the presence or absence of chromosome 5B (Viegas 1980). A wheat genotype-dependent influence of rye B-chromosomes in hybrids involving different durum wheat varieties has also been noted (Cuadrado et al. 1988). Romero and Lacadena (1980) showed that rye B-chromosomes decreased the level of homoeologous pairing in wheat–rye hybrids lacking a wheat chromosome with a suppressor effect (3A, 3D, or 5B), whereas they increased the level of pairing when a chromosome with a promoter effect (3B, 5A or 5D) was absent. Estepa et al. (1993), who analyzed wheat–rye hybrids with a different number of Bs, reported no clear effect of B-chromosomes on the level of chromosome pairing, although a greater variance in the distribution of MI associations was obtained with odd numbers of B-chromosomes. Similar results were described in diploid rye by Jones and Rees (1969). As stated by Jenkins and Jones (2004), the only general conclusion that can be drawn is that the rye Bs carry genes that interact with the pairing control genes of wheat under certain circumstances.

In a recent study, Kousaka and Endo (2012) examine the effect of the rye B-chromosome and two of its segments (namely B-9 and B-10) on homoeologous pairing in hybrids between bread wheat and *Ae. variabilis*, either in the presence or absence of chromosome 5B, and with one or two doses of chromosome 5D. When *Ph1* is absent, two rye B-chromosomes significantly decrease the chiasma frequency but do not compensate the absence of the suppressor activity of *Ph1*. A similar effect was attributed to the B-9 segment. However, both B-9 and B-10 increase the level of homoeologous pairing in hybrids carrying chromosome 5B. This study demonstrates that the effect of the Bs on homoeologous pairing is not confined to a unique region and further supports that its intensity is dose-dependent and conditioned by the numerical balance between wheat pairing regulators genes on chromosomes 5B and 5D.

Next-generation sequencing technologies have shown that B-chromosomes of rye harbor pseudogenes originated from the A chromosomes and organellar (plastid and mitochondrion) DNA sequences (Houben et al. 2013). It has been further demonstrated that some of the A-derived sequences are transcribed in a genotype-specific manner (Banaei-Moghaddam et al. 2013). This could explain the apparently contradictory findings and complex interactions pointed out on the effect of B-chromosomes on homoeologous pairing in wheat–alien combinations.

6.6.4 Cytoplasmic Effects

The influence of the cytoplasm on male fertility as well as several other morphological traits is well established in a number of plant species, including wheat (Tsunewaki et al. 2002). Analyses of meiotic pairing in wheat × rye hybrids obtained from alloplasmic wheat lines have revealed that alien cytoplasmic factors may also influence the homoeologous pairing control system of hexaploid wheat.

Wang et al. (1999) examined chromosome pairing at metaphase I in wheat × rye hybrids obtained from 16 alloplasmic lines of Chinese Spring. In each of the alloplasmic lines, the wheat cytoplasm had been replaced by the cytoplasm of a different *Triticum*or *Aegilops* species. This study revealed a significant homoeologous pairing promoter effect of the cytoplasms of *T. timopheevii*, *Ae. sharonensis*, *Ae. juvenalis*, *Ae. crassa*, *Ae. variabilis*, whereas the cytoplasms of *Ae. bicornis* and *Ae. kotschyi* decreased the level of pairing. However, such effects appeared to be nuclear genotype-dependent since Wang et al. (1991), who used a different rye parent, reported some reduction of the MI pairing by the cytoplasms of *Ae. variabilis*, *Ae. crassa*, and *Ae. juvenalis*. Besides, the cytoplasms of *Ae. kotschyi* and *Ae. crassa* showed opposite effects in hybrids with rye (ABDR) and with hexaploid triticale (AABBDR) (Wang et al. 1999).

The influence of *Ae. crassa*, *Ae. squarrosa* and *T. dicoccum* cytoplasms in wheat × rye hybrids has been examined in a more recent analysis (Kussovska et al. 2009) using alloplasmic wheat lines derived from varieties Aurora and Rusalka, which were crossed with two distinct rye inbred lines. In the four ABDR nuclear genotype combinations, the hybrids with the *Ae. crassa* cytoplasm showed a relatively high level of chiasmata compared with the euplasmic hybrids, which supports a homoeologous pairing promoter effect of the *Ae. crassa* cytoplasm. The ability of the *Ae. crassa* cytoplasm to disrupt the meiotic control system of wheat was shown also in an earlier report (Sechnyak and Simonenko 1991, cited by Simonenko et al. 1998). The effect of the cytoplasms of *Ae. squarrosa* and *T. dicoccum* is conditioned by the wheat genotype since the genetic background from cultivar Rusalka was permissive for the promoting homoeologous pairing of the alien cytoplasms while no significant deviations from the euplasmic hybrids were found in the set of hybrids derived from the Aurora alloplasmic lines (Kussovska et al. 2009).

A number of evidence support that the function of nuclear gene products can be affected by the expression of extranuclear genes (Hanson and Bentolila 2004). Interaction between nuclear and cytoplasmic genes seems a likely explanation for the disparate alien cytoplasm effect patterns reported above.

6.7 Homoeologous Pairing Control in Other Allopolyploid Species

Wheat is, by far, the polyploid species where the genetic system ensuring a diploid-like meiotic behavior has been more extensively explored and then characterized. However, there are other allopolyploids, including some important crops, whose strict homologous chromosome pairing is believed to be under genetic control. A brief summary of mostly old but some recent evidences of *Ph*-like systems fits the scope of this chapter, most examples being thoroughly reviewed by Jenczewski and Alix (2004).

Cultivated oats (*Avena sativa* and *A. bizantina*) are allohexaploid species with exclusive homologous bivalent formation at meiosis. The existence of a diploidizing genetic control system in hexaploid oat was early evidenced by the meiotic analysis of nulli-polyhaploids of *A. sativa* (Gauthier and McGinnis 1968). The complex mechanism regulating bivalency in oat, which seems to involve several chromosomes with distinct effectiveness (Gauthier and McGinnis 1968; Leggett 1977), shows a close similarity with that of wheat (reviewed in Jauhar 1977). As occurs in wheat, the homologous pairing suppression can be disrupted by certain genotypes of alien species. Rajhathy and Thomas (1972) reported a line of the diploid relative *A. longiglumis* which induces chromosome synapsis in interspecific hybrids with *A. sativa*. The promoting effect of this *A. longiglumis* genotype has indeed been successfully used for introgressive breeding of common oat (Thomas et al. 1980).

Fescues (*Festuca* ssp.) and ryegrasses (*Lolium* ssp.) form a singular complex of species with intergeneric hybrids showing a remarkable high level of homoeologous MI pairing, though the genomic relationship between parental chromosomes is distant enough for easy GISH-based cytological discrimination (e.g., Kopecký et al. 2008). The diploid-like behavior of the allohexaploid tall fescue (*Festuca arundinacea*) and other polyploid *Festuca* species seems to be controlled by homoeologous pairing suppressors that, unlike the wheat *Ph* genes, are ineffective in the haploid hemyzygous state (Jauhar 1975a, b). This has been confirmed by meiotic analyses of several *Festuca* × *Lolium* intergeneric hybrid combinations where all constituent genomes pair almost freely (see Kopecký et al. 2009 and references therein). Genotypes that suppress homoeologous pairing in interspecific and intergeneric hybrids have also been demonstrated in diploid *Lolium* species, like *L. perenne* and *L. longiflorum* (Taylor and Evans 1977; Armstead et al. 1999).

The presence of a *Ph*-like genetic control in cultivated cottons *Gossypium barbadense* and *G. hirsutum* (both being $2n=4\times=52$, AADD) was inferred from the observation that their haploids show almost no pairing at metaphase I whereas up to 11 bivalents have been reported in some interspecific hybrids between their putative diploid progenitors (Kimber 1961; Mursal and Endrizzi 1976). Some authors have questioned the need of a homoeologous pairing suppressor activity to explain the meiotic regularity of polyploid cottons. However, it seems the most substantiated hypothesis (Wendel and Cronn 2003; see Jenczewski and Alix 2004, and references therein), additionally supported by recent molecular approaches demonstrating ancient A–D homoeologous exchanges in both species (e.g., Flagel et al. 2012).

Brassica napus ($2n=4\times=38$, AACC) is an allotetraploid derived from the diploids *B. rapa* ($2n=20$, AA) and *B. oleracea* ($2n=18$, CC). Jenczewski et al. (2003) proposed the presence of a major locus regulating homoeologous pairing, which was designated as *PrBn* (for Pairing regulator in *B. napus*). This locus, genetically mapped to a C genome chromosome, is part of a complex system including several other loci with minor effect (Liu et al. 2006; Cifuentes et al. 2010). Several clues on its mode of action have been disclosed (e.g., Liu et al. 2006; Nicolas et al. 2009; Cifuentes et al. 2010; Grandont et al. 2014). A main difference relative to the wheat *Ph1* locus is that allelic variation for high and low homoeologous pairing at the *PrBn* locus greatly influences the pairing behavior of haploids but has no effect in some interspecific hybrid combinations (e.g., Leflon et al. 2006).

Jenczewski and Alix (2004) documented other important allopolyploid crops, like coffea (*Coffea arabica*; $2n=4\times=44$, CCEE) or tobacco (*Nicotiana tabacum*; $2n=4\times=48$, SSTT), for which indirect evidences suggest that their diploid-like meiotic behavior may be governed by *Ph*-like genes. Sorghum (*Sorghum bicolor*; $2n=4\times=20$, AABB) can be added to the list after the demonstration of MI pairing between its constituent genomes in interspecific hybrids with the related polyploid *S. macrospermum* (Kuhlman et al. 2008).

The potential application to *Arabidopsis suecica* ($2n=4\times=26$) of all analytical tools developed in the model plant *Arabidopsis* explains the recent research interest in this allopolyploid species (Bomblies and Madlung 2014). *A. suecica*, which originated from the hybridization between the diploids *A. thaliana* ($2n=2\times=10$) and *A. arenosa* ($2n=2\times=16$), shows a strict bivalent pairing behavior at metaphase I (Comai et al. 2003). Henry et al. (2014) have conducted a comparative analysis between *A. suecica* and synthetic amphiploids carrying *A. thaliana* and *A. arenosa* genomes which supports that the higher meiotic stability in the natural allotetraploid is not related to structural chromosomal differentiation between its constituent genomes but to some *Ph*-like mechanism that inhibits homoeologous recombination. These authors have further detected a locus on the *A. arenosa* genome whose *A. suecica* allele seems to suppress homoeologous synapsis and therefore multivalent formation.

6.8 Perspectives

The outstanding relevance of common and durum wheat as a crop in addition to the great list of chromosome variants stock lines have facilitated more advances in the knowledge of the control of meiosis in wheat than in other polyploid species. However, approaches carried out are not yet conclusive on the functional nature of *Ph* genes. Though some studies of meiosis in wheat published in the last 15 years have contributed to the general knowledge of the meiotic process in plants, most of the recent advances in this field have been obtained in diploid species, specially in *Arabidopsis*. Publication of the draft sequence of the bread wheat genome is expected to generate genomic tools that ensure the identification of genes such as *Ph1* or *Ph2*. This seems essential to decipher their mode of action. Also of capital interest is the identification of *Ph*^{I} genes as well as the molecular mechanism whereby they interact with *Ph* genes, which may open new ways of manipulating homoeologous pairing. The most relevant feature of the *Ph* and *Ph*^{I} genes, their ability to induce wheat–alien interspecific gene transfer, maintains still the potential to be applied in future breeding related programs.

Acknowledgements This work has been supported by grants AGL2012-38852 and AGL2012-38345 from Dirección General de Investigación Científica y Técnica, Ministerio de Economía y Competitividad of Spain.

References

Aghaee-Sarbarzeh M, Singh H, Dhaliwal SH (2000) *Ph*[I] gene derived from *Aegilops speltoides* induces homoeologous chromosome pairing in wide crosses of *Triticum aestivum*. J Hered 91:418–421

Aghaee-Sarbarzeh M, Ferrahi M, Singh S, Singh H, Friebe B, Gill BS (2002) *Ph*[I]-induced transfer of leaf and stripe rust-resistance genes from *Aegilops triuncialis* and *Ae. geniculata* to bread wheat. Euphytica 127:377–382

Al-Kaff N, Knight E, Bertin I, Foote T, Hart N, Griffiths S, Moore G (2008) Detailed dissection of the chromosomal region containing the *Ph1* locus in wheat *Triticum aestivum*: with deletion mutants and expression profiling. Ann Bot 101:863–872

Armstead IP, Bollard A, King IP, Forster JW, Hayward MD, Evans GM, Thomas HM (1999) Chromosome pairing in *Lolium perenne* × *L. temulentum* diploid hybrids: genetic and cytogenetic evaluation. Heredity 83:298–303

Banaei-Moghaddam AM, Meier K, Karimi-Ashtiyani R, Houben A (2013) Formation and expression of pseudogenes on the B chromosome of rye. Plant Cell 25:2536–2544. doi:10.1105/tpc.113.111856

Bayliss MW, Riley R (1972a) An analysis of temperature dependent synapsis in *Triticum aestivum*. Genet Res 20:193–200

Bayliss MW, Riley R (1972b) Evidence of premeiotic control of chromosome pairing in *Triticum aestivum*. Genet Res 20:201–212

Bhullar R, Nagarajan R, Bennypaul H, Sidhu GK, Sidhu G, Rustgi S, von Wettstein D, Gill KS (2014) Silencing of a metaphase I specific gene present in the *Ph1* locus results in phenotype similar to that of the *Ph1* mutations. Proc Natl Acad Sci U S A 111:14187–14192

Boden SA, Langridge P, Spangenberg G, Able JA (2009) TaASY1 promotes homologous chromosome interactions and is affected by deletion of *Ph1*. Plant J 57:487–497. doi:10.1111/j.1365-313X.2008.03701.x

Bomblies K, Madlung A (2014) Polyploidy in the *Arabidopsis* genus. Chromosome Res 22:117–134. doi:10.1007/s10577-014-9416-x

Ceoloni C, Donini P (1993) Combining mutations for the two homoeologous pairing suppressors genes *Ph1* and *Ph2* in common wheat and in hybrids with alien Triticeae. Genome 36:377–386

Ceoloni C, Strauss I, Feldman M (1986) Effect of different doses of group 2 chromosomes on homoeologous pairing in intergeneric wheat hybrids. Can J Genet Cytol 28:240–246

Charpentier A, Feldman M, Cauderon Y (1988) The effect of different *Agropyron elongatum* chromosomes on pairing in *Agropyron*-common wheat hybrids. Genome 30:978–983

Chen KC, Dvořák J (1984) The inheritance of genetic variation in *Triticum speltoides* affecting heterogenetic chromosome pairing in hybrids with *Triticum aestivum*. Can J Genet Cytol 26:279–287

Chen Q, Jahier J, Cauderon Y (1992) Enhanced meiotic chromosome pairing in intergeneric hybrids between *Triticum aestivum* and diploid Inner Mongolian *Agropyron*. Genome 35:98–102. doi:10.1139/g92-017

Chen PD, Tsujimoto H, Gill BS (1994) Transfer of *Ph*[I] genes promoting homoeologous pairing from *T. speltoides* to common wheat. Theor Appl Genet 88:97–101

Chhuneja P, Kaur S, Goel RK, Aghaee-Sarbarzeh M, Prashar M, Dhaliwal HS (2008) Transfer of leaf rust and strip rust resistance from *Aegilops umbellulata* Zhuk. to bread wheat (*Triticum aestivum* L.). Genet Resour Crop Evol 55:849–859. doi:10.1007/s10722-007-9289-3

Cifuentes M, Benavente E (2009a) Complete characterization of wheat-alien metaphase I pairing in interspecific hybrids between durum wheat (*Triticum turgidum* L.) and jointed goatgrass (*Aegilops cylindrica* Host). Theor Appl Genet 118:1609–1616

Cifuentes M, Benavente E (2009b) Wheat-alien metaphase I pairing of individual wheat genomes and D genome chromosomes in interspecific hybrids between *Triticum aestivum* L. and *Aegilops geniculata* Roth. Theor Appl Genet 119:805–813

Cifuentes M, Blein M, Benavente E (2006) A cytomolecular approach to assess the potential of gene transfer from a crop (*Triticum turgidum* L.) to a wild relative (*Aegilops geniculata* Roth.). Theor Appl Genet 112:657–664

Cifuentes M, Eber F, Lucas MO, Lode M, Chèvre AM, Jenczewski E (2010) Repeated polyploidy drove different levels of crossover suppression between homoeologous chromosomes in *Brassica napus* allohaploids. Plant Cell 22:2265–2276

Comai L (2005) The advantages and disadvantages of being polyploid. Nat Rev Genet 6:836–846

Comai L, Tyagi A, Lysak M (2003) FISH analysis of meiosis in *Arabidopsis* allopolyploids. Chromosome Res 11:217–226

Corredor E, Naranjo T (2007) Effect of colchicine and telocentric chromosome conformation on centromere and telomere dynamics at meiotic prophase I in wheat-rye additions. Chromosome Res 15:231–245

Corredor E, Lukaszewski AJ, Pachón P, Allen DC, Naranjo T (2007) Terminal regions of wheat chromosomes select their pairing in meiosis. Genetics 177:699–706

Cuadrado C, Torres P, Romero C (1988) The effect of rye B-chromosomes in wheat-rye hybrids. An Aula Dei 19:37–44

Dhaliwal HS (1977) The *Ph* gene and the origin of tetraploid wheat. Genetica 47:177–182

Dong C, Whitford R, Langridge P (2002) A DNA mismatch repair gene links to the *Ph2* locus in wheat. Genome 45:116–124

Dover GA, Riley R (1972a) Variation at two loci affecting homoeologous meiotic chromosome pairing in *Triticum aestivum*×*Aegilops mutica* hybrids. Nat New Biol 235:61–62

Dover GA, Riley R (1972b) Prevention of pairing of homoeologous meiotic chromosomes of wheat by an activity of supernumerary chromosomes of *Aegilops*. Nature 240:159–161

Driscoll CJ (1972) Genetic suppression of homoeologous chromosome pairing in hexaploid wheat. Can J Genet Cytol 14:39–42

Driscoll CJ (1973) Minor genes affecting homoeologous pairing in hybrids between wheat and related genera. Genetics 74:s66

Driscoll CJ, Quinn CJ (1970) Genetic variation in *Triticum* affecting the level of chromosome pairing in intergeneric hybrids. Can J Genet Cytol 12:278–282. doi:10.1139/g70-041

Driscoll CJ, Darvey NL, Barber HN (1967) Effect of colchicine on meiosis of hexaploid wheat. Nature 216:687–688

Dubcovsky J, Luo MC, Dvořák J (1995) Differentiation between homoeologous chromosomes 1A of wheat and $1A^m$ of *Triticum monococcum* and its recognition by the wheat *Phl* locus. Proc Natl Acad Sci U S A 92:6645–6649

Dvořák J (1972) Genetic variability in *Aegilops speltoides* affecting homoeologous pairing in wheat. Can J Genet Cytol 14:371–380

Dvořák J (1987) Chromosomal distribution of genes in diploid *Elytrigia elongata* that promote or suppress pairing of wheat homoeologous chromosomes. Genome 29:34–40

Dvořák J, Lukaszewski AJ (2000) Centromere association is an unlikely mechanism by which the wheat *Ph1* locus regulates metaphase I chromosome pairing between homoeologous chromosomes. Chromosoma 109:410–414

Dvořák J, Chen KC, Giorgi B (1984) The C-banding pattern of a *Ph*⁻ mutant of durum wheat. Genome 26:360–363

Dvořák J, Deal KR, Luo MC (2006) Discovery and mapping of wheat *Ph1* suppressors. Genetics 174:17–27. doi:10.1534/genetics.106.058115

Endo TR, Gill BS (1996) The deletion stocks of common wheat. J Hered 87:295–297

Estepa MA, Cuadrado C, Romero C (1993) Odd-even effect of rye B-chromosomes on homoeologous pairing in wheat-rye hybrids. Caryologia 46:17–23
Farooq S, Iqbal N, Shah TM (1990a) Intergeneric hybridization for wheat improvement. III. Genetic variation in *Triticum* species affecting homoeologous chromosome pairing. Cereal Res Commun 18:233–237
Farooq S, Iqbal N, Shah TM (1990b) Promotion of homoeologous chromosome pairing in hybrids of *Triticum aestivum* × *Aegilops variabilis*. Genome 33:825–828. doi:10.1139/g90-124
Farooq S, Iqbal N, Shah TM (1996) Intergeneric hybridization for wheat improvement. V. Production of and metaphase I chromosome analysis in F1 hybrids of wheat (*Triticum aestivum*) with *Aegilops ovata* L. Cereal Res Commun 24:155–161
Feldman M (1966) The effect of chromosomes 5B, 5D and 5A on chromosomal pairing in *Triticum aestivum*. Proc Natl Acad Sci U S A 56:1192–1199
Feldman M (1993) Cytogenetic activity and mode of action of the pairing homoeologous (*Phi*) gene of wheat. Crop Sci 33:894–897
Feldman M, Avivi L (1988) Genetic control of bivalent pairing in common wheat: the mode of *Ph1* action. In: Brandham PE (ed) Kew chromosome conference III. Proc 3rd Kew Chromosome Conference, Kew, UK. Kew Publishing, Richmond, VA, pp 269–279
Feldman M, Mello-Sampayo T (1967) Suppression of homoeologous pairing in hybrids of polyploidy wheat × *Triticum speltoides*. Can J Genet Cytol 9:307–313
Fernández-Calvín B, Orellana J (1991) Metaphase-I bound-arm frequency and genome analysis in wheat-*Aegilops* hybrids. 1. *Ae. variabilis*-wheat and *Ae. kotschyi*-wheat hybrids with low and high homoeologous pairing. Theor Appl Genet 83:264–722
Flagel L, Wendel J, Udall J (2012) Duplicate gene evolution, homoeologous recombination, and transcriptome characterization in allopolyploid cotton. BMC Genomics 13:302, http://www.biomedcentral.com/1471-2164/13/302
Friebe B, Qi LL, Liu C, Gill BS (2011) Genetic compensation abilities of *Aegilops speltoides* chromosomes for homoeologous B-genome chromosomes of polyploid wheat in disomic S(B) chromosome substitution lines. Cytogenet Genome Res 134:144–150
Gauthier FM, McGinnis RC (1968) The meiotic behaviour of a nulli-haploid plant in *Avena sativa* L. Can J Genet Cytol 10:186–189
Gill KS, Gill BS, Endo TR, Mukai Y (1993) Fine physical mapping of *Ph1*, a chromosome pairing regulator gene in polyploid wheat. Genetics 134:1231–1236
Giorgi B (1978) A homoeologous pairing mutation isolated in *Triticum durum* cv. Cappelli. Mut Breed Newsl 11:4–5
Giorgi B (1983) Origin, behaviour and utilization of a *Ph1* mutant of durum wheat *Triticum turgidum* (L) var *durum*. In: Sakamoto S (ed) Proc 6th Int Wheat Genet Symp. Kyoto University, Kyoto, pp 1033–1040
Giorgi B, Barbera F (1981a) Use of mutants that affect homoeologous pairing for introducing alien variation in both durum and common wheat. In: International Atomic Energy gency (ed) Induced mutations-a tool in plant breeding. Proc IAEA-FAO Symp Induced Mutations, Vienna, Austria. IAEA-SM, Vienna, pp 37–47
Giorgi B, Barbera F (1981b) Increase of homoeologous pairing in hybrids between *T. turgidum* L. var. *durum* and two tetraploid species of *Aegilops*: *Ae. kotschyi* and *Ae. cylindrica*. Cereal Res Commun 9:205–211
Giorgi B, Cuozzo L (1980) Homoeologous pairing in a *Ph* mutant of tetraploid wheat crossed with rye. Cereal Res Commun 8:485–490
Grandont L, Jenczewski E, Lloyd A (2013) Meiosis and its deviations in polyploid plants. Cytogenet Genome Res 140:171–184
Grandont L, Cuñado N, Coriton O, Huteau V, Eber F, Chèvre AM, Grelon M, Chelysheva L, Jenczewski E (2014) Homoeologous chromosome sorting and progression of meiotic recombination in *Brassica napus*: Ploidy does matter! Plant Cell 26:1448–1463
Greer E, Martín AC, Pendle A, Colas I, Jones AME, Moore G, Shaw P (2012) The *Ph1* locus suppresses Cdk2-type activity during premeiosis and meiosis in wheat. Plant Cell 24:152–162

Griffiths S, Sharp R, Foot TN, Bertin I, Wanous M, Reader S, Colas I, Moore G (2006) Molecular characterization of *Ph1* as a major chromosome pairing locus in polyploid wheat. Nature 439:749–752

Hanson MR, Bentolila S (2004) Interactions of mitochondrial and nuclear genes that affect male gametophyte development. Plant Cell 16:S154–S169. doi:10.1105/tpc.015966

Hao M, Luo JT, Yang M, Zhang LQ, Yan Z, Yuan Z, Zheng Y, Zhang H, Liu D (2011) Comparison of homoeologous chromosome pairing between hybrids of wheat genotypes Chinese Spring *ph1b* and Kaixian-luohanmai with rye. Genome 54:959–964. doi:10.1139/G11-062

Henry IM, Dilkes BP, Tyagi AP, Gao J, Christensen B, Comai L (2014) The *BOY NAMED SUE* quantitative trait locus confers increased meiotic stability to an adapted natural allopolyploid of *Arabidopsis*. Plant Cell 26:181–194. doi:10.1105/tpc.113.120626

Higgins JD, Osman K, Jones GH, Franklin FCH (2014) Factors underlying restricted crossover localization in barley meiosis. Annu Rev Genet 48:29–47

Hobolth P (1981) Chromosome pairing in allohexaploid wheat var. Chinese Spring. Transformation of multivalents into bivalents, a mechanism for exclusive bivalent formation. Carlsberg Res Commun 46:129–173

Holm PB (1986) Chromosome pairing and chiasma formation in allohexaploid wheat, *Triticum aestivum* analyzed by spreading of meiotic nuclei. Carlsberg Res Commun 5:239–294

Holm PB, Wang X (1988) The effect of chromosome 5B on synapsis and chiasma formation in wheat, *Triticum aestivum* cv. Chinese Spring. Carlsberg Res Commun 53:191–208

Houben A, Banaei-Moghaddam AM, Klemme S, Timmis JN (2013) Evolution and biology of supernumerary B chromosomes. Cell Mol Life Sci 71:467–478. doi:10.1007/s00018-013-1437-7

Huang S, Sirikhachornkit A, Su X, Faris J, Gill B, Haselkorn R, Gornicki P (2002) Genes encoding plastid acetyl-CoA carboxylase and 3-phosphoglycerate kinase of the *Triticum*/*Aegilops* complex and the evolutionary history of polyploid wheat. Proc Natl Acad Sci U S A 99:8133–8138

Jauhar PP (1975a) Genetic control of diploid-like meiosis in hexaploid tall fescue. Nature 254:595–597

Jauhar PP (1975b) Genetic regulation of diploid-like chromosome pairing in the hexaploid species, *Festuca arundinacea* Schreb. and *F. rubra* L. (Gramineae). Chromosoma 52:363–382

Jauhar PP (1977) Genetic regulation of diploid-like chromosome pairing in *Avena*. Theor Appl Genet 49:287–295

Jauhar PP (1992) Chromosome pairing in hybrids between hexaploid bread wheat and tetraploid crested wheatgrass (*Agropyron cristatum*). Hereditas 116:107–109

Jauhar PP, Almouslem AB, Peterson TS, Joppa LR (1999) Inter- and intragenomic chromosome pairing in haploids of durum wheat. J Hered 90:437–445

Jenczewski E, Alix K (2004) From diploids to allopolyploids: the emergence of efficient pairing control genes in plants. Crit Rev Plant Sci 23:21–45

Jenczewski E, Eber F, Grimaud A, Huet S, Lucas MO, Monod H, Chèvre AM (2003) *PrBn*, a major gene controlling homoeologous pairing in oilseed rape (*Brassica napus*) haploids. Genetics 164:645–654

Jenkins G, Jones N (2004) B chromosomes in hybrids of temperate cereals and grasses. Cytogenet Genome Res 106:314–319

Jones RN (1991) Cytogenetics of B-chromosomes in crops. In: Gupta PK, Tsuchiya T (eds) Chromosome Engineering in plants: genetics, breeding and evolution. Elsevier, Amsterdam, pp 141–157

Jones RN, Rees H (1969) An anomalous variation due to B chromosomes in rye. Heredity 24:265–271

Jubault M, Tanguy AM, Abélard P, Coriton O, Dusautoir JC, Jahier J (2006) Attempts to induce homoeologous pairing between wheat and *Agropyron cristatum* genomes. Genome 49:190–193

Kang HY, Zhang HQ, Wang Y, Jiang Y, Yuan HJ, Zhou YH (2008) Comparative analysis of the homoeologous pairing effects of *phKL* gene in common wheat × *Psathyrostachys huashanica* Keng ex Kuo. Cereal Res Commun 36:429–440. doi:10.1556/CRC.36.2008.3.7

Keeney S, Giroux CN, Kleckner N (1997) Meiosis-specific DNA double-strand breaks are catalyzed by Spo11, a member of a widely conserved protein family. Cell 88:375–384

Kempana C, Riley R (1962) Relationships between the genetic effects of deficiencies for chromosomes III and V on meiotic pairing in *Triticum aestivum*. Nature 195:1270–1273

Kimber G (1961) Basis of the diploid-like meiotic behaviour of polyploid cotton. Nature 191:98–100

Knight E, Greer E, Draeger T, Thole V, Reader S, Shaw P, Moore G (2010) Inducing chromosome pairing through premature condensation: analysis of wheat interspecific hybrids. Funct Integr Genomics 10:603–608

Kopecký D, Lukaszewski AJ, Doležel J (2008) Meiotic behaviour of individual chromosomes of *Festuca pratensis* in tetraploid *Lolium multiflorum*. Chromosome Res 16:987–998

Kopecký D, Bartoš J, Zwierzykowski Z, Doležel J (2009) Chromosome pairing of individual genomes in tall fescue (*Festuca arundinacea* Schreb.), its progenitors, and hybrids with Italian ryegrass (*Lolium multiflorum* Lam.). Cytogenet Genome Res 124:170–178. doi:10.1159/000207525

Kota RS, Dvořák J (1986) Mapping of a chromosome pairing gene and 5SRNA genes in *Triticum aestivum* L. by a spontaneous deletion in chromosome 5Bp. Can J Genet Cytol 28:266–271

Kousaka R, Endo TR (2012) Effect of a rye B chromosome and its segments on homoeologous pairing in hybrids between common wheat and *Aegilops variabilis*. Genes Genet Syst 87:1–7

Kuhlman LC, Stelly DM, Price HJ, Rooney WL (2008) Genetic recombination in *Sorghum bicolor*×*S. macrospermum* interspecific hybrids. Genome 51:749–756. doi:10.1139/G08-061

Kumar S, Friebe B, Gill BS (2010) Fate of *Aegilops speltoides*-derived, repetitive DNA sequences in diploid *Aegilops* species, wheat-*Aegilops* amphiploids and derived chromosome addition lines. Cytogenet Genome Res 129:47–54

Kussovska V, Bochev B, Bubova V (2009) Morphological characteristics and meiotic homoeologous pairing of chromosomes in alloplasmic wheat×rye hybrids. Genet Breed 38:39–45

Leflon M, Eber F, Letanneur JC, Chelysheva L, Coriton O, Huteau V, Ryder CD, Barker G, Jenczewski E, Chèvre AM (2006) Pairing and recombination at meiosis of *Brassica rapa* (AA)×*Brassica napus* (AACC) hybrids. Theor Appl Genet 113:1467–1480

Leggett JM (1977) The meiotic behaviour of aneupolyhaploids of the cultivated oat *Avena sativa* (2n=6x=42). Can J Genet Cytol 19:651–656

Lelley T (1976a) Induction of homoeologous pairing in wheat by genes of rye suppressing chromosome 5B effect. Can J Genet Cytol 18:485–489

Lelley T (1976b) The effect of supernumerary chromosomes of rye on homoeologous pairing in hexaploid wheat. Z Pflanzenzuchtg 77:281–285

Li L, Jean M, Belzile F (2006) The impact of sequence divergence and DNA mismatch repair on homeologous recombination in *Arabidopsis*. Plant J 45:908–916

Li H, Gill BS, Wang X, Chen P (2011) A Tal-Ph^I wheat genetic stock facilitates efficient alien introgression. Genet Resour Crop Evol 58:667–678

Liu DC, Luo MC, Yang JL, Yan J, Lan XJ, Yang WY (1997) Chromosome location of a new paring promoter in natural populations of common wheat. SW China J Agric Sci 10:10–15

Liu DC, Zheng YL, Yan ZH, Zhou YH, Wei YM, Lan XJ (2003) Combination of homoeologous pairing gene *phKL* and *Ph2*-deficiency in common wheat and its mitotic behaviors in hybrids with alien species. Acta Bot Sin 45:1121–1128

Liu Z, Adamczyk K, Manzanares-Dauleux M, Eber F, Lucas MO, Delourme R, Chèvre AM, Jenczewski E (2006) Mapping *PrBn* and other quantitative trait loci responsible for the control of homeologous chromosome pairing in oilseed rape (*Brassica napus* L.) haploids. Genetics 174:1583–1596

Lloyd AH, Milligan AS, Langridge P, Able JA (2007) *TaMSH7*: a cereal mismatch repair gene that affects fertility in transgenic barley (*Hordeum vulgare* L.). BMC Plant Biol 7:67. doi:10.1186/1471-2229-7-67

Luo MC, Dubcovsky J, Dvořák J (1996) Recognition of homeology by the wheat *Ph1* locus. Genetics 144:1195–1203

Maestra B, Naranjo T (1997) Homoeologous relationships of *Triticum sharonense* chromosomes to *T. aestivum*. Theor Appl Genet 94:657–663

Maestra B, Naranjo T (1998) Homoeologous relationships of *Aegilops speltoides* chromosomes to bread wheat. Theor Appl Genet 97:181–186

Maestra M, de Jong JH, Shepherd KW, Naranjo T (2002) Chromosome arrangement and behaviour of two rye homologous telosomes at the onset of meiosis in disomic wheat-5RL addition lines with and without the *Ph1* locus. Chromosome Res 10:655–667

Marais GF, Bekker TA, Eksteen A, McCallum B, Fetch T, Marais AS (2010) Attempts to remove gametocidal genes co-transferred to common wheat with rust resistance from *Aegilops speltoides*. Euphytica 171:71–85

Marcussen T, Sandve SR, Heier L, Spannagle M, Pfeifer M, IWGSC, Jakobsen KS, Wulff BBH, Steuernagel B, Mayer KFX, Olsen OA (2014) Ancient hybridizations among the ancestral genomes of bread wheat. Science 345:1250092

Martín AC, Shaw P, Phillips D, Reader S, Moore G (2014) Licensing MLH1 sites for crossover during meiosis. Nat Commun 5:4580. doi:10.1038/ncomms5580

Martínez M, Cuñado N, Carcelén N, Romero C (2001a) The *Ph1* and *Ph2* loci play different roles in the synaptic behaviour of hexaploid wheat *Triticum aestivum*. Theor Appl Genet 103:398–405

Martínez M, Naranjo T, Cuadrado C, Romero C (2001b) The synaptic behaviour of *Triticum turgidum* with variable doses of the *Ph1* locus. Theor Appl Genet 102:751–758

Martinez M, Cuadrado C, Laurie DA, Romero C (2005) Synaptic behaviour of hexaploid wheat haploids with different effectiveness of the diploidizing mechanism. Cytogenet Genome Res 109:210–214. doi:10.1159/000082402

Martínez-Pérez E, Shaw P, Reader S, Aragón-Alcaide L, Miller T, Moore G (1999) Homologous chromosome pairing in wheat. J Cell Sci 112:1761–1769

Martínez-Pérez E, Shaw P, Moore G (2001) The *Ph1* locus is needed to ensure specific somatic and meiotic centromere association. Nature 411:204–207

Martínez-Pérez E, Shaw P, Aragón-Alcaide L, Moore G (2003) Chromosomes form into seven groups in hexaploid and tetraploid wheat as a prelude of meiosis. Plant J 36:21–29

McGuire PE, Dvořák J (1982) Genetic regulation of heterogenetic chromosome pairing in polyploid species of the genus *Triticum* sensu lato. Can J Genet Cytol 24:57–82

Mello-Sampayo T (1971a) Genetic regulation of meiotic chromosome pairing by chromosome 3D of *Triticum aestivum*. Nat New Biol 230:22–23

Mello-Sampayo T (1971b) Promotion of homoeologous pairing in hybrids of *Triticum aestivum*×*Aegilops longissima*. Genet Iber 23:1–9

Mello-Sampayo T, Canas AP (1973) Suppressors of meiotic chromosome pairing in common wheat. In: Sears ER, Sears LMS (eds) Proc 4th Int Wheat Genet Symp. University of Missouri, Columbia, MO, pp 675–684

Mikhailova EI, Naranjo T, Shepherd K, Wennekes-van Eden J, Heyting C, de Jong JH (1998) The effect of the wheat *Ph1* locus on chromatin organisation and meiotic chromosome pairing analysed by genome painting. Chromosoma 107:339–350

Mochizuki A (1964) Further studies on the effect of accessory chromosomes on chromosome pairing in *Aegilops mutica*. Jpn J Genet 39:356–362

Morais L, Queiroz A, Viegas W, Roca A, Naranjo T (1992) Synaptonemal complex formation and metaphase I bond distribution at low temperature in nullisomic 5D-tetrasomic 5B hexaploid wheat. Genome 35:498–502

Motsny II, Simonenko VK (1996) The influence of *Elymus sibiricus* L. genome on the diploidization system of wheat. Euphytica 91:189–193

Mursal IEJ, Endrizzi JE (1976) A reexamination of diploid-like meiotic behaviour of polyploid cotton. Theor Appl Genet 47:171–178

Naranjo T (1992) The use of homoeologous pairing in the identification of homoeologous relationships in Triticeae. Hereditas 116:219–223

Naranjo T, Corredor E (2008) Nuclear architecture and chromosome dynamics in the search of the pairing pattern in meiosis in plants. Cytogenet Genome Res 120:320–330
Naranjo T, Maestra B (1995) The effect of *ph* mutations on homoeologous pairing in hybrids of wheat with *Triticum longissimum*. Theor Appl Genet 91:1265–1270
Naranjo T, Roca A, Goicoechea PG, Giráldez R (1987) Arm homoeology of wheat and rye chromosomes. Genome 29:873–882
Naranjo T, Roca A, Goicoechea PG, Giráldez R (1988) Chromosome structure of common wheat: genome reassignment of chromosomes 4A and 4B. In: Miller TE, Koebner RMD (eds) Proc 7th Int Wheat Genet Symp. Cambridge University, Cambridge, pp 115–120
Nicolas SD, Leflon M, Monod H, Eber F, Coriton O, Huteau V, Chèvre AM, Jenczewski E (2009) Genetic regulation of meiotic cross-overs between related genomes in *Brassica napus* haploids and hybrids. Plant Cell 21:373–385
Okamoto M (1957) Asynaptic effect of chromosome V. Wheat Info Serv 5:6
Ortega S, Prieto I, Odajima J, Martin A, Dubus P, Sotillo R et al (2003) Cyclin-dependent kinase 2 is essential for meiosis but not for mitotic cell division in mice. Nat Genet 35:25–31
Ozkan H, Feldman M (2001) Genotypic variation in tetraploid wheat affecting homoeologous pairing in hybrids with *Aegilops peregrina*. Genome 44:1000–1006. doi:10.1139/g01-100
Petersen G, Seberg O, Yde M, Berthelsen K (2006) Phylogenetic relationships of *Triticum* and *Aegilops* and evidence for the origin of the A, B, and D genomes of common wheat (*Triticum aestivum*). Mol Phylogenet Evol 39:70–82
Rajhathy T, Thomas H (1972) Genetic control of chromosome pairing in hexaploid oats. Nat New Biol 239:217–219
Riley R (1966a) Genotype-environmental interaction affecting chiasma frequency in *Triticum aestivum*. Chrom Today 1:57–65
Riley R (1966b) The genetic regulation of meiotic behaviour in wheat and its relatives. In: MacKey J (ed) Proc 2nd Int Wheat Genet Symp, Lund, Sweden. Hereditas 2(Suppl):395–408
Riley R, Chapman V (1958) Genetic control of the cytologically diploid behaviour of hexaploid wheat. Nature 182:713–715
Riley R, Chapman V (1964) The effect of the deficiency of the long arm of chromosome 5B on meiotic pairing in *Triticum aestivum*. Wheat Info Serv 17:12–15
Riley R, Chapman V (1967) Effect of 5BS in suppressing the expression of altered dosage of 5BL on meiotic chromosome pairing in *Triticum aestivum*. Nature 216:60–62
Riley R, Kempana C (1963) The homoeologous nature of the non-homologous meiotic pairing in *Triticum aestivum* deficient for chromosome V (5B). Heredity 18:287–306
Riley R, Kimber G, Chapman V (1961) Origin of genetic control of diploid-like behaviour of polyploid wheat. J Hered 52:22–25
Riley R, Chapman V, Young RM, Belfield AM (1966) Control of meiotic chromosome pairing by the chromosomes of homoeologous group 5 of *Triticum aestivum*. Nature 212:1475–1477
Riley R, Chapman V, Johnson R (1968) The incorporation of alien disease resistance in wheat by genetic interference with the regulation of meiotic chromosome synapsis. Genet Res Camb 12:199–219
Roberts MA, Reader SM, Dalgliesh C, Miller TE, Foot TN, Fish LJ, Snape JW, Moore G (1999) Induction and characterization of *Ph1* wheat mutants. Genetics 153:1909–1918
Romero C, Lacadena JR (1980) Interaction between rye B-chromosomes and wheat genetic systems controlling homoeologous pairing. Chromosoma 80:33–48
Roothan M, Sybenga J (1976) No 5B compensation of rye B-chromosomes. Theor Appl Genet 48:63–66
Sánchez-Morán E, Benavente E, Orellana J (2001) Analysis of karyotypic stability of homoeologous-pairing (*ph*) mutants in allopolyploid wheats. Chromosoma 110:371–377
Sánchez-Morán E, Santos JL, Jones GH, Franklin FCH (2007) ASY1 mediates AtDMC1-dependent interhomolog recombination during meiosis in *Arabidopsis*. Genes Dev 21:2220–2233
Sears ER (1954) The aneuploids of common wheat. MO Agric Exp Stn Res Bull 572:58
Sears ER (1976) Genetic control of chromosome pairing in wheat. Annu Rev Genet 10:31–51

Sears ER (1977) An induced mutant with homoeologous pairing in common wheat. Can J Genet Cytol 19:585–593

Sears ER (1982) A wheat mutation conditioning an intermediate level of homoeologous chromosome pairing. Can J Genet Cytol 24:715–719

Sears ER (1984) Mutations in wheat that raise the level of meiotic chromosome pairing. In: Gustafson JP (ed) Gene manipulation in plant improvement. Proc 16th Stadler Genet Symp. Plenum Press, New York, NY, pp 295–300

Sears ER, Okamoto M (1958) Intergenomic chromosome relationship in hexaploid wheat. In: Proc 10th Int Congr Genetics. Toronto, ON: University of Toronto Press, 258–259

Sechnyak AL, Simonenko VK (1991) The crossability of *Aegilops crassa* Boiss. with *Triticum aestivum* L. and meiosis features in the hybrids. Nauch Techn Bull BSGI 2:10–14, Russian

Sidhu GK, Rustgi S, Shafqat MN, von Wettstein D, Gill KS (2008) Fine structure mapping of a gene-rich region of wheat carrying *Ph1*, a suppressor of crossing over between homoeologous chromosomes. Proc Natl Acad Sci U S A 105:5815–5820

Simonenko VK, Motsny II, Sechnyak AL, Kulbida MP (1998) The use of wheat-alien and *Aegilops*-rye amphiploids for introgression of genetic material to wheat. Euphytica 100:313–321

Sutton T, Whitford R, Baumann U, Dong C, Able JA, Langridge P (2003) The *Ph2* pairing homoeologous locus of wheat (*Triticum aestivum*): identification of candidate meiotic genes using a comparative genetics approach. Plant J 36:443–456

Tam SM, Hays JB, Chetelat RT (2011) Effects of suppressing the DNA mismatch repair system on homeologous recombination in tomato. Theor Appl Genet 123:1445–1458

Taylor IB, Evans GM (1977) The genotypic control of homoeologous chromosome association in *Lolium temulentum* × *Lolium perenne* interspecific hybrids. Chromosoma 62:57–67

The International Wheat Genome Sequencing Consortium (IWGSC) (2014) A chromosome-based draft sequence of the hexaploid bread wheat (*Triticum aestivum*) genome. Science 345:1251788

Thomas H, Powell W, Aung T (1980) Interfering with regular meiotic behaviour in *Avena sativa* as a method of incorporating the gene for mildew resistance from *A. barbata*. Euphytica 29:635–640

Tsunewaki K, Wang GZ, Matsuoka Y (2002) Plasmon analysis of *Triticum* (wheat) and *Aegilops*. 2. Characterization and classification of 47 plasmons based on their effects on common wheat phenotype. Genes Genet Syst 77:409–427

Upadhya MD, Swaminathan MS (1967) Mechanism regulating chromosome pairing in *Triticum*. Biol Zentralbl Suppl 86:239–255

Valenzuela NT, Perera E, Naranjo T (2012) Dynamics of rye chromosome 1R regions with high and low crossover frequency in homology search and synapsis development. PLoS One 7(4):e36385. doi:10.1371/journal.pone.0036385

Vardi A, Dover GA (1972) The effect of B chromosomes on meiotic and pre-meiotic spindles and chromosome pairing in *Triticum*/*Aegilops* hybrids. Chromosoma 38:367–385

Viegas WS (1980) The effect of B-chromosomes of rye on the chromosome association in F_1 hybrids *Triticum aestivum* × *Secale cereale* in the absence of chromosomes 5B or 5D. Theor Appl Genet 56:193–198

Viera A, Rufas JS, Martínez I, Barbero JL, Ortega S, Suja JA (2009) CDK2 is required for proper homologous pairing, recombination and sex-body formation during male mouse meiosis. J Cell Sci 122:2149–2159

Wall AM, Riley R, Chapman V (1971a) Wheat mutants permitting homoeologous chromosome pairing. Genet Res 18:311–328

Wall AM, Riley R, Gale MD (1971b) The position of a locus on chromosome 5B of *Triticum aestivum* affecting homoeologous meiotic pairing. Genet Res 18:329–339

Wang G, Xue X, Wang T, Xu X, Li J (1991) Study on obtaining and cytogenetics of octoploid triticales with different alien cytoplasms. Chinese J Genet 15:235–244

Wang T, Liu W, He C, Zhang G, Shi R, Xu X, Li J (1999) Effect of alien cytoplasms on meiosis in the F1 hybrids between alloplasmic lines of *Triticum aestivum* cv. Chinese Spring and rye or triticale. Cytologia 64:285–291

Wang RRC, Li XM, Hu ZM, Zhang JY, Larson SR, Zhang XY, Grieve CM, Shannon M (2003) Development of salinity-tolerant wheat recombinant lines from a wheat disomic addition line carrying a *Thinopyrum junceum* chromosome. Int J Plant Sci 164:25–33

Wendel JF, Cronn RC (2003) Polyploidy and the evolutionary history of cotton. Adv Agron 78:139–186

Whitford R, Baumann U, Sutton T, Gumaelius L, Wolters P, Tingey S, Able JA, Langridge P (2007) Identification of transposons, retroelements, and a gene family predominantly expressed in floral tissues in chromosome 3DS of the hexaploid wheat progenitor *Aegilops tauschii*. Funct Integr Genomics 7:37–52

Xiang ZG, Liu DC, Zheng YL, Zhang LQ, Yan ZH (2005) The effect of *phKL* gene on homoeologous pairing of wheat-alien hybrids is situated between gene mutants of *Phl* and *Ph2*. Hereditas-Beijing 27:935–940. Chinese

Yang GH, Yang XM, Wang RH, Gao AN, Li LH, Liu WH (2010) The inhibiting effect of 1·4 recombinant P chromosome of wheat-*Agropyron cristatum* addition line on the *Ph* gene. Chinese Sci Bull 55:153–157

Yousafzai FK, Al-Kaff N, Moore G (2010) Structural and functional relationship between the *Ph1* locus protein 5B2 in wheat and CDK2 in mammals. Funct Integr Genomics 10:157–166

Zhang H, Bian Y, Gou X, Dong Y, Rustgi S, Zhang B, Xu C, Li N, Qi B, Han F, von Wettstein D, Liu B (2013) Intrinsic karyotype stability and gene copy number variations may have laid the foundation for tetraploid wheat formation. Proc Natl Acad Sci U S A 110:19466–19471

Chapter 7
Introgressions Between Wheat and Rye

Adam J. Lukaszewski

7.1 Introduction

Rye is a distant relative of wheat; it is estimated that the spilt from the common ancestor took place about 3.5 million years ago (Middleton et al. 2014). Within the genus Secale, *S. cereale*, or common rye, is the only widely cultivated species, especially in Central and Northern Europe where it is the crop of choice for light soils. It is valued for its hardiness and tolerance to many biotic and abiotic stresses. While its general quality/utility is nowhere near to that of wheat, it can be reliably grown on much more demanding soils and in harsher environments. As such, it has always been viewed with much envy by wheat breeders and many efforts have been made to utilize its gene pool for wheat improvement. Perhaps the most comprehensive of these efforts is the creation of triticale, a man-made amphiploid combining entire genomes of diploid rye and either two (AB) or three (ABD) genomes of hexaploid wheat. Tetraploid triticales, combining one genome each of wheat and rye, and produced in several different ways, have been created in several places but do not appear to offer a perspective as a crop. Of the three ploidy levels, hexaploid triticale, genomes AABBRR with possible modifications, is successful, with yields often exceeding wheat, at least on lands marginal for wheat; production. It is an excellent feed crop with several still unresolved issues in the context of human consumption.

This chapter, however, does not deal with issues of introgressions of entire genomes, and hence, triticale is beyond its scope. The chapter deals with introgressions on a sub-genome scale, such as chromosomes and chromosome segments. The author does not have the ambition of presenting a comprehensive literature review on the topic but rather to present and discuss general concepts and the state

A.J. Lukaszewski (✉)
University of California, Riverside, CA, USA
e-mail: adam.lukaszewski@ucr.edu

M. Molnár-Láng et al. (eds.), *Alien Introgression in Wheat*,
DOI 10.1007/978-3-319-23494-6_7

of the art when it comes to chromosome manipulations. The body of literature is very large and this chapter cannot possibly aim to mention every contribution ever made. It attempts, however, to present the most important points.

Specific potential advantages of rye in wheat breeding, allocated to individual chromosomes and even chromosome arms, were well reviewed and summarized by Zeller and Hsam (1983). Since then the number of identified characteristics useful in wheat breeding must have doubled, many of those were allocated to chromosomes, chromosome arms, and even chromosome segments. The list is now long and it includes resistance to various diseases and pests such as leaf, stem, and yellow rust, powdery mildew (Ren et al. 2009; Yang et al. 2009; An et al. 2013) barley yellow dwarf virus (Nkongolo et al. 1992), nematodes (Asiedu et al. 1990), several destructive aphid species (Lukaszewski et al. 2001; Crespo-Herrera et al. 2013), Hessian fly (Friebe et al. 1990; Mukai et al. 1993), better zinc and copper efficiency (Schlegel et al. 1991b, 1998), restoration of male fertility in *timopheevi* cytoplasm (Curtis and Lukaszewski 1993), and many more, even including potential weed control (Bertholdsson et al. 2012). A very specific class of newly identified parameters are the characteristics of the root system, such as those encoded by loci on rye chromosome arms 1RS and 2RS (Ehdaie et al. 2003; Waines and Ehdaie 2007; Sharma et al. 2011). The 1RS arm has already made an immense contribution to wheat production worldwide, even though it has only recently been recognized as such (Waines and Ehdaie 2005).

In most cases, the term "alien introgression," as awkward as it is, is reserved for chromatin transfers into cultivated crops from its wild relatives. Depending on the genetic distance and the isolation barriers separating the donor and the recipient, the issue may be as simple as the cross-backcross approach combined with selection for the desired allele, or as complicated and tedious as enforcement of crossing over between distantly related chromosomes, selection of translocation breakpoints in the desired intervals and assembly of intercalary alien segments in recipient wheat chromosomes. Because donors are wild, undomesticated species, the issue of the so-called linkage drag is always present; the transfers have to involve as little chromatin as possible to exclude any flanking loci with potential negative effects such as weediness. In this sense, rye, a cultivated species itself, would appear to be an easier source, as domestication has already removed most of its undesirable wild characteristics and one could perhaps assume that the precision required for the introgressions would be less. Unfortunately, experience so far shows that this is not always the case.

While it is generally believed that the transfers should be as small as possible to minimize linkage drag, linkage drag is not always bad, and unexpected benefits may appear from larger introgressions. At times, unknown and unrecognized alien loci linked to the targeted ones, in the background of cultivated wheat may produce astounding effects. It is more than likely that the poster child of alien introgressions into a crop, the wheat–rye translocation 1RS.1BL in wheat, appeared under selection pressure for disease resistance (three cereal rust and one powdery mildew resistance loci on the 1RS arm). However, the translocation really took off globally well after some or all these resistance genes had broken down. We now know that the introgressed rye chromosome arm also carries a locus affecting root biomass in wheat (Ehdaie et al. 2003) and its presence positively affects yield, especially under drought stress (Rajaram et al. 1983; Villareal et al. 1998). A similar yield effect was observed in wheat with an introgression from an Agropyron species (Singh et al. 1998), where the responsible genes have already been identified and cloned (Placido et al. 2013).

The techniques for transfers of chromatin from related species into crops are not new (see Jiang et al. 1994) but only crossing over is capable of producing compensating chromosome translocations with predictable regularity. It requires certain minimum lengths and levels of DNA sequence homology so it is unlikely to take place between unrelated chromosomes or in non-corresponding positions. However, the effect of the *Ph1* locus, or, rather, its absence when homoeologous recombination is induced, is the unknown and unquantified factor. Crossing over in the absence of *Ph1*, especially that involving homoeologues, may be different not only quantitatively but also qualitatively from the very stringent exchanges taking place with *Ph1* present. Be it as it may, crossing over still appears as the most reliable approach to production of alien introgressions. E.R. Sears who invented the approach considered it the easiest (Sears 1972) but it might have been overly optimistic. The approach is very demanding in terms of labor and resources. Because pairing and recombination of alien chromosomes with their wheat homoeologues is infrequent, and because linkage drag demands crossover points in specific, narrow intervals, there is no way around large populations. Regardless of what technique is used, whether it is cytogenetic or DNA marker, whether plants are bulked or analyzed individually, thousands must be screened to assure proper precision. Only our Chinese colleagues are lucky to find perfect centric translocation from monosomic additions, and introgressions of small intercalary segments without ever manipulating chromosome pairing (Ren et al. 2009; Fu et al. 2010). For everybody else, the prospect of sifting through large populations is intimidating and so over the years many other options have been tried. These include random chromosome fragmentation by irradiation or gametocidal chromosomes (Massoudi-Nejad et al. 2002) and even karyotypic aberrations occurring in tissue culture (Lapitan et al. 1984). Unfortunately, these almost always end up with predictable results: because the points of breakage (and then fusion) are random the resulting translocations are not compensating. They may create pretty pictures but not particularly useful products with acceptable agronomic value. There is hope that new molecular techniques will alleviate the problem (Wulf and Moscou 2014), but this may be long in coming. For the time being the GMO label is not welcome in many regions of the world and even if and when it starts to be ignored, there will still be the issue of identification, isolation, and the ownership of specific genes in alien species (relative to wheat).

Chromosome engineering is commonly associated with methods of cytology and long hours over the microscope. In fact, all chromatin manipulations as described here can be performed just as well, if not better, using the techniques of molecular biology, but usually far more expensively. Until we learn how to manipulate crossing over for practical purposes, population sizes will remain large regardless of techniques used to screen them.

7.2 Addition Lines

If introgression is understood as insertion of alien chromatin into the genome of another species, here in a crop such as wheat, addition lines do not entirely fit into the picture. Addition lines, as the name implies, are sets of lines where single chromosomes of a donor are ADDED to the genome of the recipient. Addition lines can be monosomic or disomic, depending on the dosage of the added chromosome. Added chromosomes in disomic additions are transmitted to progeny with sufficient fidelity for easy maintenance, but insufficient fidelity to be of any agronomic importance. In research, they are routinely tested for chromosome numbers or chromosome constitution to make sure that the alien pair is indeed present; in large scale grow-outs they have a tendency to quickly lose added chromosomes.

That addition of alien chromosomes to the wheat genome is technically possible was first demonstrated by O'Mara (1940) who created several disomic additions of cv. 'Imperial' rye to wheat. This set was completed only many years later (Driscoll and Sears 1971) and it became a standard tool in wheat and rye genetics. A wonderful tool it is indeed: it separates the entire genetic contents of a donor species into individual linkage groups (chromosomes). Using it, chromosome location of specific loci can be determined and specific individual chromosomes can then be manipulated to generate more precise transfers into the genomes of crops. Once the chromosome location of a locus of interest is determined, the addition lines are a convenient starting point for the introgression work, to create substitution, translocation or recombinant lines. So, while addition lines are not introgressions in the true sense, they are often the first step in the process of creating introgressions and hence appear worthy of some space here. In a feat surely never anticipated by O'Mara or Sears, such added chromosomes can now be sorted by flow cytometry, creating large samples of DNA from specific chromosomes or chromosome arms.

In the standard O'Mara approach, addition lines are created by backcrossing a wheat–alien species hybrid (F_1 or an amphiploid) to wheat to create a heptaploid where the entire alien genome is present in a single dose. For a wheat–rye situation the crossing scheme is: AABBDD×RR=ABDR, chromosome doubling to AABBDDRR, backcross to wheat to create a heptaploid AABBDDR. This heptaploid is backcrossed to wheat again, plants with 42+1 chromosomes (single rye chromosomes present) selected and self-pollinated to create disomic additions. There are two difficult steps in this procedure, one of which appears to have been solved by technology. The first is identification of 42+1 plants with each of the seven rye chromosome present at least once. It used to be a very demanding task; it was simplified once various chromosome identification techniques entered the field (cytological or DNA markers). The second difficult step is selection of disomic additions among progenies of monosomic additions. In wheat, monosomic (unpaired) chromosomes are included into gametes with ca. 25 % frequency (Sears 1954). Competition for fertilization on the male side strongly favors balanced, 21-chromosome gametes. As a consequence, the recovery rate of disomics is only in the range of 1 %. For a 95 % probability of recovering a disomic, ca. 300 progeny have to be screened for each chromosome, or 2100 for the entire set (for a seven-chromosome genome). Needless to say, there are few volunteers for this job and for many years the number of complete sets of wheat–rye addition lines was very low: Chinese Spring-Imperial, Chinese Spring-King II, Holdfast-King II and that was about it.

Attempts were made to circumvent the gametic selection process. E.R. Sears recommended avoiding pollen competition by pollination with single pollen grains but never tried it himself and did not find volunteers to try it for him. Pollen competition can be minimized by increasing the proportion of pollen with abnormal chromosome constitution: rather than selecting disomic additions from monosomic additions, progenies of multiple monosomic additions can be screened or the first backcross can be made using the heptaploid (here AABBDDR) as male in a backcross to the amphiploid. In each case the proportion of abnormal gametes (21 + 1 chromosomes) will be much higher, and the proportion of euploid gametes will much lower, than among those produced by a monosomic addition, and they would be at a lower competitive disadvantage. Progenies are screened and plants with pairs of individual chromosomes are selected and grown. After that it is only a matter of time (generations) before monosomic chromosomes are eliminated, leaving the selected disomes. This system seriously reduces the required amount of labor but at the same time demands precise chromosome monitoring as unpaired chromosomes are susceptible to centric misdivision and may form translocations. Several sets of additions were produced in this manner by the author, including BH1146-Blanco (BH1146 is a spring wheat; from Brazil known for high resistance to soil aluminum), CS-Blanco and CS-*Haynaldia villosa* (Lukaszewski 1988).

A different way of mitigating the effects of gametic competition is via haploidization. If an assumption is made that the presence of an alien chromosome (here a rye chromosome) does not affect gamete's performance in gametogenesis, a quarter of haploids recovered from a monosomic addition would have to carry the added chromosome; upon chromosome doubling, disomic additions are expected. The author is not aware of any serious attempts to generate complete sets of additions in this fashion; an attempt to generate a specific one was unsuccessful, apparently because the added chromosome did affect gamete's behavior (B. Friebe pers. comm.).

There is no general system of keeping track of wheat–rye addition sets available around the world right now but the number must have grown to at least a dozen. The author himself created six such sets (CS-Blanco, BH1146-Blanco, Pavon-Salvo, Pavon-Presto, Henika-Salvo, Henika-ANOAS; the last four using hexaploid triticales as the starting point); recently a set of additions from *S. africanum* was described (Lei et al. 2013).

7.3 Chromosome Substitutions

Single chromosome substitutions are usually the first step in introgressions of specific segments of chromatin into the wheat genome. The author is not aware of any whole chromosome substitutions in commercial wheats, but this may only reflect

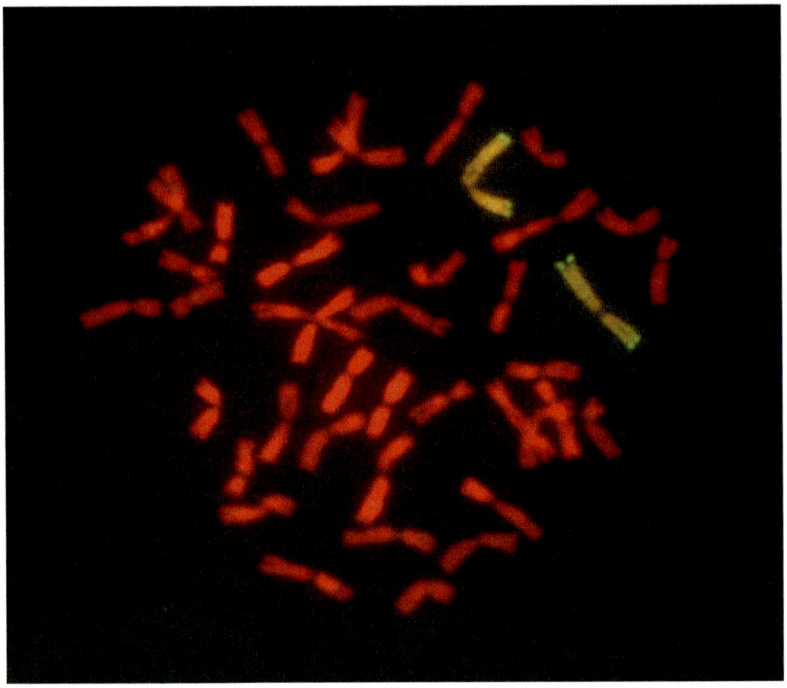

Fig. 7.1 Substitution of rye chromosome 2R for wheat chromosome 2B in hexaploid wheat. Here, 2R was reconstituted from two centric translocations, 2RS.2BL and 2BS.2RL. Rye chromatin labeled *green*; wheat chromatin *red*

his ignorance. The history of the 1RS.1BL translocation includes a cultivar with the 1R(1B) substitution (Zeller and Hsam 1983) but apparently this was not good enough to withstand the selection pressures in wheat breeding. This is perhaps because substitution of the long arm of rye 1R for any of the group-1 long arms of wheat reduces seed set and makes plants appear weaker, thinner, more slender with clearly smaller grains and this effect extends to whole chromosome substitutions in homoeologous group 1 (Kim et al. 2004).

Chromosome substitutions (Fig. 7.1) helped allocate individual chromosomes of rye to wheat homoeologous groups. In most cases, a rye chromosome compensates (that is, is homoeologous to) even if incompletely, for just one homoeologous group of wheat, but within that group it may show more affinity to one chromosome than to another (Sears 1968).

Exceptions are rye chromosomes 4 and 7 where each one of them shows some affinity to both wheat; homoeologous groups 4 and 7 (Zeller and Hsam 1983). Chromosome rearrangements during rye evolution after the wheat–rye split differentiated the ryegenome from those of wheat to a considerable extent. DNA markers made mapping of these rearrangements possible and a detailed map of individual segments' affinities is available (Devos et al. 1993).

Whole chromosome substitutions can be generated in several different ways. If disomic additions are available, they can be crossed as male to monosomics or nullisomics of wheat, creating either double monosomics (one rye and one wheat chromosome present in the targeted homoeologous group), or monosomic substitutions of the alien chromosome. The latter are easier to work with as upon self-pollination they quite readily generate disomic substitutions (about 25 % frequency, consistent with the 25 % inclusion rate of monosomics in the gametes and strong selection for euploid gametes on the male side, Sears 1954). Unfortunately, only 12 nullisomics of wheat are available, and almost all of them in cv. Chinese Spring. If double monosomics are created (one alien chromosome and one wheat homoeologue), the issue of competition for fertilization between the two chromosomes comes into play and with few exceptions (the so-called cuckoo chromosomes), alien chromosomes tend to lose. No matter, this approach to production of alien chromosome substitutions can be called "direct" as specific chromosomes are placed into specific positions in one of the three genomes of wheat, one chromosome at a time. A less direct approach is by intercrossing, say, hexaploid triticale, genomic constitution AABBRR or nearby, with hexaploid wheat, creating a hexaploid hybrid AABBDR. Upon self-pollination a series of chromosome substitutions can be created, with both single and multiple substitutions possible. The bulk of substitutions will be for the D-genome chromosomes, but exceptions to this are possible and not infrequent. The success rate for any individual rye chromosome depends on its compensating ability for the corresponding D-genome chromosome. Sets of substitutions were developed in this manner by Shchapova and Kravtsova (1982), Merker (1984), Friebe and Larter (1988), Alkhimova et al. (1999), and Silkova et al. (2006); more are probably available but not published on. This process of producing substitutions could perhaps be simplified by backcrossing the F_1 as male to triticale, selection of plants disomic for specific rye chromosomes followed by several generations of self-pollination to eliminate monosomics. A somewhat hybrid approach is by intercrossing sets of monosomics/nullisomics of wheat with wheat–rye amphiploids (say, an octoploid AABBDDRR) and sorting out chromosome constitutions after several cycles of self-pollination. For any specific targeted substitution, the original cross would create a double monosomic (one wheat and one rye chromosome) and with sufficient homoeology (compensating ability) of the rye chromosome, a substitution is possible. Again, this approach demands reliable chromosome identification and patience. An added advantage using monosomics/nullisomics is that should any centric translocations occur, they should involve wheat chromosomes from the donor plant (here the octoploid) and not the nullisomic/monosomic parent. In this sense, involvement of chromosomes of Chinese Spring, not the most desirable genetic background in wheat breeding, can be easily avoided. The issue is not trivial: once a centric translocation is formed, the proximal part of the wheat chromosome arm will remain unchanged by crossing over, as crossing over in wheat is limited to distal halves of the arms.

7.4 Centric Translocations

A note on terminology: it has been proposed to describe centric translocations in wheat by listing the recipient arm first and the donor second, separated by a ".". Following this rule, the most common wheat–rye translocation in wheat should be denoted as T1BL.1RS, and often is. This system of notation is background dependent: when a wheat–rye translocation is transferred from wheat to triticale or to rye, the background changes and the order of arms should be flipped, to T1RS.1BL. In other words, the notation depends not on the arm composition of the translocation but on the genetic background in which it is present. Since the author works with both wheat and triticale and encountered this problem numerous times, in this chapter the notation will list the short arms first, followed by the long arms that is, 1RS.1BL. This still leaves unresolved the issue of the homoeo-isochromosomes, such as, say, 1BL.1RL.

The poster child of rye introgressions into wheat is the 1RS.1BL translocation. It involves entire chromosome arms, 1RS of rye and 1BL of wheat. With the breakpoint in the centromere region it must have formed by misdivision of univalents and fusion of telocentric chromosomes. Its history has been presented by Zeller and Hsam (1983) and recently reviewed by Mujeeb-Kazi et al. (2013) with heavy emphasis on the contribution from CIMMYT, so it will be outlined here only briefly. The translocation appeared among progenies of hexaploid triticale × wheat hybrids selected for "triticale-like resistance to wheat diseases." It involves a B-genome chromosome of wheat and not 1D, as the pedigree would dictate, but such shifts have been noted and may be a consequence of reduced chromosome pairing in hybrids. The rye chromosome arm (1RS) carried into wheat four loci for resistance to wheat fungal diseases: *Lr26*, *Yr9m*, *Sr31*, and *Pm9*. The translocation has spread to breeding programs and commercial wheats all over the world (Rabinovich 1998). In compendium assembled by Schlegel (http://www.rye-gene-map.de/rye-introgression/index.html) among lines released since 2000 (not counting germplasm releases and experimental lines), ca. 30 % carry the 1RS.1BL translocation. Among US wheat tested in recent nurseries, about 12–13 % carry the 1RS.1BL translocation (http://www.ars.usda.gov/Research/docs.htm?docid=11932). This is almost 50 % increase from 7.1 % of such lines in wheat yield test nurseries in 1989 (Lukaszewski 1990). It is interesting that most of the spread of the translocation occurred when, or after, the rye resistance genes have broken down and contributed little value in breeding. Measurable yield effects have been associated with this translocation, even in absence of disease pressure (Rajaram et al. 1983) but the results of individual trials, in different environments and in different sets of wheats were at times confusing, with some experiments showing a clear grain yield benefit of the translocation (Carver and Rayburn 1994; Villareal et al. 1997, 1998; Kumlay et al. 2003) while others showed no such effect (McKendry et al. 1996; Bullrich et al. 1998; Singh et al. 1998; Espita-Rangel et al. 1999a, b; Lelley et al. 2004). Eventually, it was discovered that the rye chromosome arm 1RS carries a locus for increased root biomass in wheat (Ehdaie et al. 2003; Waines and Ehdaie 2005,

2007) and this may explain confusing results of individual experiments: the yield effect was present, or was more pronounced only in some, usually more demanding environments (Villareal et al. 1997, 1998) and in spring rather than winter wheats. Perhaps for this reason the spread of the translocation has accelerated after the involvement of CIMMYT (Rajaram et al. 1983) as it introduced the translocation to spring wheats with a global reach and its benefits became more pronounced. The translocation has also been transferred to durum wheat, by backcrosses (Friebe et al. 1987) where it appears to have a similar positive effect on grain yield (Villareal et al. 1997).

While the positive effect of the 1RS.1BL translocation on grain yield is indisputable, so is its negative effect on bread making quality. It is not entirely clear whether this negative effect is due strictly to the presence of the rye secalin locus *Sec-1*on 1RS, or the absence of wheat storage protein loci on 1BS (*Glu-B1*, *Glu-B3* plus some others) or the combination of both factors. In pairs of isogenic lines that differ by the presence/absence of the translocation, the presence of *Sec-1* is always associated with the absence of all 1BS-located wheat storage protein loci, and vice versa. Hence, the actual contribution of each locus cannot be measured. However, tests of sets of isogenic lines indicated that the negative impact of rye secalin (*Sec-1*) presence is greater than that of the absence of wheat loci (Kumlay et al. 2003) and of the three possible positions of 1RS in the wheat genome, that in the D-genome (translocation 1RS.1DL) was the most detrimental to bread making quality while 1RS.1AL was the least. This position effect of course measures relative contribution of the storage protein loci on the short arms of group-1 chromosomes to the overall bread making quality, as the *Sec-1* locus is always the same.

Sebesta and Wood (1978) created another centric translocation, 1RS.1AL, to transfer into wheat a locus for resistance to greenbug, a serious pest of wheat in the South-Eastern USA. While the translocation was officially produced by irradiation, the fact that it is centric makes it much more likely to be a product of centric misdivision and fusion, very much of the same nature as the 1RS.1BL translocation (Zeller and Fuchs 1983). This translocation was first released in cv. Amigo; since then it has spread to many soft wheats in the USA. Among the 1RS wheats in nurseries listed at http://www.ars.usda.gov/Research/docs.htm?docid=11932 about one half are 1RS.1AL. In the compendium of Schlegel (http://www.rye-gene-map.de/rye-introgression/index.html) about 10 % of cultivars released since 2000 carry 1RS.1AL and it appears that it has begun to spread beyond the USA. This translocation also appears to enhance grain yield in wheat (Villareal et al. 1996; Kumlay et al. 2003; Kim et al. 2004) apparently also by increasing the root biomass (Ehdaie et al. 2003; Waines and Ehdaie 2007). Its impact on breadmaking quality appears to be less severe than that of the 1RS.1BL translocation (Espita-Rangel et al. 1999a, b).

Translocation 1RS.1BL served for many cytogenetic exercises, including untranslocating the two arms, moving the 1RS arm to each of the three wheat genomes and re-translocating it again to each of the long arms of group-1 homoeologues in Pavon 76 wheat. In the end, a set of isogenic lines was produced (Fig. 7.2) where the same chromosome arm 1RS, as it was originally derived from rye cv. Petkus (Zeller

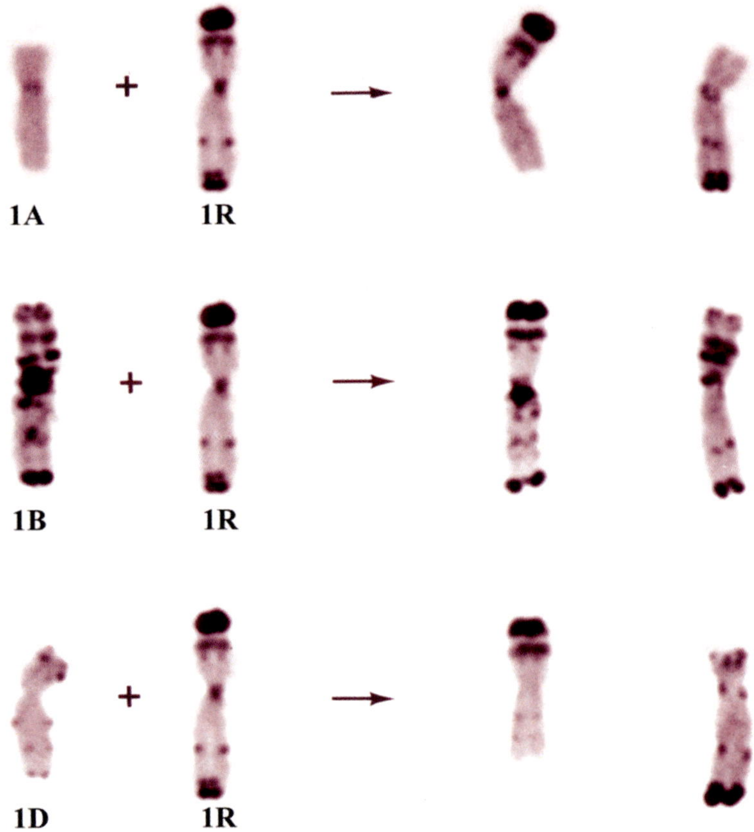

Fig. 7.2 A complete set of centric translocation made from group-1 homoeologues of wheat cv. Pavon 76 and rye chromosome 1R reconstructed from translocation 1RS.1BL of the Aurora/Kavkaz origin and 1BS.1RL

and Hsam 1983), was placed in translocations 1RS.1AL, 1RS.1BL (1BL arm of Pavon 76, hence different from that in the original translocation as in cv. Aurora and Kavkaz) and 1RS.1DL (Lukaszewski 1993, 1997). This did not affect the disease resistance profiles of the recipient wheat, but affected root biomass: the highest effect was associated with the 1RS.1AL arm combination, less so with the 1RS.1DL combination, and the least with the original arm composition 1RS.1BL. It appears that the combination of loci on 1RS, 1BS, and 1DS produces the best effect as far as root biomass is involved (Ehdaie et al. 2003; Waines and Ehdaie 2007).

The historical account of the origins of the 1RS.1BL translocation of wheat (Zeller and Hsam 1983) shows that originally several different sources of the 1R chromosome were available (however, Schlegel and Korzun (1997) suggest that in fact these sources originated from a common stock). Over time only the Aurora/Kavkaz source spread to breeding programs and commercial wheat the world over. It seems a question worth asking if this specific chromosome arm carries the most desirable alleles, such as disease resistance early on and root biomass to this day, and no better alleles are present in rye. The 1RS from cv. Petkus (in 1RS.1BL from Aurora/Kavkaz) does not carry the greenbug resistance locus present in 1RS from cv. Insave rye in the 1RS.1AL Amigo translocations, and the two 1RS arms carry different rust and powdery mildew resistance loci (Hsam and Zeller 1997; Graybosch et al. 1999). Centric translocation 1RS.1DL produced from the cv. Imperial rye chromosome 1R carries a *Sr* locus allelic to *Sr31* of the Aurora/Kavkaz translocation which still offers good protection against stem rust, including Ug99 (Anugrahwati et al. 2008). Attempts have been made to map/combine some of these alleles into single translocations (Lapitan, personal communication; Hsam et al. 2000). Given the level of heterozygosity in rye, it would appear sensible to assume that more allelic variation for various resistance and the root biomass locus/loci must exists but testing for their presence is somewhat complicated, as it requires sets of wheat–rye additions from wide sources.

Several different experiments with 1RS arms from different sources translocated to different positions in the wheat genome (Ehdaie et al. 2003; Kim et al. 2004; Waines and Ehdaie 2007) suggest that the Amigo-originating 1RS arm (from Insave rye), as well as an arm from a different wheat line apparently unrelated to the Aurora/Kavkaz source (CIMMYT line E12165) produce larger increases in root biomass under water stress than the 1RS arm from the Aurora/Kavkaz (rye cv. Petkus). In addition, there appear to be significant position effects involved, with 1RS.1AL having the best positive effect on root biomass and the least negative effect on the end use quality of wheat. On the other hand, Karki et al. (2014) suggest that the 1RS.1AL translocation where the 1RS arm originates from Aurora/Kavkaz produces larger root biomass only under normal moisture levels, and this increase may come at the cost to grain yield.

There are several 1RS.1AL translocations with 1RS from different sources apart from Amigo (including the original Petkus source, triticale cv. Rhino and the E12165 wheat line from CIMMYT) but these are yet to be tested for root characteristics. The 1RS.1BL configuration involves more sources of 1RS, including six in the possession of the author, one of which includes 1RS from *S. montanum*. Yang et al. (2009) generated a new 1RS.1BL translocation with 1RS from *S. africanum*; another new 1RS.1BL was created from a Korean rye (Ko et al. 2002) and an entire swarm appears to have occurred spontaneously in China and even recombined themselves to produce small segment introgressions (Fu et al. 2010). Ren et al. (2009) generated a new 1RS.1BL translocation with a different profile of disease resistance, somewhat surprisingly as the donor was an inbred of Petkus, the source of the original 1RS.1BL translocation.

Rapid spread of the 1RS.1BL Aurora/Kavkaz translocation clearly points to its agronomic value but this does not mean that this specific 1RS arm carries the most beneficial alleles possible. Only careful screening may uncover the range of available variation. Moreover, research of Waines (Ehdaie et al. 2003; Waines and Ehdaie 2007) suggests that as far as root characteristics are concerned, translocation of 1RS to 1BL may not be the best possible and in this sense the true potential of introgressions of rye 1RS chromosome arm has not yet been completely exploited.

Many other centric wheat–rye translocations have been produced in wheat but none seems to have made it into commercial production even when they offer a good set of disease resistance genes, such as translocations of the long arm of chromosome 2R (Friebe et al. 1990; Hysing et al. 2007). The author himself has all possible centric translocation of rye 1R (both arms, all wheat homoeologues, Fig. 7.2), both 2B-2R and 2AS.2RL (the latter obtained from ER Sears), both 3R-3B, 7DL.4RL, 5BS.5RL, 6BS.6RL and two of 7RL presumably to 7D. Several other translocations, including those with similar structure to the above but from unrelated sources have been described in the literature (Friebe et al. 1996). The old 2AS.2RL translocation with the long arm originating from Imperial rye is an interesting one, as it was shown to reduce yield reduction under water stress (Lahsaiezadeh et al. 1983).

The two widespread wheat–rye translocations, 1RS.1BL and 1RS.1AL are centric, that is, they were formed by misdivision of centromeres and fusion of two chromosome arms broken at the centromere. Both were lucky occurrences; they were not planned or designed. By design, centric translocations can be produced from any two chromosomes as long as each one has a tendency to break at the centromere when present as a univalent in meiosis. The process does not require homology, homoeology, affinity, similarity. Any arm combination can be produced from any two chromosomes; the process of arm fusion is random and so compensating translocations (short arm-long arm, S.L, and long-arm-short arm, L.S) are produced as frequently as the non-compensating combinations (S.S or L.L). Of course, only compensating translocations have a chance of surviving breeding pressures; others can be maintained for experimental purposes, such as homoeo-isochromosomes tested for the mode of action of the *Ph1* locus in wheat (Dvorak and Lukaszewski 2000). Because of sister chromatid cohesion in the first meiotic anaphase, breakpoints are not always located in the centromere itself (understood as the region of the chromosome underlying the kinetochore) but may occur in the regions flanking the centromere (Lukaszewski 2010). Cytogenetic exercises with un-translocating and re-translocating chromosome arms in centric translocation led to introduction of rye centromeres into wheat chromosomes (Zhang et al. 2001); these do not appear to have any detectable effect on the behavior of the chromosomes or plants carrying them.

The two widespread translocations in wheat, 1RS.1BL and 1RS.1AL, occurred in progenies of triticale × wheat hybrids but both appear to be spontaneous events. Both involve wheat chromosomes that should not have been univalent to misdivide. Counting on such occurrences may be disappointing so luck usually needs a bit of help. Centric translocations are relatively easy to produce, if suitable aneuploid lines are available and targeted chromosomes misdivide. It is not clear at all what makes a chromosome susceptible to misdivision, and considerable variation in susceptibility is present. Among wheat group-1 homoeologues in cv. Pavon 76, chromosome 1B misdivides the most frequently; chromosome 1D the least, hinting that chromosome length may be a factor. Chromosomes reconstructed from centric translocations tended to misdivide more often than their original counterparts (Lukaszewski 1997) and over 20 % of screened progeny carried centric misdivision products. In arm-flip exercises with chromosomes 2R and 2B, 35 % of progeny from double monosomics carried at least one misdivision product (Brunell et al. 1999) while for 3R and 3B, this frequency was 6.1 % and only 11 centric translocations (including all possible arm combinations) were recovered from among 1891 plants screened. Chromosome 3B misdivided three times less frequently than chromosome 3R (Lukaszewski unpublished). On the far end of the spectrum are several wheat chromosomes that either do not misdivide in any detectable frequency, or one of their misdivision products cannot be recovered. The same chromosomes from other sources misdivide with usual frequencies. For this reason, E.R. Sears was forced to import into Chinese Spring several telocentrics which could not be produced from Chinese Spring chromosomes (Sears and Sears 1978). Lukaszewski attempted to generate original Chinese Spring telocentrics for these imports but so far, only 6AS was created. Not a single misdivision product was observed among over 500 progeny of a 5D monosomic screened but when 5D from a different source was tested, both telocentrics and isochromosomes of both arms were recovered from a much smaller sample.

7.5 Translocations Produced by Chromosome Fragmentation

If the targeted alien (rye?) chromosome arms are syntenic with their wheat counterparts, and introgressions are produced by controlled means such as centric breakage-fusion of designated chromosomes or by homoeologous recombination, the resulting translocations are genetically balanced and offer a chance of agronomic benefit. Unfortunately, cytogenetic stocks suitable for controlled production of centric translocations are available only in very few genetic backgrounds and these are not necessarily much desired in wheat breeding. Homoeologous recombination, which assures proper placement of alien chromatin in the wheat genome, is even more difficult to handle and always requires screening large populations. For this reason, in times of need, many people resort to random processes. Their advantage is the ease of use; the disadvantage is randomness. While there is a statistical chance of placing the desired segment of alien chromatin in its proper position in the wheat genome, this chance is low. In most cases, the targeted segment is inserted in a random position in the wheat genome, it is non-compensating, and the degree of non-compensation depends on the length of the inserted segment as well as the effect of the absence, if any, of a part of the wheat genome.

Random insertions are produced by fragmentation of the wheat genome together with the alien donor chromosome; DNA repair mechanisms stitch broken pieces together, almost never in the proper order. Therefore, random translocations can only differ in their degree of non-compensation. If done on a large enough scale, there is a chance that the transferred segment will be small and placed in a position in the wheat genome where little harm is done. As a result, the benefit of the introgression may outweigh its possible problems. The first one of such random introgressions was produced by E.R. Sears, to transfer a leaf rust resistance locus from *Aegilops umbellulata* (Sears 1956). This translocation offered substantial agronomic benefit thereby setting the trend. Sears (1993) presented a complete protocol for irradiation-induced chromatin transfers into wheat; unfortunately, many attempts that followed were not nearly as precise and as successful as the original one of Sears (1956).

Two quite popular transfers were nominally by irradiation but ended up as centric translocations. The first was the already mentioned translocation 1RS.1AL in cv. Amigo (Sebesta and Wood 1978) the other is a transfer of a segment of a *Haynaldia villosa* chromosome to wheat by irradiation that started from a centric translocation of an entire short arm of chromosome 6V (Chen et al. 1995). In both cases it is more than likely that the translocations originated by centric misdivision; perhaps irradiation was responsible in some degree for poor chromosome pairing but it is unlikely that it produced the desired chromosome breaks. Follow-up irradiation reduced the *H. villosa* segment to only a fraction of the arm (Chen et al. 2008) but the first results of a recombination study indicate that the remaining small fragment of *H. villosa*chromatin, while residing on the correct wheat chromosome arm, may not be in its correct position (Lukaszewski, unpublished data).

In most cases, however, transfers by random means are truly random and noncompensating, and have little chance for acceptance in commercial agriculture. Many such translocations have been produced in wheat; many are reviewed and illustrated by Friebe et al. (1996). The irradiation approach is usually undertaken to avoid large samples required for transfers via homoeologous recombination but the savings may be illusory. E.R. Sears (1956) selected the *Ae. umbellulata* transfer from among 6000 progeny of irradiated plants. The initial size of the progeny population was not given for the *Pm21* transfer from *H. villosa* but 100 resistant progeny were screened cytologically and several with the centric translocation were identified. For chromosomes with normal misdivision frequency the author usually screens ca. 100 progeny to identify all possible centric fusion combinations and almost always finds them, especially for such readily misdividing chromosomes as 1R or 6V. On the other hand, the 6VS arm recombines with 6A of wheat with ca. 3 % frequency and screening of ca. 1000 progeny of *ph1b*plants yielded 32 recombinants of the *H. villosa* arm. In this sense, standard cytogenetic approaches are not necessarily more laborious.

An interesting translocation, also created by irradiation, was Transec (Driscoll and Anderson 1967). It was created to introduce leaf rust and powdery mildew resistance from rye into wheat but was also hoped to offer a system of hybrid wheat production. Since the translocation removes chromosome arm 4BS carrying a locus critical for male fertility, translocation homozygotes are male sterile. Unfortunately, no reliable system of hybrid wheat production could ever be established based on Transec.

Fragmentation of the donor alien chromosome can also be accomplished by the so-called gametocidal chromosomes (Endo 2003). These chromosomes appear to cut other chromosomes indiscriminately and in this sense they may be no different than irradiation. Therefore, chances of recovery of compensating balanced translocations are about the same. Long series of chromosome deficiencies have been created using the gametocidal chromosomes, not only for wheat but also rye and barley chromosomes in wheat (Massoudi-Nejad et al. 2002; Endo 2003), but the author is not aware of any agricultural significance of the translocations that might have been produced.

Given the predictably high amount of work needed to produce translocations via chromosome pairing and recombination, many keep trying shortcuts, with irradiation being the means of choice. At times it does succeed but the issue brings on an interesting question: what is the minimum acceptable level of chromosome compensation in agricultural practice? Incomplete compensation created by unbalanced chromosome translocation may affect grain yield; what level of grain yield loss can be accepted in exchange for a defined benefit, such as resistance, tolerance or similar? It is this author's experience from years of interactions with breeders that the ultimate criterion deciding of cultivar release/acceptance is grain yield and tests are usually performed in stress-free environments. As a general rule, breeders do not accept yield reduction in exchange for a promise of greater yield stability in the event of an infection, heat wave or similar calamity. This in itself is interesting as in traditional agriculture yield stability was far more important to farmers than a promise of large yields under favorable conditions (Denison 2012).

7.6 Induced Homoeologous Recombination

To those who deal with interspecific introgressions it is obvious that only crossing over has the capacity to deliver the desired segment of alien chromatin into its proper position in the crop's genome. No method based on chromosome fragmentation, physical or genetic, can reliably produce a genetically balanced introgression. Given genome sizes and an almost infinite number of possible break positions, it is practically impossible to place the desired piece of alien chromatin in its correct position in the wheat genome while removing the corresponding portion of the wheat genome to avoid any gene dosage issues. All transfers made by random fragmentation are, by definition, non-compensating and differ only by the degree of non-compensation. The bad ones are rejected outright; the few good ones with minimal adverse effects may be tolerated when the benefit they offer outweigh problems. However, the first alien chromosome engineering effort in wheat was done by irradiation (that is, chromosome fragmentation) and it was accepted in agriculture to considerable benefit (Sears 1956).

Introgressions via chromosome fragmentation seem quick, cheap and effective, but most of the time they generate stocks unsuitable for production agriculture. On the other hand, introgressions via homoeologous recombination are guaranteed laborious and can be very time consuming but once certain assumptions are met, they guarantee the desired outcome. In wheat, the approach via homoeologous recombination requires a method of disabling the *Ph1* system which enforces strictly homologous chromosome pairing, by imposing some criteria of chromosome affinity.

At times, even pairing of homologous chromosomes, such as those in inter-cultivar hybrids, may be restricted by *Ph1* (Dvorak and McGuire 1981).

The *Ph1* system is disabled either by its removal, with two options available, or by its suppression, by specific chromosomes of *Aegilops speltoides* (Feldman and Mello-Sampayo 1967; Dvorak 1972; Chen et al. 1994). The *Ph1* removal options are either nullisomy for chromosome 5B (ditelosomy 5BS would be even better except that it has not yet been produced) or by the *ph1b* mutation (Sears 1975, 1984), which in fact is an intercalary deletion of less than 3 Mb of DNA including the *Ph1* locus (Gill et al. 1993). The other locus of the system, *Ph2* on 3DS, has a minor effect on chromosome pairing and a combination of two mutations, *ph1b* and *ph2b*, does not appear to improve the frequency of rye homoeologous paring in wheat (Ceoloni and Donini 1993), so it does not increase the chances of success while adding an additional degree of complication to the system.

Once the *Ph* system is disabled, homoeologous chromosomes are free to pair and recombine. Their pairing frequencies appear to depend on two factors: the level of affinity and their structural similarity. Chromosomes 1, 3 and 5 from *T. monococcum* pair and recombine with their wheat homoeologous (1A, 3A, 5A, respectively) only rarely in the presence of *Ph1*, but behave like normal homologues in its absence (Luo et al. 1996, 2000). Chromosomes from more distantly related species, such as *S. cereale* rye, do not pair with wheat homoeologues at all when the *Ph* system is present and operational; in its absence individual chromosome arms show surprisingly large differences in pairing (and recombination), ranging from as high as ca. 12–13 % for 1RL or 2RL to essentially zero for chromosome 4R (Naranjo and Fernandez-Rueda 1996). Of the 14 rye chromosome arms, only five (1RS, 1RL, 2RL, 3RS, and 5RS) (Naranjo and Fernandez-Rueda 1991; Devos et al. 1993) are colinear with wheat; the rest are translocated. Presence of such structural changes must severely hinder pairing and recombination to the point where no recombinant chromosomes can be recovered (Dundas et al. 2001; Lukaszewski et al. 2001). Even relatively minor structural differences, such as the presence/absence of telomeric C-bands on rye chromosomes, reduce pairing frequencies (Naranjo and Lacadena 1980; Naranjo and Fernandez-Rueda 1991). It would be an interesting exercise in cytogenetics to test if removal of the translocated segment of an alien homoeologue would permit pairing with its homoeologous segment of a wheat chromosome, so that structural chromosome differences would no longer hinder chromosome engineering in wheat. Such removal could be accomplished by chromosome fragmentation using the gametocidal chromosomes (Endo 2003). The original chromosome structure would then have to be restored by recombination with a normal chromosome, however difficult it could be given the effect of heterozygosity for deficiencies on MI pairing (Curtis et al. 1991).

The system of interspecific introgression by homoeologous recombination was created by E.R. Sears (1981) and is used to this day, however sparingly. The donor and recipient chromosomes are set up as monosomic and *Ph1*is disabled. Recombination of the two chromosomes creates recombinants in two configurations: wheat chromosomes with terminal segments of the alien chromosome; and the alien chromosome with terminal segments of the wheat chromosome. Homoeologous

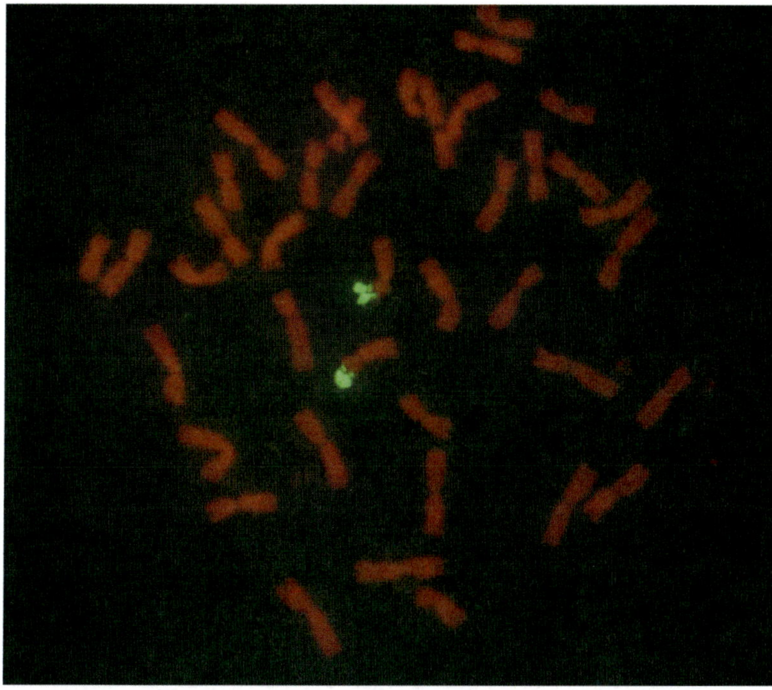

Fig. 7.3 Hexaploid wheat homozygous for a recombinant wheat–rye chromosome involving the distal end of rye chromosome 5RS and wheat chromosome 5DS. Rye segment labeled *green*; wheat chromatin *red*

recombination is based primarily on single crossovers (Fig. 7.3); doubles are so rare that they may just as well be ignored. Since each recombinant chromosome originates from a different event, their translocation breakpoints are different. All are screened for the presence of the desired locus and the position of the translocation breakpoint. For the second step of the procedure one chromosome from each configuration is selected, with the desired locus present and the closest available breakpoint on the proper side of the locus. The two are combined in a single plant with *Ph1* present, and allowed to recombine. *Ph1* permits only homologous recombination and the only segment of homology in the two primary recombinants is between the two translocation breakpoints. A crossover in that segment produces two chromosomes: a wheat chromosome with an intercalary alien introgression and a perfectly normal alien chromosome.

The size of the introgressed segment depends on the positions of the two breakpoints selected for the second round of recombination. Therefore, the precision of the entire exercise is determined by the number of primary recombinants recovered: the more are isolated the better the chance of finding two flanking the target locus in immediate vicinity. The number of primary recombinants in turn depends on the recombination frequency of the donor and target chromosome arms and the size of the screened population. If the recombination rate of the two arms is taken as a constant, the size of the screened population determines the precision of a transfer. The calculation is simple (metaphase I pairing rates for each rye chromosome in wheat are known, Naranjo and Fernandez-Rueda 1996): with MI pairing rate of 2 % (recombination rate of 1 %) and with no more than 1 cM of alien chromatin to remain on either side of the targeted locus, ca. 30,000 progeny would have to be screened for a 95 % probability of success (the formula is $n=\text{Ln}(1-p)/\text{Ln}(1-x)$ where n is the populations size, p the desired probability of success and x the event frequency). Care needs to be taken to use the recombination rate of the donor and recipient arms, and not the listed general rates for the donor arms. Recombination fidelity of the donor arms is never perfect; quite often they recombine with non-targeted arms. This reduces the yield of primary recombinant chromosomes by as much as one-third and the screening populations need to be increased accordingly.

The general perception is that chromosome engineering by homoeologous recombination requires very large populations and in most cases is not feasible. This may not be entirely true (see above). Several such transfers have been successfully made in wheat; more so with more closely related species than rye. With rye, the first demonstration of feasibility was by Koebner and Shepherd (1985) using the chromosome arm 1RL. This was followed by engineering of the 1RS arm in the 1RS.1DL translocation. The 1RS arm in this translocation originates from Imperial rye and the translocation was produced to introduce a very potent locus for stem rust resistance. Judging by the publication dates, the effort took over 20 years but after three rounds of chromosome engineering (Koebner and Shepherd 1986; Rogovsky et al. 1991; Anugrahwati et al. 2008) the stem rust locus (*Sr50*) was separated from the *Sec-1* locus and so it is now available without the penalty to bread making quality.

A somewhat different approach to and engineering the 1RS arm was taken for the 1RS.1BL and 1RS.1DL translocations (Lukaszewski 2000, 2006). The presence of the rye arm was deemed beneficial; the only identified problem of the translocation is the quality defect (see discussion above). Rather than to extract a specific locus from 1RS for a transfer to wheat, the starting concept was to replace *Sec-1* with *Gli-B1*/*Glu-B3* of wheat, leaving as much rye chromatin present as possible. To do this as quickly as possible, 103 primary breakpoints involving 1RS and 1BS were isolated (and another 40+ involving 1AS and 1DS). Tests of these recombinant chromosomes quickly showed that wheat and rye storage protein loci were located in non-corresponding positions and were separated by a block of resistance genes. This dictated creation of a four translocation breakpoint chromosome arm 1RS, with two small inserts from wheat: the proximal one to remove the *Sec-1* locus, and the distal one inserting the block of wheat storage protein loci. The two are separated by a segment of rye chromatin with four disease resistant loci: *Pm8*, *Lr26*, *Yr9*, and *Sr31*. It took screening of ca. 17,000 progeny to identify all necessary primary breakpoints to accomplish the intended task and another ca. 8000 to produce the secondary and tertiary recombinants 1RS-1BS (Lukaszewski 2000). The irony of the entire exercise was that when it was all done and finished (Fig. 7.4), it became apparent that a locus responsible for increased root biomass in the 1RS.1BL wheats is closely linked to the *Sec-1* locus (Sharma et al. 2009, 2011), and it was removed

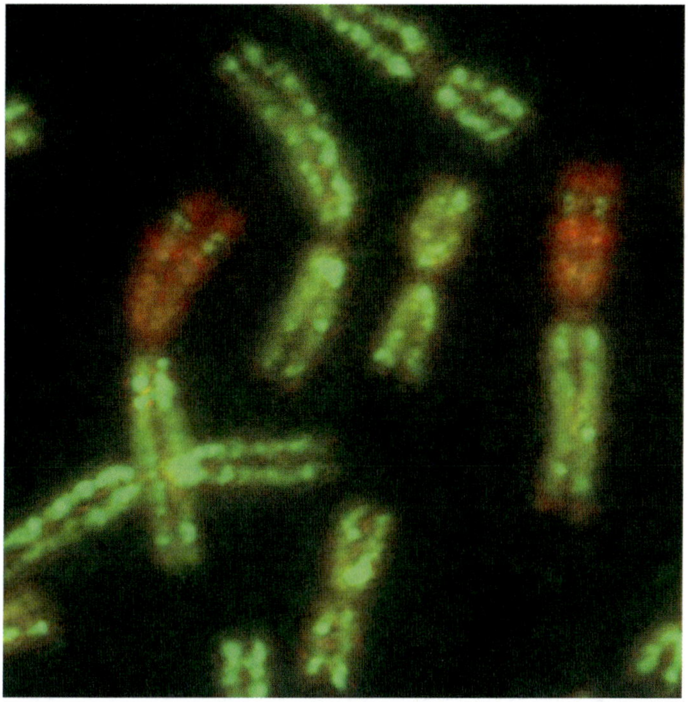

Fig. 7.4 A pair of engineered translocations 1RS.1BL of the Aurora/Kavkaz origin. Wheat chromatin labeled *green*/rye chromatin *red*. The 1RS arms of the translocation carry two wheat inserts: the proximal one removes the *Sec-1* locus of rye; the distal one introduces wheat loci *Gli-B1* and *Glu-B3*. The rye segment separating the two inserts contains four relevant resistance loci

from the engineered chromosome with *Sec-1* (Howell et al. 2014). Given that the resistance loci are now of little value, the entire exercise which required cytological screening of well over 25,000 plants, remedied the translocation problems but also removed its main advantage. Now new primary recombinants have to be isolated to find a breakpoint between the *Sec-1* and the root biomass locus.

Induced homoeologous recombination exercises always require fair amounts of labor; the amount depends on the desired level of precision and the affinity between the engineered chromosomes. However, at times it does not work no matter how much effort is invested. Failed attempts are not routinely reported but several have been published. Dundas et al. (2001) failed to recover any recombinant chromosomes in an attempt to transfer a cyst nematode resistance locus from the long arm of rye chromosome 6R to a wheat homoeologue among 3786 progeny screened. Screening was done using DNA markers so perhaps some recombinants involving non-designated chromosomes were present but not identified. Lukaszewski et al. (2001) screened 3563 progeny in an attempt to transfer from rye chromosome arm 4RL a locus for resistance to Russian wheat aphid. Screening was done using cytological markers in terminal positions so any single crossover event should have been detected, no matter which wheat homoeologues were involved. Two recombinants were recovered and confirmed by probing with labeled genomic DNA but both were associated with poor vigor of carrier plants, and the effort was abandoned. Both failures were probably due to structural differences between wheat and rye chromosomes: the 4RL arm is composed of segments homoeologous to wheat 7S and 6S groups while 6RL is composed of segments homoeologous to wheat homoeologous groups 6L, 3L, and 7L (Devos et al. 1993). It is not known in which segments of the 4RL and 6RL arms the targeted loci were located. Perhaps terminal deficiencies such as those recovered by Dundas et al. (2001) and those produced by gametocidal chromosomes (Endo 2003) can be used to recombine the carrier segment with its corresponding homoeologous segment in the wheat genome, without interference from segments constituting the structural difference and homoeologous to different wheat chromosomes that very likely disturb meiotic pairing and make the transfer impossible. This approach has never been tried, and it would be a complex one. Not only proper deficient chromosomes would have to be created, the transfer by induced homoeologous recombination affected, but the normal structure of the recipient chromosome would have to be restored at the end of the exercise. The author once attempted to transfer a dwarfing gene from 4DS to 4AL (chromosome 4A is inverted relative to other group-4 chromosomes of wheat and on L it carries segments originating from two other chromosomes) but the effort lost in competition for microscopy time with projects offering more immediate payoffs.

Once collections of recombined wheat–alien chromosomes are generated, usually with much effort, they can be used to transfer onto the alien segments, again by crossing over, specific desired alleles from an alien species. In this fashion Cainong et al. (2010) introduced an allele for Hessian fly resistance into wheat using preexisting recombinant chromosomes 2BL-2RL. This is not necessarily as easy a process as it appears, even though only homologous recombination is involved (between the alien segment already translocated to a wheat chromosome and a donor alien chromosome). The success rate, or, rather, the size of the progeny population needed to recover the desired product, depends on the distance between the translocation breakpoint and the locus of interest. Differences in chromosome configurations may severely limit the crossover rate between such two chromosomes.

7.7 Introgressions from Wheat to rye

While rye has been a popular donor of specific loci to wheat, the reverse has never been seriously considered. Of course, transfers from a diploid to polyploid can now be called routine, no matter how complicated and labor intensive they may be; transfers into diploids are inherently more difficult as diploids poorly tolerate some types of chromosome aberrations. However, such transfers can be done, and much easier at the tetraploid level than diploid. Probably the first attempt at introgression from wheat to rye was by Schlegel et al. (1991a) who introgressed wheat chromosome 5B into diploid rye. Lukaszewski (unpublished) introduced each of the B-genome chromosomes of wheat into tetraploid rye, and some disomics were produced. This led to studies on the effects of the *Ph1*locus on pairing of rye chromosomes (Lukaszewski and Kopecky 2010; Oleszczuk et al. 2014). Not surprisingly, the *Ph1* locus from wheat does affect rye chromosomes in their native environment but it also appears to interfere with the chromosome paring control system of rye. With two system overlapping, it is not clear what criteria are used in selecting pairs of homologues for bivalent pairing (in an otherwise an autotetraploid).

To improve breadmaking quality of hexaploid triticale Lukaszewski and Curtis (1992) created rye chromosome 1R where a segment of the long arm with the storage protein locus *Sec-3* was replaced with a corresponding segment from wheat 1DL carrying the *Glu-D1* locus, the most important locus for bread making quality in wheat. Once it was created, the chromosome was transferred to diploid rye (Fig. 7.5) and a population of homozygotes was established (Lukaszewski et al. 2000). This introgression had no apparent adverse effects on grain yield but it significantly changed the dough properties of rye. By the same approach, once the short arm of 1RS was engineered for bread making quality in hexaploid triticale (Lukaszewski 2006), it was transferred to diploid rye and populations of homozygotes were created. These populations have sets of wheat storage protein loci

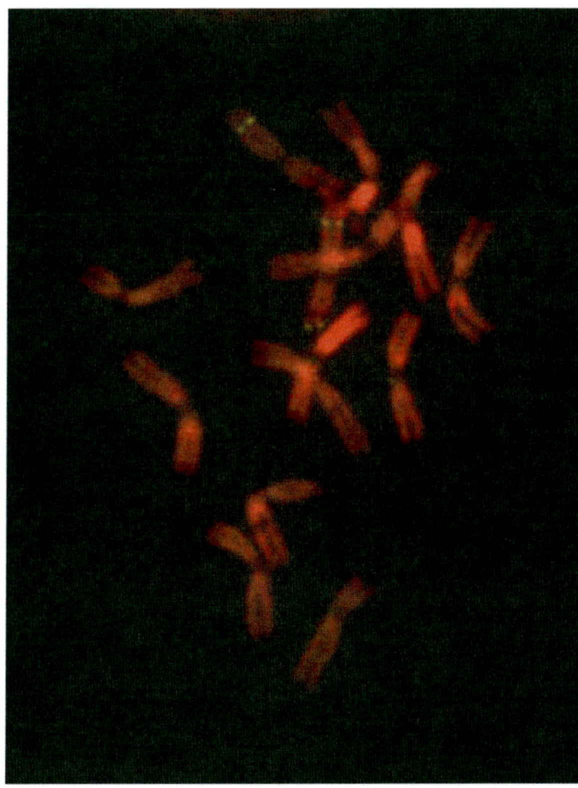

Fig. 7.5 Diploid rye homozygous for a segment of wheat chromatin on the long arm of chromosome 1R, replacing rye *Sec-3* locus with wheat *Glu-D1*, and heterozygous for a terminal segment on the short arm of 1R introducing wheat loci *Gli-D1*/*Glu-D3*. Wheat chromatin labeled green; rye chromosomes stained *red*

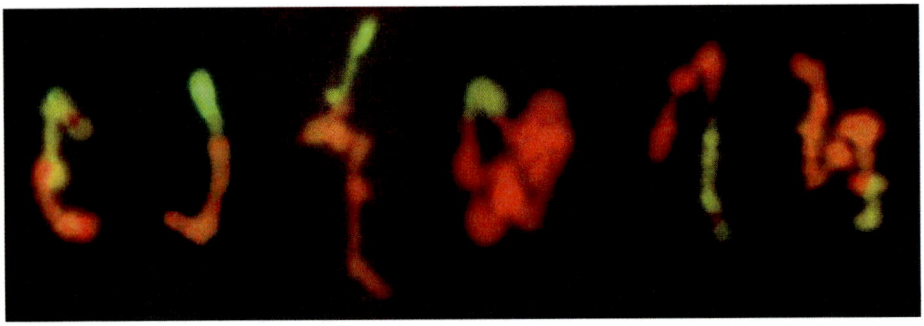

Fig. 7.6 Wheat B-genome chromosomes (*green*) paring with their rye homoeologues in tetraploid rye. From *left* to *right*: 1B, 2B, 3B, 4B, 6B, and 7B

equivalent to a single genome of wheat (Lukaszewski, unpublished). Again, as judged by seed set, the introgressions do not appear to affect fertility; their effect on bread making quality is as yet unknown.

The two exercises illustrate that chromatin transfers from wheat to rye are feasible and can have beneficial effects. Many more wheat loci could be transferred to rye for the benefit of rye breeding. Perhaps the most tempting, and the easiest to accomplish, is *Rht8*, from chromosome 2D as 2D-2R recombinant and centric translocation chromosomes are already available (Lukaszewski et al. 2004 and unpublished data). Interestingly, cursory observations of the behavior of wheat's B-genome monosomics in tetraploid rye suggest that wheat–rye chromosome pairing in rye may be much more frequent than in wheat. With the exception of chromosome 5B, each of the remaining six chromosomes was observed involved in pairing, quite often both arms at the same time (Fig. 7.6). Perhaps it is the difference in chromosome pairing control, which in rye appears to favor differentiated chromosomes, or perhaps it is the residual effect of the *Ph2* locus in wheat, but wheat–rye translocations are often recovered in tetraploid rye (Lukaszewski, unpublished; Wiśniewska et al. 2013).

References

Alkhimova AG, Heslop-Harrison JS, Schapova AI, Vershinin AV (1999) Rye chromosome variability in wheat-rye addition and substitution lines. Chromosome Res 7:205–212

An D, Zheng Q, Zhou Y, Ma P, Lu Z, Li L et al (2013) Molecular cytogenetic characterization of a new wheat–rye 4R chromosome translocation line resistant to powdery mildew. Chromosome Res 21:419–432

Anugrahwati DR, Shepherd KW, Verlin DC, Zhang P, Mirzaghaderi G, Walker E, Francki MG, Dundas IS (2008) Isolation of wheat-rye 1RS recombinants that break the linkage between the stem rust resistance gene *SrR* and secalins. Genome 51:341–349

Asiedu R, Fisher JM, Driscoll CJ (1990) Resistance to *Heterodera avenae* in the rye genome of triticale. Theor Appl Genet 79:331–336

Bertholdsson NO, Andersson SC, Merker A (2012) Allelopathic potential of Triticum spp., Secale spp. and Triticosecale spp. and use of chromosome substitutions and translocations to improve weed suppression ability in winter wheat. Plant Breed 131:75–80

Brunell MS, Whitkus R, Lukaszewski AJ (1999) Development of arm specific RAPD markers for rye chromosome 2R in wheat. Crop Sci 39:1702–1706

Bullrich L, Tranquilli G, Pfluger LA, Suarezm EY, Barneix AJ (1998) Bread-making quality and yield performance of 1BL/1RS wheat isogenic lines. Plant Breed 117:119–122

Cainong J, Zavatsky L, Chen M, Johnson JW, Friebe B, Gill BS, Lukaszewski AJ (2010) Wheat-rye T2BS.2BL-2RL recombinants with resistance to Hessian fly (H21). Crop Sci 50:920–925

Carver B, Rayburn AL (1994) Comparison of related wheat stocks possessing 1B or 1RS.1BL chromosomes: agronomic performance. Crop Sci 34:1505–1510

Ceoloni C, Donini P (1993) Combining mutations for the two homoeologous pairing suppressor genes *Ph1* and *Ph2* in common wheat and in hybrids with alien Triticeae. Genome 36:377–386

Chen PD, Tsujimoto H, Gill BS (1994) Transfer of Ph^I genes promoting homoeologous pairing from *Triticum speltoides* to common wheat. Theor Appl Genet 88:97–101

Chen PD, Qi LL, Zhou B, Zhang SZ, Liu DJ (1995) Development and molecular cytogenetic analysis of *wheat-Haynaldia villosa* 6VS/6AL translocation lines specifying resistance to powdery mildew. Theor Appl Genet 91:1125–1128

Chen PD, Chen SW, Cao AZ, Xing LP, Yang XM, Zhang SZ, Wang XE, Qi LL and Liu DJ (2008) Transferring, mapping, cloning of powdery mildew resistance gene of Haynaldia villosa and its utilization in common wheat. In: Appels R, Eastwood R, Lagudah E et al. (eds.), Proc 11th Int Wheat Genet Symp, Brisbane, QLD, pp 1–3

Crespo-Herrera L, Smith CM, Singh R, Ahman I (2013) Resistance to multiple cereal aphids in wheat–alien substitution and translocation lines. Arthropod Plant Interact 7:535–545

Curtis CA, Lukaszewski AJ (1993) Localization of genes in rye that restore male fertility to hexaploid wheat with *timopheevi* cytoplasm. Plant Breed 111:106–112

Curtis CA, Lukaszewski AJ, Chrzastek M (1991) Metaphase I pairing of deficient chromosomes and genetic mapping of deficiency breakpoints in common wheat. Genome 34:553–560

Denison RF (2012) Darwinian Agriculture. How understanding evolution can improve agriculture. Princeton University Press, Princeton and Oxford

Devos KM, Atkinson MD, Chinoy CN, Francis HA, Harcourt RL, Koebner RMD, Liu CJ, Masojc P, Xie DX, Gale MD (1993) Chromosomal rearrangements in the rye genome relative to wheat. Theor Appl Genet 85:673–680

Driscoll CJ, Anderson LM (1967) Cytogenetic studies of Transec-a wheat-rye translocation line. Can J Genet Cytol 9:375–380

Driscoll JC, Sears ER (1971) Individual addition of the chromosomes of Imperial rye to wheat. Agron Abst, ASA, Madison WI, p 6

Dundas IS, Frappell DE, Crack DM, Fisher JM (2001) Deletion mapping of a nematode resistance gene on rye chromosome 6R in wheat. Crop Sci 41:1771–1778

Dvorak J (1972) Genetic variability in *Aegilops speltoides* affecting homoeologous pairing in wheat. Can J Genet Cytol 14:317–380

Dvorak J, Lukaszewski AJ (2000) Centromere association is an unlikely mechanism by which the wheat *Ph1* locus regulates metaphase I pairing between homoeologous chromosomes. Chromosoma 109:410–414

Dvorak J, McGuire P (1981) Nonstructural chromosome differentiation among wheat cultivars with special reference to differentiation of chromosomes in related species. Genetics 97:391–414

Ehdaie B, Whitkus RW, Waines JG (2003) Root biomass, water-use efficiency and performance of wheat-rye translocations of chromosomes 1 and 2 in spring bread wheat ‘Pavon’. Crop Sci 43:710–717

Endo TR (2003) Wheat stocks carrying alien chromosomal segments induced by the gametocidal system. In: Pogna NE, Romano M, Pogna EA, Galterio G (eds.), Proc 10th Int Wheat Genet Symp, Paestum, Italy, pp 69–72

Espita-Rangel E, Baenziger PS, Graybosch RA, Shelton DR, Moreno-Sevilla B, Peterson CJ (1999a) Agronomic performance and stability of 1A vs 1AL.1RS genotypes derived from winter wheat 'Nekota'. Crop Sci 39:643–648

Espita-Rangel E, Baenziger PS, Graybosch RA, Shelton DR, Moreno-Sevilla B, Peterson CJ (1999b) End-use quality performance and stability of 1A vs 1AL.1RS genotypes derived from wheat 'Nekota'. Crop Sci 39:649–654

Feldman M, Mello-Sampayo T (1967) Suppression of homoeologous pairing in hybrids of polyploid wheat x *Triticum speltoides*. Can J Genet Cytol 9:307–313

Friebe B, Larter EN (1988) Identification of a complete set of isogenic wheat/rye D-genome substitution lines by means of Giemsa C-banding. Theor Appl Genet 76:473–479

Friebe B, Zeller FJ, Kunzmann R (1987) Transfer of the 1BL/1RS wheat-rye translocation from hexaploid bread wheat to tetraploid durum wheat. Theor Appl Genet 74:423–425

Friebe B, Hatchett JH, Sears RG, Gill BS (1990) Transfer of Hessian fly resistance for 'Chaupon' rye to hexaploid wheat via a 2BS/2RL wheat-rye chromosome translocation. Theor Appl Genet 79:385–389

Friebe B, Jiang J, Raupp WJ, McIntosh RA, Gill BS (1996) Characterization of wheat-alien translocations conferring resistance to diseases and pests: current status. Euphytica 91:59–87

Fu S, Tang Z, Ren Z, Zhang H (2010) Transfer to wheat (*Triticum aestivum*) of small chromosome segments from rye (*Secale cereale*) carrying disease resistance genes. J Appl Genet 51:115–121

Gill KS, Gill BS, Endo TR, Mukai Y (1993) Fine mapping of *Ph1*, a chromosome pairing regulator gene in polyploid wheat. Genetics 134:1231–1236

Graybosch RA, Lee J-H, Peterson CJ, Porter DR, Chung OK (1999) Genetic agronomic and quality comparisons of two 1AL.1RS wheat rye chromosomal translocations. Plant Breed 118:125–130

Howell T, Hale I, Jankuloski L, Bonafede M, Gilbert M, Dubcovsky J (2014) Mapping a region within the 1RS.1BL translocation in common wheat affecting grain yield and canopy water status. Theor Appl Genet 127:2695–2709

Hsam SKL, Zeller FJ (1997) Evidence of allelism between genes Pm8 and Pml7 and chromosomal location of powdery mildew and leaf rust resistance genes in the common wheat cultivar 'Amigo'. Plant Breed 116:119–122

Hsam SKL, Mohler V, Hartl L, Wenzel G, Zeller FJ (2000) Mapping of powdery mildew and leaf rust resistance genes on the wheat-rye translocated chromosome T1BL.1RS using molecular and biochemical markers. Plant Breed 119:87–89

Hysing SC, Hsam SL, Singh RP, Huerta-Espino J, Boyd LA, Koebner R et al (2007) Agronomic performance and multiple disease resistance in T2BS. 2RL wheat-rye translocation lines. Crop Sci 47:254–260

Jiang J, Friebe B, Gill BS (1994) Recent advances in alien gene transfer in wheat. Euphytica 73:199–212

Karki D, Wyant W, Berzonsky WA, Glover KD (2014) Investigating physiological and morphological mechanisms of drought tolerance in wheat (*Triticum aestivum* L.). Lines with 1RS translocation. Am J Plant Sci 5:1936–1944. doi:10.4236/ajps.2014.513207

Kim W, Johnson JW, Baenziger PS, Lukaszewski AJ, Gaines SC (2004) Agronomic effect of wheat-rye translocation carrying rye chromatin (1R) from different sources. Crop Sci 44:1254–1258

Ko JM, Seo BB, Suh DY, Do GS, Park DS, Kwack YH (2002) Production of a new wheat line possessing the 1BL.1RS wheat-rye translocation derived from Korean rye cultivar Paldanghomil. Theor Appl Genet 104:171–176

Koebner RMD, Shepherd KW (1985) Induction of recombination between rye chromosome 1RL and wheat chromosomes. Theor Appl Genet 71:208–215

Koebner RMD, Shepherd KW (1986) Controlled introgression to wheat of genes from rye chromosome 1RS by induction of allosyndesis. I. Isolation of recombinants. Theor Appl Genet 73:197–208

Kumlay AM, Baenziger PS, Gill BS, Shelton DR, Graybosch RA, Lukaszewski AJ, Wesenberg DM (2003) Understanding the effect of rye chromatin in bread wheat. Crop Sci 43:1643–1651
Lahsaiezadeh M, Ting IP, Waines JG (1983) Drought resistance in Chinese Spring wheat/Imperial rye additions and substitution lines. In: Sakamoto S (ed) Proc 6th Int. Wheat Genet Symp. Kyoto, Japan, pp 945–950
Lapitan NLV, Sears RG, Gill BS (1984) Translocations and other karyotypic structural changes in wheat x rye hybrids regenerated from tissue culture. Theor Appl Genet 68:547–554
Lei MP, Li GR, Zhou L, Li CH, Liu C, Yang ZJ (2013) Identification of wheat-*Secale africanum* chromosome $2R_{af}$r introgression lines with novel disease resistance and agronomic characteristics. Euphytica 194:197–205
Lelley T, Eder C, Grausgruber H (2004) Influence of 1BL.1RS wheat-rye chromosome translocation on genotype by environment interaction. J Cereal Sci 39:313–320
Lukaszewski AJ (1988) A comparison of several approaches in the development if disomic alien addition lines of wheat. In: Miller TE, Koebner RMD (eds.), Proc 7th Int Wheat Genet Symp, Cambridge, UK, pp 363–367
Lukaszewski AJ (1990) Frequency of 1RS.1BL and 1RS.1AL translocations in US wheats. Crop Sci 30:1151–1153
Lukaszewski AJ (1993) Reconstruction in wheat of complete chromosomes 1B and 1R from the 1RS.1BL translocation of 'Kavkaz' origin. Genome 36:821–824
Lukaszewski AJ (1997) Further manipulation by centric misdivision of the 1RS.1BL translocation in wheat. Euphytica 94:257–261
Lukaszewski AJ (2000) Manipulation of the 1RS.1BL translocation in wheat by induced homoeologous recombination. Crop Sci 40:216–225
Lukaszewski AJ (2006) Cytogenetically engineered rye chromosomes 1R to improve bread-making quality of hexaploid triticale. Crop Sci 46:2183–2194
Lukaszewski AJ (2010) Behavior of centromeres in univalents and centric misdivision in wheat. Cytogenet Genome Res 129:97–109
Lukaszewski AJ, Curtis CA (1992) Transfer of the *Glu-D1* gene from chromosome 1D of bread wheat to chromosome 1R in hexaploid triticale. Plant Breed 109:203–210
Lukaszewski AJ, Kopecky D (2010) The *Ph1* Locus from wheat controls meiotic chromosome pairing in autotetraploid rye (*Secale cereale* L.). Cytogenet Genome Res 129:124–132
Lukaszewski AJ, Brzezinski W, Klockiewicz-Kaminska E (2000) Transfer of the *Glu-D1* locus encoding high molecular weight glutenin subunits from bread wheat to diploid rye. Euphytica 115:49–57
Lukaszewski AJ, Porter DR, Baker CA, Rybka K, Lapinski B (2001) Attempts to transfer Russian wheat aphid resistance from a rye chromosome in Russian triticales to wheat. Crop Sci 41:1743–1749
Lukaszewski AJ, Rybka K, Korzun V, Malyshev SV, Lapinski B, Whitkus R (2004) Genetic and physical mapping of homoeologous recombination points involving wheat chromosome 2B and rye chromosome 2R. Genome 47:36–45
Luo MC, Dubcovsky J, Dvorak J (1996) Recognition of homoeology by the wheat *Ph1* locus. Genetics 144:1195–1203
Luo MC, Yang ZL, Kota RS, Dvorak J (2000) Recombination of chromosomes $3A^m$ and $5A^m$ of *Triticum monococcum* with homoeologous chromosomes 3A and 5A of wheat: distribution of recombination across chromosomes. Genetics 154:1301–1308
Massoudi-Nejad A, Nasuda S, McIntosh RA, Endo TR (2002) Transfer of rye chromosome segments to wheat by gametocidal system. Chromosome Res 10:349–357
McKendry AL, Tagues DN, Miskin KE (1996) Effect of 1BL.1RS on agronomic performance of soft red winter wheat. Crop Sci 36:844–847
Merker A (1984) The rye genome in wheat breeding. Hereditas 100:183–191
Middleton CP, Senerchia N, Stein N, Akhunov ED, Keller B et al (2014) Sequencing of chloroplast genomes from wheat, barley, rye and their relatives provides a detailed insight into the evolution of the Triticeae tribe. PLoS One 9(3):e85761. doi:10.1371/journal.pone.0085761

Mujeeb-Kazi A, Gul Kazi A, Dundas I, Rasheed A, Ogbonnaya F et al (2013) Genetic diversity for wheat improvement as a conduit to food security. In: Sparks DL (ed) Advances in agronomy, vol 122. Academic, Burlington, VT, pp 179–257
Mukai Y, Friebe B, Hatchett JH, Yamamoto M, Gill BS (1993) Molecular cytogenetic analysis of radiation-induced wheat-rye terminal and intercalary chromosomal translocations and the detection of rye chromatin specifying resistance to Hessian fly. Chromosoma 102:88–92
Naranjo T, Fernandez-Rueda P (1991) Homoeology of rye chromosome arms to wheat. Theor Appl Genet 82:577–586
Naranjo T, Fernandez-Rueda P (1996) Pairing and recombination between individual chromosomes of wheat and rye in hybrids carrying the *ph1b* mutation. Theor Appl Genet 93:242–248
Naranjo T, Lacadena JR (1980) Interaction between wheat chromosomes and rye telomeric heterochromatin on meiotic pairing of chromosome pair 1R of rye in wheat-rye derivatives. Chromosoma 81:249–261
Nkongolo KK, Armstrong KC, Comeau A, St-Pierre CA (1992) Identification of rye chromosomes involved in tolerance to barley yellow dwarf virus disease in wheat x triticale hybrids. Plant Breed 109:123–129
O'Mara JG (1940) Cytogenetic studies on *Triticale*: I. A method for determining the effects of individual Secale chromosomes on Triticum. Genetics 25:401–408
Oleszczuk S, Tyrka M, Lukaszewski AJ (2014) The *Ph1* locus of wheat does not discriminate between identical and non-identical homologues in rye. Cytogenet Genome Res 142:293–298
Placido DF, Begcy K, Sato SJ, Nersesian N et al (2013) The LATERAL ROOT DENSITY gene regulates drought tolerance response and promotes grain productivity in wheat. Plant Phys 161:1806–1819
Rabinovich SV (1998) Importance of wheat-rye translocations for breeding modern cultivars of *Triticum aestivum* L. Euphytica 100:323–340
Rajaram S, Mann CE, Ortiz-Ferrara G, Mujeeb-Kazi A (1983) Adaptation, stability and high yield potential of certain 1B/1R CIMMYT wheats. In: Sakamoto S (ed) Proc 6th Int. Wheat Genet Symp, Kyoto, Japan pp, pp 613–621
Ren TH, Yang ZJ, Yan BJ, Zhang HQ et al (2009) Development and characterization of a new 1BL.1RS translocation line with resistance to stripe rust and powdery milder of wheat. Euphytica 169:207–213
Rogovsky PM, Gudet FLY, Langride P, Shepherd KW, Koebner RMD (1991) Isolation and characterization of wheat-rye recombinants involving chromosome arm 1DS of wheat. Theor Appl Genet 82:537–544
Schlegel R, Korzun V (1997) About the origin of 1RS.1BL wheat-rye chromosome translocations from Germany. Plant Breed 116:537–540
Schlegel R, Boerner A, Thiele V, Melz G (1991a) The effect of the *Ph1* gene in diploid rye, *Secale cereale* L. Genome 34:913–917
Schlegel R, Werner T, Hugelnhof E (1991b) Confirmation of a 4BL/5RL Wheat-rye chromosome translocation line in the wheat cultivar 'Viking' showing high copper efficiency. Plant Breed 107:226–234
Schlegel R, Camak I, Torun B, Eker S et al (1998) Screening for zinc efficiency among wheat relatives and their utilization for alien gene transfer. Euphytica 100:281–286
Sears ER (1954) The aneuploids of common wheat, vol 572, Missouri Agricultural Experiment Station Research Bulletin. University Archives of the University of Missouri, Columbia, MO
Sears ER (1956) The transfer of leaf-rust resistance from *Aegilos umbellulata* to wheat. Brookhaven Symp Biol 9:1–22
Sears ER (1968) Relationships of chromosomes 2A, 2B and 2D with their rye homoeologue. In: Finlay KW, Shepher KW (eds.), Proc 3rd Int Wheat Genet Symp Aust Acad Sci, Canberra, CT, pp 163–171
Sears ER (1972) Chromosome engineering in wheat. Proc. 4th Stadler Symp. Columbia, MO, pp 23–38

Sears ER (1975) An induced homoeologous pairing mutant in wheat. Genetics 80:s74
Sears ER (1981) Transfer of alien genetic material to wheat. In: Evans LT, Peacock WJ (eds) Wheat science – today and tomorrow. Cambridge University Press, Cambridge, pp 75–89
Sears ER (1984) Mutations in wheat that raise the level of meiotic chromosome pairing. In: Gustafson JP (ed) Gene manipulation in plant improvement. Proc 16th Stadler Genetics Symp, Columbia, MO. Plenum Press, New York, NY, pp 295–300
Sears ER (1993) Use of radiation to transfer alien chromosome segments to wheat. Crop Sci 33:897–901
Sears ER, Sears LMS (1978) The telocentric chromosomes of common wheat. In: Ramanujam S (ed) Proc 5th In Wheat Genet Symp New Delhi, India. Indian Agric. Res. Institute, New Delhi, pp 380–407
Sebesta EE, Wood EA (1978) Transfer of greenbug resistance from rye to wheat with X-rays. Agron Abstr ASA, Madison WI, pp 61–62
Sharma S, Bhat P, Ehdaie B, Close TJ et al (2009) Integrated genetic map and genetic analysis of a region associated with root traits on the short arm of rye chromosome 1 in wheat. Theor Appl Genet 119:783–793
Sharma S, Xu S, Ehdaie B, Hoops A et al (2011) Dissection of QTL effects for root traits using a chromosome arm-specific mapping population in bread wheat. Theor Appl Genet 122:759–769
Shchapova AI, Kravtsova LA (1982) The production of wheat-rye substitution lines by using the Giemsa staining technique. Cereal Res Commun 10:33–39
Silkova OG, Dobrovolskaya OB, Dubovets NI, Adonina IG et al (2006) Production of wheat–rye substitution lines and identification of chromosome composition of karyotypes using C-banding, GISH, and SSR markers. Rus J Genet 42:645–653
Singh RP, Huerta-Espino J, Rajaram S, Crossa J (1998) Agronomic effects from chromosome translocations 7DL.7Ag and 1BL.1RS in spring wheat. Crop Sci 38:27–33
Villareal RL, del Toro E, Rajaram S, Mujeeb-Kazi A (1996) The effect of chromosome 1AL/1RS translocation on agronomic performance of 85 F2-derived F6 lines from three *Triticum aestivum* L crosses. Euphytica 89:363–369
Villareal RL, Banuelos O, Mujeeb-Kazi A (1997) Agronomic performance of related durum (*Triticum turgidum* L) stocks possessing the chromosome substitution T1BL.1RS. Crop Sci 37:1735–1740
Villareal RL, Banuelos O, Mujeeb-Kazi A, Rajaram S (1998) Agronomic performance of chromosome 1B and T1BL.1RS near-isolines in the spring breadwheat Seri M82. Euphytica 103:195–202
Waines JG, Ehdaie B (2005) Optimizing root characters and grain yield in wheat. Czech J Genet Plant Breed 41:326–330
Waines JG, Ehdaie B (2007) Domestication and crop physiology: roots of Green Revolution wheat. Ann Bot 100:991–998
Wiśniewska H, Kwiatek M, Kulak-Książczyk S, Apolinarska B (2013) Introgression of A- and B-genome of tetraploid triticale chromatin into tetraploid rye. J Appl Genet 54:435–440
Wulf BBH, Moscou MJ (2014) Strategies for transferring resistance into wheat: from wide crosses to GM cassettes. Front Plant Sci 5:692. doi:10.3389/fpls.2014.0692
Yang ZJ, Li GR, Jia JQ, Zeng X et al (2009) Molecular cytogenetic characterization of wheat–*Secale africanum* amphiploids and derived introgression lines with stripe rust resistance. Euphytica 167:197–202
Zeller FJ, Fuchs E (1983) Cytologie und Krankheitsresistenz einer 1A/1R- un mehrerer 1B/1R-Weizen-Roggen-Translokationsorten. Z Pflanzenzuchtg 90:285–296
Zeller FJ, Hsam SLK (1983) Broadening the genetic variability of cultivated wheat by utilizing rye chromatin. In: Sakamoto S (ed) Proc 6th Int Wheat Genet Symp. Japan, Kyoto, pp 161–173
Zhang P, Friebe B, Lukaszewski AJ, Gill BS (2001) The centromere structure in Robertsonian translocation chromosomes indicates that centric breakage-fusion can occur at various locations within the primary constriction. Chromosoma 110:335–344

Chapter 8
Alien Introgressions from wild *Triticum* species, *T. monococcum*, *T. urartu*, *T. turgidum*, *T. dicoccum*, *T. dicoccoides*, *T. carthlicum*, *T. araraticum*, *T. timopheevii*, and *T. miguschovae*

George Fedak

8.1 *Triticum monococcum–T. urartu*

The diploid einkorn group consists of four "species." The wild hulled form encompasses *T. boeoticum* Boiss, (A^b genome) commonly called wild einkorn, and *T. urartu* Tumansian ex Gandilyam (A^u genome) commonly called wild urartu einkorn. There is no evidence for the differentiation of the genomes of the two species. The cultivated group includes the hulled *T. monococcum* L. (A^b genome) and the free threshing form *T. sinskayae* A. Filat. & Kurkiev (A^b) genome otherwise known as Sinskaya einkorn. Diploid cultivated wheat is one of the most ancient crops domesticated in the Middle East (Harlan 1971). Distribution in the wild of *T. boeoticum* occurs in the area of the Taurus-Zagros arc of southeastern Turkey, northeastern Iraq, and eastern Iran (Harlan and Zohary 1966).

T.monococcum can be crossed directly to hexaploid wheat then followed by 1–2 backcrosses to restore fertility. From personal experience we found that such crossing was more difficult than anticipated and backcrossing was essential to restore fertility. Progeny of such crosses show minimal linkagedrag. Other methods to introgress traits from *T. monococcum* to bread wheat are to induce triploid bridges using durum or using the autotetraploid form of *T. monococcum* in crosses to durum. Crosses to tetraploid wheat are somewhat more difficult such that triploid bridging was necessary (Gerechter-Amitai et al. 1971).

G. Fedak (✉)
Eastern Cereal and Oilseed Research Centre, Agriculture and Agri-Food Canada,
960 Carling Ave, Ottawa, ON, Canada, K1A 0C6
e-mail: george.fedak@agr.gc.ca

M. Molnár-Láng et al. (eds.), *Alien Introgression in Wheat*,
DOI 10.1007/978-3-319-23494-6_8

8.1.1 *Construction of Genetic Maps*

Whereas the levels of polymorphism between cultivars of *Triticum aestivum* for RFLP mapping are very low, it is very high in *T. monococcum* (85 %). On eRFLP map was constructed in *T. monococcum* having 335 markers (Dubcovsky et al. 1996). It was found that the linkage maps of barley and *T. monococcum* are very highly conserved. Of the 328 mapped loci in the latter (detected with DNA probes), 18 % were duplicated, 5.9 % were triplicated, 4.6 % were quadruplicated, and 1.55 were present five times. Another genetic linkage map was constructed on a set of 93R1L lines obtained from the cross of *T. boeoticum* × *T. monococcum* (Singh et al. 2007b). The levels of polymorphism between parents were 73 % for SSR markers and 50 % of the RFLP markers. A total of 188 polymorphic loci were mapped. About 58 loci showed distorted segregation with most of these mapping to chromosome $2A^m$. Chromosome $1A^m$ of the two parents showed a small pericentric inversion relative to the A genome of hexaploid wheat.

8.1.2 *Disease Resistance*

8.1.2.1 Stem Rust Resistance

A list of current *Sr* genes introgressed from *T. monococcum* is shown in Table 8.1. As for deployment of introgressed genes; *Sr21* has not been deployed yet, but has the potential for limited use especially in pyramids. *Sr22* (RL5244) has received limited use thus far. *Sr35* has not been deployed yet, but is being stacked into pyramids. *Sr35* is effective against the three TTKS variants plus TRTTF a virulent race from Yemen. *Sr35* has been mapped to chromosome 3AL in *T. monococcum* and maps to the distal portion of chromosome 3BL in the hexaploid map (Zhang et al. 2010). *T. monococcum* appears to be a valuable reservoir for additional unique *Sr* genes. It

Table 8.1 Sources of stem rust resistance genes in *Triticum monococcum*

Gene	Source	Chromosome location	Reference
Sr21	*T. monococcum*	2A	The (1973)
			Gerechter-Amitai et al. (1971)
			Kerber and Dyck (1973)
Sr22	*T. monococcum*	7A	Kerber and Dyck (1973)
Sr35	*T. monococcum*	3AL	McIntosh et al. (1984)
	1062 accessions screened (78 % resistant to TTKSK)		Rouse and Jin (2011a)
	2 new genes resistant to TTKSK		Rouse and Jin (2011b)
	102 accessions screened with 2 races		Vallega (1979)

was found that of 1062 gene bank accessions of *T. monococcum*, 78.7 % was found to be resistant to TTKSK (Ug99) and four of its variants (Rouse and Jin 2011b). Some accessions were resistant to multiple races. Genetic tests were conducted with some of the resistant accessions (Rouse and Jin 2011b). They found two genes with resistance to TTKSK that were different from *Sr21*, *Sr22* and *Sr35*. One of the accessions was resistant to all races tested.

205 gene bank accessions of *T. urartu* were screened for resistance to TTKSK and some of its variants. It was found that 93 % of the accessions were resistant to TTKSK (Rouse and Jin 2011a). The observed resistance is likely due to new genes. Thus far no stem rust resistance genes have been introgressed into wheat from *T. urartu*. In earlier screening studies, 102 accessions of *T. monococcum* were screened with 2 Italian races of stem rust and 48 and 60 % of accessions were resistant (Vallega 1979). A new stem rust resistance gene from *T. monococcum* has been transferred to hexaploid wheat (Valkoun et al. 1989). On a much smaller scale, in testing of 10 accession of *T. monococcum* obtained of from M. Trottet of INRA, one line was found to carry unique genes for resistance to TTKSK (Fedak 2000).

8.1.2.2 Leaf Rust Resistance

Triticum monococcum has not made a significant contribution to leaf rust resistance to date, however there appears to be considerable potential for a source of leaf rust resistance genes. For example Vallega (1979) screened up to 102 accessions of *T. monococcum* with five different accessions of leaf rust and all accessions were resistant to all races. A study of 49 accessions in the University of Saskatchewan collection revealed 2 independent dominant genes for leaf rust resistance (Bai et al. 1998). Two accessions were also shown to carry *Sr22* and *Sr35* for stem rust resistance. *Triticum monococcum* accession number 14087 was shown to have resistance to leaf rust at the seedling and adult plant stage plus adult plant resistance to stripe rust (Dhaliwal et al. 2005). Resistance to leaf rust, stripe rust, and karnal bunt was subsequently transferred to hexaploid wheat. Prehaustorial resistance to the leaf rust fungus, *Puccinia triticina* has been detected in some accessions of *T. monococcum* (Anker and Niks 2001). This is a non-hypersensitive reaction and acts before the formation of the haustoria by the fungus. In this study, 84 % of 182 accessions of *T. monococcum* were found to be resistant to leaf rust, but only three showed prehaustorial resistance. A number of leaf and stripe rust resistance genes have been mapped in *T. monococcum* and transferred to bread wheat (Singh et al. 2007a). A dominant gene for leaf rust resistance was introgressed from *T. monococcum* to hexaploid wheat and located on chromosome 3A (Valkoun et al. 1986). One leaf rust resistance gene from *T. monococcum* var. *monococcum* and two from *T. monococcum* var. *boeoticum* were transferred to hexaploid wheat and their race specificities were confirmed (Hussein et al. 1997). The gene *Lr63* from *T. monococcum* was introgressed onto chromosome 3AS of hexaploid wheat (Kolmer et al. 2010).

A large number of accessions of diploid and tetraploid wheats were screened for resistance to a number of insects and fungal pathogens (Gill et al. 1983). Sixty eight

Fig. 8.1 A field plot of *T. monococcum*-derived line M321 showing leaf rust resistance. The adjacent plot of AC Barrie is susceptible to leaf rust

of 198 accessions of *T. urartu* tested were resistant to leaf rust. No resistance to Hessian fly, greenbug, or WSMV was detected. A novel leaf rust resistance gene from *T. monococcum* has been introgressed onto chromosome 5AS and is independent of a FHB QTL also located on that chromosome arm (Fedak et al. 2006, 2011). Since there are no other leaf rust resistance genes mapped on that chromosome, this gene must be unique (McCallum p.c.). The introgressed line designated as M321 is shown in Fig. 8.1 as a field plot. There is no obvious linkage drag associated with this introgression. In Fig. 8.1 M321 is shown adjacent to a plot of AC Barrie showing the difference in leaf rust infection. AC Barrie is a widely grown Canadian cultivar of hard red spring wheat that has been recently overcome by leaf rust.

8.1.2.3 Stripe Rust Resistance

Although *T. monococcum* is a good source of genes for resistance to leaf rust, it has not provided many genes for resistance to stripe rust. Adult plant stripe rust resistance was detected in strain pau14087 of *T. monococcum* and strain pau5088 of *T. boeoticum* (Chhuneja et al. 2008). A mapping population was created from a hybrid between the two strains. The resistance locus in *T. monococcum* mapped to chromosome 2A and to chromosome 5A in *T. boeoticum*. Selected *RIL* were used to transfer the resistance to hexaploid wheat using *T. durum* as a bridge. The B genome of durum suppressed the resistance in F1 plants but this was overcome by several backcrosses. The resistance gene from *T. boeoticum* was successfully transferred to hexaploid wheat (Chhuneja et al. 2008). Seedlings of 75 accessions of *T. boeoticum*, 12 of *T. monococcum*, 16 of *T. urartu*, 230 of durum wheat and 128 amphiploids AAAABB were evaluated for stripe rust resistance in field and greenhouse conditions (Ma et al. 1997). Resistant reactions were observed on 10 accessions of *T. boeoticum*, 19 accessions of durum, and 32 amphiploids. Suppressions of resistance were common in the A and AB genomes and were resistance gene specific.

8.1.2.4 Powdery Mildew Resistance

There are very few examples of variation for mildew resistance in *Triticum monococcum*. A new powdery mildew resistant allele at the *Pm4* locus was introgressed from that species (Schmolke et al. 2012). Powdery mildew resistance was found in one accession of *T. urartu*. Through conventional crossing and backcrossing a single dominant gene was transferred and mapped to chromosome 7AL (Qui et al. 2005). The gene is temporarily designated as *PmU*. Two dominant and different powdery mildew resistance genes, designated as *Mlm* 2033 and Mlm80, were identified in two different accessions of *T. monococcum*. They both mapped to chromosome 7AL. Further tests indicated that they are probably allelic to each other and mapped close to the *Pm1* locus (Yao et al. 2007). Mildew resistance locus *Pm1b* was transferred from *T. monococcum* to hexaploid wheat and mapped to chromosome 7AL (Hsam et al. 1998). Similarly *Pm25* was introgressed from *T. boeoticum* into hexaploid wheat and mapped to chromosome 1AL (Shi et al. 1998). Two powdery mildew resistance genes from *T. boeoticum* were introgressed into bread wheat (Chhuneja et al. 2012). They mapped to chromosome 7AL. One gene may be allelic to *Pm1* whereas the other is probably a new gene.

8.1.2.5 Fusarium Head Blight (FHB) Resistance

Fusarium head blight has now become a serious disease of cereals in all temperate grain growing regions of the world. Effects of the disease are manifest as direct yield reduction by kernel shrivelling plus the deposition of a vomitoxin in the seed rendering it as a dangerous component in food and feed. One of the reasons for the slow rate of progress in breading for resistance is the oligogenic control of resistance. Additional genes/QTL for resistance are continually being sought, even among wild species. Several reviews have been written on the identification of FHB resistance QTL in alien species (Cai et al. 2005). *Triticum monococcum* is not listed in those reviews. *T. monococcum* was one of the species screened for FHB resistance in our studies. To begin with, 200 accessions of *T. monococcum* were obtained from M. Trottet of INRA. After repeated screening, line 10-1 was identified as having a fair level of FHB resistance (Fig. 8.2) (Fedak et al. 1997a, b, 2003; Fedak 2000; 2007). Line 10-1 was crossed to the spring wheat cultivar AC Domain. After repeated backcrossing and selection line M321 was selected. The values for percent infected florets following point inoculation were 8 % as compared to 4 % for the resistant check Sumai3, and 32 % for Roblin the susceptible check. M321 was crossed to AC Domain and a doubled haploid mapping population of 80 lines was produced by the maize pollination method (Fedak et al. 1997a, b). The population was phenotyped by point inoculation and genotyped with SSR markers. A QTL for FHB resistance was located on chromosome 5A, linked to the marker *Xwme705* (Fedak et al. 2011).

The agronomic characteristic of M321 is listed in Table 8.2. Line M321 compares favorably with check cultivars in terms of agronomic traits such as plant

Fig. 8.2 Resistance to Fusarium head blight in an accession of *Triticum monococcum*. Disease symptoms developed at 21 days after artificial inoculation

Table 8.2 Agronomic characteristics of FHB resistant lines obtained from *T. monococcum* (M321) and *Ae. speltoides* (S184)

Genotypes	Yield (kg/ha)	TSTWT[a] (kg/hl)	HT[b] (cm)	Protein (%)	Flour yield (%)	DON (ppm)
Sumai3	2895	–	88.5	13.0	–	2.1
M 321	3272	79.3	76	13.9	57.5	5.5
S 184	3246	80.3	86	13.3	67.2	3.4
AC Barrie	3304	80.5	79	13.7	66.8	6.5
Roblin	–	–	–	–	–	17.2

[a]Test weight
[b]Height

height, yield, TKW, protein content, and even flour yield. The yield of this line is reasonable compared to AC Barrie a check cultivar. The data in Table 8.2 indicated that there is minimal linkage drag in M321. The lowered DON content relative to the checks is a useful attribute for improvement of disease resistance of wheat. FHB symptoms on the spike of the parental cultivar 10-1, scored at 21 days after inoculation are shown in Fig. 8.2 and a field plot of M321 is shown in Fig. 8.1.

FHB resistance was also sought in *Ae. speltoides*. In this case, 50 accessions were screened and line S184 selected (Fedak et al. 2004, 2006). It has been previously shown that different accessions of *Ae. speltoides* can lead to different levels of meiotic chromosome pairing in F1 hybrids with wheat. The hybrid between wheat and the resistant selection in our studies showed 3–4 bivalents at meiosis, i.e., an accession that induced fairly high meiotic chromosome pairing. Despite this, three backcrosses were required to restore fertility in the progeny. The agronomic characteristics of line S184 are shown in Table 8.2. For most agronomic traits such as yield, plot yield, plant height, protein content, and even flour yield, the values for S184 compare favorably to the check cultivars. This may be attributed to the fact that large populations were screened at each backcross generation and the most vigorous plants were selected. Probably the most important attribute of this line is the lower DON content following inoculation.

8.1.2.6 Karnal Bunt Resistance

Karnal bunt resistance genes have been identified in *T. monococcum*, transferred to hexaploid wheat and subsequently tagged with molecular markers (Vasu et al. 2000).

8.1.3 Other Traits

8.1.3.1 Virus Resistance

A large number of diploid and tetraploid wild wheat accessions, currently being maintained at the University of California Riverside, and Kansas State University were screened for resistance to a number of insects and fungal diseases. Nine of 233 accessions of *T. boeoticum* screened showed resistance to Hessian fly and nine of 237 accessions were resistant to wheat streak mosaic virus.

8.1.3.2 Cold Hardiness

A cluster of 11 CBF genes have been mapped to the frost hardiness locus (*Fr.*A^m*2*) on chromosome 5A (Miller et al. 2006) of *T. monococcum*. CBF genes (C-repeat binding factor) are transcription factors that activate cold-regulated genes, whose proteins confer cold resistance to plants. *Fr* A^m2is a complex locus that contains genes for whole plant frost survival plus controls transcript levels of cold induced genes (Knox et al. 2008). Orthologous loci have been identified in barley and wheat. The locus in *T. monococcum* is being used to dissect the components of frost hardiness in plants.

8.1.3.3 Storage Proteins

Triticum monococcum contains allelic variation for storage proteins and unique loci that could be used to enhance baking quality of bread wheat and cookies (Tranquilli et al. 2002). The main locus is located on chromosome $1A^m$. Whereas the high molecular weight glutenins of *T. monococcum* reduced cookie diameters and cookie quality and increased sedimentation volumes, the low molecular weight glutenins had opposite effects. Recombinant substitution lines between chromosome $1A^m$ and bread wheat chromosome 1A have been produced (Dubcovsky et al. 1997).

Puroindoline genes *Pina* and *Pinb* are mapped to the kernel hardness locus in chromosome $5A^m$. These loci when transferred to hexaploid wheat (Bonafede et al. 2007) will result in softer-textured kernels which require less energy in milling and reduced starch damage. The translocated segment has been defined by STS and microsatellite markers. The hardness locus has been sequenced (Chantret et al. 2004). A grain texture-related locus in *T. monococcum* was further characterized physically and genetically (Tranquilli et al. 1999). A new *Mr* (Low molecular weight glutenin) locus has been mapped to chromosome 7AL in *T. monococcum* (Dubcovsky et al. 1997).

8.2 Emmer Wheat *T. dicoccum*, *T. dicoccoides*, *T. carthlicum*, *T. araraticum*, and *T. durum*

Wild emmer was first described by Aaronsohn (1910) after its discovery in Israel. Its habitat is believed to be centered near Sanliurba in Turkey. The wild hulled form is designated as *T. dicoccoide*s Körn otherwise known as wild emmer. The cultivated hulled forms are designated as *T. dicoccum* Schrank, commonly known as emmer and *T. ispahanicum* Heslot commonly known as Ispahan wheat. As for the cultivated free threshing forms, seven species are listed according to the Dorofeyev system of classification. The six species are *T. aethiopicum* Jakubz., *T. polonicum* L., *T. turanicum* Jakubz., *T. carthlicum* Nevski, *T. jakubzineri* Udacz. and Shakhm., and *T. turgidum* L. (or Poulard wheat). All of the above species are thought to have originated as hybrids between *T. urartu* and a B genome donor such as *Ae. speltoides* or *Ae. searsii*, and all share the A and B genome. The centre of diversity of the wild forms is considered to be in the Fertile Crescent around the Mediterranean.

8.2.1 Molecular Maps

The first molecular map was constructed by Peng et al. (2000) in a population derived from crossing *T. dicoccoides* with the durum variety Langdon. The total map length exceeded 3000 cM. Genetic differentiation between parental genomes was controlled by the B genome as determined by AFLP markers. Segregation distortion markers were mapped to chromosomes 4A, 5A, and 5B.

8.2.2 Disease Resistance

8.2.2.1 Stem Rust Resistance

The first recorded introgression of stem rust resistance from an emmer to hexaploid wheat took place over 80 years ago (McFadden 1930). This source of resistance has been exploited regularly ever since. *Sr2* is a resistance gene that was introgressed from "Yaroslav" emmer into cultivars Hope and H44-24. It is effective against race TTKSK. It is carried by cultivars such as Selkirk, Redman, and Renown. Seven alleles are known for *Sr9*, located in chromosome 2B, as shown in Table 8.3. This gene is not effective against TTKSK. *Sr9d* and *9e* are not effective against race 15B, whereas *Sr9g* is a useful gene. The genes *Sr11*, *Sr12*, and *Sr17* have been widely used and deployed, whereas *Sr13* and *Sr14* are potentially useful and have not been deployed extensively as yet. 359 accessions of [*T. turgidum l. subsp. dicoccum* (Shrank) Thell] were evaluated for resistance to race TTKSK (Ug99) of stem rust (Oliver et al. 2011). It was found that 31.8 % of accessions were resistant at the seedling stage. Genetic studies on five accessions indicated that resistance was controlled by single genes. Resistance to TTTTF was also identified in some

Table 8.3 Sources of stem rust resistance in emmer wheat

Gene	Source	Chromosome location	Reference
Sr2	*T. turgidum* var *dicoccoides cv. Yaroslav*	3BS	McFadden (1930)
Sr9	*T. dicoccum* (7 alleles are known)	2B	Knott and Anderson (1956)
Sr9a	Frontiere		Green et al. (1960)
Sr9b	Kenyan wheat		Green et al. (1960)
Sr9d (*Sr1*)	*T. turgidum cv. Yaroslav*		McFadden (1930)
Sr9e (*SrV*)	*T. turgidum cv. Bernalemmer*		McIntosh and Luig (1973)
Sr9f	*T. turgidum var. durum*		Loegering (1975)
Sr9g	*T. turgidum var. durum*		McIntosh and Luig (1973)
Sr11(*Kc2*)	*T. turgidum var. durum cv. Gaza*	6B	Knott and Anderson (1956)
Sr12	*T. turgidum var. durum cv. Iumillo*	3B	Sheen and Snyder (1964)
Sr13	*T. turgidum var. dicoccum cv. Khapli*	6AL	Knott (1962)
Sr14	*T. turgidum var. dicoccum cv. Khapli*	1BL	Knott (1962)
Sr17	*T. turgidum var. dicoccum cv. Yaroslav*	7BL	McIntosh (1988)
Srdp2	*T. turgidum*	6AS	Rondon et al. (1966)

Note: of the above genes, only *Sr2* and *Sr14* are effective against Ug99

Table 8.4 Leaf rust resistance genes derived from emmer wheat

Lr gene	Chr. location	Origin	Tester source	Tester line	Gene reference	Marker reference
Lr14a	7BL	*T. turgidum*	Selkirk	RL6013	Dyck and Samborski (1968)	Herrera-Foessel et al. (2008)
Lr23	2BS	*T. turgidum*	Gabo	RL6012	McIntosh and Dyck (1975)	Nelson et al. (1997; 1975)
Lr53	6BS	*T. dicoccoides*	*T. dicoccoides*		Marais et al. (2005)	
Lr61	6BS	*T. turgidum*	Guayacan 1N1A	Guayacan 1N1A	Herrera-Foessel et al. (2008)	Herrera-Foessel et al. (2008)
Lr64	6AL	*T. dicoccoides*	*T. dicoccoides*	RL6149	JA Kolmer et al. (2010)	

of the accessions. It is expected that some of the resistance genes are novel. A gene for stem rust resistance has also been introgressed into common wheat from *T. araraticum* (Dyck 1992).

8.2.2.2 Leaf Rust Resistance

A total of 67 leaf rust resistant genes in wheat have now been defined (McCallum et al. 2012). Of this total 35 have been introduced from alien sources including primary and secondary gene pools. Five of these genes were introgressed from *T. turgidum* or *T. dicoccoides* (Table 8.4).

There is another example of a leaf rust resistance gene having been transferred to bread wheat from *T. dicoccoides* (Marais et al. 2005). Suppression of the expression of alien leaf rust resistance genes in a wheat background has always been a problem. Some of these suppressors are being mapped (Nelson et al. 1997). Markers have been assigned to three of the five genes shown in Table 8.4. In general the alien *Lr* genes are not widely deployed in wheat breeding programs (McCallum et al. 2012).

8.2.2.3 Stripe Rust Resistance

Aaronsohn (1910) who first discovered *T. dicoccoides* predicted that it would be a useful source of rust resistance. It was shown that many populations of wild emmer were valuable sources of stripe resistance, and one selection in particular, G-25 was resistant to all 21 races of the fungus (Gerechter-Amitai and Stubbs 1970).

One dominant major gene from this accession (*Yr15*) was subsequently transferred to durum and hexaploid wheat (Gerechter-Amitai et al. 1989). Some of the accessions of wild emmer carried genes that were temperature sensitive in their reaction to the fungus; for example the resistance in 10 of 26 lines was temperature sensitive (Gerechter-Amitai and Van Silfhout 1984).

The gene bank at the Punjab Agricultural University, Ludhiana, India maintains about 1000 accessions of wild *Triticum* species (Dhaliwal et al. 1993). After screening, resistance to stripe rust, leaf rust and karnal bunt was found in accessions of durum, *T. urartu*, *T. boeoticum*, *T. dicoccoides*, and *T. araraticum*. They provide a summary of the genes that were transferred to cultivated wheat from *T. dicoccoides*. *Yr15* and at least 8 other *Yr* genes were transferred. More than 60 genes for stripe rust resistance have been identified in wheat and its wild relatives. Unfortunately most of the genes have now been defeated by new races of the pathogen. In the last 10 years, more than 140 QTL have been identified for *Yr* resistance; many of these could be identical genes. 850 samples of wild emmer populations collected in Israel were evaluated for resistance to stripe rust. About 10 % of the collection was resistant to Israeli isolates of the pathogen (van Silfhout et al. 1989a, b). Fifty eight accessions of *T. dicoccoides* from 32 collection sites in Israel were screened with 21 races of the stripe rust fungus. Two accessions G7 and G25 were found to be resistant to all 21 races (Gerechter-Amitai and Stubbs 1970). A single dominant gene for resistance was identified in accession G25 and transferred to durum wheat (Gerechter-Amitai and Grama 1974). A total of 541 accessions of *T. dicoccoides* from 32 locations in Israel were screened with one accession of each of stripe rust, stem rust, and leaf rust (The et al. 1993). Stripe rust resistance showed a wide range of variability indicative of a number of genes for resistance. Only moderate levels of stem rust resistance were observed. This level of resistance was not considered adequate to transfer to commercial cultivars. *Yr15* was assigned by aneuploid analysis to chromosome 1BS and found to be linked to *Yr10* (McIntosh et al. 1996). Flanking markers were defined for *Yr15* (Sun et al. 1997).

It has been shown that a relatively small number of microsatellites can be used to estimate genetic diversity in wild accessions of *T. dicoccoides* (Fahima et al. 1998). In this case 20 stripe rust resistant accessions were screened with 23 microsatellites to obtain estimates of variability within the sample. In total, 230 alleles were detected located on 14 different chromosomes and 23 chromosome arms. A further refinement of the screening is the application of GWAM (Genome-wide association mapping). In this way three putative new *Yr* gene regions were identified (Sela et al. 2014) after screening 128 accessions of *T. dicoccoides*. Genome-Wide Association mapping (GWAM) using the wheat 9K Infinium chip was used on 128 accessions of *T. dicoccoides* to find association with stripe rust resistance loci (Sela et al. 2014). The model produced significant association with putative resistance loci on chromosome 1BS, 1BL, and 3AL. The 1BS locus was located in a region known to contain stripe rust resistance genes. In one mapping effort, the gene *YrH52* derived from *T. dicoccoides* was mapped to chromosome 1B (Peng et al. 1999).

The high temperature adult plant stripe rust resistance gene *Yr36* was mapped to chromosome 6B and found to be very tightly linked to the grain protein content locus (Uauy et al. 2005). This study was made possible by the use of recombinant substitution lines of chromosome 6B of *T. dicoccoides* in a Langdon background. Thus far, 140 QTL for stripe rust resistance in wheat have been identified, spread over 49 chromosome regions on wheat maps (Rosewarne et al. 2013). Many of these QTL are likely identified genes. Many of these genes have already been defeated by alterations in the organism. It is therefore necessary to search for additional resistance genes. The wild emmer wheat *Triticum turgidum* spp. *dicoccoides* is proving to be one of these sources of new resistance genes.

8.2.2.4 Powdery Mildew Resistance

All of the resistant reactions shown in Table 8.5 are controlled by single genes. Some genes such as *Pm4* are effective at the seedling and adult plant stages. Some loci have numerous alleles such as *Pm4* with 4 and *Pm5* with 5. According to Dhaliwal et al. (1993), the *T. dicoccoides*-derived genes *Pm4a*, *Pm5*, *Pm16* were all integrated into cultivated wheat. Molecular markers have now been assigned to four of the emmer-derived *Pm* genes shown in Table 8.5. A dominant powdery mildew resistance gene *PmAS846* was transferred from *T. dicoccoides* to hexaploid

Table 8.5 Powdery mildew resistance genes introgressed from wild emmer or durum

Gene/allele	Chromosome location	Cultivar/Line	Source	Reference
Pm3h	1AS	Abessi	*T. durum*	Zeller and Hsam (1998)
Pm4a[a]	2AL	Khapli	*T. dicoccum*	The et al. (1979)
Pm5a	7 BL	Hope	*T. dicoccum*	Law and Wolfe (1973)
Pm16	4A	Noman rec. line	*T. dicoccoides*	Reader and Miller (1991)
Pm26	2BS	TTD140	*T. dicoccoides*	Rong et al. (2000)
Pm30[a]	5BS	C20	*T. dicoccoides*	Liu et al. (2002)
Pm31[a]	6AL	G-305-M/781//Jin411[a]3	*T. dicoccoides*	Xie et al. (2003)
*MlZeo*l	2BL	Zecoi-1	*T. dicoccoides*	Mohler et al. (2005)
				Hua et al. (2009)
Pm42	2BS	P63	*T. dicoccoides*	Piarulli et al. (2012)
Pm50	2AL	K2	*T. dicoccum*	Mohler et al. (2013)
MlIW72[a]	7A	IW72	*T. dicoccoides*	Ji et al. (2008)

[a]Indicates that genes have been mapped

wheat and mapped to a gene-rich region on chromosome 5BL (Xue et al. 2012). An incompletely dominant powdery mildew resistance gene that gave resistance at the seedling and adult plant stages was detected in *T. dicoccoides* (Liu et al. 2012). It was mapped to the distal region of chromosome 2BS. According to Dhaliwal et al. (1993) the powdery mildew resistance genes *Pm4a*, *Pm5* and *Pm16* from *T. dicoccum* were transferred to cultivated wheat. As were *Pm*46 from *T. carthlicum*, *Pm10* & *Pm11* from *T. spelta*, *Pm15* from *T. compactum*, and *Pm36* from *T. sphaerococum*. The emmer wheats are a good source of resistance to powdery mildew. For example, 233 *T. dicoccoides*accessions were collected at ten sites in Israel and inoculated with four cultures of the fungus (Moseman et al. 1984). It was determined that 49 % of the accessions were resistant. The conclusion was that the number of resistance sources in *T. dicoccoides* was unlimited. It has been stated (Huang and Röder 2004) that up to that time, 32 loci with 48 genes/alleles for mildew resistance have been identified and located on various chromosomes. Eleven of those resistance genes are shown in Table 8.5.

8.2.2.5 Fusarium Head Blight Resistance

Although wild emmer wheats, *T. dicoccoides* are good sources of resistance to such diseases as stem rust, there is limited variability for resistance to Fusarium head blight in this species. For example, 151 *T. dicoccoides* genotypes originating from 16 habitats in Israel and one habitat in Turkey were screened by point inoculation in greenhouse conditions. Most of the accessions proved to be highly susceptible and only a few were moderately resistant (Buerstmayr et al. 2003). Similar results were obtained by several other labs when screening *T. dicoccoides* accessions for FHB resistance. Wan et al. (1997) screened 21 *T. dicoccoides* accessions and found that five were moderately resistant while the remainder were susceptible to highly susceptible. Among 97 tetraploids screened by Gilbert and Tekauz (2000), six were rated as moderately resistant. A total of 282 *T. dicoccoides* accessions were screened by Miller et al. (2006) and most were found to be highly susceptible, with only a few lines rating as moderately resistant.

In another study, 416 accessions of *T. dicoccoides* were screened over multiple seasons and environments (Oliver et al. 2005, 2007). Two accessions PI 481521 and PI 478742 had consistently low FHB scores across four seasons. Substitution lines were developed with chromosomes from these lines in a Langdon background. QTL for resistance in the latter were located on chromosomes 7A and 7B (Stack et al. 2002) while in the former, resistance QTL were located on chromosome 1A, 3A, 5B, and 7A (Stack et al. 2003). In continued screening of tetraploid wheat for FHB resistance (Oliver et al. 2008) a total of 376 accessions of five subspecies of *T. turgidum* including subsp. *carthlicum*, subsp. *dicoccum*, subsp. *polonicum*, subsp. *turanicum*, and subsp. *turgidum* were evaluated. Point inoculation was applied for three seasons and grain inoculation at two locations. Sixteen *T. turgidum* subsp. *carthlicum* and four subsp. *dicoccum* accessions consistently gave resistant or moderately resistant reactions. It was shown that the main FHB resistance QTL in tetraploid wheat maps to chromosome 3AS and is not homologous to the locus on chromosome 3BS of Sumai3 (Chen et al. 2007). A novel FHB resistance QTL was mapped to chromosome 7A in tetraploid wheat (Kumar et al. 2007). This is only the second major FHB QTL mapped in tetraploid wheat. Its R^2 value was 19 %. Tetraploid wheat accessions from Tunisia were a source of additional resistance to FHB. Twenty such accessions were haplotyped with markers for known FHB loci (Huhn et al. 2012). Based on these results there were indications that those accessions likely carry novel sources of Type II FHB resistance. Moderate FHB resistance was detected in one accession of *T. carthlicum* (4×), designated as "Blackbird" (Fig. 8.3). This accession was crossed to Strongfield and a DH mapping population of 85 lines was produced (Somers et al. 2006; Han and Fedak 2003). Two QTL for FHB resistance were reported in this population. A QTL on chromosome 6BS was derived from Blackbird and one on 2BL from Strongfield (Fig. 8.4). The QTL on chromosome 6BS was coincident with the known FHB resistance QTL *Fhb2* derived from Sumai3. Two other loci located on chromosomes 5AS and 2AL had on epistatic effect on the locus on 6BS (Singh et al. 2008).

Three mapping populations derived from crosses of a resistant *T. dicoccum* donor line with three Austrian recipient cultivars were produced (Buerstmayr et al. 2012). Resistant QTL were detected on chromosomes 3B, 4B, 6A, 6B, and 7B; all QTL except for 3B were contributed by *T. dicoccum*. All QTL except for 6A mapped to genomic regions where FHB QTL have previously been reported in hexaploid

Fig. 8.3 FHB resistance expressed on spike of strain "Blackbird" of *T. carthlicum* at 21 days after inoculation

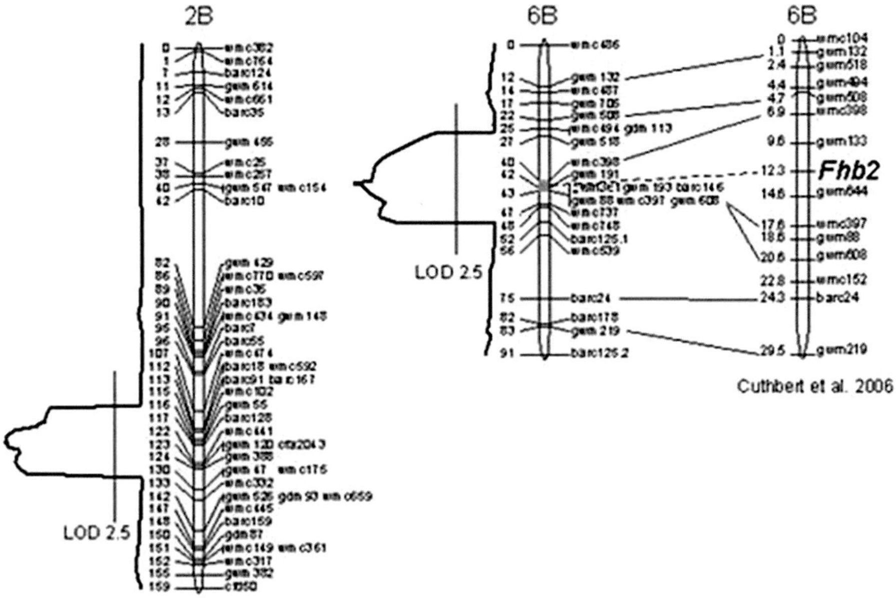

Fig. 8.4 FHB resistance QTL from *T. carthlicum* mapping population. The 6B QTL is from *T. carthlicum* and 2BL from Strongfield, the other parent in the mapping population

wheat. The QTL on 3B and 6B were coincident with *Fhb1* and *Fhb2*. The 4B QTL collocated with a major QTL for plant height and the 7B QTL overlapped with a QTL for flowering time. A complete series of *T. dicoccoides* substitution lines were produced in Langdon durum by L.R. Joppa of USDA, Fargo ND (Joppa and Williams 1988). After point inoculation with FHB it was found that the substitution line LDN (DIC-3A) was consistently less susceptible than Langdon and another line LDN (DIC2A) was more susceptible (Stack et al. 2002). Several other substitution lines such as 2A and 6B also showed lower percentages of infected kernels indicating that these chromosomes also carried QTL for FHB resistance. The QTL on *T. dicoccoides* chromosome 3A was found to account for 37 % of the phenotypic variability plus occupied a 29.3 cM region between the marker *Xmwg14* to *Xbcd828* (Otto et al. 2002). Using the Langdon substitution lines produced by Leonard it was found that *T. dicoccoides* substitution lines LDN (DIC 3A) and LDN (DIC 3B) had about the same level of FHB resistance as the parental cultivar Langdon. However the substitution line for chromosome 3A completely suppressed the bleaching of the heads caused by *F. graminearum*. It was concluded that chromosome 3A of *T. dicoccoides* carries gene(s) for resistance to head bleaching following Fusarium infection. In none of the studies did any accessions approach the level of Sumai3 for FHB resistance. However the FHB resistance in durum cultivars is so low, that some of the variability observed in the screening mentioned above, might be useful in providing limited incremental improvements in FHB resistance. Despite the paucity of large numbers of effective FHB QTL in tetraploid wheat, incremental progress is being made in improving the FHB resistance of durum wheat (Clarke et al. 2004; Comeau et al. 2008).

8.2.3 Other Factors

8.2.3.1 Virus Resistance

A large collection of wild diploid and tetraploid wheat was screened for resistance to a number of insect pests and fungal pathogens (Gill et al. 1983). Four out of 25 *T. dicoccoides* accessions tested were resistant to WSMV. No resistance was detected to Hessian fly or greenbug.

8.2.3.2 Leaf Spotting

The set of Langdon–*Triticum dicoccoides* durum substitution lines were screened with three components of leaf spotting disease complex (Singh et al. 2006). The location of genes for resistance to tan spot, caused by *Pyrenophora tritici-repentis* was on chromosomes 5B and 3B. The resistance to toxin Ptr Tox A of Stagonospora blotch was located on chromosome 5A.

8.2.3.3 Storage Proteins

One of the important factors in durum wheat production is that of protein content. Wild emmer wheat has the potential to transfer this trait to durum wheat. High grain protein content (GPC) was first reported in *T. turgidum var. dicoccoides* in 1978 (Avivi 1978). Three QTL for GPC were identified on chromosome 5B of *T. turgidum var. dicoccoides* (Gonzalez-Hernandez et al. 2004). In an earlier report, a high GPC QTL was described on chromosome 6B of the same species (Chee et al. 2001). In addition to high GPC, some dicoccoides accessions also had novel glutamic subunits and gliadins, located on chromosome 1B, that could be used to alter storage protein patterns of durum wheat (Xu et al. 2004). Loci controlling levels of storage proteins in *T. dicoccoides* were determined by studying *T. dicoccoides* disomic substitution lines in a Langdon background (Joppa and Cantrell 1990). Disomic substitution lines of *dicoccoides* chromosomes 6B, 2A and 5B had lower protein levels than Langdon, while substitution lines 3B, 4B and 7B had higher levels. A gene for high protein concentration was mapped on chromosome 5B of *T. dicoccoides* (Gonzalez-Hernandez et al. 2004). A high grain protein locus from *T. dicoccoides* was mapped close to the centromere on chromosome 6BS (Joppa et al. 1997). PCR markers were developed for the 6BS QTL (Khan et al. 2000).

In tetraploid and hexaploid wheat, the low molecular weight glutenin loci (*Mr*) were mapped at the *X glu-3* locus in the distal regions of chromosomes 1AS, 1BS, and 1DS. A new *Mr* locus mapped to chromosome 7AL in *T. monococcum* (Dubcovsky et al. 1997). Glaucousness inhibitor *IW3*gene has been identified in *Triticum dicoccoides* (Wang et al. 2014). *IW3* inhibits glaucousness formation by altering wax composition. The gene has been mapped to chromosome 1BS.

8.2.3.4 Insect Resistance

Hessian Fly Resistance

Resistance to Hessian fly was found in 65 of 147 accessions of *T. araraticum* screened (Gill et al. 1983). No resistance to greenbug or WSMV was detected in this collection. A number of Hessian fly resistance genes have been transferred to hexaploid wheat from *T. tauschii*. One such resistance gene was transferred from *T. turgidum ssp. dicoccum* and mapped to the short arm of chromosome 1A (Liu et al. 2005). This is the first emmer-derived insect resistance gene to be mapped and characterized. To date, a total of 14 *Hf* resistance genes have been transferred from *T. turgidum* ssp. durum to hexaploid wheat. More recently the gene *Hdic* has been identified in *T. dicoccum* and transferred to hexaploid wheat (Liu et al. 2005). This introgressed gene mapped to wheat chromosome 1A in the same region as *H9*, *H10*, and *H11*.

8.3 *Triticum timopheevii*

The wild hulled form of *T. timopheevii* is designated as *Triticum timopheevii* (Zhuk.) Zhuk subsp. armeniacum (Jakubz) Slageren (Traditionally known as *Triticum araraticum* Jakubz). The domesticated hulled form is classified as *Triticum timopheevii* (Zhuk.) Zhuk subsp. *timopheevii*. The cultivated free threshing form is designated as *T. militinae* Zhuk & Migush. In the wild the species is found, according to some accounts, in southern zones of *T. dicoccoides* (AABB), distribution which occurs in the Fertile Crescent. The genome of *T. timopheevi* is designated as AAGG. Its origin is not entirely clear, but appears to be a double mutation from an AABB genotype that controls chromosome asynapsis that lead to reproductive isolation, *T. timopheevii* has a diploidizing mechanism similar to the *Ph1b* system in hexaploid wheat (Martinez et al. 1996). There are not many reports of gene transfer from *T. timopheevii* to wheat. It is initiated by normal crossing and embryo rescue followed by up to three backcrosses to restore fertility. Interest in *T. timopheevii* as a source of disease resistance for bread wheat goes back about 70 years. Wheat-like progeny with resistance to stem rust, leaf rust and mildew were reported by Pridham (1939) and Shands (1941). Semeniuk (1947) reported detailed cytogenetic studies of advanced generations of stem and leaf rust resistant progeny obtained from the hybrid of

Steinwedel (a soft white wheat) and *T. timopheevii*. Allard (1949), provided detailed cytogenetic studies of progenies of wheat and *T. timopheevii* hybrids in the process of transfer of disease resistance to hexaploid wheat. In addition to resistance to stem rust and mildew in his progenies, Allard (1949) also reported bunt resistance.

Subsequent screening results report a high frequency of accessions carrying resistance to six different pests. A total of 301 accessions of *T. timopheevii* var. *araraticum* were screened for resistance to leaf rust, stem rust, stripe rust, powdery mildew, tan spot, wheat curl mite, and Hessian fly (Brown-Guedira et al. 1996). A large number of accessions were recognized with R ratings for five of the pests. For stripe rust, only MR levels were observed.

8.3.1 Disease Resistance

8.3.1.1 Stem Rust Resistance

Three wheat lines with stem rust derived from *T. timopheevii* have been reported. One strain was CI12632 (W1656) with resistance introgressed to chromosome 2B (Nyquist 1957). Line W (W3563) was derived from selections in W1906/*T. timopheevii* W 1899/2/3 Steinwedel W199 (McIntosh and Gyarfas 1971). CI 13005 was another *T. timopheevii*-derived line that gave moderate resistance to race 15B (Atkins 1967). *Timopheevii* with stem rust resistance is another wheat strain derived from *T. timopheevii* (Watson and Luig 1958). In a subsequent report the *timopheevii-derived* lines CI 12632 and CI 12633 were described as having adult plant resistance to leaf rust, stem rust and mildew (Allard and Shands 1954). CI12633 was also reported to be highly resistant to loose smut (Table 8.6). The timopheevii wheats were screened for resistance to rusts and powdery mildew and variability for both traits was observed (Tomar et al. 1988).

8.3.1.2 Leaf Rust Resistance

A QTL for adult plant leaf rust resistance was transferred to wheat from a synthetic allopolyploidderived from *T. timopheevii*/*T. tauschii*. The gene originated in chromosome 2G and was mapped to chromosome 2B with microsatellite markers. It explained 31 % of the total variance (Leonova et al. 2007).

8.3.1.3 Powdery Mildew

Powdery mildew resistant gene *Pm27* was introgressed from chromosome 6G, of *T. timopheevii* elite wheat and mapped to chromosome 6BL by RFLP and microsatellite analysis (Järve et al. 2000). Two unlinked dominant genes for mildew resistance were introgressed from *T. timopheevii* into winter wheat line TP114.

One of the genes was identical to *Pm2*. The other was a unique gene located on chromosome 2BL and designated as *Pm6* (Jorgensen and Jensen 1973). *Pm6* has since been mapped to chromosome 2BL with DNA markers. The linked marker BCD135 has been used to detect *Pm6* in various backgrounds (Tao et al. 2000).

According to Dhaliwal et al. (1993) the *T. timopheevii* derived mildew resistance gene *Pm6* was transferred to hexaploid wheat.

8.3.1.4 Fusarium Head Blight Resistance

Fusarium head blight resistance was discovered in accession PI 343447 of *T. timopheevii* (Cao et al. 2009). This accession was crossed as a pollen parent to the variety Crocus which has the three crossability genes *Kr1*, *Kr2*, *Kr3* (dominant *Kr* genes). The F1 and BC1 embryos were excised and raised on an artificial medium. A population of 535 BC1F7 lines were established by SSD. One hundred lines were selected from this population based on plant fertility and agronomic traits. The lines were evaluated for the FHB reaction in field and greenhouse trials in four replicates and two seasons. From this trial line TC67 was selected, based on its superiority in FHB reaction and agronomic traits (Fig. 8.5) (Table 8.7).

Fig. 8.5 Resistance to Fusarium head blight expressed in TC67 an introgression from *T. timopheevii*. Disease symptoms expressed at 21 days after inoculation. Roblin is the susceptible check

Table 8.6 Stem rust resistance genes from *T. timopheevii*

Gene	Source	Chromosome location	Reference
Sr36 (*SrTt1*)	*T. timopheevii*	2B	McIntosh (1988)
Sr37 (*SrTt2*)	*T. timopheevii*	4B	McIntosh and Gyarfas (1971)
SrTt3	*T. timopheevii*	2B	Gyarfas (1978)
CI12632 (W1656)	*T. timopheevii*	2B	Nyquist (1957)
Linew (W3563)	*T. timopheevii*		Nyquist (1957)
CI13005	*T. timopheevii*		Atkins (1967)
	T. timopheevii		Watson and Luig (1958)
CI12632	*T. timopheevii*		Allard and Shands (1954)
CI12633	*T. timopheevii*		Allard and Shands (1954)

Table 8.7 Means of Fusarium head blight (FHB) incidence, FHB severity, Fusarium-damaged kernels (FDK) and deoxynivalenol (DON) for three spring wheat checks and the line TC67 (derived from *T. timopheevii*) in the field FHB nursery

Line	Incidence (%)	Severity (%)	FDK (%)	DON (mgL^{-1})
Sumai3	14.4	6.3	9.5	6.6
HY644	43.6	20.0	35.1	25.3
Roblin	90.7	81.9	66.1	35.0
TC67	29.4	16.9	14.5	7.7

The FHB reaction of line TC67 is shown in Table 8.7, compared to three check cultivars with highly tolerant (Sumai3), partially tolerant (HY644), and susceptible (Roblin) reactions. As shown in Table 8.7 the reaction of TC67 to four parameters of FHB evaluation was slightly lower than Sumai3, but exceeded the other two checks. Compared to Sumai3, TC67 had somewhat higher values for Incidence and Severity and FDK. One of the most important factors in FHB resistance is the DON content of the grain following inoculation. For this parameter, the values for TC67 at 7.7 ppm are only slightly higher than Sumai3 at 6.6 ppm. In order to map the resistance of TC67, it was crossed to AC Brio a susceptible cultivar and a mapping population of 235 lines were produced by SSD (single seed descent). This population was grown in replicated trails in field plots at two locations for two seasons plus replicated trails in greenhouses. A major QTL for resistance was mapped on chromosome 5A (Malihipour et al. 2015).

8.4 *Triticum macha* (AABBDD Genome)

There are not many reports of variability for disease resistance in *T. macha*. Two reports were found describing FHB resistance in that species. In one report (Steed et al. 2005), type I FHB resistance was mapped to the short arm of chromosome 4A.

In another study (Buerstmayr et al. 2011) resistance QTL in *T. macha* were mapped to chromosomes 2A, 2B, 5A, and 5B. The 5A QTL mapped very tightly to the spelt-type locus.

8.5 *Triticum miguschovae*

Triticum miguschovae is a synthetic hexaploid (AAGGDD) derived from crossing *T. militinae* Zhuk and Migush (genome formula A^bG) with *Aegilops squarossa* (Zhirov 1980). The amphiploid is stable and fully fertile. It has overly thick outer glumes as shown in Fig. 8.6 and is difficult to thresh. In our screening tests, it was found to have a degree of resistance to FHB as shown in Fig. 8.6 (Fedak et al. 2011). This resistance is currently being mapped. With backcrossing and selection the resistance was carried over to BC_2. Resistance to leaf rust at the adult plant stage was detected in the amphiploid *T. miguschovae* and transferred to bread wheat (Davoyan and Ternovskaya 1996). The resistance was controlled by a pair of complementary genes located on chromosomes 7B and 1D. The latter resistance would have come from *Ae. squarrosa* and the former from *T. militinae*. The amphiploid also had resistance to stem rust, yellow rust, and mildew. Powdery mildew resistance was introgressed into bread wheat from *T. militinae* (Jakobson et al. 2012). The QTL responsible for the resistance were mapped to chromosomes 4AL, 5AL, and 7AL. The resistance was non race specific and effective at seedling and adult plant stages.

Fig. 8.6 Resistance to Fusarium head blight expressed on spike of *Triticum miguschovae* (AGD) (R) at 21 days after inoculation. Roblin (L) is the susceptible check

8.6 Conclusions

There is extensive variability for resistance to disease and pests and value added traits in wild wheats and species related to them. Even in the primary and secondary gene pools covered in this text there is extensive variability. There is much more in the tertiary gene pool. The variability in the primary and secondary gene pool is readily accessible. Genes can be transferred by simple crossing then backcrossing. Embryo rescue needs to be used in many combinations. Of the totals of genes that have been identified and transferred to wheat relatively few have been deployed. The deployment is currently being accelerated. This may be attributed to more effort in developing molecular markers for the genes and using these to "pyramid" the genes. Technologies such as doubled haploids are being employed to permit selection of multiple genes in reduced population sizes. Pyramids might consist of multifunctional genes such as *Lr34*, *Sr33*, *Sr42*, *Fhb1* (Zhang et al. 2014) or multiple genes for one function such as *Sr33*, *Sr36*, *Sr42*, and *Sr43* (Jin et al. 2014). In the case of the latter, it was produced to give prolonged resistance to stem rust race Ug99. Three of the four genes involved in that pyramid have been introgressed from alien sources.

Acknowledgements The author wishes to acknowledge the excellent input and assistance provided by Dawn T. Chi in typing, formatting the manuscript and arranging the figures and tables.

References

Aaronsohn A (1910) Agricultural and botanical explorations in Palestine, vol 180, Bull. USDA Bureau of Plant industry, Washington, DC

Allard RW (1949) A cytogenetic study dealing with the transfer of genes from *Triticum timopheevi* to common wheat by backcrossing. J Agric Res 78:33–64

Allard RW, Shands RG (1954) Inheritance of resistance to stem rust and powdery mildew in cytologically stable spring wheat derived from *Triticum timopheevii*. Phytopathology 44:266–274

Anker CC, Niks RE (2001) Prehaustorial resistance to the wheat leaf rust fungus, *Puccinia triticina* in *Triticum monococcum*(s.s.). Euphytica 117:209–215

Atkins RE (1967) Inheritance of reaction to race 15B of wheat stem rust in derivatives of Frontana and *Triticum timopheevi* Zhuk. Iowa State J Sci 41:305–311

Avivi L (1978) High protein content in wild tetraploid *Triticum dicoccoides* Körn. In: Ramanujam S (ed) Proceedings 5th International Wheat Genetics Symposium. Indian Soc Genet and Plant Breed, New Delhi, pp 372–380

Bai D, Knott DR, Zale JM (1998) The inheritance of leaf and stem rust resistance in *Triticum monococcum* L. Can J Plant Sci 78:223–226

Bonafede M, Kong L, Tranquilli G, Ohm H, Dubcovsky J (2007) Reduction of a *Triticum monococcum* chromosome segment carrying the softness genes *Pina* and *Pinb* translocated to bread wheat. Crop Sci 47:821–828

Brown-Guedira GL, Gill BS, Bockus WW, Cox TS, Hatchett JH, Leath S, Peterson CJ, Thomas JB, Zwer PK (1996) Evaluation of a wild collection of *timopheevi* wheat for resistance to disease and arthropod pests. Plant Dis 80:928–933

Buerstmayr H, Stierschneider M, Steiner B, Lemmens M, Griesser M, Nevo E, Fahima T (2003) Variation for resistance to head blight caused by *Fusarium graminearum* in wild emmer (*Triticum dicoccoides*) originating from Israel. Euphytica 130:17–23

Buerstmayr M, Lemmens M, Steiner B, Buerstmayr H (2011) Advanced backcross QTL mapping of resistance to Fusarium head blight and plant morphological traits in a *Triticum macha* x *T. aestivum* population. Theor Appl Genet 123:293–306

Buerstmayr M, Huber K, Heckmann J, Steiner B, Nelson JC, Buerstmayr H (2012) Mapping of QTL for Fusarium head blight resistance and morphological and developmental traits in three backcross populations derived from *Triticum dicoccum* x *Triticum durum*. Theor Appl Genet 25:1751–1765

Cai X, Chen PD, Xu SS, Oliver RE, Chen X (2005) Utilisation of alien genes to enhance Fusarium head blight resistance in wheat: a review. Euphytica 142:309–318

Cao W, Fedak G, Armstrong K, Xue A, Savard ME (2009) Registration of spring wheat germplasm TC67 resistant to Fusarium head blight. J Plant Registr 3:104–106

Chantret N, Cenci A, Sabot F, Anderson O, Dubcovsky J (2004) Sequencing of the *Triticum monococcum* hardness locus reveals good microcolinearity with rice. Mol Genet Genomics 271:377–386

Chee PW, Elias EM, Anderson JA, Kianian SF (2001) Evaluation of a high grain protein QTL from *Triticum turgidum* L. var *didoccoides* in an adopted durum wheat background. Crop Sci 41:295–301

Chen XF, Faris JD, Hu JG, Stack RW, Adhikari T, Elias EM, Kianian SF, Cai XW (2007) Saturation and comparative mapping of a major Fusarium head blight resistance QTL in tetraploid wheat. Mol Breed 19:113–124

Chhuneja P, Kaur S, Garg T, Ghai M, Kaur S, Prashar M, Bains NS, Goel RK, Keller B, Dhaliwal HS, Singh K (2008) Mapping of adult plant stripe rust resistance genes in diploid A genome wheat species and their transfer to bread wheat. Theor Appl Genet 116:313–324

Chhuneja P, Kumar K, Stirnweis D, Hurni S, Keller B, Dhaliwal HS, Singh K (2012) Identificaation and mapping of two powdery mildew resistance genes in *Triticum boeoticum* L. Theor Appl Genet 124:1051–1058

Clarke J, Thomas J, Fedak G, Somers D, Gilbert J, Pozniak C, Fernandez M, Comeau A (2004) Progress in improvement of Fusarium resistance of durum wheat. In: Canty SM, Boring T, Wardwell J, Ward RW (eds) Proceedings of the 2nd International Symposium on Fusarium head blight; incorporating the 8th European Fusarium Seminar, Orlando, FL, 11–15 December 2004. Michigan State University Publisher, East Lansing, MI, p 43

Comeau A, Langevin F, Caetano VR, Haber S, Savard ME, Voldeng HD, Fedak G, Dion Y, Rioux S, Gilbert J, Somers DJ, Martin RA (2008) A systemic approach for the development of FHB-resistant germplasm accelerates progress. Cereal Res Commun 36:5–9

Davoyan RD, Ternovskaya TK (1996) Use of a synthetic hexaploid *Triticum miguschovae* for transfer of leaf rust resistance to common wheat. Euphytica 89:99–102

Dhaliwal HS, Singh H, Gill KS, Randhawa HS (1993) Evaluation and cataloguing of wheat germplasm for disease resistance and quality. In: Damania AB (ed) Biodiversity and wheat improvement. Wiley, London, pp 123–140

Dhaliwal HS, Chhuneja P, Singh I, Ghai M, Goel RK, Garg M, Keller B, Roder M, Singh K (2005) Triticum monococcum a novel source of transfer and exploitation of disease resistance in wheat. In: Pogna NE, Romeno M, Pogna EA, Galterio G (eds.), Proceedings of the 10th International Wheat Genetics Symposium, Paestum, Italy, pp 346–349

Dubcovsky J, Luo MC, Zhong GY, Bransteitter R, Desai A, Kilian A, Kleinhofs A, Dvoràk J (1996) Genetic map of diploid wheat, *Triticum monococcum L.*, and its comparison with maps of *Hordeum vulgare* L. Genetics 143: 983–999

Dubcovsky J, Echaide M, Giancola S, Rousset M, Luo MC, Joppa LR, Dvorak J (1997) Seed-storage-protein loci in RFLP maps of diploid, tetraploid, and hexaploid wheat. Theor Appl Genet 95:1169–1180

Dyck PL (1992) Transfer of a gene for stem rust resistance from *Triticum araraticum* to hexaploid wheat. Genome 35:788–792

Dyck PL, Samborski DJ (1968) Host-parasite interactions involving two genes for leaf rust resistance in wheat. In: Findley FW, Shepherd KW (eds) Proceedings of the Third International Wheat Genetics Symposium. Australian Academy of Science, Canberra, ACT, pp 245–250

Fahima T, Röder MS, Grama A, Nevo E (1998) Microsatellite DNA polymorphism divergence in *Triticum dicoccoides* accessions highly resistant to yellow rust. Theor Appl Genet 96:187–195

Fedak G (2000) Sources of resistance to Fusarium head blight. In: Raupp WJ, Ma Z, Chen PD, Liu DJ (eds.), Proceedings of the International Symposium on Wheat Improvement for Scab Resistance, Suzhou and Nanjing, China, p 4

Fedak G, Burvill M, Voldeng HD (1997a) A comparison of anther cultivar and maize pollination for haploid production in wheat. J Appl Genet 38:407–414

Fedak G, Armstrong KC, Sinha RC, Gilbert J, Procunier JD, Miller JD, Pandeya R (1997b) Wide crosses to improve Fusarium head blight resistance. Cereal Res Commun 25:651–654

Fedak G, Han F, Cao W, Burvill M, Kritenko S, Wang L (2003) Identification and characterization of novel sources of resistance to FHB In: Pogna Ne E, Romano M, Pogna EA, Galterio G et al. (eds.), Proceedings of the 10th International Wheat Genetics Symposium, Paestum, Italy, September 1–6, 2003, pp 354–356

Fedak G, Cao W, Han F, Savard MJ, Gilbert J, Xue A (2004) Germplasm enhancement for FHB resistance in spring wheat through alien introgression. In: Canty SM, Boring T, Versdahl K, Wardwell J, Ward RW (eds.), Proceedings of the 2nd International Symposium on Fusarium head Blight, Incorporating the 8th European Fusarium Seminar, Wyndham Orlando Resort, Orlando, FL, USA, December 11–15, 2004, p 56

Fedak G, Cao W, Xue A, Savard MJ, Gilbert J, Clarke J, Somers D (2006) Fusarium head blight resistance from wide crosses in bread wheat and durum. In: Ban T, Lewis JM, Phipps EE (eds.), The Global Fusarium Initiative for International Collaboration; a strategy planning workshop held at Cimmyt-El Batan, Mexico, March 14–17, 2006, pp 20–24

Fedak G, Cao W, Xue A, Savard, MJ, Clarke J, Somers DJ (2007) Enhancement of Fusarium head blight resistance in bread wheat and durum by means of wide crosses. In: Buck HT, Nisi JE, Salomon N (eds.), Wheat Production in Stressed Environments. Proc 7th Int Wheat Conference, 27 November–2 December, 2005, Mar del Plata, Argentina, pp 91–95

Fedak G, Cao W, Chi D, Somers D, Miller S, Ouellet T, Xue A, Gilbert J, Savard M, Voldeng H (2011) New sources of resistance to Fusarium head blight and their mode of action. In: Canty S, Clark A, Anderson-Scully A, Van Sanford D (eds.), Proceedings of the 2011 National Fusarium Head Blight Forum, East Lansing, MI, pp 19–22

Gerechter-Amitai ZK, Grama A (1974) Inheritance of resistance to stripe rust (*Puccinia striformis*) in crosses between wild emmer (*Triticum dicoccoides*) and cultivated tetraploid and hexaploid wheat. I. Triticum durum. Euphytica 23:387–392

Gerechter-Amitai ZK, Stubbs RW (1970) A valuable source of yellow rust resistance in Israeli populations of wild emmer, *Triticum dicoccoides* körn. Euphytica 19:12–21

Gerechter-Amitai ZK, Van Silfhout CH (1984) Resistance to powdery mildew in wild emmer (*Triticum dicoccoides* Körn). Euphytica 33:273–280

Gerechter-Amitai ZK, Wahl I, Vardi A, Zohary D (1971) Transfer of stem rust seedling resistance from wild diploid einkorn to tetraploid durum by means of a triploid hybrid bridge. Euphytica 20:281–285

Gerechter-Amitai ZK, Van Silfhout CH, Grama A, Kleitman F (1989) *Yr*15 - a new gene for resistance to *Puccinia striiformis* in *Triticum dicoccoides* sel. *G*-25. Euphytica 43:187–190

Gilbert J, Tekauz A (2000) Review: recent developments in research on Fusarium head blight of wheat in Canada. Can J Plant Pathol 22:1–8

Gill BS, Browder LE, Hatchett JH, Harvey TL, MartinTJ, Raupp WJ, Sharma HC, Waines JG (1983) Disease and insect resistance in wild wheats. In: Sakamoto S (ed.), Proc 6th Int Wheat Genetics Symposium, Kyoto, Japan, November 28–December 3, 1983, pp 785–792

Gonzalez-Hernandez JL, Elias EM, Kianian SF (2004) Mapping genes for grain protein concentration and grain yield on chromosome 5B of *Triticum turgidum* (L.) var. *dicoccoides*. Euphytica 139:217–225

Green GJ, Knott DR, Watson IA, Pugsley AT (1960) Seedling reactions to stem rust of lines of Marquis wheat with substituted genes for rust resistance. Can J Plant Sci 40:524–538

Gyarfas J (1978) Transference of disease resistance from Triticum timopheevii to Triticum aestivum. MSc thesis, University of Sydney, NSW, Australia.

Han F, Fedak G (2003) Molecular characterization of partial amphiploids from *Triticum durum* x tetraploid *Thinopyrum elongatum* as novel sources of resistance to wheat Fusarium head blight In: Pogna NE, Romano M, Pogna EA, Galteno G (eds.), Proc 10th Int Wheat Genet Symposium III, Paestum, Italy, pp 1148–1150

Harlan JR (1971) Agricultural origins: centers and noncenters. Science 174:468–474

Harlan JR, Zohary D (1966) Distribution of wild wheats and barleys. Science 153:1074–1080

Herrera-Foessel SA, Singh RP, Huerta-Espino J, William HM, Garcia V, Djurle A , Yuen J (2008) Identification and molecular characterization of leaf rust resistance gene *Lr14a* in durum wheat. Plant Dis 92:469–473

Hsam SLK, Huang XQ, Ernst F, Hartl L, Zeller FJ (1998) Chromosomal location of genes for resistance to powdery mildew in common wheat (*Triticum aestivum* L. em Thell.). 5. Alleles at the *Pm*1 locus. Theor Appl Genet 96:1129–1134

Hua W, Liu Z, Zhu J, Xie C, Yang T, Zhou Y, Duan X, Sun Q, Liu Z (2009) Identification and genetic mapping of *Pm42*, a new recessive wheat powdery mildew resistance gene derived from wild emmer (*Triticum turgidum var. dicoccoides*). Theor Appl Genet 119:223–230

Huang X-Q, Röder MS (2004) Molecular mapping of powdery mildew resistance genes in wheat: a review. Euphytica 137:203–223

Huhn M, Elias EM, Ghavami F, Kianian SF, Chao SM, Zhong SB, Alamri MS, Yahyaoui A, Mergoum M (2012) Tetraploid Tunisian wheat germplasm as a new source of *Fusarium* head blight resistance. Crop Sci 52:136–145

Hussein T, Bowden RL, Gill BS, Cox TS, Marshall DS (1997) Performance of four new leaf rust resistance genes transferred to common wheat from *Aegilops tuaschii* and *Triticum monococcum*. Plant Dis 81:582–586

Jakobson I, Reis D, Tiiderma A, Peusha H, Timofejeva L, Valarik M, Kladivova M, Simkova H, Doležel J, Järve K (2012) Fine mapping, phenotypic characterization and validation, of non-race-specific resistance to powdery mildew in a wheat-*Triticum militinae* introgression line. Theor Appl Genet 25:609–623

Järve K, Peusha HO, Tsymbalova J, Tamm S, Devos KM, Enno TM (2000) Chromosomal location of a *Triticum timopheevii* - derived powdery mildew resistance gene transferred to common wheat. Genome 43:377–381

Ji X, Xie C, Ni Z, Yang T, Nevo E, Fahima T, Liu Z, Sun Q (2008) Identification and genetic mapping of a powdery mildew resistance gene in wild emmer (*Triticum dicoccoides*) accession IW72 from Israel. Euphytica 159:385–390

Jin X, Chi D, Zhang B, Xue A, Cao W. Fedak G (2014) Wheat germplasm development by gene pyramiding for resistance to Ug99. In: Proceedings of 2nd Canadian Wheat Symposium, Saskatoon, SK, June 8–11, 2014, poster 19

Joppa LR, Cantrell RG (1990) Chromosomal location of genes for grain protein content of wild tetraploid wheat. Crop Sci 30:1059–1064

Joppa LR, Williams ND (1988) Langdon durum disomic substitution lines and aneuploidy analysis in tetraploid wheat. Genome 30:222–228

Joppa LR, Du C, Hart GE, Hareland GA (1997) Mapping gene (s) for grain protein in tetraploid wheat (*Triticum turgidum L*) using a population of recombinant inbred chromosome lines. Crop Sci 37:1586–1589

Jorgensen JH, Jensen CJ (1973) Gene *Pm6* for resistance to powdery mildew in wheat. Euphytica 22:423

Kerber ER, Dyck PL (1973) Inheritance of stem rust resistance transferred from diploid wheat (*Triticum monococcum*) to tetraploid and hexaploid wheat and chromosome location of the gene involved. Can J Genet Cytol 15:397–409

Khan IA, Procunier JD, Humphreys DG, Tranquilli G, Schlatter AR, Marcucci-Poltri S, Frohberg R, Dubcovsky J (2000) Development of PCR-based markers for high grain protein content gene from *Triticum turgidum* ssp. *dicoccoides* transferred to bread wheat. Crop Sci 40:518–524

Knott DR (1962) The inheritance of rust resistance IX. The inheritance of resistance to races 15B and 56 of stem rust in the wheat variety Khapstein. Can J Plant Sci 42:415–419
Knott DR, Anderson RG (1956) The inheritance of rust resistance I. The inheritance of stem rust resistance in ten varieties of common wheat. Can J Agri Sci 36:174–195
Knox AK, Li C, Vágùjfalvi A, Galiba G, Stockinger EJ, Dubcovsky J (2008) Identification of candidate *CBF* genes for the frost tolerance locus *Fr-A^{m}2* in *Triticum monococcum*. Plant Mol Biol 67:257–270
Kolmer JA, Anderson JA, Flor JM (2010) Chromosome location, linkage with simple sequence repeat markers and leaf rust resistance conditioned by gene *Lr63* in wheat. Crop Sci 50:2392–2395
Kumar S, Stack RW, Friesen TL, Faris JD (2007) Identification of a novel Fusarium head blight resistance quantitative trait locus on chromosome 7A in tetraploid wheat. Phytopathology 97:592–597
Law CN, Wolfe MS (1966) Location of genetic factors for mildew resistance and ear emergence time on chromosome 7B of wheat. Can J Genet Cytol 8:462–479
Leonova IN, Laikova LI, Popova OM, Unger O, Borner A, Roder MS (2007) Detection of quantitative trait loci for leaf rust resistance in wheat-*T.timopheevi*/*T.tauschii* introgression lines. Euphytica 155:79–86
Loegering WO (1975) An allele for low reaction to *Puccinia graminis tritici* in Chinese Spring Wheat. Phytopathology 65:925
Liu Z, Sun Q, Ni Z, Nevo E, Yang T (2002) Molecular characterization of a novel powdery mildew resistance gene *Pm30* in wheat originating from wild emmer. Euphytica 123:21–29
Liu XM, Brown-Guidera GL, Hatchett JH, Owuoche JO, Chen MS (2005) Genetic characterization and molecular mapping of a Hessian fly resistance gene transferred from *T. turgidum ssp dicoccum* to wheat. Theor Appl Genet 111:1308–1315
Liu Z, Zhu J, Cui Y, Liang Y, Wu H, Song W, Liu Q, Yang T, Sun Q, Liu Z (2012) Identification and comparative mapping of a powdery mildew resistance gene derived from wild emmer (*Triticum turgidum var. dicoccoides*) on chromosome 2BS. Theor Appl Genet 124:1041–1049
Ma H, Singh RP, Mujeeb-Kazi A (1997) Resistance to stripe rust in durum wheats, A genome diploids and their amphiploids. Euphytica 94:279–286
Malihipour A, Gilbert J, Fedak G, Brule-Babel A, Cao W (2015) Molecular mapping of quantitative trait loci for Fusarium head blight resistance in a population of wheat with *Triticum timopheevi*i background (submitted to Genome)
Marais GF, Pretorius ZA, Wellings CR, McCallum B, Marais AS (2005) Leaf rust and stripe rust resistance genes transferred to common wheat from *Triticum dicoccoides*. Euphytica 143:115–123
Martinez M, Naranjo T, Cuadrado C, Romero C (1996) Synaptic behaviour of the tetraploid wheat *Triticum timopheevii*. Theor Appl Genet 93:1139–1144
McCallum B, Hiebert C, Huerta-Espino J, Cloutier S (2012) Wheat Leaf Rust. In: Sharma I (ed.), Disease Resistance in Wheat Cab International 2012, pp 33–62
McFadden ES (1930) A successful transfer of emmer characters to vulgare wheat. J Am Soc Agron 22:1020–1034
McIntosh RA (1988) The Catalogue of gene symbols in wheat. In: Miller TE, Koebner RMD (eds.), Proc 7th Int Wheat Genetics Symposium, vol 2, Institute of Plant Science Research, Cambridge, pp 1225–1323
McIntosh RA, Dyck PL (1975) Cytological studies in wheat. VII Gene *Lr23* for reaction to *Puccinia recondita* in Gabo and related cultivars. Aust J Biol Sci 28:201–211
McIntosh RA, Gyarfas J (1971) *Triticum timopheevii* as a source of resistance to wheat stem rust. Z Pflanzenzüchtg 66:240–248
McIntosh RA, Luig NH (1973) Linkage of genes for reaction to *Puccinia graminis F.* sp. *Tritici* and *P. recondita* in Selkirk wheat and related cultivars. Aus J Biol Sci 26:1145–1152
McIntosh RA, Dyck PL, The TT, Cusick JE, Milne DL (1984) Cytogenetical studies in wheat XIII. *Sr35* – a third gene from *Triticum monococcum* for resistance to *Puccinia graminis*. Z Pflanzenzüchtg 92:1–14

McIntosh RA, Silk J, The TT (1996) Cytogenetic studies in wheat. XVII. Monosomic analysis and linkage relationships of gene *Yr15* for resistance to stripe rust. Euphytica 89:395–399

Miller AK, Galiba G, Dubcovsky J (2006) A cluster of 11 *CBF* transcription factors is located at the frost tolerance locus Fr-A^{m}2 in *Triticum monococcum*. Mol Genet Genomics 275:193–203

Mohler V, Zeller FJ, Wenzel G, Hsam SLK (2005) Chromosomal location of genes for resistance to powdery mildew in common wheat (*Triticum aestivum* L. em Thell.) 9. Gene MlZec1 from *Triticum dicoccoides* - derived wheat line Zecoi-1. Euphytica 142:161–167

Mohler V, Bauer C, Schweizer G, Kempf H, Hartl L (2013) *Pm50*: a new powdery mildew resistance gene in common wheat derived from cultivated emmer. J Appl Genet 54:259–263

Moseman JG, Nevo E, El Morshidy MA, Zohary D (1984) Resistance of *Triticum dicoccoides* to infection with *Erysiphe graminis tritici*. Euphytica 33:41–47

Nelson JC, Singh RP, Autrique JE, Sorrells ME (1997) Mapping genes conferring and suppressing leaf rust resistance in wheat. Crop Sci 37:1928–1935

Nelson DR and Sukkestad DR (1975) Normal and branched alkanes from cast skins of the grasshopper Schistocerca vaga (Scudder). J. Lipid Res 16:12–18

Nyquist WE (1957) Monosomic analysis of stem rust resistance of a common wheat strain derived from *Triticum timopheevi*. Agron J 49:22–23

Oliver RE, Cai X, Xu SS, Chen X, Stack RW (2005) Wheat-alien species derivatives: a potential source of novel resistance to Fusarium head blight in wheat. Crop Sci 45:1353–1360

Oliver RE, Stack RW, Miller JD, Cai X (2007) Reaction of wild emmer wheat accessions to Fusarium head blight. Crop Sci 47:893–899

Oliver RE, Cai X, Friesen TL, Halley S, Stack RW, Xu SS (2008) Evaluation of Fusarium head blight resistance in tetraploid wheat (*Triticum turgidum L.*). Crop Sci 48:213–222

Oliver PD, Jin Y, Badelo A, Xu S, Klindworth D (2011) Resistance to race TTKSK of *Puccinia graminis f.sp. tritici* in tetraploid wheat. Phytopathology 101:132–135

Otto CD, Kianian SF, Elias EM, Stack RW, Joppa LR (2002) Genetic dissection of a major Fusarium head blight QTL in tetraploid wheat. Plant Mol Biol 48:625–632

Peng JH, Fahima T, Roder MS, Li YC, Dahn A, Grama A, Ronin YI, Korol AB, Nevo E (1999) Microsatellite tagging of the stripe-rust resistance gene *YrH52* derived from wild emmer wheat, *Triticum dicoccoides*, and suggestive negative crossover interference on chromosome 1B. Theor Appl Genet 98:862–872

Peng UJ, Korol AB, Fahima T, Roder M, Ronin YI, Youchum CL, Nevo E (2000) Molecular genetic maps in wild emmer wheat *Triticum dicoccoides*: genome-wide coverage, massive negative interference and putative quasi-linkage. Genome Res 10:1509–1531

Piarulli L, Gadaleta A, Mangini G, Signorile MA, Pasquini M, Blanco A, Simeone R (2012) Molecular identification of a new powdery mildew resistance gene on chromosome 2BS from *Triticum turgidum* ssp. *dicoccum*. Plant Sci 196:101–106

Pridham JT (1939) A successful cross between *T. vulgare* and *T. timopheevi*. J Aus Inst Agric Sci 5:160–161

Qui YC, Zhou RH, Kong XY, Zhang SS, Jia JZ (2005) Microsatellite mapping of a *Triticum urartu* Tum. derived powdery mildew resistance gene transferred to common wheat (*Triticum aestivum* L.). Theor Appl Genet 111:1524–1531

Reader SM, Miller TE (1991) The introduction into bread wheat of a major gene for resistance to powdery mildew from wild emmer. Euphytica 53:57–60

Rondon MR, Gough FJ and Williams ND (1966) Inheritance of stem rust resistance in *Triticum aestivum* ssp. *vulgare* 'Reliance' and P.I. 94701 of *Triticum durum*. Crop Sci. 6:177–179

Rong JK, Millet E, Manisterski J, Feldman M (2000) A new powdery mildew resistance gene: introgression from wild emmer into common wheat and RFLP-based mapping. Euphytica 115:121–126

Rosewarne GM, Herrera-Foessel SA, Singh RP, Huerta-Espino J, Lan CX, He ZH (2013) Quantitative trait loci of stripe rust resistance in wheat. Theor Appl Genet 126:2427–2449

Rouse MN, Jin Y (2011a) Stem rust resistance in A-genome diploid relatives of wheat. Plant Dis 95:941–944

Rouse MN, Jin Y (2011b) Genetics of resistance to race TTKSK of *Puccinia graminis* f. sp. *tritici* in *Triticum monococcum*. Phytopathology 101:1418–1423

Schmolke M, Mohler V, Hartl L, Zeller FS, Hsam SLK (2012) A new powdery mildew resistance allele at the Pm4 wheat locus transferred from einkorn (*Triticum monococcum*). Mol Breed 29:449–456

Sela H, Ezrati S, Ben-Yehuda P, Manisterski J, Akhunov E, Dvorak J, Breiman A, Korol A (2014) Linkage disequilibrium and association analysis of stripe rust resistance in wild emmer wheat (*Triticum turgidum* ssp. *dicoccoides*) population in Israel. Theor Appl Genet 127:2453–2463

Semeniuk W (1947) Chromosome stability in some rust resistant derivatives from *a T.vulgare* x *T. timopheevi* cross. Sci Agric 27:7–20

Shands RG (1941) Disease resistance of *Triticum timopheevi* transferred to common winter wheat. J Am Soc Agron 33:709–712

Sheen SJ, Snyder LA (1964) Studies on the inheritance of resistance to six stem rust cultures using chromosome substitution lines of a Marquis wheat selection. Can J Genet Cytol 6:74–82

Shi AN, Leath S, Murphy JP (1998) A major gene for powdery mildew resistance transferred to common wheat from wild einkorn wheat. Phytopathology 88:144–147

Singh AK, Knox RE, Clarke FR, Clarke JM, Fedak G, Singh A, DePauw R (2008) Fusarium head blight mapping in durum wheat and *T. carthlicum* sources of resistance. In: Appels R, Eastwood R, Lagudah E, Langridge P, McKay M, McIntyre L, Sharpe P (eds.), Eleventh International Wheat Genetics Symposium, Brisbane, QLD, Australia, August 24-29, 2008, poster 161

Singh K, Chhuneja P, Ghai M, Kaur S, Goel RK, Bains NS, Keller B, Dhaliwal HS (2007a) Molecular mapping of leaf and stripe rust resistance genes in *Triticum monococcum* and their transfer to hexaploid wheat. In: Buck H, Nisi JE, Solomon N (eds) Wheat production in stressed environments. Springer, Netherlands, pp 779–786

Singh K, Ghai M, Garg M, Chhuneja P, Kaur P, Schnurbusch T, Keller B, Dhaliwal HS (2007b) An integrated molecular linkage map of diploid wheat based on a *Triticum boeoticum* x *T. monococcum* RIL population. Theor Appl Genet 115:301–312

Singh RP, Hodson DP, Jin Y, Huerta-Espino J, Kinyua, MG, Wanyera R, Njau P & Ward RW (2006) Current status, likely migration and strategies to mitigate the threat to wheat production from race Ug99 (TTKS) of stem rust pathogen. CAB Reviews: Perspectives in Agriculture, Veterinary Science, Nutrition and Natural Resources 54: 1–13

Somers DJ, Fedak G, Clarke J, Cao W (2006) Mapping of FHB resistance QTLs in tetraploid wheat. Genome 49:1586–1593

Stack RW, Elias EM, Mitchell-Fetch J, Miller JD, Joppa LR (2002) Fusarium head blight reaction of Langdon durum-*Triticum dicoccoides* chromosome substitution lines. Crop Sci 42:637–642

Stack RW, Miller JD, Joppa LR (2003) A wild emmer having multiple genes for resistance to Fusarium head blight. In: Pogna NE, Romano M, Pogna EA, Galteno E (eds.), Proc 10th Int Wheat Genet Symp I., Paestum, Italy, pp 1257–1259

Steed A, Chandler E, Thomsett M, Gosman N, Fuare S, Nicholson P (2005) Identification of Type I resistance to Fusarium head blight controlled by a major gene located on chromosome 4A of *Triticum macha*. Theor Appl Genet 111:521–529

Sun GL, Fahima T, Korol AB, Turpeinen T, Grama A, Ronin YI, Nevo E (1997) Identification of molecular markers linked to the *Yr*15 stripe rust resistance gene of wheat originated in wild emmer wheat, *Triticum dicoccoides*. Theor Appl Genet 95:622–628

Tao W, Liu D, Liu J, Feng Y, Chen P (2000) Genetic mapping of the powdery mildew resistance gene *Pm*6 in wheat by RFLP analysis. Theor Appl Genet 100:564–568

The TT (1973) Chromosome location in genes conditioning stem rust resistance transferred from diploid to hexaploid wheat. Nat New Biol 241:256

The TT, McIntosh RA, Bennett FGA (1979) Cytogenetical studies in wheat. 1X. Monosomic analysis, telocentric mapping and linkage relationships of genes *Sr21*, *Pm4* and *Mle*. Aust J Biol Sci 32:115–125

The TT, Nevo E, McIntosh RA (1993) Response of Israeli wild emmer to selected Australian pathotypes of *Puccinia* species. Euphytica 71:75–81

Tomar SMS, Kochumadhavan M, Nambisan PNN (1988) Evaluation of timopheevi wheats for resistance to rusts and powdery mildew. Ind J Genet 48:69–73
Tranquilli G, Lijavetsky D, Muzzi G, Dubcovsky J (1999) Genetic and physical characterization of grain texture-related loci in diploid wheat. Mol Gen Genet 262:846–850
Tranquilli G, Heaton J, Chicaiza O, Dubcovsky J (2002) Substitutions and deletions of genes related to grain hardness in wheat and their effect on grain texture. Crop Sci 42:1812–1817
Uauy C, Brevis JC, Chen X, Khan I, Jackson L, Chicaiza O, Distelfeld A, Fahima T, Dubcovsky J (2005) High-temperature adult-plant (HTAP) stripe rust resistance gene *Yr36* from *Triticum turgidum* ssp. *dicoccoides* is closely linked to the grain protein content locus *Gpc-B1*. Theor Appl Genet 112:97–105
Valkoun J, Kucerova D, Bartos P (1986) Transfer of leaf rust resistance from *Triticum monococcum* to hexaploid wheat. Z Pflanzenzüchtg 96:271–278
Valkoun J, Kucerova D, Bartos P (1989) Transfer of a new gene for stem rust resistance from *Triticum monococcum* L. to hexaploid wheat *T. aestivum* L. Genetika Selektsiya 25:209–212
Vallega V (1979) Search for useful characters in diploid Triticum ssp. In: Ramanujam S (ed.), Proc 5th Int. Wheat Genet Symp, New Delhi, pp 156–162
Van Silfhout CH, Grama A, Gerechter-Amitai ZK and Kleitman F (1989a) Resistance to yellow rust in Triticum dicoccoides I. Crosses with susceptible Triticum durum. Netherlands J. Plant Path. 95 (1989) 73-–78
Van Silfhout CH, Grama A, Gerechter-Amitai ZK, Kleitman F (1989b) Resistance to yellow rust in *Triticum dicoccoides I.* Crosses with susceptible *Triticum durum*. Neth J Plant Pathol 95:73–78
Vasu K, Hayit-Singh S, Chhuneja P, Singh S, Dhaliwal HS (2000) Molecular tagging of Karnal bunt resistance genes of *Triticum monococcum* transferred to *Triticum aestivum* L. Crop Improv 27:33–42
Wan YF, Yen C, Yang JL (1997) Sources of resistance to head scab in Triticum. Euphytica 94:31–36
Wang J, Li W, Wang W (2014) Fine mapping and metabolic and physiological characterization of the glume glaucousness inhibitor *Iw3* derived from wild wheat. Theor Appl Genet 127:831–841
Watson IA, Luig NH (1958) *Timvera-A Steinwedel* x *T. timopheevi* derivative. Agron J 50:644
Xie C, Sun Q, Ni Z, Yang T, Nevo E, Fahima T (2003) Chromosomal location of a *Triticum dicoccoides-* derived powdery mildew resistance gene in common wheat by using microsatellite markers. Theor Appl Genet 106:341–345
Xu SS, Khan K, Klindworth DL, Faris JD, Nygard G (2004) Chromosomal location of genes for novel glutenin subunits and gliadins in wild emmer wheat (*Triticum turgidum* L. var. *dicoccoides*). Theor Appl Genet 108:1221–1228
Xue F, Wang G, Zhang H, Yang B (2012) High density mapping and marker development for the powdery mildew resistance gene *PmAS846* derived from wild emmer wheat (*Triticum tiugidum var. dicoccoides*). Theor Appl Genet 124:1549–1560
Yao G, Zhang J, Yang L, Xu H, Jiang Y, Xiong L, Zhang C, Zhengzhi Z, Ma Z, Sorrells ME (2007) Genetic mapping of two powdery mildew resistance genes in einkorn (*Triticum monococcum* L.) accessions. Theor Appl Genet 114:351–358
Zeller FJ, Hsam SLK (1998) Progress in breeding for resistance to powdery mildew in commom wheat (*Triticum aestivum L.*). In Proceedings of the Ninth International Wheat Genetics Symposium, Saskatoon, Saskatchewan, 1:178–180
Zhang W, Olson E, Saintenac C, Rouse M, Abate Z, Jin Y, Akhumov D, Pumphrey M, Dubcovsky J (2010) Genetic maps of stem rust resistance gene *Sr35* in diploid and hexaploid wheat. Crop Sci 50:2464–2474
Zhang B, Chi D, Hiebert C, Fetch T, Xue A, Cao W, Fedak G (2014) Pyramiding genes for resistance to stem rust, leaf rust and FHB. In: Proceedings of the 2nd Canadian Wheat Symposium, Saskatoon SK, June 8–11, 2014, Poster 18
Zhirov EG (1980) Synthesis of a new hexaploid wheat, Trudi ProPrikladnoi Botanike. Genetika i Selectsiya 68:14–16

Chapter 9
Wheat–*Aegilops* Introgressions

Peng Zhang, Ian S. Dundas, Robert A. McIntosh, Steven S. Xu, Robert F. Park, Bikram S. Gill, and Bernd Friebe

9.1 Introduction

In the tribe Triticeae, *Aegilops* is the most closely related genus to *Triticum* and represents the largest portion of the secondary gene pool for cultivated wheat (van Slageren 1994). Based on the botanical classification by van Slageren (1994), the genus *Aegilops* comprises ten diploid, eight tetraploid, two tetraploid-hexaploid, and two hexaploid species, and five nontypical varieties with six diverse genomes including C, D, M, N, S, and U (Table 9.1). Among the diploid *Aegilops* species, *Ae. tauschii* Coss. ($2n=2\times=14$, DD) is the D genome donor of hexaploid wheat and *Ae. speltoides* Tausch ($2n=2\times=14$, SS) is believed to be the donor of B genome of tetraploid (*T. turgidum* L., $2n=4\times=28$, AABB) and hexaploid wheat. In addition to *Ae. tauschii* and *Ae. speltoides*, five polyploid species carry the D genome, and four diploid and three polyploid species have the S genome (Table 9.1).

P. Zhang • R.A. McIntosh • R.F. Park
Plant Breeding Institute, University of Sydney,
107 Cobbitty Road, Camden, NSW 2570, Australia

I.S. Dundas
School of Agriculture, Food and Wine, Waite Campus, The University of Adelaide,
PMB 1, Glen OsmoND SA 5064, Australia

S.S. Xu
USDA-ARS Cereal Crops Research Unit, Northern Crop Science Laboratory,
Fargo, ND 58102-2765, USA

B.S. Gill • B. Friebe (✉)
Department of Plant Pathology, Wheat Genetic Resources Center, Kansas State University,
Manhattan, KS 66506-5502, USA
e-mail: friebe@ksu.edu

M. Molnár-Láng et al. (eds.), *Alien Introgression in Wheat*,
DOI 10.1007/978-3-319-23494-6_9

Table 9.1 Classification of *Aegilops* speciesaccording to van Slageren (1994)

Section and Species	Ploidy	Genome
Section *Aegilops* L.		
Aegilops biuncialis Vis.	4×	U^bM^b
Aegilops columnaris Zhuk.	4×	U^cM^c
Aegilops geniculata Roth	4×	M^gU^g
Aegilops kotschyi Boiss.	4×	S^kU^k
Aegilops neglecta Req. ex Bertol.	4×, 6×	U^nM^n, $U^nM^nN^n$
Aegilops peregrina (Hackel in J. Fraser) Maire & Weiller	4×	S^pU^p
var. *peregrina*		
var. *brachyathera* (Boiss.) Maire & Weiller		
Aegilops triuncialis (L.) Á. Löve	4×	C^tU^t
var. *triuncialis*		
var. *persica* (Boiss.) Eig		
Aegilops umbellulata Zhuk.	2×	U
Section *Comopyrum* (Jaub. & Spach) Zhuk		
Aegilops comosa Sm. in Sibth. & Sm.	2×	M
var. *comosa*		
var. *subventricosa* Boiss.		
Aegilops uniaristata Vis.	2×	N
Section *Cylindropyrum* (Jaub. & Spach) Zhuk.		
Aegilops markgrafii (Greuter) Hammer	2×	C
Aegilops cylindrica Host	4×	C^cD^c
Section *Sitopsis* (Jaub. & Spach) Zhuk.		
Aegilops bicornis (Forsskål) Jaub. & Spach	2×	S^b
var. *bicornis*		
var. *anathera* Eig		
Aegilops longissima (Schweinf. & Muschl. in Muschl.) Eig	2×	S^l
Aegilops searsii Feldman & Kislev ex K. Hammer	2×	S^s
Aegilops sharonensis Eig	2×	S^{sh}
Aegilops speltoides Tausch	2×	S
var. *speltoides*		
var. *ligustica* (Savig.) Fiori		
Section *Vertebrata* Zhuk. emend. Kihara		
Aegilops crassa Boiss.	4×, 6×	D^cM^c, $D^cD^cM^c$
Aegilops juvenalis (Thell). Eig	6×	$D^jM^jU^j$
Aegilops tauschii Cosson	2×	D
Aegilops vavilovii (Zhuk.) Chennav.	6×	$D^vM^vS^v$
Aegilops ventricosa Tausch	4×	D^vN^v

9.2 Early History

Aegilops species areexcellent sources of resistance to many biotic (e.g., rusts, powdery mildew, cereal cyst nematode, root-knot nematode, Hessian fly, and greenbug) and abiotic (drought, cold, heat, and salinity) stresses (Sharma and Gill 1983;

Monneveux et al. 2000; Schneider et al. 2008; Stoyanov 2014). The close relationship of *Aegilops*to *Triticum* species makes it feasible to transfer the desirable genes from the alien genomes of *Aegilops* species into wheat genomes. The earliest reported successful artificial hybridization between wheat and an *Aegilops* species is that of DA Godron (1863, cited in Roberts 1929), although a report on spontaneous intergeneric hybrids between wheat and *Aegilops neglecta* Req. ex Bertol. and *Ae. triuncialis* (L.) Á. Löve was published several years earlier (Godron 1854, cited in Andersson and de Vicente 2010). The vast majority of *Aegilops*introgressions into wheat have taken place during the past century and mostly involved intentional transfer of genes for resistance to diseases and pests (Friebe et al. 1996; Schneider et al. 2008).

9.3 Hybrid Production

The first step in the transfer of a potentially useful gene from a wild *Aegilops* species to wheat is the production of an F_1 hybrid that combines the parental genomes. Difficulties are often encountered in this transfer process due to (a) inability to hybridize, (b) lethality of the hybrid, (c) inability of chromosomes from the two species to pair and exchange chromosome segments, (d) instability of the incorporated alien chromosome resulting in their elimination over generations, and (e) the resulting resistant line being unacceptable for agriculture. McIntosh et al. (2013) stated that the high crossability of some wheat genotypes, particularly of Chinese origin (e.g., Chinese Spring), with *Aegilops* was determined by additive recessive *kr* genes. When two species cannot be hybridized directly, a third species that is cross-compatible with one has often been used as a bridging species for gene transfer. For instance, several of the diploid relatives of common wheat cannot be hybridized easily with the hexaploid species but can be readily crossed with tetraploid durum wheat (*T. turgidum* ssp. *durum*). The derived amphiploid can then be crossed with common wheat. Examples where resistance genes derived from *Aegilops* have been transferred to common wheat via durum wheat include genes transferred from *Ae. ventricosa* Tausch (Doussinault et al. 1983; Garcia-Olmedo et al. 1984; Delibes et al. 1987).

If cross-pollination between wheat and *Aegilops* species results in a hybrid, in vitro embryo culture may be required to rescue F_1 plantlets. The F_1 hybrids are sterile and at approximately 4 weeks of age, the seedlings are treated for 18 h in an aerated solution of 0.03 % colchicine to double the chromosome number of dividing cells, which may result in the later production of amphidiploid tillers. Selfed seeds on these late-developing amphidiploid tillers or BC_1 progeny (using wheat pollen) may ensure the continuity of the hybrids. These hybrids/amphidiploids have no value to agriculture but after further backcrossing with wheat may result in alien chromosome addition lines carrying whole or telocentric *Aegilops* chromosomes. These addition lines may be studied to confirm that the gene(s) of interest is expressed in a wheat background and then determine the homoeologous group of the *Aegilops* chromosome carrying that character using molecular markers (Sharp et al. 1988; see also review by Qi et al. 2007; Dundas et al. 2008).

9.4 Chromosome Translocations

Wheat–*Aegilops* chromosome addition lines have no practical application in agriculture. The *Aegilops* chromosome segment carrying the gene of interest must be transferred to a wheat chromosome. This can be achieved by spontaneous or radiation translocations, or homoeologous recombination induced either by removal of the *Ph1* gene on wheat 5B chromosome or use of the *Ph1* suppressor in some *Aegilops* accessions.

The most common type of spontaneous translocation is the centric fusion or Robertsonian translocation. Sears (1972) demonstrated that in wheat plants that are monosomic for two chromosomes, both monosomes can misdivide at their centromeres at meiosis and rejoin to give chromosomes carrying arms from each original chromosome. Shepherd and Islam (1988) listed spontaneous translocations in common wheat involving chromosomes from *Ae. markgrafii* (Greuter) Hammer (*Ae. caudata* L.), *Ae. comosa* Sm. in Sibth. & Sm., and *Ae. longissima* (Schweinf. & Muschl. in Muschl.) Eig. The first report of using irradiation to transfer a disease resistance gene was that of Sears (1956) who translocated a segment of an *Ae. umbellulata* Zhuk. chromosome carrying the leaf rust resistance gene *Lr9* to common wheat.

9.5 Homoeologous Recombination

Chromosome pairing behavior is under genetic control. However, it can be manipulated to induce homoeologous recombination in order to achieve gene transfer between chromosomes that normally do not pair. Meiotic pairing between homoeologous chromosomes of the three wheat genomes (A, B, and D) is normally prevented by the action of several genes, the principal one being *Ph1* on chromosome arm 5BL (Okamoto 1957; Riley and Chapman 1958). Later, it was recognized that, in addition to the gene on 5BL, several other genes on chromosomes 3D, 3A, and 4D acted to reduce homoeologous pairing. But genes also existed on other wheat chromosomes (5BS, 5D, 5A, 3AL, 3BL, 3DL, and 2AS) that promoted homoeologous pairing (Sears 1976). Two additional pairing suppressors and nine promoters on other chromosomes of wheat were later reported (Gale and Miller 1987).

Wheat lines lacking chromosome 5B or possessing the mutant gene *ph1b* (Sears 1977) (an interstitial chromosome deletion of the *Ph1* gene) showed homoeologous pairing between wheat chromosomes of different genomes and also with other species in the tribe Triticeae. Transfers of disease resistance into wheat using the *ph1b* mutant were reported from *Ae. longissima* (Ceoloni 1984; Ceoloni et al. 1988, 1992), *Ae. triuncialis* (Fan et al. 1995), and *Ae. cylindrica* Host (Bai et al. 1995). In addition, the *ph1b* mutant was employed to reduce the size of *Ae. speltoides* chromosome segments (see many examples listed below). Wheat nullisomic for the entire 5B chromosome (carrying the *Ph1* gene) was used to transfer disease resistance from *Ae. markgrafii* into wheat (Dyck et al. 1990).

An alternative to the *ph1b* mutant for inducing homoeologous recombination in wheat is to use *Ph* suppressors in *Ae. speltoides*. Riley et al. (1968a, b) transferred stripe rust resistance gene *Yr8* from *Ae. comosa* chromosome 2M to wheat using a *Ph1*-suppressing line of *Ae. speltoides* and produced the line known as Compair. Chen et al. (1994) reported the transfer of a gene influencing homoeologous chromosome pairing from *Ae. speltoides* to a wheat chromosome.

In this chapter, we review all the formally named genes transferred to common wheat from *Aegilops* species, except those from *Ae. tauschii*. Species in *Sitopsis* section, such as *Ae. speltoides* and *Ae. longissima*, have high pairing and low pairing genotypes that require different transfer strategies.

9.6 *Ae. speltoides*

Wells and coworkers (Lay et al. 1971; Wells et al. 1973, 1982) used radiation treatment to transfer *Gb5*, conferring resistance to greenbug (*Schizaphis graminum* Rond.), and *Lr47* conferring resistance to leaf rust (caused by *Puccinia triticina* Eriks.) from a group-7 *Ae. speltoides* Tausch ($2n=2x=14$, SS) chromosome to 7A of wheat (Friebe et al. 1991; Dubcovsky et al. 1998). Genomic in situ hybridization (GISH) and meiotic pairing analyses suggested that the wheat–*Ae. speltoides* recombinant chromosomes present in lines CI 17883, CI 17884, and CI 17885 consisted of the complete long arm of 7S#1, most of the short arm of 7S#1, and a small distal segment derived from the short arm of wheat chromosome 7A, and therefore it was designated as T7AS-7S#1S•7S#1L (Friebe et al. 1991, 1996; Dubcovsky et al. 1998). Dubcovsky et al. (1998) used induced homoeologous recombination to shorten the *Ae. speltoides* segment and recovered two interstitial recombinant chromosomes. Ti7AS-7S#1S-7AS•7AL conferring resistance to leaf rust and Ti7AS•7AL-7S#1L-7AL conferring resistance to greenbug. This placed *Lr47* on the short arm and *Gb5* on the long arm of the *Ae. speltoides* chromosome 7S#1.

Miller et al. (1988) used induced homoeologous recombination and transferred leaf rust resistance gene *Lr28* from *Ae. speltoides* to wheat. *Lr28* is present in the lines 2A/2M#4/2 and 2D/2M#3/8 (McIntosh et al. 1982) that were actually selected for stripe rust resistance (*Yr8* from *Ae. comosa*). Chromosome C-banding analysis suggested that the wheat–*Ae. speltoides* recombinant chromosome in these lines consists of the short arm of chromosome 4A of wheat, most of the long arm of 4A, and a distal segment derived from *Ae. speltoides* (Friebe et al. 1996). Chromosome 4A of *T. turgidum*and *T. aestivum* is highly rearranged and the distal region of the long arm is actually homoeologous to group-7S arms (Miftahudin et al. 2004). Therefore, the *Ae. speltoides* segment transferred to the 4AL arm most likely is homoeologous to the group-7S and the recombinant chromosome should be described as T4AS•4AL-7S#2S.

ER Sears used induced homoeologous recombination to transfer *Sr32* conferring resistance to stem rust (*Puccinia graminis* Pers.: Pers f. sp. *tritici* Eriks. & Henn.) from an *Ae. speltoides* chromosome 2S#1 to wheat chromosomes 2A, 2B, and 2D

(McIntosh 1991). C-banding analysis suggested that the wheat–*Ae. speltoides* recombinant chromosome in germplasm C95.24 (TA3945) consisted of the complete short arm of chromosome 2S#1, part of the long arm of 2S#1, and a distal segment derived from 2AL of wheat, and it was designated as T2AL-2S#1L•2S#1S (Friebe et al. 1996). This study further indicated that germplasm C82.1 (TA3944) had the recombinant chromosome T2BL-2S#1S and germplasm C82.2 (TA3947) had the recombinant chromosome T2DL-2S#1L•2S#1S. The translocation line C82.2 was crossed with Sears' *ph1b* mutant and a set of wheat–*Ae. speltoides* recombinant lines were produced (Mago et al. 2013). By studying these recombinants it was determined that *Sr32* was located on the short arm of the 2S#1 chromosome and an additional stem rust resistance gene was located on the long arm of this chromosome.

Kerber and Dyck (1990) transferred the adult-plant leaf rust resistance gene *Lr35* and the seedling stem rust resistance gene *Sr39* from an *Ae. speltoides* chromosome, 2S#2, to wheat chromosome 2B using induced homoeologous recombination (Table 9.2). C-banding analysis indicated that the wheat–*Ae. speltoides* recombinant chromosome T2B-2S#2 consisted of an *Ae. speltoides* segment from both arms (Friebe et al. 1996). This translocation line was crossed with Sears' *ph1b* mutant and a set of wheat–*Ae. speltoides* recombinants carrying shortened segments of the 2S#2 chromosome were produced (Mago et al. 2009). By studying these recombinants, *Sr39* was located on an interstitial region of the chromosome 2S#2S. Niu et al. (2011) also described the production of wheat–*Ae. speltoides* 2S/2B recombinants involving *Sr39*, some of which carried extremely small *Ae. speltoides* chromosome segments.

A durum wheat–*Ae. speltoides*chromosome translocation line DAS15 carrying stem rust resistance gene *Sr47* was developed by LR Joppa (USDA-ARS) through *ph1b*-induced homoeologous recombination. Although two wheat–*Ae. speltoides* translocation chromosomes are present in the line, only one T2BL-2SL•2SS harbors *Sr47*. The distal 2BL segment comprised less than 10 % of the long arm (Faris et al. 2008). Even though *Sr47* is derived from chromosome 2S, its pathotype specificity, rust infection type and map location showed that it was different from genes *Sr32* and *Sr39*, which were also transferred into common wheat from 2S (Faris et al. 2008; Klindworth et al. 2012). Homoeologous pairing was induced and allosyndetic recombinants with stem rust resistance were recovered for both arms of 2S (Klindworth et al. 2012). The *Ae. speltoides* segment in DAS15 carried two stem rust resistance genes, *Sr47* was located on chromosome 2BL, while the gene on chromosome 2BS may be the same as *Sr39*. Interstitial translocation lines carrying *Sr47* characterized as Ti2BL-2SL-2BL•2BS retained various sized *Ae. speltoides* segments in the subtelomeric region of chromosome 2BL.

Dvorak and Knott (Dvorak 1977; Dvorak and Knott 1980, 1990) transferred leaf rust resistance from *Ae. speltoides* to wheat chromosomes 1B and 6B. The gene present in the germplasm 2-9-2 was located on the wheat–*Ae. speltoides* recombinant chromosome T6BL•6BS-6S#2S and was designated as *Lr36* (McIntosh 1991), and *Lr51* present in the germplasm F-7-3 was mapped on an interstitial wheat–*Ae. speltoides* recombinant chromosome Ti1BS•1BL-1S#1L-1BL by Helguera et al. (2005).

Table 9.2 Genes transferredfrom *Aegilops* species into wheat

Aegilops species	Gene of interest	Germplasm	Type	Chromosome constitution	Origin	Reference
Ae. speltoides	*Gb5*/*Lr47*	CI17883, CI17884, CI17885	REC	T7AS-7S#1S•7S#1L	I	Lay et al. (1971); Wells et al. (1973, 1982); Friebe et al. (1991); Dubcovsky et al. (1998)
	Lr47		REC	Ti7AS-7S#1S-7AS•7AL	HR	Dubcovsky et al. (1998)
	Gb5		REC	Ti7AS•7AL-7S#1L-7AL	HR	Dubcovsky et al. (1998)
	Lr28	2A/2M#4/2	REC	T4AS•4AL-7S#2S	HR	McIntosh et al. (1982); Miller et al. (1988)
		2D/2M#3/8	REC			Friebe et al. (1996)
	Sr32	C95.24 (TA3945)	REC	T2AL-2S#1L•2S#1S	HR	McIntosh (1991); Friebe et al. (1996)
		C82.1 (TA3944)	REC	T2BL-2S#1S		
		C82.2 (TA3947)	REC	T2DL-2S#1L•2S#1S		
	Sr39, *Lr35*	RL5711	REC	T2B-2S#2	HR	Kerber and Dyck (1990)
	Sr47	DAS15	REC	T2BL-2SL•2SS	HR	Faris et al. (2008)
		RWG35 (0406), RWG36 (0696)	REC	Ti2BL-2SL-2BL•2BS	HR	Klindworth et al. (2012)
	Lr36	2/9/2002	REC	T6BL•6BS-6S#2S	HR	Dvorak (1977); Dvorak and Knott (1980, 1990); McIntosh (1991); Helguera et al. (2005)
	Lr51	F-7-3	REC	Ti1BS•1BL-1S#1L-1BL	HR	
	Lr66	S13 (line 8029)	REC	T3A-3S	HR	Marais et al. (2003)
			REC		HR	Marais et al. (2010c)
	Pm12	#31	REC	T6BS-6S#1S•6S#1L	HR	Miller et al. (1987); Jia et al. (1996)
	Pm32	L501	ROBT	T1BL•1S#2S	S	Hsam et al. (2003)
Ae. longissima	*Pm13*	R1A	REC	T3BL•3BS-3S^{l}#1S	HR	Ceoloni et al. (1988, 1992); Donini et al. (1995); Biagetti et al. (1998)
		R1D	REC	T3DL•3DS-3S^{l}#1S	HR	

Table 9.2 (continued)

Aegilops species	Gene of interest	Germplasm	Type	Chromosome constitution	Origin	Reference
Ae. sharonensis	*Lr56*, *Yr38*	0352-4	REC	T6AL-6S^{sh}#1L•6S^{sh}#1S	S	Marais et al. (2006)
		#39, #157, #175	REC	Ti6AL-6S^{sh}#1L-6AL•6AS	HR	Marais et al. (2010a, b, c)
Ae. searsii	*Sr51*	TA5619	ROBT	T3AL•3S^{s}#1S	MU	Liu et al. (2011a)
		TA5620	ROBT	T3BL•3S^{s}#1S		
		TA5621	ROBT	T3DL•3S^{s}#1S		
Ae. umbellulata	*Lr9*	Transfer	T	T6BS•6BL-6U#1L	I	Sears (1956); Friebe et al. (1995)
Ae. comosa	*Yr8*, *Sr34*	Compair	REC	T2DS-2M#1L•2M#1S	HR	Riley et al. (1968a, b); McIntosh et al. (1982); Miller et al. (1988); Nasuda et al. (1998)
		2A/2M#4/2	REC	T2AS-2M#1L•2M#1S		
		2D/2M#3/8	REC	T2DS-2M#1L•2M#1S		
Ae. peregrina	*Lr59*	306	ROBT	T1AS•1^{p} #1L	S	Marais et al. (2008)
		#10, #21, #25, #29, #36, #101, #144, #151	REC	T1AS•1AL-1^{p}#1L	HR	Marais et al. (2010a, b, c)
	Rkn2 (*Rkn-mn -1*)	X8	N/A	3B	S	Yu et al. (1995); Barloy et al. (2000)
Ae. kotschyi	*Lr54*, *Yr37*	S14	ROBT	T2DS•2^{k}#1L	MU	Marais et al. (2005)
Ae. geniculata	*Lr57*, *Yr40*	TA5599	REC REC	T5DL-5M^{g}#1L•5M^{g}#1S	HR	Kuraparthy et al. (2007b, 2009)
		TA5601, TA5602		T5DL•5DS-5M^{g}#1S		
	Sr53	TA5599	REC REC	T5DL-5M^{g}#1L•5M^{g}#1S	HR	Liu et al. (2011b)
		TA5630		Ti5DS•5DL-5M^{g}#1L-5DL		
	Pm29	Pova	N/A	7D	S	Friebe and Heun (1989); Zeller et al. (2002)

Ae. triuncialis	*Lr58*	TA5605	REC	T2BS•2BL-2^t#1L	S	Kuraparthy et al. (2007a)
	Cre7	TR-353	N/A	unknown	S	Romero et al. (1998)
	H30	TR-3511	N/A	unknown	S	Martin-Sanchez et al. (2003)
Ae. ventricosa	*Pch1*	H-93-70	REC	T7D-7D^v#1	S	Doussinault et al. (1983); Jahier et al. (1979, 1989, 2001)
	Yr17, *Lr37*, *Sr38*, *Cre5*, *Pch1*	VPM1	REC	T2AL•2AS-2N^v#1/6N^v#1	S	Bariana and McIntosh (1993, 1994); Tanguy et al. (2005); Badaeva et al. (2008)
	Cre2	H93-8	N/A	unknown	S	Delibes et al. (1993)
	Cre6	H-93-35	N/A	5N^v#1	S	Ogbonnaya et al. (2001)
	H27	H-93-33	N/A	4M^v#1(4D)	S	Delibes et al. (1997)
	Rkn3	VPM1, Lassik (PI 653535)	REC	T2AL•2AS-2N^v#1/6N^v#1		Williamson et al. (2013)
Ae. neglecta	*Lr62*, *Yr42*	03M119-71A	REC	T6AL-6^n#1L•6^n#1S	HR	Marais et al. (2009)

Chromosome designations follow the nomenclature of Raupp et al. (1995) where "T" indicates a terminal translocation, "Ti" indicates an interstitial translocation, a "•" indicates the location of the centromere, a "-" indicates an interstitial breakpoint, "S" identifies the chromosome short arm, "L" identifies the chromosome long arm, numbers are used to identify the homoeologous group followed by a letter indicating the genomic origin followed by a superscript identifying the donor species, and the "#" sign is used to distinguish between homologous chromosomes derived from different donor accessions. In cases where the genomic origin is unknown only the homoeologous group followed by a superscript indicating the donor species is used; REC = recombinant chromosome, ROBT = Robertsonian translocation, I = ionizing radiation, HR = homoeologous recombination, S = spontaneous, MU = misdivision of univalents

The *Ae. speltoides* introgression line S13, with genes for resistance to all three rust pathogens, transferred by Marais et al. (2003) involves the complete 3AS of wheat and a proximal region of 3AL. However, it could not be used commercially due to associated gametocidal gene (*Gc*) with strong effects on plant development, fertility and seed quality. Marais et al. (2010c) recovered interstitial recombinants in which the leaf rust resistance gene *Lr66* in S13 was separated from the *Gc* gene using homoeologous recombination. Unfortunately, the stripe rust and stem rust resistance genes in S13 were lost during selection of the line with *Lr66*.

Miller et al. (1987) transferred *Pm12*, conferring resistance to powdery mildew (caused by *Blumeria graminis* DC E. O. Speer f. sp. *tritici* Em. Marchal), from *Ae. speltoides* to wheat chromosome 6B using induced homoeologous recombination and the recombinant chromosome can be described as T6BS-6S#1S•6S#1L (Jia et al. 1996).

Lapochkina and coworkers (Hsam et al. 2003) transferred powdery mildew resistance gene *Pm32* from *Ae. speltoides* to wheat in the form of a Robertsonian translocation (ROBT) T1BL•1S#2S using the process of univalent misdivision during meiosis followed by the centric fusion of the broken chromosome arms.

In addition to these formally named stem and leaf rust resistance genes, many more introgressions/additions of *Ae. speltoides* chromosome segments carrying temporarily designated genes are listed in Mujeeb-Kazi et al. (2013).

9.7 *Ae. longissima*

Coeloni and coworkers (Ceoloni et al. 1988, 1992; Donini et al. 1995; Biagetti et al. 1998) used induced homoeologous recombination to transfer the powdery mildew resistance gene *Pm13* from *Ae. longissima* (Schweinf. & Muschl. in Muschl.) Eig ($2n=2x=14$, S^lS^l) to wheat chromosome arms 3BS and 3DS. The wheat–*Ae. longissima* recombinant chromosomes present in the two germplasm lines R1A and R1D were characterized as T3BL•3BS-3S^l#1S and T3DL•3DS-3S^l#1S, respectively.

9.8 *Ae. sharonensis*

Ae. sharonensis Eig ($2n=2x=14$, $S^{sh}S^{sh}$) is the source of leaf rust and stripe rust (caused by *Puccinia striiformis* f. sp. *tritici* Westend.) resistance genes *Lr56* and *Yr38* (Marais et al. 2006). Genetic and molecular marker analyses suggested that the wheat–*Ae. sharonensis* recombinant chromosome in germplasm 0352-4 was characterized as T6AL-6S^{sh}#1L•6S^{sh}#1S, which consists of a distal segment of wheat chromosome 6AL, part of the long arm of the *Ae. sharonensis* chromosome 6S^{sh}#1, and the complete short arm of 6S^{sh}#1. Marais et al. (2010b) used induced homoeologous recombination to reduce the *Ae. sharonensis* chromatin in this translocation chromosome and obtained three leaf rust and stripe rust resistant lines (#39, #157, and #175) with recombinant chromosome Ti6AL-6S^{sh}#1L-6AL•6AS, which carries smaller interstitial *Ae. sharonensis* segments inserted into the 6AL arm.

Ae. sharonensis is also the source of the gametocidal gene *Gc2*, which was mapped to chromosome 4S^{sh}#1 (Miller et al. 1982). *Gc2* is available in the form of a spontaneous wheat–*Ae. sharonensis* recombinant chromosome T4BS•4BL-4S^{sh}#1L (TR Endo, pers. comm.). Previous attempts to produce a complete set of wheat–*Ae. sharonensis* chromosome addition lines failed, and only chromosome 4S^{sh} was recovered from eight different *Ae. sharonensis* accessions (B Friebe, N Tuleen et al., unpublished). The failure to produce a complete set of chromosome addition lines suggested that most *Ae. sharonensis* accessions are homozygous for the *Gc2* gene, making gene transfer from this species into wheat difficult. Friebe et al. (2003) used EMS treatment to produce a knockout mutation at the *Gc2* locus, allowing the introgression of all *Ae. sharonensis* chromosomes (besides 4S^{sh}) into wheat, and may make the gene transfer from this species more efficient (Olivera and Steffenson 2009; Millet et al. 2014).

9.9 *Ae. searsii*

Ae. searsii Feldman & Kislev ex Hammer ($2n = 2\times = 14$, S^{s}S^{s}) is the source of stem rust resistance gene *Sr51* that was transferred to wheat as centric fusion or Robertsonian translocation (Sears 1952; Friebe et al. 2005). *Sr51* is located on the short arm of the *Ae. searsii* chromosome 3S^{s}#1 and is available in the form of the Robertsonian translocations T3AL•3S^{s}#1S, T3BL•3S^{s}#1S, and T3DL•3S^{s}#1S (Liu et al. 2011a). The level of resistance conferred by the T3AL•3S^{s}#1S germplasm is lower than that of ROBTs involving wheat chromosomes 3B and 3D, indicating that the 3AS chromosome may have a modifier that affects the expression of the *Sr51* resistance.

9.10 *Ae. umbellulata*

Sears (1956) was the first to use radiation to transfer a leaf rust resistance gene, *Lr9*, from *Ae. umbellulata* Zhuk. ($2n = 2\times = 14$, UU) to wheat. Notably, this is the only success case of transferring genes by irradiating pollen, all the others irradiated seeds. Radiation treatment causes random chromosome breakage and fusion of broken ends resulting in translocation chromosomes. Although 17 different wheat–*Ae. umbellulata* translocation chromosomes were produced, only one, designated as T4 or Transfer, showed normal male and female transmission and was used in wheat improvement. The translocation chromosome T6BS•6BL-6U#1L in Transfer consists of wheat chromosome 6BS, most of 6BL, and a small distal segment derived from the long arm of *Ae. umbellulata* chromosome 6U#1 (Friebe et al. 1995). *Ae. umbellulata* has a highly nonsymmetrical karyotype, caused by a large number of chromosomal rearrangements, and the distal region of chromosome arm 6U#1L is actually homoeologous to the distal region of group-1L chromosomes (Zhang et al. 1998; Danilova et al. 2014). Because the missing 6BL segment in the Transfer translocation is replaced by a nonhomoeologous group-1 long arm segment, which causes duplications/deficiencies, the resulting Transfer translocation is a noncompensating type. Because radiation treatment induces chromosome breaks at random, most of the resulting translocations are between nonhomoeologous chromosomes and are noncompensating. Stringent selection among noncompensating translocations needs to be used to recover transfers with acceptable agronomic performance.

9.11 *Ae. comosa*

Ae comosa Sm. in Sibth. & Sm. ($2n=2\times=\mathrm{MM}$) is the source of the stripe rust resistance gene *Yr8* and the stem rust resistance gene *Sr34* that were transferred to wheat by induced homoeologous recombination using a high-pairing line of *Ae. speltoides* (Riley et al. 1968a, b). The wheat–*Ae. comosa* recombinant chromosome in the genotype Compair consists of a distal segment of chromosome 2DS, most of the long arm of 2M#1, and the complete short arm of 2M#1, and it was designated T2DS-2M#1L•2M#1S. Miller and coworkers (Miller et al. 1988) produced independent 2D/2M#3/8 and 2A/2M#4/2 *Yr8*/*Sr34* transfers. The 2D/2M#3/8 transfer is similar to the wheat–*Ae. comosa* recombinant chromosome present in Compair, and the 2A/2M#4/2 recombinant chromosome was identified as T2AS-2M#1L•2M#1S (Nasuda et al. 1998). The structures of these wheat–*Ae. comosa* recombinant chromosomes suggested that chromosome 2M#1 was rearranged and that the distal region of the long arm was actually derived from 2M#1S. Molecular marker analysis confirmed this rearrangement and revealed that both translocations 2D/2M and 2A/2M are noncompensating (Nasuda et al. 1998). The advantage of using induced homoeologous recombination to transfer an alien gene to wheat is that usually recombination events involve homoeologous chromosome regions and, therefore, are more likely to be agronomically useful, compensating type. However, this is only the case in which the wheat and alien chromosomes are not structurally rearranged. Any rearrangement, as in the case of chromosome 2M#1 and most cereal rye chromosomes, will result in noncompensating transfers. This is the most likely reason that the *Yr8*/*Sr34* transfer was not successfully used in wheat improvement.

9.12 *Ae. peregrina*

Ae. peregrina (Hack. in J. Fraser) Maire & Weller ($2n=4\times=28$, $\mathrm{U^pU^pS^pS^p}$) is the source of the leaf rust resistance gene *Lr59* that was transferred to wheat by Marais et al. (2008). The *Lr59* transfer occurred spontaneously and monosomic, meiotic pairing, and microsatellite analyses suggested that the translocation present in germplasm 0306 is a ROBT consisting of the complete short arm of wheat chromosome 1A and the long arm of a group-1 *Ae. peregrina* chromosome, T1AS•1^p#1L. The exact breakpoint in this translocation and the genomic affinity of the translocated *Ae. peregrina* segment remain to be determined. Marais et al. (2010a) used induced homoeologous recombination to shorten the *Ae. peregrina* segment in this translocation. Several recombinants T1AS•1AL-1^p#1L with shortened *Ae. peregrina* segments with *Lr59* were obtained (Table 9.2). However, molecular marker analysis suggested that the transferred *Ae. peregrina* segment is structurally rearranged.

Ae. peregrina is also the source of the root-knot nematode (*Meloidogyne naasi* Franklin) resistance gene *Rkn2* (*Rkn-mn1*) that was transferred to the long arm of wheat chromosome 3B (Yu et al. 1995; Barloy et al. 2000).

9.13 *Ae. kotschyi*

Leaf rust resistance gene *Lr54* and stripe rust resistance gene *Yr37* were transferred to wheat from *Ae. kotschyi* Boiss. ($2n=4\times=28$, $U^kU^kS^kS^k$) using the centric breakage-fusion mechanism of univalents at meiosis (Marais et al. 2005). Meiotic pairing indicated that the leaf rust and stripe rust resistance genes in germplasm S14 are located on the ROBT T2DS•2^k#1L. GISH analysis failed to detect the 2^k#1L chromatin in this translocation and, therefore, the exact breakpoint in this translocation and its genomic origin remain to be determined.

9.14 *Ae. geniculata*

Ae. geniculata Roth ($2n=4\times=28$, $U^gU^gM^gM^g$) is the source of powdery mildew resistance gene *Pm29*, leaf rust resistance gene *Lr57*, stripe rust resistance gene *Yr40* (Kuraparthy et al. 2007b, 2009), and stem rust resistance gene *Sr53* (Liu et al. 2011b).

Zeller et al. (2002) transferred powdery mildew resistance gene *Pm29* from a Poros-*Ae. geniculata* chromosome addition line (Friebe and Heun 1989) to wheat. Meiotic pairing and molecular marker analyses mapped the *Pm29* resistance gene on wheat chromosome 7D.

Sr53, conferring resistance to stem rust including race Ug99, is located in the long arm of the *Ae. geniculata* chromosome $5M^g$#1 and is available in wheat in the form of a spontaneous distal translocation T5DL-$5M^g$#1L•$5M^g$#1S (TA5599) and an interstitial recombinant chromosome Ti5DS•5DL-$5M^g$#1L-5DL (TA5630). The interstitial *Ae. geniculata* segment in Ti5DS•5DL-$5M^g$#1L-5DL corresponds to approximately 20 % of the 5DL-$5M^g$#1L-5DL arm (Liu et al. 2011b).

Lr57 and *Yr40* are located in the distal region of the short arm of the *Ae. geniculata* chromosome $5M^g$#1 and are available in the form of the wheat–*Ae. geniculata* recombinant chromosomes T5DL-$5M^g$#1L•$5M^g$#1S (TA5599) and T5DL•5DS-$5M^g$#1S (TA5601 and TA5602). The breakpoint in the germplasm TA5601 is

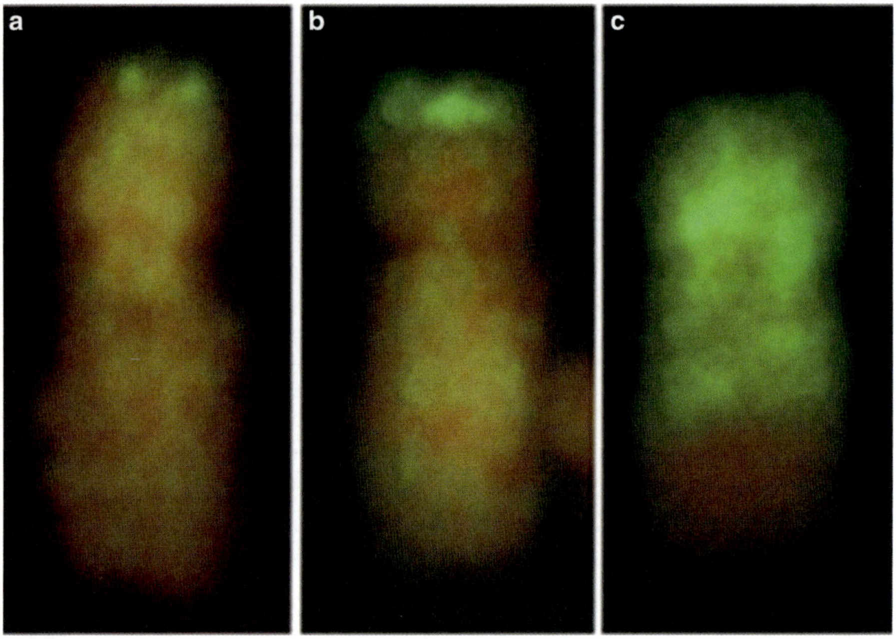

Fig. 9.1 Genomic in situ hybridization (GISH) pattern of wheat–*Aegilops geniculata* introgression lines resistant to leaf rust (*Lr57*), stripe rust (*Yr40*), and stem rust (*Sr53*). *Ae. geniculata* chromatin is visualized by *green* FITC fluorescence and wheat chromosomes were counterstained with PI and fluoresce *red*. The wheat–*Ae. geniculata* recombinant chromosome in TA5599 (**c**) with *Lr57*, *Yr40*, and *Sr53* consists of the distal part of the long arm of 5DL of wheat, the proximal part of the long arm of 5M^g#1L, and the complete short arm of chromosome 5M^g#1S (T5DL-5M^g#1L.5M^g#1S), whereas the wheat–*Ae. geniculata* recombinant chromosome in TA5601 (**b**) and TA5602 (**a**) with only *Lr57* and *Yr40* consists of the long arm of 5DL, the proximal part of 5DS and the distal part derived from 5M^g#1S (T5DL•5DS-5M^g#1S). The breakpoint in the recombinant chromosome present in TA5602 was determined to be at FL 0.95 and assumed to be below the detectability of GISH, which was the reason why this recombinant chromosome was characterized as cryptic. However, careful adjustment of probe and blocking DNA detected the presence of *Ae. geniculata* chromatin at the telomere of the short arm of T5DL•5DS-5M^g#1S in germplasm TA5602

located in the 5DS arm at a fraction length of FL 0.75, whereas in TA5602 it is located at FL 0.95. Molecular marker data suggested that the size of the *Ae. geniculata* segment in TA5602 is less than 3.5 % of the 5DS chromosome arm. GISH failed to detect the terminal *Ae. geniculata* segment in T5DL•5DS-5M^g#1S present in TA5602, which suggested that this transfer is below the detection limit of GISH and, was therefore designated as a "cryptic" transfer (Kuraparthy et al. 2007b, 2009). However, after careful adjustment of the ratio of probe to blocker, it was feasible to visualize the *Ae. geniculata* segment in the telomeric region of chromosome 5DS (Fig. 9.1) (T Danilova et al., unpublished).

For transfer of *Lr57* and *Yr40* to wheat, a disomic wheat–*Ae. geniculata* substitution line DS5M^g#1(5D) was crossed with the Chinese Spring Ph^I stock, and the F_1 was backcrossed to the wheat cultivar WL711 (Aghaee-Sarbarzeh et al. 2002). The Ph^I stock promotes homoeologous recombination (Chen et al. 1994) and, therefore, both the *Lr57* and *Yr40* transfers were assumed to be induced by homoeologous recombination. On the contrary, the *Sr53* transfer occurred spontaneously in the presence of *Ph1*, which ensures that in hexaploid wheat only homologous chromosomes pair and recombine. Chromosome 5M^{g}#1 not only escapes the diploid pairing control of wheat but also freely recombines in proximal chromosome regions where crossing over is usually suppressed (Liu et al. 2011b). We further analyzed this phenomenon by GISH, which revealed that a very small distal region of the short arm of chromosome 5M^{g}#1 is replaced by 5DS (3 %) and that this chromosome is actually a spontaneous wheat–*Ae. geniculata* recombinant chromosome and can be described as T5DS-5M^{g}#1S•5M^{g}#1L (D-H Koo et al., unpublished). This translocation presumably occurred during the development of the substitution line. Chromosome 5M^{g}#2, from a different *Ae. geniculata* accession, is a complete *Ae. geniculata* chromosome that does not have a wheat segment attached to the distal region of the short arm, and, as a result, does not recombine with wheat chromosomes in the presence of *Ph1*. The presence of a very small distal wheat segment in heterozygous condition not only allows recombination in this small region of homology, but also permits crossing over in distal and proximal homoeologous regions where recombination is usually suppressed in the presence of *Ph1* (D-H Koo et al., unpublished).

9.15 *Ae. triuncialis*

Ae. triuncialis L. ($2n=4x=28$, U^{t}U^{t}C^{t}C^{t}) is the source of the leaf rust resistance gene *Lr58*. The *Lr58* transfer occurred spontaneously in the progeny of a cross between the wheat cultivar WL711 and *Ae. triuncialis* accession TA10438. Molecular marker analysis mapped *Lr58* to a wheat–*Ae. triuncialis*recombinant chromosome, T2BS•2BL-2^{t}#1L, where approximately 5 % of the distal end of the long arm was derived from the long arm of a group-2 *Ae. triuncialis* (2^{t}#1L) chromosome (Kuraparthy et al. 2007a). The T2BS•2BL-2^{t}#1L translocation present in germplasm TA5605 was designated as cryptic because the *Ae. triuncialis* segment in this translocation was cytologically undetectable.

Ae. triuncialis is also the source of cereal cyst nematode (*Heterodera avenae* Woll.) resistance gene *Cre7* (Romero et al. 1998) and Hessian fly (*Mayetiola destructor* Say.) resistance gene *H30* (Martin-Sanchez et al. 2003). Both transfers occurred spontaneously. Preliminary data suggested that the *H30* gene was derived from a group-4 U^{t}-genome chromosome and was transferred to wheat chromosome 4D. However, further work is necessary to confirm this.

9.16 *Ae. ventricosa*

Doussinault et al. (1983) transferred a gene conferring resistance to eyespot [caused by *Pseudocercosporella herpotrichoides* (Fron) Dreighton], *Pch1*, from *Ae. ventricosa* Tausch ($2n=4x=28$, D^{v}D^{v}N^{v}N^{v}) to wheat chromosome 7D (Jahier et al. 1979, 1989;

Worland et al. 1988). The wheat–*Ae. ventricosa* recombinant chromosome T7D-7D^{v}#1 is present in the germplasms H-93-70 and VPM1, and has also been transferred to other wheat cultivars. In addition to *Pch1*, VPM1 has the *Ae. ventricosa*-derived leaf rust, stripe rust, and stem rust resistance genes *Lr37*, *Yr17*, and *Sr38*, and cereal cyst nematode resistance gene *Cre5* present on the wheat–*Ae. ventricosa* recombinant chromosome T2AL•2AS-2N^{v}#1/6N^{v}#1 (Bariana and McIntosh 1993, 1994; Bonhomme et al. 1995; Jahier et al. 2001; Tanguy et al. 2005; Badaeva et al. 2008).

Ae. ventricosa is also the source of cereal cyst nematode resistances genes *Cre2* (Delibes et al. 1993) and *Cre6* (Ogbonnaya et al. 2001). Delibes et al. (1997) also transferred the Hessian fly resistance gene *H27* to wheat in the form of a 4M^{v}#1 chromosome addition line and a 4M^{v}#1(4D) substitution line (H-93-33) (Delibes et al. 1997).

9.17 *Ae. neglecta*

Marais et al. (2009) transferred leaf rust resistance gene *Lr62* and the stripe rust resistance gene *Yr42* from *Ae. neglecta* Req. ex Bertol. ($2n=6x=42$, U^{n}U^{n}X^{n}X^{n}N^{n}N^{n}) using induced homoeologous recombination. Meiotic pairing and microsatellite marker analyses suggested that the wheat–*Ae. neglecta* recombinant chromosome T6AL-6^{n}#1L•6^{n}#1S in germplasm 03M119-71A consists of a distal segment of wheat chromosome 6AL, a proximal segment of the long arm, and complete short arm of a group-6 *Ae. neglecta* chromosome. The exact breakpoint and the homoeologous and genomic origin of the *Ae. neglecta* chromatin in this translocation remain to be determined.

9.18 Concluding Remarks

The genus *Aegilops* has been the most favorable genetic source for wheat improvement through alien gene introgression due to its close relatedness to wheat and rich sources of unique genes for resistance to various biotic and abiotic stresses. During the boom era of wide hybridization in wheat in the twentieth century, most of the *Aegilops* species were successfully crossed with wheat (Sharma and Gill 1983; Ozkan et al. 2001; Schneider et al. 2008), despite various problems preventing transfer of genes from *Aegilops* species to wheat, such as crossability barriers, gametocidal genes, and high vs. low pairing variants. Some were not successful because they were either too large or noncompensating. Much of the more recent work has focused on their improvement by further round of homoeologous pairing and use of markers to facilitate in shortening them.

A large number of wheat–*Aegilops* amphiploids and chromosome addition, substitution, translocation, and introgression lines have been produced (Schneider et al. 2008). A total of 41 formally named resistance genes have been transferred from

Aegilops species into wheat through chromosome translocation or homoeologous recombination, and several of the genes such as *Lr9*, *Pch1*, *Yr17*, *Lr37*, and *Sr38* have been used in wheat production (Jahier et al. 1989; Ambrozková et al. 2002). However, in most cases, these were defeated by new variants of the respective pathogens. If such genes transferred from alien species are to be used in the future, they should be used responsibly in gene combinations that may extend the durability and compensate for the considerable effort in their development.

Based on this review, we are aware of 41 unique resistance genes that were transferred to wheat from approximately 30 accessions representing 12 *Aegilops* species. Among the transferred genes, *Ae. speltoides* alone accounted for the origins of 12 genes, approximately one-third of the total genes transferred. Most of the thousands of *Aegilops* accessions maintained in various genetic resource centers worldwide (Monneveux et al. 2000) remain untapped for wheat improvement. Until now the genetic resources in the primary gene pool of wheat have been used in wheat breeding; increased emphasis is now needed to transfer more genes for resistance to pathogens and pests, tolerance to abiotic stresses, and increased yield potential from *Aegilops* species of the secondary gene pool into wheat for future wheat breeding programs.

Alien gene introgression has not been efficiently used in developing breeding-ready germplasm because selection of new homoeologous recombinants has been a tedious and laborious task in the past (Niu et al. 2011). However, the high throughput marker technologies that are recently available have greatly increased the efficiency for the precise identifications of desirable recombinants from large populations (Niu et al. 2011; Tiwari et al. 2014). We anticipate that the availability of genomic tools and rich genetic resources (i.e., wheat–*Aegilops* amphiploids and chromosome addition, substitution, and translocation lines) will facilitate the future endeavors of wheat–*Aegilops* introgressions for wheat improvement.

References

Aghaee-Sarbarzeh M, Ferrahi M, Singh S, Singh H, Friebe B, Gill BS (2002) Ph^I-induced transfer of leaf and stripe rust resistance genes from *Aegilops triuncialis* and *Ae. geniculata* to bread wheat. Euphytica 127:377–382. doi:10.1023/A:1020334821122

Ambrozková M, Dedryver F, Dumalasová V, Hanzalová A, Bartoš P (2002) Determination of the cluster of wheat rust resistance genes *Yr17*, *Lr37* and *Sr38* by a molecular marker. Plant Protect Sci 38:41–45

Andersson MS, de Vicente MC (2010) Gene flow between crops and their wild relatives. The Johns Hopkins University Press, Baltimore, MD, p 514

Badaeva ED, Dekova OS, Koenig J, Bernard S, Bernard M (2008) Analysis of the introgression of *Aegilops ventricosa* Tausch genetic material in a common wheat background using C-banding. Theor Appl Genet 117:803–811. doi:10.1007/s00122-008-0821-4

Bai D, Scoles GJ, Knott DR (1995) Rust resistance in *Triticum cylindricum* Ces. (4x, CCDD) and its transfer into durum and hexaploid wheats. Genome 38:8–16. doi:10.1139/g95-002

Bariana HS, McIntosh RA (1993) Cytogenetic studies in wheat. XV. Location of rust resistance genes in VPM1 and its genetic linkage with other disease resistance genes in chromosome 2A. Genome 36:476–482. doi:10.1139/g93-065

Bariana HS, McIntosh RA (1994) Characterization and origin of rust resistance and powdery mildew resistance in VPM1. Euphytica 76:53–61. doi:10.1007/BF00024020

Barloy D, Lemoine J, Dedryer F, Jahier J, McIntosh RA (2000) Molecular markers linked to the *Aegilops variabilis*-derived root-knot nematode resistance gene *Rkn-mn1* in wheat. Plant Breed 119:169–172. doi:10.1046/j.1439-0523.2000.00451.x

Biagetti M, Vitellozzi F, Ceoloni C (1998) Physical mapping of wheat-*Aegilops longissima* breakpoints in mildew-resistant recombinant lines using FISH with highly repeated and low-copy DNA probes. Genome 42:1013–1019. doi:10.1139/gen-42-5-1013

Bonhomme A, Gale MD, Koebner RMD, Nicolas P, Jahier J, Bernard M (1995) RFLP analysis of an *Aegilops ventricosa* chromosome that carries a gene conferring resistance to leaf rust (*Puccinia recondita*) when transferred to hexaploid wheat. Theor Appl Genet 90:1042–1048. doi:10.1007/BF00222919

Ceoloni C (1984) Transfer of a mildew resistance gene from *Triticum longissimum* to common wheat by induced homoeologous recombination. Genetica Agraria 38:326–327

Ceoloni C, Del Signore G, Pasquini M, Testa A (1988) Transfer of mildew resistance from *Triticum longissimum* into wheat by *ph1b* induced homeologous recombination. In: Miller TE, Koebner RMD (eds) Proceedings of the 7th International Wheat Genetics Symposium. The Institute of Plant Science Research, Cambridge, UK, pp 221–226

Ceoloni C, Del Signore G, Ercoli L, Donini P (1992) Locating the alien chromatin segment in common wheat-*Aegilops longissima* mildew resistant transfers. Hereditas 116:239–245. doi:10.1111/j.1601-5223.1992.tb00148.x

Chen PD, Tsujimoto H, Gill BS (1994) Transfer of Ph^I genes promoting homoeologous pairing from *Triticum speltoides* to common wheat. Theor Appl Genet 88:97–101. doi:10.1007/BF00222400

Danilova TV, Friebe B, Gill BS (2014) Development of a wheat single gene FISH map for analyzing homoeologous relationship and chromosomal rearrangements within the Triticeae. Theor Appl Genet 127:715–730. doi:10.1007/s00122-013-2253-z

Delibes A, Lopez-Brana I, Mena M, Garcia-Olmedo F (1987) Genetic transfer of resistance to powdery mildew and of an associated biochemical marker from *Aegilops ventricosa* to hexaploid wheat. Theor Appl Genet 73:605–608. doi:10.1007/BF00289201

Delibes A, Romero D, Aguaded S, Duce A, Mena M, Lopez-Brana I, Andres MF, Martin-Sanchez JA, Garcia-Olmedo F (1993) Resistance to cereal cyst nematode (*Heterodera avenae* Woll.) transferred from the wild grass *Aegilops ventricosa* to hexaploid wheat by a 'stepping stone' procedure. Theor Appl Genet 87:402–408. doi:10.1007/BF01184930

Delibes A, Del Moral J, Martin-Sanchez JA, Mejias A, Gallego M, Casado D, Sin E, Lopez-Brana I (1997) Hessian fly-resistance gene transferred from Chromosome $4M^v$ of *Aegilops ventricosa* to *Triticum aestivum*. Theor Appl Genet 94:858–864. doi:10.1007/s001220050487

Donini P, Koebner RMD, Ceoloni C (1995) Cytogenetic and molecular mapping of the wheat-*Aegilops longissima* chromatin breakpoints in powdery mildew-resistant introgression lines. Theor Appl Genet 91:738–743. doi:10.1007/BF00220952

Doussinault G, Delibes A, Sanchez-Monge R, Garcia-Olmedo F (1983) Transfer of a dominant gene for resistance to eyespot disease from a wild grass to hexaploid wheat. Nature 303:698–700. doi:10.1038/303698a0

Dubcovsky J, Lukaszewski AJ, Echaide M, Antonelli EF, Porter DR (1998) Molecular characterization of two *Triticum speltoides* interstitial translocations carrying leaf rust and greenbug resistance genes. Crop Sci 38:1655–1660. doi:10.2135/cropsci1998.0011183X003800060040x

Dundas IS, Verlin DC, Park RF et al. (2008) Rust resistance in *Aegilops speltoides* var. *ligustica*. In: Appels R et al. (eds) Proceedings of the 11th International Wheat Genetics Symposium, vol 1, Brisbane, Australia, Vol., pp 200–202

Dvorak J (1977) Transfer of leaf rust resistance from *Aegilops speltoides* to *Triticum aestivum*. Can J Genet Cytol 19:133–141. doi:10.1139/g77-016

Dvorak J, Knott DR (1980) Chromosome location of two leaf rust resistance genes transferred from *T. speltoides* to *T. aestivum*. Can J Genet Cytol 22:281–289. doi:10.1139/g80-047

Dvorak J, Knott DR (1990) Location of a *Triticum speltoides* chromosome segment conferring resistance to leaf rust in *Triticum aestivum*. Genome 33:892–897. doi:10.1139/g90-134
Dyck PL, Kerber ER, Martens JW (1990) Transfer of a gene for stem rust resistance from *Aegilops caudata* to common wheat. Can J Plant Sci 70:931–934. doi:10.4141/cjps90-114
Fan L, Han JY, Deng JY (1995) Transfer of gene for resistance to powdery mildew from *Ae. triuncialis* to common wheat by using Chinese Spring Ta1 kr *ph1b* plants. In: Li ZS, Xin ZY (eds) Proceedings of the 8th International Wheat Genetics Symposium, Beijing, China, 20–25 July, 1993, vol 1. China Agricultural Scientech Press, Beijing, pp 489–493
Faris JD, Xu SS, Cai X, Friesen TL, Jin Y (2008) Molecular and cytogenetic characterization of a durum wheat-*Ae. speltoides* chromosome translocation conferring resistance to stem rust. Chromosome Res 16:1097–1105. doi:10.1007/s10577-008-1261-3
Friebe B, Heun M (1989) C-banding pattern and powdery mildew resistance of *Triticum ovatum* and four *T. aestivum-T. ovatum* chromosome addition lines. Theor Appl Genet 78:417–424. doi:10.1007/BF00265306
Friebe B, Mukai Y, Dhaliwal H, Martin TJ, Gill BS (1991) Identification of alien chromatin specifying resistance to wheat streak mosaic virus and greenbug in wheat germ plasm by C-banding and in situ hybridization. Theor Appl Genet 81:381–389
Friebe B, Jiang J, Tuleen N, Gill BS (1995) Standard karyotype of *Triticum umbellulatum* and the characterization of derived chromosome addition and translocation lines in common wheat. Theor Appl Genet 90:150–156. doi:10.1007/BF00221010
Friebe B, Jiang J, Raupp WJ, McIntosh RA, Gill BS (1996) Characterization of wheat-alien translocations conferring resistance to diseases and pests: current status. Euphytica 91:59–87. doi:10.1007/BF00035277
Friebe B, Zhang P, Nasuda S, Gill BS (2003) Characterization of a knock-out mutation at the *Gc2* locus in wheat. Chromosoma 111:509–517. doi:10.1007/s00412-003-0234-8
Friebe B, Zhang P, Linc G, Gill BS (2005) Robertsonian translocations in wheat arise by centric misdivision of univalent at anaphase I and rejoining of broken centromeres during interkinesis of meiosis II. Cytogenet Genome Res 109:293–297. doi:10.1159/000082412
Gale MD, Miller TE (1987) The introduction of alien genetic variation in wheat. In: Lupton FGH (ed) Wheat breeding, its scientific basis. Chapman and Hall, London, New York, pp 173–210
Garcia-Olmedo F, Delibes A, Sanchez-Monge R (1984) Transfer of resistance to eyespot disease from *Aegilops ventricosa* to wheat. Vortrage fur Pflanzenzuchtung 6:156–168
Helguera M, Vanzetti L, Soria M, Khan IA, Kolmer J, Dubcovsky J (2005) PCR markers for *Triticum speltoides* leaf rust resistance gene *Lr51* and their use to develop isogenic hard red spring wheat lines. Crop Sci 45:728–734. doi:10.2135/cropsci2005.0728
Hsam SLK, Lapochkina IF, Zeller FJ (2003) Chromosomal location of genes for resistance to powdery mildew in common wheat (*Triticum aestivum* L. em Thell.) 8. Gene *Pm32* in a wheat-*Aegilops speltoides* translocation line. Euphytica 133:367–370. doi:10.1023/A:1025738513638
Jahier J, Doussinault G, Dosba F, Bourgois E (1979) Monosomic analysis of resistance to eyespot in the variety 'Roazon'. In: Ramanujam S (ed) Proceedings of the 5th International Wheat Genetics Symposium. Indian Society of Genetics & Plant Breeding, New Delhi, India, pp 437–440
Jahier J, Tanguy AM, Doussinault G (1989) Analysis of the level of eyespot resistance due to genes transferred to wheat from *Aegilops ventricosa*. Euphytica 44:55–59. doi:10.1007/BF00022599
Jahier J, Abelard P, Tanguy AM, Dedryver F, Rivoal R, Khathar S, Bariana HS (2001) The *Aegilops ventricosa* segment on chromosome 2AS of the wheat cultivar VPM1 carries the cereal cyst nematode resistance gene *Cre5*. Plant Breed 120:125–128. doi:10.1046/j.1439-0523.2001.00585.x
Jia J, Devos KM, Chao S, Miller TE, Reader SM, Gale MD (1996) RFLP-based maps of homoeologous group-6 chromosomes of wheat and their application in the tagging of *Pm12*, a powdery mildew resistance gene transferred from *Aegilops speltoides* to wheat. Theor Appl Genet 92:559–565. doi:10.1007/BF00224558
Kerber ER, Dyck PL (1990) Transfer to hexaploid wheat of linked genes for adult plant leaf rust and seedling stem rust resistance from an amphiploid of *Aegilops speltoides* x *Triticum monococcum*. Genome 33:530–537. doi:10.1139/g90-079

Klindworth DL, Niu Z, Chao S, Friesen TL, Jin Y, Faris JD, Cai X, Xu SS (2012) Introgression and characterization of a goatgrass gene for a high level of resistance to Ug99 stem rust in tetraploid wheat. G3 2:665–673. doi:10.1534/g3.112.002386

Kuraparthy V, Sood S, Chhuneja P, Dhaliwal HS, Kaur S, Bowden RL, Gill BS (2007a) A cryptic wheat-*Aegilops triuncialis* translocation with leaf rust resistance gene *Lr58*. Crop Sci 47:1995–2003. doi:10.2135/cropsci2007.01.0038

Kuraparthy V, Chhuneja P, Dhaliwal HS, Kaur S, Bowden RL, Gill BS (2007b) Characterization and mapping of cryptic alien introgressions from *Aegilops geniculata* with new leaf rust and stripe rust resistance genes *Lr57* and *Yr40* in wheat. Theor Appl Genet 114:1379–1389. doi:10.1007/s00122-007-0524-2

Kuraparthy V, Sood S, Gill BS (2009) Molecular genetic description of the cryptic wheat–*Aegilops geniculata* introgression carrying rust resistance genes *Lr57* and *Yr40* using wheat ESTs and synteny with rice. Genome 52:1025–1036. doi:10.1139/G09-076

Lay CL, Wells DG, Gardner WAS (1971) Immunity from wheat streak mosaic virus in irradiated *Agrotricum* progenies. Crop Sci 1:431–432. doi:10.2135/cropsci1971.0011183X001100030037x

Liu WX, Jin Y, Rouse M, Friebe B, Gill BS, Pumphrey MO (2011a) Development and characterization of wheat–*Ae. searsii* Robertsonian translocations and a recombinant chromosome conferring resistance to stem rust. Theor Appl Genet 122:1537–1545. doi:10.1007/s00122-011-1553-4

Liu W, Rouse M, Friebe B, Jin Y, Gill BS, Pumphrey MO (2011b) Discovery and molecular mapping of a new gene conferring resistance to stem rust, *Sr53*, derived from *Aegilops geniculata* and characterization of spontaneous translocation stocks with reduced alien chromatin. Chromosome Res 19:669–682. doi:10.1007/s10577-011-9226-3

Mago R, Zhang P, Bariana HS, Verlin DC, Bansal UK, Ellis JG, Dundas IS (2009) Development of wheat lines carrying stem rust resistance gene *Sr39* with reduced *Aegilops speltoides* chromatin and simple PCR markers for marker-assisted selection. Theor Appl Genet 119:1441–1450. doi:10.1007/s00122-009-1146-7

Mago R, Verlin D, Zhang P, Bansal U, Bariana H, Jin Y, Ellis J, Hoxha S, Dundas I (2013) Development of wheat-*Aegilops speltoides* recombinants and simple PCR-based markers for stem rust resistance genes on the 2S#1 chromosome. Theor Appl Genet 126:2943–2955. doi:10.1007/s00122-013-2184-8

Marais GF, Pretorious ZA, Marais AS, Wellings CR (2003) Transfer of rust resistance genes from *Triticum* species to common wheat. S Afr J Plant Soil 20:193–198. doi:10.1080/02571862.2003.10634934

Marais GF, McCallum B, Snyman JE, Pretorious ZA, Marais AS (2005) Leaf rust and stripe rust resistance genes *Lr54* and *Yr37* transferred to wheat from *Aegilops kotschyi*. Plant Breed 124:538–541. doi:10.1111/j.1439-0523.2005.01116.x

Marais GF, McCallum B, Marais AS (2006) Leaf rust and stripe rust resistance genes derived from *Aegilops sharonensis*. Euphytica 149:373–380. doi:10.1007/s10681-006-9092-9

Marais GF, McCallum B, Marais AS (2008) Wheat leaf rust resistance gene *Lr59* derived from *Aegilops peregrina*. Plant Breed 124:538–541. doi:10.1111/j.1439-0523.2008.01513.x

Marais F, Marais A, McCallum B, Pretorious Z (2009) Transfer of leaf rust and stripe rust resistance genes *Lr62* and *Yr42* from *Aegilops neglecta* Req. ex Bertol. to common wheat. Crop Sci 49:871–879. doi:10.2135/cropsci2008.06.0317

Marais GF, Kotze L, Eksteen A (2010a) Allosyndetic recombinants of the *Aegilops peregrina-derived Lr59* translocation in common wheat. Plant Breed 129:356–361. doi:10.1111/j.1439-0523.2009.01713.x

Marais GF, Badenhorst PE, Eksteen A, Pretorious ZA (2010b) Reduction of *Aegilops sharonensis* chromatin associated with resistance genes *Lr56* and *Yr38* in wheat. Euphytica 171:15–22. doi:10.1007/s10681-009-9973-9

Marais GF, Bekker TA, Eksteen A, McCallum B, Fetch T, Marais AS (2010c) Attempts to remove gametocidal genes co-transferred to common wheat with rust resistance from *Aegilops speltoides*. Euphytica 171:71–85. doi:10.1007/s10681-009-9996-2

Martin-Sanchez JA, Gomez-Colmenarejo M, Del Moral J, Sin E, Montes MJ, Gonzalez-Belinchon C, Lopez-Brana I, Delibes A (2003) A new Hessian fly resistance gene (*H30*) transferred from the wild grass *Aegilops triuncialis* to hexaploid wheat. Theor Appl Genet 106:1248–1255. doi:10.1007/s00122-002-1182-z

McIntosh RA (1991) Alien sources of disease resistance in bread wheats. In: Sasakuma T, Kinoshita T (eds) Nuclear and organellar genomes of wheat species. Proceedings of Dr. Kihara Memorial International Symposium on Cytoplasmic Engineering in Wheat. Yokohama, Japan, pp 320–332

McIntosh RA, Miller TE, Chapman V (1982) Cytogenetical studies in wheat XII. *Lr28* for resistance to *Puccinia recondita* and *Sr34* for resistance to *P. graminis tritici*. Z Pflanzenzüchtg 89:295–306

McIntosh RA, Yamazaki Y, Dubcovsky J, Rogers J, Morris C, Appels R, Xia XC (2013) Catalogue of gene symbols for wheat. In: 12th International Wheat Genetics Symposium, 8–13 September, 2013, Yokohama, Japan. http://www.shigen.nig.ac.jp/wheat/komugi/genes/download.jsp

Miftahudin RK, Ma XF, Mahmoud AA, Layton J, Rodriguez M, Chikmawati T et al. (2004) Analysis of EST loci on wheat chromosome group 4. Genetics 168:651–663. doi:10.1534/genetics.104.034827

Miller TE, Hutchinson J, Chapman V (1982) Investigation of a preferentially transmitted *Aegilops sharonensis* chromosome in wheat. Theor Appl Genet 61:27–33. doi:10.1007/BF00261506

Miller TE, Reader SM, Ainsworth CC, Summers RW (1987) The introduction of a major gene for resistance to powdery mildew of wheat, *Erisiphe graminis* f. sp. *tritici*, from *Aegilops speltoides* into wheat *T. aestivum*. In: Jorna ML, Slootmaker LAJ (eds) Cereal breeding related to integrated cereal production. Proceedings of Eucarpia Conference, Wageningen, The Netherlands, pp 179–183

Miller TE, Reader SM, Singh D (1988) Spontaneous non-Robertsonian translocations between wheat chromosomes and an alien chromosome. In: Miller TE, Koebner RMD (eds) Proceedings of the 7th International Wheat Genetics Symposium. UK, Cambridge, pp 387–390

Millet E, Manisterski J, Ben-Yehuda P, Distelfeld A, Deek J, Wan A, Chen X, Steffenson BJ (2014) Introgression of leaf rust and stripe rust resistance from Sharon goatgrass (*Aegilops sharonensis* Eig) into bread wheat (*Triticum aestivum* L.). Genome 57:309–316. doi:10.1139/gen-2014-0004

Monneveux P, Zaharieva M, Rekika D (2000) The utilisation of *Triticum* and *Aegilops* species for the improvement of durum wheat. In: Royo C, Nachit M, Di Fonzo N, Araus JL (eds.) Seminar on Durum Wheat Improvement in the Mediterranean Region: New Challenges. Zaragoza, Spain, 12–14 April, 2000, pp 71–81

Mujeeb-Kazi A, Kazi AG, Dundas I, Rasheed A, Ogbonnaya F, Kishii M, Bonnett D, Wang RR-C, Xu S, Chen P, Mahmood T, Bux H, Farrakh S (2013) Genetic diversity for wheat improvement as a conduit to food security. Adv Agron 4:179–257. doi:10.1016/B978-0-12-417187-9.00004-8

Nasuda S, Friebe B, Bush W, Kynast RG, Gill BS (1998) Structural rearrangement in chromosome 2M of *Aegilops comosa* has prevented the utilization of the Compair and related wheat-*Ae. comosa* translocations in wheat improvement. Theor Appl Genet 98:780–785. doi:10.1007/s001220050802

Niu Z, Klindworth DL, Friesen TL, Chao S, Jin Y, Cai X, Xu SS (2011) Targeted introgression of a wheat stem rust resistance gene by DNA marker-assisted chromosome engineering. Genetics 187:1011–1020. doi:10.1534/genetics.110.123588

Ogbonnaya FC, Seah S, Delibes A, Jahier J, Lopez-Brana I, Eastwood RF, Lagudah E (2001) Molecular-genetic characterization of a new nematode resistance gene in wheat. Theor Appl Genet 102:623–629. doi:10.1007/s001220051689

Okamoto M (1957) Asynaptic effect of chromosome V. Wheat Info Serv 5:6

Olivera PD, Steffenson BJ (2009) *Aegilops sharonensis*: origin, genetics, diversity, and potential for wheat improvement. Botany 87:740–756. doi:10.1139/B09-040

Ozkan H, Levy AA, Feldman M (2001) Allopolyploidy-induced rapid genome evolution in the wheat (*Aegilops–Triticum*) group. Plant Cell 13:1735–1747. doi:10.1105/tpc.13.8.1735

Qi L, Friebe B, Zhang P, Gill BS (2007) Homoeologous recombination, chromosome engineering and crop improvement. Chromosome Res 15:3–19. doi:10.1007/s10577-006-1108-8

Raupp WJ, Friebe B, Gill BS (1995) Suggested guidelines for the nomenclature and abbreviation of the genetic stocks of wheat, *Triticum aestivum* L. em Thell., and its relatives. Wheat Info Serv 81:51–55

Riley R, Chapman V (1958) Genetic control of the cytologically diploid behaviour of hexaploid wheat. Nature 182:713–715. doi:10.1038/182713a0

Riley R, Chapman V, Johnson R (1968a) The incorporation of alien disease resistance in wheat by genetic interference with the regulation of meiotic chromosome synapsis. Genet Res Camb 12:198–219. doi:10.1017/S0016672300011800

Riley R, Chapman V, Johnson R (1968b) Introduction of yellow rust resistance of *Aegilops comosa* into wheat by genetically induced homoeologous recombination. Nature 217:383–384. doi:10.1038/217383a0

Roberts HF (1929) Plant hybridization before Mendel. Princeton University Press, Princeton, pp 123–129

Romero MD, Montes MJ, Sin E, Lopez-Brana I, Duce A, Martin-Sanchez JA, Andres MF, Delibes A (1998) A cereal cyst nematode (*Heterodera avennae* Woll.) resistance gene transferred from *Aegilops triuncialis* to hexaploid wheat. Theor Appl Genet 96:1135–1140. doi:10.1007/s001220050849

Schneider A, Molnar I, Molnar-Lang M (2008) Utilization of *Aegilops* (goatgrass) species to widen the genetic diversity of cultivated wheat. Euphytica 163:1–19. doi:10.1007/s10681-007-9624-y

Sears ER (1952) Misdivision of univalents in common wheat. Chromosoma 4:535–550. doi:10.1007/BF00325789

Sears ER (1956) The transfer of leaf rust resistance from *Aegilops umbellulata* to wheat. Brookhaven Symp Biol 9:1–22

Sears ER (1972) Chromosome engineering in wheat. In: Stadler Symposium, vol. 4, University of Missouri, Columbia, pp 23–38

Sears ER (1976) Genetic control of chromosome pairing in wheat. Annu Rev Genet 10:31–51. doi:10.1146/annurev.ge.10.120176.000335

Sears ER (1977) An induced mutant with homoeologous pairing in common wheat. Can J Genet Cytol 19:585–593. doi:10.1139/g77-063

Sharma HC, Gill BS (1983) Current status of wide hybridization in wheat. Euphytica 32:17–31. doi:10.1007/BF00036860

Sharp PJ, Desai S, Chao S, Gale MD (1988) Isolation, characterization, and applications of a set of 14 RFLP probes identifying each homoeologous chromosome arm in the Triticeae. In: Miller TE, Koebner RMD (eds) Proceedings of the 7th International Wheat Genetics Symposium, Cambridge, UK, 13–19 July, 1988. Institute of Plant Science Research, Cambridge, UK, pp 639–646

Shepherd KW, Islam AKMR (1988) Fourth compendium of wheat-alien chromosome lines. In: Miller TE, Koebner RMD (eds) Proceedings of the 7th International Wheat Genetics Symposium, Cambridge, UK, 13–19 July, 1988. Institute of Plant Science Research, Cambridge, UK, pp 1373–1398

Stoyanov H (2014) Representatives of genus *Aegilops* as a source of pathogens resistance. AgroLife Sci J 3:139–148

Tanguy A-M, Coriton O, Abelard P, Dedryver F, Jahier J (2005) Structure of *Aegilops ventricosa* chromosome $6N^v$, the donor of wheat genes *Yr17*, *Lr37*, *Sr38*, and *Cre5*. Genome 48:541–546. doi:10.1139/g05-001

Tiwari VK, Wang S, Sehgal S, Vrána J, Friebe B, Kubaláková M, Chhuneja P, Doležel J, Akhunov E, Kalia B, Sabir J, Gill BS (2014) SNP discovery for mapping alien introgressions in wheat. BMC Genomics 15:273. doi:10.1186/1471-2164-15-273 van Slageren MW (1994) Wild wheats: a monograph of *Aegilops* L. and *Amblyopyrum* (Jaub. & Spach) Eig (Poaceae). Wageningen Agricultural University Papers, Wageningen, The Netherlands. No. 94-7, pp. xiii + 512 pp

Wells DG, Sze-Chung R, Lay CL, Gardner WA, Buchenau GW (1973) Registration of C.I. 15092 and C.I. 15093 wheat germplasm. Crop Sci 13:776. doi:10.2135/cropsci1973.0011183X001300060078x

Wells DG, Kota RS, Sandhu HS, Gardner WS, Finney KF (1982) Registration of one disomic substitution line and five translocation lines of winter wheat germplasm resistant to wheat streak mosaic virus. Crop Sci 22:1277–1278. doi:10.2135/cropsci1982.0011183X002200060083x

Williamson VM, Thomas V, Ferris H, Dubcovsky J (2013) An *Aegilops ventricosa* translocation confers resistance against root-knot nematodes to common wheat. Crop Sci 53:1412–1418. doi:10.2135/cropsci2012.12.0681

Worland AJ, Law CN, Hollins TW, Koebner RMD, Guira A (1988) Location of a gene for resistance to eyespot (*Pseudocercosporella herpotrichoides*) on chromosome 7D of bread wheat. Plant Breed 101:43–51. doi:10.1111/j.1439-0523.1988.tb00265.x

Yu MQ, Jahier J, Person-Dedryer F (1995) Chromosomal location of a gene (*Rkn-mn1*) for resistance to the root-knot nematode transferred into wheat from *Aegilops variabilis*. Plant Breed 114:358–360. doi:10.1111/j.1439-0523.1995.tb01250.x

Zeller FJ, Kong L, Hartl L, Mohler V, Hsam SLK (2002) Chromosomal location of genes for resistance to powdery mildew in common wheat (*Triticum aestivum* L. em Thell.) 7. Gene *Pm29* in line Pova. Euphytica 123:187–194. doi:10.1023/A:1014944619304

Zhang H, Jia J, Gale MD, Devos KM (1998) Relationships between the chromosomes of *Aegilops umbellulata* and wheat. Theor Appl Genet 96:69–75. doi:10.1007/s001220050710

Chapter 10
Aegilops tauschii Introgressions in Wheat

Andreas Börner, Francis Chuks Ogbonnaya, Marion S. Röder, Awais Rasheed, Sambasivam Periyannan, and Evans S. Lagudah

10.1 Introduction

Progenitor species that contributed the constituent triple genomes, designated AA, BB and DD found in the allohexaploid genome of common wheat (*Triticum aestivum*) have to a large extent been defined. *Aegilops tauschii* has been established as the D genome donor to common wheat (Kihara 1944; McFadden and Sears 1946). In contrast to the narrow geographic distribution of the other progenitor species, *Ae. tauschii* extends over a wide geographic range from eastern Turkey to China. Early taxonomic classifications based on spike morphology placed *Ae. tauschii* into two subspecies-ssp *tauschii* and ssp. *strangulata* (Eig 1929). Under subspecies *tauschii*, there were three varietal taxa classified as *anathera*, *meyeri* and *typica* whereas in subspecies *strangulata* a monotypic varietal taxa also referred

A. Börner • M.S. Röder
Leibniz Institute of Plant Genetics and Crop Plant Research (IPK Gatersleben), Corrensstraße 3, Stadt Seeland/OT Gatersleben 06466, Germany

F.C. Ogbonnaya
Grains Research and Development Corporation, Level 1, 40 Blackall St, Barton, Canberra, ACT 2600, Australia

A. Rasheed
Institute of Crop Science, National Wheat Improvement Centre, Chinese Academy of Agricultural Sciences (CAAS), 12 Zhongguancun South Street, Beijing 100081, China

International Maize and Wheat Improvement Centre (CIMMYT) China Office, c/o CAAS, 12 Zhongguancun South Street, Beijing 100081, China

S. Periyannan • E.S. Lagudah (✉)
Commonwealth Scientific and Industrial Research Organisation, Agriculture, Clunies Ross St., Canberra, ACT 2601, Australia
e-mail: evans.lagudah@csiro.au

M. Molnár-Láng et al. (eds.), *Alien Introgression in Wheat*,
DOI 10.1007/978-3-319-23494-6_10

to as *strangulata* was adopted. Some of the early studies in comparative D genomes of common wheat and accessions of the diploid progenitor led to an overly simplified conclusion that the *strangulata* taxa was the source of the D genome in the allohexaploid.

With increasing advances in genotyping, from AFLP analysis to single nucleotide polymorphisms (SNPs) where thousands of genes and DNA fragments across the entire D genome have been analysed, it has become evident that the morphological classification is inadequate in describing the specific source of the wheat D genome. A genetic classification that surpasses the morphological groupings revealed two lineages of the *Ae. tauschii* genepool, designated lineage 1(L1) and lineage 2 (L2) (Mizuno et al. 2010; Wang et al. 2013). While the evolutionary lineage classifications has some parallels with the morphological groupings, L2 contains the taxa *strangulata*, *meyeri* and some accessions of *anathera* and typica. These two lineages are further subdivided with sublineage 1W located in eastern Turkey, Armenia, Azerbaijan, and western Iran and sublineage 1E was distributed from central Iran to China. Sublineage 2W was found in Armenia and Azerbaijan, and sublineage 2E was located in Caspian Azerbaijan and Caspian Iran (Wang et al. 2013). On the basis of the SNP data, a population within L2E in the southwestern and southern Caspian was shown to be the main source of the wheat D genome, whereas L1 contributed as little as 0.8 % of the wheat D genome (summarized in Fig. 10.1). It has been postulated that recurrent hybridisation and introgression between *Ae tauschii* and common wheat aided by tetraploid wheats as a bridging species may have contributed to the origin of D genome diversity in wheat.

Interest in *Ae. tauschii* introgressions stems from the observation that the diploid D genome progenitor possesses a higher genetic diversity compared to bread wheat cultivars and landraces (Reif et al. 2005). *Ae. tauschii* was used to introgress specific traits that include diverse resistance genes (Olson et al. 2013; Mandeep et al. 2010; Leonova et al. 2007; Miranda et al. 2006; Ma et al. 1993; Eastwood et al. 1994), bread-making quality (Li et al. 2012), pre-harvest sprouting tolerance (Gatford et al. 2002; Imtiaz et al. 2008), yield (Gororo et al. 2002) and also morphological characters (Watanabe et al. 2006) into breeding material and cultivars of bread wheat. Since the mid-twentieth century directed efforts at *Ae tauschii* introgressions into wheat has come from two avenues. Firstly, the more common approach of artificial hexaploid wheat synthesis that is generated by crossing tetraploid wheats with *Ae tauschii* and then doubling the triploid chromosome set by colchicine treatment or spontaneous doubling arising from unreduced gamete formation. Numerous reports on synthetic hexaploids have been reviewed by Ogbonnaya et al. (2013). Secondly, the process of direct introgression which involves *Ae tauschii* crosses with bread wheat and through repeated backcrosses to recover a stable bread wheat derivative (Gill and Raupp 1987), where recombinant chromosomes between the diploid and hexaploid D genomes are produced. Introgression approaches that occur via synthetic hexaploids are not limited to the D genome but also involve the A and B genomes. With the recent whole genome shotgun sequence of the synthetic hexaploid, W7984 and derivatives from recombi-

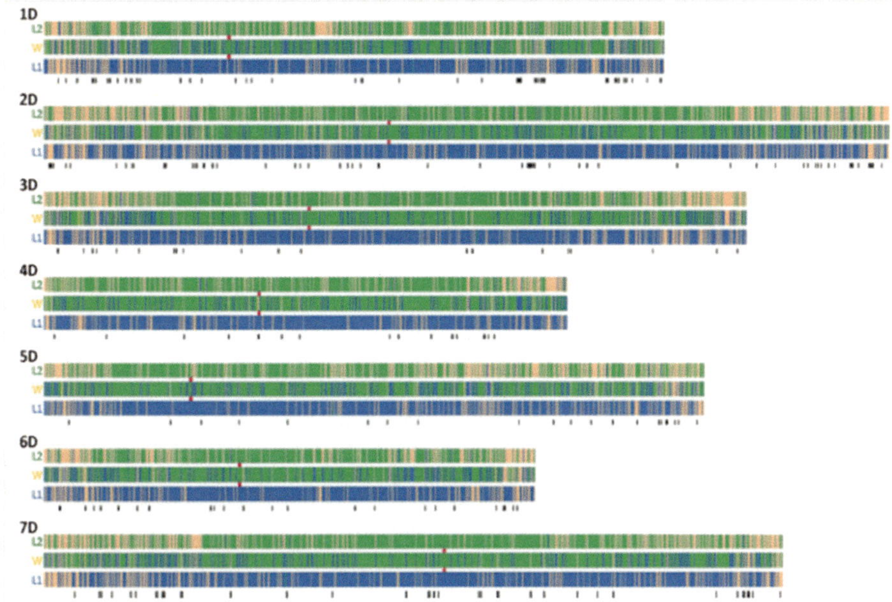

Fig. 10.1 SNP haplotypes along wheat chromosomes1D-7D (W) and *Aegilops tauschii* lineage1 (L1) and lineage2 (L2). The *pink coloured* regions are polymorphic sites whereas the monomorphic regions derived from L1 are shown in *blue* and L2 in *green*. Centromeric regions are marked in *red* and putative haplotypes from L1 introgressed into wheat are marked in *black*. Figure taken from Wang et al. (2013)

nant inbred lines (Chapman et al. 2015), it provides opportunities to assess the effects of selected sequence characterized genomic intervals introgressed into wheat. Further refinements towards achieving introgressions specifically derived from *Ae tauschii* are reported in the next section.

10.2 Development of Introgression Lines

As an alternative approach, wheat chromosome substitution lines carrying different chromosomes of *Ae. tauschii* were used in generating a set of well-characterized *Trtitcumaestivum–Ae. tauschii* introgression lines (Pestsova et al. 2001, 2006). Substitution lines of 'Chinese Spring' ('CS') in which single chromosomes of the D-genome had been replaced by homologous chromosomes of a synthetic wheat were developed by Law and Worland (1973). 'Synthetic 6x' was obtained from a cross of tetraploid emmer and the wild grass *Ae. tauschii* (*T. dicoccoides* var. *spontaneovillosum*×*Ae. squarrosa* ssp. *eusquarrosa*) (McFadden and Sears 1947), i.e. the D-genome substitution lines represent *T.aestivum*/*Ae. tauschii* replacements.

The D-genome substitution lines were backcrossed twice with 'CS'. In order to select a set of homozygous introgression lines representing the whole *Ae. tauschii* genome, 450 BC_2-plants were genotyped with microsatellite markers (Röder et al. 1998) and 60 were selected and selfed for development of homozygous lines. In total, 84 different homozygous 'CS (Synthetic 6x)' ('CS (Syn)') introgression lines were developed from BC_1 and BC_2 progenies (Pestsova et al. 2006). The advantage of this approach is that with a limited effort introgression lines with clean background can be generated; the disadvantage is that the approach is only applicable for varieties with existing chromosome substitution lines.

10.3 Utilisation of Wheat–*Ae tauschii* Chromosome Substitution Lines for Genetic Mapping

The set of 'CS (Syn)' D genome introgression lines has been used to identify chromosomal regions responsible for a range of agronomic traits including biotic and abiotic stress response. Both enhancing and detrimental effects were related to the *Ae. tauschii* segments incorporated into the bread wheat genome. Several examples are described below.

10.3.1 Yield and Related Characters

Yield associated traits including flowering time, plant height and single ear characters such as ear length, spikelet number and grain weight per ear were investigated under greenhouse (Pestsova et al. 2001) and field (Pestsova et al. 2006) conditions. Large effects were detected on the long arm of chromosome 5D due to the effect of the vernalisation response gene *Vrn-D1* (Fig. 10.2). The sensitivity of the synthetic wheat to vernalisation causes a delay in flowering time of at least 14 days. This delay in flowering was associated with an increased tillering and a higher spikelet number. The traits spike fertility (number of grains divided by the number of spikelets per ear) and grain weight per ear, however, were reduced. Two genomic regions appeared to have favourable alleles derived from the *Ae. tauschii* segment. Loci on chromosome 2DS contributed to earliness and ear length. The detected QTL coincided with the location of the photoperiodic insensitivity gene *Ppd-D1*. A second region was detected on chromosome arm 3DL. Favourable alleles from the *Ae. tauschii* introgressed segment for spikelet number per ear and grain weight per ear were present. The reason for finding only a few favourable alleles may be the rather long size of introgressions. It is also possible that beneficial alleles of *Ae. tauschii* were masked by many deleterious alleles located on the same chromosomal segment (Pestsova et al. 2006).

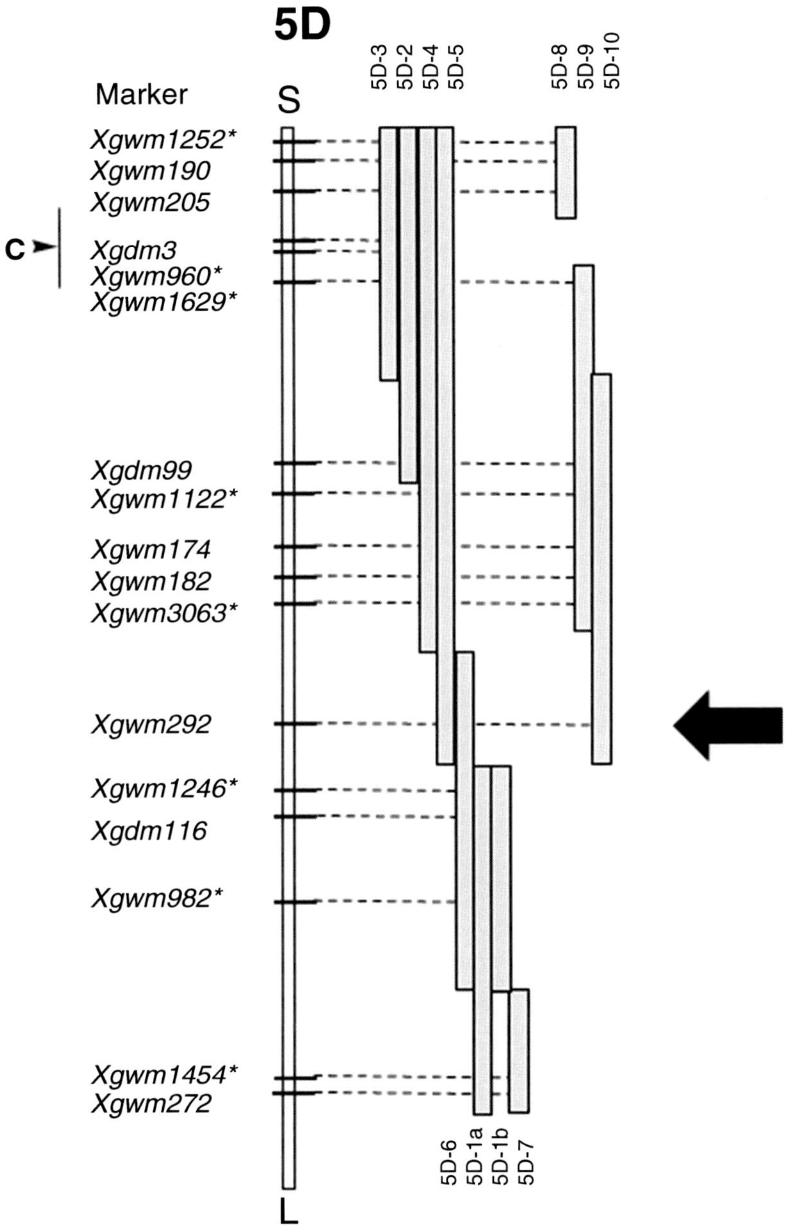

Fig. 10.2 Wheat/*Ae. tauschii*chromosome 5D introgression lines. The *black arrow* indicates the region of *Vrn-D1*.showing large effects on yield-related characters. *c* centromere position, *L* long arm, *S* short arm (Pestsova et al. 2006)

*10.3.2 Resistance to Septoria Tritici Blotch (*Mycosphaerella graminicola*)*

The analysis of different sets of single chromosome substitution lines, including those derived from the 'CS' x 'Synthetic 6x' identified chromosome 7D of the synthetic wheat to carry a gene(s) encoding near complete resistance to two virulent Argentinian isolates of the foliar fungal disease septoria tritici blotch, caused by the pathogen *Mycosphaerella graminicola* (Fuckel) Schroeter in Cohn (Simón et al. 2001, 2005). In subsequent studies, Simón et al. (2007) explored the 13 chromosome 7D introgression lines, along with the parental lines 'CS' and 'CS (Syn 7D)'. Both seedling and adult plant disease tests were performed by inoculation with the two isolates 'IPO 92067' and 'IPO 93014', which had been selected on the basis of prior observations made of their reactions in the 'CS (Syn)' single chromosome substitution lines.

Results are presented in Fig. 10.3. The introgression lines differed significantly in disease reaction from 'CS' in at least two (grey vertical bars) or four (black

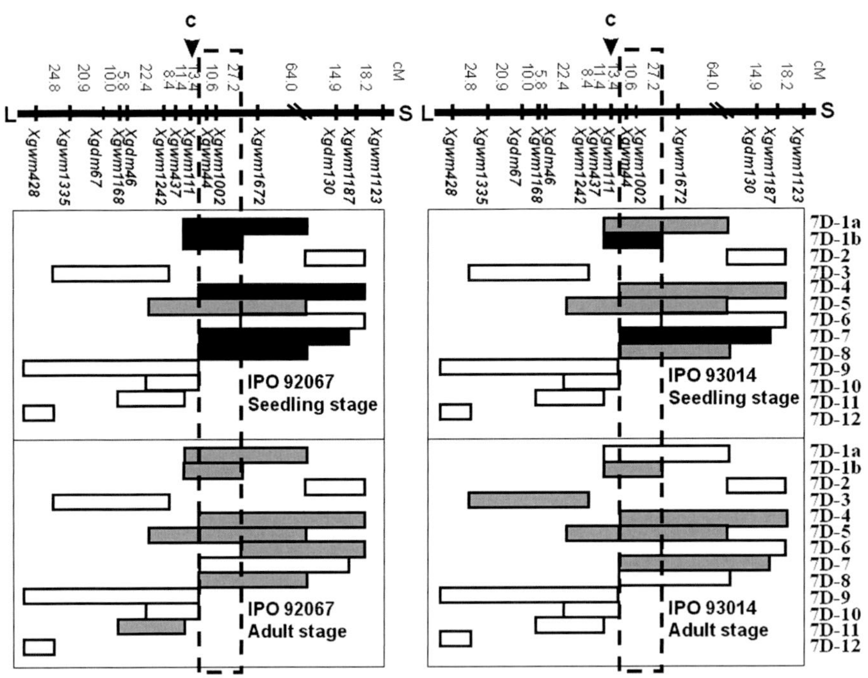

Fig. 10.3 Wheat/*Ae. tauschii* chromosome 7D introgression lines inoculated with septoria tritici blotch isolates IPO 92067 and IPO 93014 at the seedling and adult plant stages. *Lines* differing significantly from CS in two and four independent experiments are indicated by, respectively, *grey* and *black colour. Boxes* in *broken lines* indicate the position of the resistance locus. *c* centromere position, *L* long arm, *S* short arm (Simón et al. 2007)

vertical bars) independent experiments, and showed that the disease-resistance locus mapped to the centromeric region of chromosome arm 7DS. The locus identified by the 'introgression mapping' approach was consistent with that described by Arraiano et al. (2001), who investigated single chromosome recombinant lines developed from the 'CS (Syn 7D)' substitution line. The gene was designated *Stb5*. It was concluded that *Stb5* confers resistance against *M. graminicola* isolates from both Europe (Portugal, the Netherlands) and South America (Argentina).

10.3.3 Aluminium Tolerance

Aluminium (Al) toxicity is a major constraint to crop productivity in acidic soils. Although lime application can be used to neutralize soil acidity, and hence reduce the level of soluble Al, the development of wheat cultivars able to tolerate Al toxicity is considered a more sustainable approach. To identify loci responsible for Al tolerance Navakode et al. (2009) investigated the seven 'CS (Syn)' D genome chromosomes substitution lines as well as the 84 introgression lines. Of the whole chromosome substitution lines it was shown that 'CS (Syn 4D)' was most sensitive compared to 'CS'.

Among the 4D introgressions, lines '4D-1' and '4D-8' were tolerant, but the others were all sensitive. Genotyping of the 4D introgression lines using SSR markers enabled the delineation of the region of the 'CS' chromosome in which a gene(s) for tolerance is present (Fig. 10.4). The introgression lines derived from the remaining D genome chromosomes showed no significant deviation from control parent 'CS'. Therefore, it was concluded that the *Ae. tauschii* donor used to create the introgression lines carried no positive factors for Al tolerance. However, in a subsequent study, Ryan et al. (2010) reported identification of genetic variation for this trait in *Ae. tauschii*. A major gene for Al tolerance, *ALMT1*, located on chromosome 4DL encodes an anion channel protein which releases malate ions that bind to Al^{3+} to protect the sensitive root apex. While Al tolerance is largely determined by expression levels of ALMT1 governed by tandem repeats in the promoter, the reported variation of Al tolerance in *Ae. tauschii* is suggested to occur outside the promoter region. The demarcated region shown in Fig. 10.4 where the *Ae tauschii* introgressed segment has replaced CS, most likely carries the *ALMT1* gene.

10.3.4 Seed and Seedling Characters: Dormancy, Germination, Vigour and Longevity

Seed and seedling-related characters were investigated by Lohwasser et al. (2005) and Landjeva et al. (2008). For dormancy testing fresh seeds were germinated under two different temperature conditions: at 20 °C for 7 days and at 10 °C for 14 days.

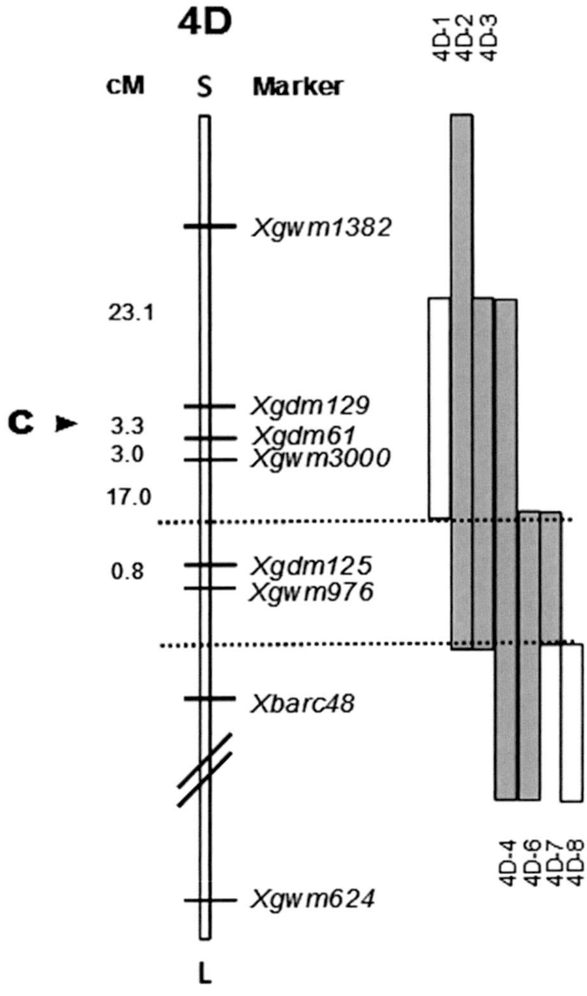

Fig. 10.4 Wheat/*Ae. tauschii* chromosome 4D introgression lines investigated for Al tolerance. *White bars*: tolerant lines, *grey bars*: sensitive lines. The area within the *dotted lines* indicates the location of positive factors in 'Chinese Spring' and/or negative factors in *Ae. tauschii*. *c* centromere position, *L* long arm, *S* short arm (Navakode et al. 2009)

Dormancy index following Strand (1965) was calculated. One major QTL (LOD>3.0) and one minor QTL (LOD>1.5<2.0) were detected on chromosomes 6DL and 6DS, respectively (Lohwasser et al. 2005).

A comprehensive study that examined germination, seed vigour and longevity, and early seedling growth revealed QTL that mapped to chromosomes 1D, 2D, 4D, 5D, and 7D (Landjeva et al. (2008)). Most of the QTL for germination per se clustered on chromosome 1DS whereas chromosome 7DS harboured loci controlling the development of normal seedlings. Seed vigour-related QTL were present on chromosome 5DL. Loci for seed longevity were coincident with those for germination or seed vigour on chromosomes 1D or 5D. Finally, QTL for seedling growth were identified on chromosomes 4D and 5D. In summary, *Ae. tauschii* contributed alleles allowing earlier and faster germination, whereas CS alleles were responsible for improved germination capacity, and increased synchronicity of germination. In addition, CS contributed favourable alleles for seedling growth, seed vigour and longevity.

10.4 Molecular Genetics of Disease-Resistance Gene Introgressions

Several reviews have documented the wide range of pest and disease-resistance sources found in *Ae. tauschii* and introgressed into wheat (Gill et al. 1985; Friesen et al. 2008; Halloran et al. 2008). A few of the underlying genes for these resistance sources have been cloned and many others are in the pipeline towards their eventual isolation. Two of the genes cloned, *Lr21* and *Sr33*, provide some insights into the loci found in their respective introgressed segments. *Lr21* introgressions into wheat has come from different avenues; these include direct crosses that involved the accessions TA1649 into the background of the cultivar Wichita in the Unites States, the use of the synthetic hexaploid RL5406 (Tetra Canthatch x *Ae. tauschii* RL5289) crossed into Canadian and Australian wheats. *Lr21* encodes a nucleotide-binding leucine rich repeat (NB-LRR) protein (Huang et al. 2003), and analysis of the corresponding locus in common wheat revealed sequence variants that included SNPs and insertion/deletion events that account for at least ten haplotypes (Fu et al. 2010). Apart from the haplotype that characterizes *Lr21* resistance, it was unclear whether the other haplotypes carried functional alleles. Nevertheless intragenic recombination involving non-functional haplotypes that reconstituted a functional *Lr21* haplotype was reported in a single plant selection from over 5000 F_2 progeny (Huang et al. 2009).

A stem rust-resistance gene, *Sr45*, was also present on the same chromosome arm (1DS) that harboured *Lr21* in the *Ae. tauschii* accession RL5289. Using the single chromosome substitution line where chromosome 1D from RL5289 had replaced the corresponding homologue in Chinese Spring (CS), the progeny from CS × CS (1DRL5289) enabled a high resolution mapping of the *Sr45* locus. By combining diagnostic markers for *Sr45* and *Lr21*, it became evident that selections from backcross derivatives between RL5406 and Australian wheats carried different *Ae. tauschii* introgressed segments; some had segments with both *Lr21* and *Sr45* while some had retained only the *Sr45* carrying segment (Periyannan et al. 2014).

Another Chinese Spring chromosome 1D substitution line from the synthetic hexaploid RL5405 (Tetra-Canthatch x *Ae. tauschii* RL5288), the source of *Sr33*, was used to generate an EMS mutagenized population to inactivate *Sr33*. These mutants were pivotal in validating the candidate *Sr33*-resistance gene from a cluster of the *Mla* (barley powdery mildew-resistance gene family)-related gene sequences at the *Sr33* locus (Periyannan et al. 2013). In addition to the *Mla* gene family, the

Sr33 locus carried an NB-LRR gene with an unusual C terminal domain that possessed an exocyst70 subunit (NB-LRR-Exo70). There is increasing evidence that adjacent pairs of distantly related NB-LRR genes where often one member within the pair with additional domains are all required for functional activity in defense against pathogens (see review by Cesari et al. 2014). While *Sr33* was confirmed to be the result of a functional diversification of an *Mla* gene member that recognize stem rust, further investigation is needed to establish whether NB-LRR-Exo70 is involved in defense to any other wheat pathogens other than stem rusts. The role of exocyst 70 in defense against bacterial and fungal pathogens has been reported in the model plant *Arabidopsis* (Pecenkova et al. 2011). It is also possible that genes with such unusual domain fusions occurs randomly through insertion, recombination or deletion events that facilitates the birth of new disease specificities under increased fitness from pathogens. Thus the introgressed *Sr33* segment and the corresponding locus in wheat of which Chinese Spring also carries an allelic variant of NB-LRR-Exo70 (IWGS wheat survey sequence) may provide new opportunities in deciphering the functional variability associated with this gene fusion.

10.5 Introgressions via Artificial Hexaploid Wheat Synthesis

Synthetic hexaploid wheat (SHW) that combines genes from the tetraploid wheat *Triticum turgidum* L. and wild ancestor *Ae. tauschii*, are arguably the most preferred and widely exploited wheat genetic resource as sources of new variation for the improvement of bread wheat. Breeding improvements are commonly achieved when the desirable gene from the primary SHW are introgressed into bread wheat via advanced derivatives (synthetic backcross-derived lines-SBLs) or synthetic derivatives (SYN-DER) (Table 10.1). As pointed out earlier, this approach introduces introgressed segments from the D genome of *Ae. tauschii* as well as the AB genomes of tetraploid wheat. The research focus during the past two decades were largely on characterizing primary SHW for various economic traits, and to proposing their putative usefulness in wheat breeding. However, there is now increasing trends to exploit characterized SHW through the introgression of desirable genes via SBLs for wheat improvement. Recently, Ogbonnaya et al. (2013) presented a comprehensive review and analysis of research on SHW that covered the important historical landmarks on their development, characterization and exploitation in wheat improvement. The current efforts will focus on impacts of successful introgressions from SHW post Ogbonnaya et al. (2013).

The largest collection of SHW in the world was developed at CIMMYT during 1988 to 2010 with 1300 SHW produced using about 50 improved durum genotypes and 900 *Ae. tauschii* accessions. What remains unclear is how many of these *Ae. tauschii* accessions are unique, given that the same accession can have different identification tags from the germplasm banks where they were sourced. Of these, about 100 SHW were developed using wild accessions of *Triticum dicoccoides* and *Ae. tauschii*. Additional collections of SHW were also developed at CIMMYT and

Table 10.1 Selected examples of recent advances in the characterization, and exploitation of *Ae. tauschii*, SHW and SYN-DER

Group	Trait	Germplasm	Genotyping	Phenotyping	Summary	Reference
Biotic stresses	Stem rust	456 *Ae. tauschii* accessions	–	UG-99 races	12 accessions resistant to all Ug99 races	Rouse et al. (2011)
	Stem rust	267 *Ae. tauschii* accessions	–	Seedling	239 resistant to local Sr races	Vikas et al. (2014)
	Yellow rust	SHW	SNP 9K	Seedling, adult (2 locations)	Nine genomic regions identified	Zegeye et al. (2014)
	Pest resistance	SHW	DArT	Field trials	26 QTLs identified for resistance to five major pests.	Joukhadar et al. (2013)
	Pest resistance	SHW	–	Field trials	15 resistant to Hessian fly, 1 resistant to Russian wheat aphid, and 21 resistant to Sunn pest	El Bouhssini et al. (2013)
	Leaf rust	267 *Ae. tauschii* accessions	–	Seedling	All resistant to local *Lr* races	Vikas et al. (2014)
	Bird cherry-oat aphid and green bug	SYN-DER RILs	GBS	Evaluations for resistance	QTL on 2DL	Crespo-Herrera et al. (2014)
	Spot blotch	SYN1xOcoroni86	GBS and SSRs	Two years field screening	3 QTLs identified on 1B, 3B and 5A contributed by SYN1	Zhu et al. (2014)
Yield	TKW	231 SHW	DArT	Grain size and weight	31 loci identified for grain size, weight and shape	Rasheed et al. (2014)
	TKW, GS	4 SYN-DER F2 populations	DArT	Grain size and weight	18 QTLs for grain size and shape identified	Okamoto et al. (2013)
	Yield	SYN-DER	–	Field trials	SYN-DER cultivars prove superior to bread wheat cultivars	Tang et al. (2014)
	Wax	210 *Ae. tauschii* and 3 SYN-DER mapping populations	Genotyping *Iw2* locus	–	–	Nishijima et al. (2014)

(continued)

Table 10.1 (continued)

Group	Trait	Germplasm	Genotyping	Phenotyping	Summary	Reference
Abiotic	Drought	210 SYN-DER	*TaCwi-5D*	Relative water contents and field trials	2 new haplotypes identified	Khalid et al. (Unpublished)
	Drought	F2 SYN-DER	SSR and *TaABA8O′H1*	Seedling traits	QTL on 6D	Iehisa et al. (2014b)
	Drought	SYN-DER	–	Relative water contents	SYN-DER outperformed bread wheat cultivars	Ali et al. (2014)
	Sprouting	SYN-DER	sequencing *amy1* gene		QTLs on group 6 chromosomes	Yang et al. (2014)
	Boron	SHW	DArT	Relative root length at seedling stage	3 QTLs on 1A, 4A and 5B	Emebiri and Ogbonnaya (2015)
	Boron	SHW	–	Screening at seedling and adult plant stage	17 SHW tolerant	Ilyas et al. (2015)
	Boron	*Ae. tauschii*, SHW	sequencing *Bot1* gene	–	*Bot-D5* gene identified	Pallotta et al. (2014)
	Heat	24 SHW	SSRs	Field trials over two years	3 tolerant and 9 medium tolerant	Sharma et al. (2014)
Genotyping		*Ae.tauschii*	10K	–	–	Wang et al. (2013)
		Ae. tauschii	30K	–	–	Iehisa et al. (2014a)
		Ae. tauschii, SHW	*Ppd-D1*	–	–	Jones et al. (2014)
		SHW	*Rht-D1*, *Ppd-D1* and *Vrn-D1*	–	Modified Eco-TILLING approach to identify natural variation at D-genome	Rasheed et al. (Unpublished)

USA using winter durum wheat and *Ae. tauschii* accessions (Hanif et al. 2014) to facilitate incorporation of desirable traits into winter wheat breeding programs. In the 1980s, L.R. Joppa developed a number of spontaneous SHW from partially fertile hybrids between 'Langdon' durum and different *Ae. tauschii* accessions. The advantage of the latter is that they were developed with only one durum parent thus much easier to attribute potential source of new and desirable genes. Both of these two SHW collections are globally distributed and extensively exploited in genetics and pre-breeding research. Similarly, durum wheat cv. 'Langdon' was also used to develop 82 SHW from 69 different *Ae. tauschii* accessions and have been widely used in Japan (Takumi et al. 2009; Kajimura et al. 2011; Nishijima et al. 2014). Recently, 86 SHW were developed by Sichuan Agriculture University, China (Lianquan Zhang, Personal communication) using 23 *Ae. tauschii* accessions and 54 accessions of tetraploids comprising of *T. dicoccon*, *T. dicoccoides* and *T. turgidum* durum accessions. The other collections include limited number of Australian SHWs developed by University of Melbourne and Department of Primary industry, Victoria. A limited number were also produced at ICARDA. In recent years, additional synthetic hexaploids have been produced at the National Institute of Agricultural Botany (NIAB) in the UK.

10.6 Impact of SHWs and SBLs

10.6.1 Biotic Stresses

There is a long history of the evaluation of *Ae. tauschii* and SHW as sources of resistance to biotic stresses and their introgressions into wheat. These have been well presented by van Ginkel and Ogbonnaya (2007) and Mujeeb-Kazi et al. (2008). In the review by Ogbonnaya et al. (2013) a comprehensive catalogue of all major genes identified in *Ae. tauschii* and SHW have been documented. In Australia, successful examples of discovery of useful genes in SHW and their transfer in elite germplasm that are now routinely used in breeding programs include cereal cyst and root lesion nematodes (*Pratylenchus neglectus* and *P. thornei*), Fusarium crown rot (*Fusarium pseudograminearum*), yellow leaf spot, Septoria nodorum blotch (SNB, *Parastagonospora nodorum*) and Septoria tritici blotch (STB, *Zymoseptoria tritici* (syn. *Mycosphaerella graminicola*) and leaf, stem and stripe rusts. The delivery of elite germplasm with introgressed genes in adapted genetic background to Australian breeding companies has been underpinned by development and use of robust phenotyping (e.g. managed environmental facility) that maximizes gene expression, development and provision of appropriate tools such as linked and/or diagnostic molecular markers accompanied by detailed knowledge and protocols for their deployment in breeding.

The delivery to breeders of new genes identified in SHW and transferred into elite Australian adapted backgrounds, along with markers is resulting in rapid introgression into breeding populations, minimising the time needed to deploy them in commercial varieties.

The prevalence and severity of loss by wheat rusts make them one of the most important threats to wheat production and priority objective in all wheat breeding programs globally. The initial screening of 456 non-duplicated *Ae. tauschii* accession for resistance to stem rust Ug99 races was an important step for subsequent identification of novel genes and their transfer to bread wheat (Rouse et al. 2011). The results suggested that 22 % of the *Ae. tauschii* accessions screened were resistant to Ug99 races. Similarly, Vikas et al. (2014) reported the results from the screening of D-genome accessions of India for leaf and stem rust resistance and found that 90% were resistant to stem rust. Recently, Periyannan et al. (2013; 2014) identified two genes *Sr33* and *Sr45* from *Ae. tauschii* and developed diagnostic markers for use in marker assisted selection in wheat.

Resistance to stripe rust was directly evaluated in 181 primary SHWs at seedling and adult plant stages in Ethiopia using virulent Kusba/Attila isolates. The SHWs were genotyped with 9K infinium SNP array and used in GWAS to identify loci linked to stripe rust resistance (Zegeye et al. 2014). They identified nine genomic regions influencing stripe rust resistance with a novel QTL on 6DS. These results provide further stimulus to exploit resynthesized SHWs as a rich source of new stripe rust resistance that may be useful in choosing SHWs and incorporating diverse yellow rust-resistance loci into locally adapted wheat cultivars.

Spot blotch caused by *Cochliobolus sativus* (anamorph: Bipolaris sorokiniana) is an important disease of wheat in warmer wheat growing regions like eastern India, southeast Asia, Latin America and sub-Saharan Africa (Dubin and Duveiller 2000) and can substantially reduce yields (Sharma and Duveiller 2006). SHW have been known to be the major resistance source to spot blotch (Mujeeb-Kazi et al. 2001) that have been widely used in CIMMYT's breeding activities producing many resistant derivative lines. However, there were no published reports on mapping of resistance from SHWs. Recently, Zhu et al. (2014) identified three major loci for resistance to spot blotch and favourable alleles of these loci were contributed by the SYN-DER parent.

Wheat streak mosaic virus (WSMV) is found throughout the Great Plains of North America (Burrows et al. 2009) and throughout the world, where wheat is grown (Ellis et al. 2003). Crop losses due to WSMV ranged from a trace to 13 % in Kansas from 1976 to 2000 (Bockus et al. 2001); however, complete field losses have been reported. Unfortunately, only a single dominant resistance gene, *Wsm2* (Haley et al. 2002), and minor resistance has been found in bread wheat. There are also genes for resistance to wheat curl mite which is an alternate method to reduce the incidence of WSMV through control of the vector (Martin et al. 1984). Recently, 412 SHW were screened for WSMV and 30 were found to be resistant including 4 SHW with very high level of resistance (Rupp et al. 2014). This initial finding provides new sources of resistance to WSMV and will help identify new genes and facilitate their transfer to bread wheat.

Resistance to root lesion nematodes (RLN, *Pratylenchus neglectus* and *P. thornei*) and cereal cyst nematodes (CCN) are widely acknowledged to be economically important biotic constraints in rainfed wheat production regions of Australia, USA, China, India and several countries in West Asia and North Africa (Nicol and Rivoal 2008). GWAS was conducted using 332 SHW to identify genomic regions associated with resistance to both nematode stresses using DArT markers. Seventeen DArT marker loci were found to be significantly associated with CCN and twelve to *P. neglectus* resistance. The novel QTL on chromosomes 1D, 4D, 5B, 5D and 7D for resistance to CCN and 4A, 5B and 7B for resistance to CCN are suggested to represent new sources of genes which could be deployed in further wheat improvement against these two important root diseases of wheat (Mulki et al. 2013). Similarly, Lindsell et al. (2014) identified eight QTL associated with *P. thornei* resistance in a DH population from a cross between the synthetic-derived wheat Sokoll and an Australia wheat cultivar Krichauff. Three QTL were identified on chromosome 2B, two on chromosome 6D, and a single QTL on each of chromosomes 2A, 2D and 5D. The QTL on chromosomes 2BS and 6DS mapped to locations previously identified to be associated with Pratylenchus resistance. Together, the QTL on 2B (QRlnt.sk-2B.1–2B.3) and 6D (QRlnt.sk-6D.1 and 6D.2) explained 30 and 48 % of the genotypic variation, respectively. Flanking PCR-based markers based on SSRs and SNPs were developed for the major QTL on 2B and 6D and are being used by Australian wheat breeding entities as a cost-effective high-throughput tool for marker-assisted breeding of wheat with improved *P. thornei* resistance.

Another important finding recently reported is on the identification of resistance to Hessian fly, Russian wheat aphid, and Sunn pest (El Bouhssini et al. 2013) using SHW, which are important due to the prevalence of these pests which cause economically significant damage in many wheat producing areas. About 914 SHWs were screened for resistance to these pests; fifteen SHWs showed high levels of resistance to Hessian fly and four showed moderate resistance. A SHW derived from the cross with (*T. dicoccoides*) also showed a high level of resistance to Hessian fly. The level of resistance to RWA in SHW was considerably lower; only one SHW and one durum wheat 'Altar 84' exhibited a high level of resistance, while four SHW were moderately resistant. Twenty one SHWs and one durum wheat 'Langdon' were identified to be resistant to Sunn pest at the vegetative stage. Crosses between these potentially novel resistance sources and elite bread wheat were initiated. Genetic and genomic studies using these accessions are on-going to identify and characterize the resistance genes. This will be useful in breeding programs to develop wheat germplasm with multiple resistances to these pests. In a subsequent study, Joukhadar et al. (2013) carried out GWAS with the SHW and identified 54 DArT markers which were significantly associated with 26 different QTLs conferring resistance to five insect pests (Hessian fly (HF), Russian wheat aphid (RWA), Sunn pest (SP), wheat stem saw fly (WSSF) and cereal leaf beetle (CLB)). This was the first study to utilize GWAS to identify markers linked to many insect pest resistances. The DArT markers linked to QTLs for resistance to CLB on 7DS (wPt- 66406) and 3BL (wPt-73166) were highly significant and explained up to 33 and 43 % of the variation for resistance respectively, which is quite high, suggesting that both are major QTLs and perhaps even major genes for CLB resistance. These QTLs are likely to be novel, being the first reported identification of QTLs on 3B and 7D for resistance to CLB in wheat.

10.6.2 Grain Yield Enhancement

Number of grains per m^2 and thousand kernel weight (TKW) are two important components determining grain yield. In the past four decades, improvement of grain yield has come from increased grains per m^2, due to the utilization of *Rht* genes in wheat breeding (Rebetzke et al. 2011). However, improvement of TKW is considered to be equally as important for further improving yield potential in various parts of world (Rasheed et al. 2014; Tang et al. 2014).

Significant variation in grain yield and its component traits have been reported for SHWs and for SBLs (summarized in Ogbonnaya et al. 2013). Yield advantages of SBLs over elite cultivars of bread wheat have been reported to be as high as 30% in northern Australia and 11 % in southern Australia (Dreccer et al. 2007; Ogbonnaya et al. 2007). In southern Australia, Gororo et al. (2002) found that a set of SBLs yielded similarly to their bread wheat recurrent parent in high-yielding environments, but up to 49 % more than the recurrent parent in low-yielding environments. They found that significant improvements to grain yield from one SHW were achieved through increases to the number of grain produced per m^2. SHW that exhibited significant variation for grain weight compared to bread wheat and TKW of up to 67 g have been reported in Mexico (Calderini and Reynolds 2000).

Cooper et al. (2012, 2013) examined the yield potential of SHWs under rain-fed field conditions over years of consecutive experiments and concluded that grain weight is the most heritable trait in SHW; even some lines with higher number of spikes and higher number of grains per spike maintained their grain size and weight. Recently, Tang et al. (2014) evaluated three SBLs and five bread wheat cultivars consecutively for 3 years under field conditions in Sichuan, China. The SBLs cultivars showed on average an 11.5 % or 951 kg ha^{-1} yield increase compared to bread wheat cultivars. This yield gain was mainly attributed to increases in both grain number per m^2 (5.7 %) and TKW (5.9 %). Other superior phenotypes associated with SBLs cultivars include higher rate of above-ground dry matter accumulation in the early growth stages, better partitioning to the grain, relatively compact and erect plant type with medium and upper leaves having a mean EC45° increase of 8.4 % over the non SBL cultivars at 20 days after flowering.

In a very comprehensive study, Talbot (2011) investigated the potential of SHWs to increase the grain yield of an Australian bread wheat cultivar, Yitpi, under water stressed conditions. In the study, grain yield and its major components were measured in 27 families of BC1 synthetic-derived lines under five drought stressed environments in southern Australia. Fourteen SHWs were donor parents to SBLs families with significantly ($P<0.05$) higher grain yields compared to the recurrent Australian bread wheat Yitpi parent. These lines produced the highest significant grain yield improvements under the lowest grain-yielding environments. The grain yield component responsible for these increases was grain weight under the highest-yielding environments, whereas grain per m^2 was commonly responsible under the lowest-yielding environments. The study showed that many, but not all SHWs can be used to increase the grain yield of an Australian bread wheat cultivar, particularly in low-yielding moisture-limiting environments of southern Australia.

Similarly, the yield performance of SBLs against the recurrent parent, Cham-6 and other reputed high-yielding cultivars, including Attila-7 and Wyalkatchem was conducted in nine site year by location in Mediterranean environment of Syria and Lebanon. Five SBLs had superior average yield compared to the best check, Wyalkatchem, and seventeen SBLs had superior average yield compared to the parent, Cham6. Further, two SBLs were also the first two winning genotypes in most of the nine environments using AMMI model (results not shown). The coefficient of variation (CV) from Francis and Kannenberg (1978) was used to assess a genotype stability by plotting the CV-FK against mean grain yield. The result indicated that some SBLs (e.g. 69, 9, 66, 8, 9, etc) not only possessed higher grain yield but were also stable across environments (Fig. 10.5). The grain yield was positively associated with the number of kernel per meter square, harvest index, early ground cover and vigour, and NDVI at the beginning of grain filling. The increase in grain yield was mainly attributed to two components, number of grains per m^2 and TKW.

Recently, Trethowan (2014) reviewed the contributions of wheat genetic resources for drought tolerance and concluded that 30 % yield advantage is associated with SHWs under drought stress. Perhaps, yield advantages in SHW are due to favourable alleles underpinning important yield related traits that are preferentially retained in SBLs (McIntyre et al. 2014). Similar results were also reported by Ali et al. (2014) which demonstrated the superiority of SBLs, over improved bread wheat cultivars. Nishijima et al. (2014) reported on sequence polymorphism of *Iw2* gene controlling glaucous appearance in *Ae. tauschii* and SBLs segregating populations. Glaucous appearance is associated with drought tolerance, prevents non-stomatal water loss, inhibits organ fusion during development, protects from UV radiation damage, and imposes a physical barrier against pathogenic infection. However, application of this information in practical wheat improvement is yet to be evaluated.

Collectively, kernel size, shape and TKW are relatively new yield-related traits that can be targeted to get more genetic gains for grain yield, following the example of yield advantages achieved in rice by enhancing kernel size (Gegas et al. 2010). Rasheed et al. (2014) conducted a GWAS for grain size, shape and TKW in a collection of SHW and identified two important loci on 3D and 6D chromosomes consistently associated with kernel length, width and TKW. Similarly, Okamoto et al. (2013) and Williams and Sorrells (2014) conducted QTL mapping in populations derived from SHW and identified several loci that underpin these traits and may have an impact on enhancing grain yield without compromising the number of grains m^2.

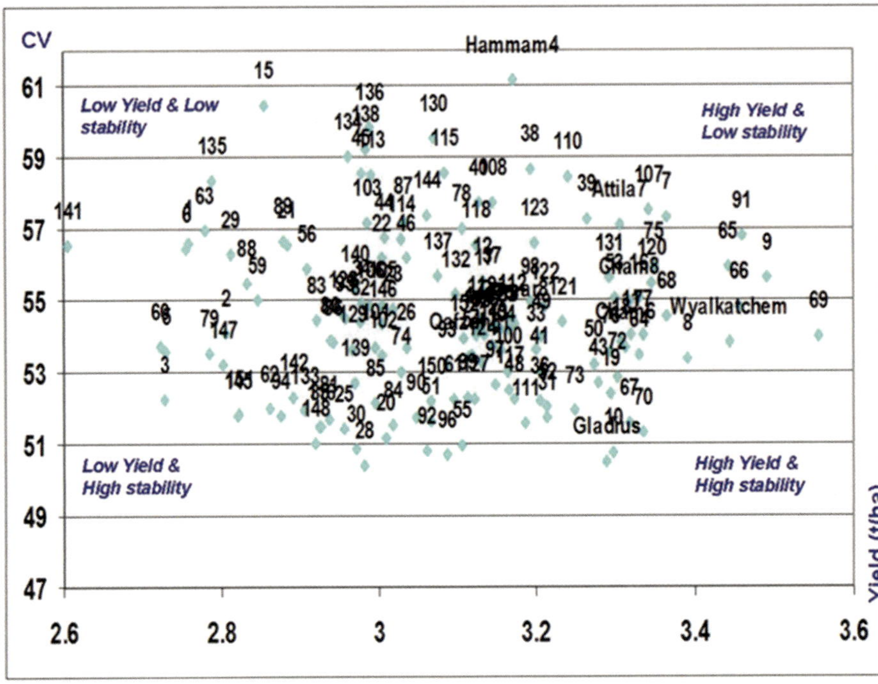

Fig. 10.5 Stability plot of CV-FK against Mean GY K) for a BC_2F_7 synthetic backcross-derived lines

Recently, several genes have been cloned in wheat influencing kernel size and weight using rice-wheat orthologue information (Valluru et al. 2014), out of which *TaCKX-D1* and *TaGS-D1* are present on D-genome. Zhang et al. (2012) identified five haplotypes in *Ae. tauschii* for *TaCKX-D1*, while only two haplotypes were observed in modern wheat cultivars and landraces and concluded that severe domestication bottleneck appeared to be involved in loss of alleles. It could also be argued that unconscious and indirect selection of favoured haplotypes at this and other loci may have contributed to narrowing genetic base for grain yield. Introgressions from SHW may be a preferred strategy to introduce new allelic variation at loci influencing grain yield, especially TKW, kernel size and kernel weight (Rasheed et al. 2014). Whilst SHWs possess favourable disease- and insect-resistance traits, it is now evident that SHWs and its derived SBLs contribute to improved yield potential in favourable environments as well as semi-arid and hot environments. Several authors have also demonstrated that improved water extraction of SBLs relative to respective recurrent parents was due to a greater distribution of root biomass deeper in the soil profile and better water use efficiency.

10.6.3 Impact on Providing Protection from Abiotic Stresses

Drought, heat and salinity are the major abiotic stresses that affect wheat production worldwide. Other important stresses include freezing tolerance (reproductive and vegetative), and soil toxicities including boron toxicity. The use of SHWs is seen as important genetic resources in identifying superior traits associated with some tolerance to these abiotic stresses. Salinity is a severe problem, affecting more than 800 million hectares of land worldwide that accounts for more than 6% of the global land mass (Munns and Tester 2008). It is well known that hexaploid bread wheat generally shows higher salt tolerance than its tetraploid progenitor, *Triticum turgidum*. We evaluated Na+ exclusion in a set of 150 genotypes including SHWs, elite ICARDA germplasm and focused identification of germplasm strategy (FIGs) genotypes. Amongst the SHWs, genotypes significantly varied in Na+ blade concentrations from 56 to 1216 μmol Na+ g^{-1} leaf blade dry weight (BDW) (exceeding 21-fold) for Aus-34453 and SHW-860 respectively. The differences between the lowest nine genotypes in term of Na+ blade conc. and the standard salt-tolerant Indian genotype (KHARCHIA 65 = 166 μmol Na+ g^{-1} BDW) were significant, and approximated threefold. This is similar to earlier results reported by Dreccer et al. (2004) who observed a threefold range of Na+ concentrations in SHW compared to hexaploid wheat control used in the study. Ogbonnaya et al. (2013) demonstrated the successful transfer of salinity tolerance in SHW measured as Na+ exclusion into an elite Australian common wheat cultivar, Yitpi with some of the SBLs showing significantly enhanced Na+ exclusion compared to either the SHW or the recurrent common wheat cultivar (Fig. 10.6). This was also confirmed by an independent study where the SBL genotype ranked 3rd out of 150 lines evaluated for salinity tolerance using a hydroponic system at ICARDA.

Boron toxicity is a major problem in many parts of world, especially in Australia, limiting wheat production. Previously, a major locus *Bo1* on chromosome 7BL was identified to be contributing tolerance to boron toxicity. Dreccer et al. (2003) reported high levels of B tolerance in SHWs. Tolerance to boron toxicity was also evaluated in 45 SHW derived from the susceptible durum cultivar 'Decoy' and 16 SHW were identified as tolerant which may be derived from the *Ae. tauschii* D-genome givven the susceptibility of the durum parent used in the study (Ilyas et al. 2015). In a recent study, Emebiri and Ogbonnaya (2015) used a genome-wide scan with DArT markers to identify regions that might harbour novel genetic loci that confer enhanced boron tolerance in SHWs than currently available in bread wheat. They showed that the SHWs were uniformly more tolerant to boron toxicity than the sensitive check, Meering, and 25 showed tolerance levels that were superior ($P \leq 0.05$) to that of Halberd, the most tolerant wheat check cultivar. At a threshold of $-\log(P) \geq 2.8$, a mixed linear model association mapping identified DArT markers on chromosomes 1A, 4A and 5B, but only the 4A region is known to harbour genes for boron toxicity tolerance. The chromosomes 1A and 5B loci represent novel regions, which when validated will increase the options of achieving tolerance beyond that conferred by *Bo1* and *Bo4* in breeding programs.

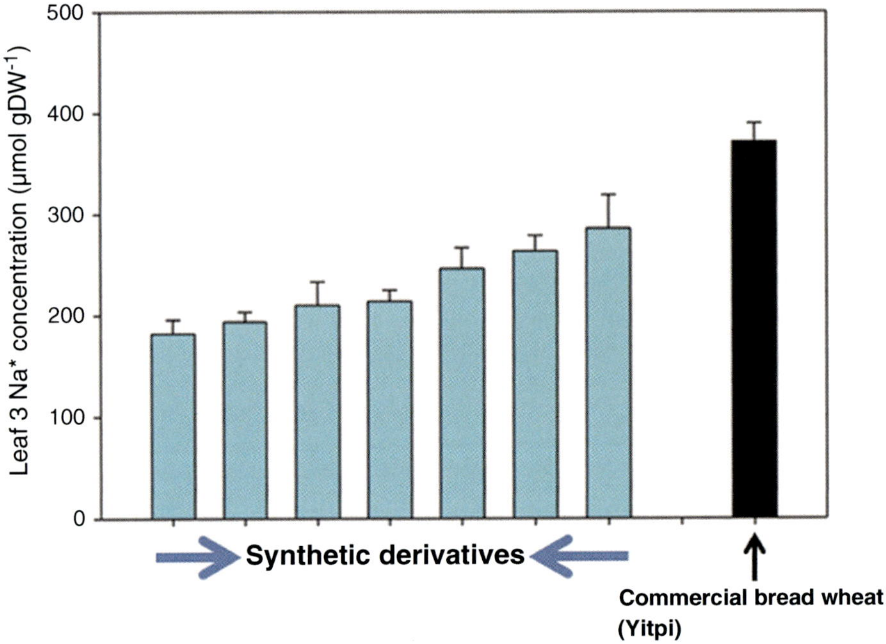

Fig. 10.6 Synthetic backcross-derived lines with lower Na^+ exclusion than the recurrent elite Australian wheat cultivar, "Yitpi" from Ogbonnaya et al. 2013

In most of winter wheat producing area, susceptibility to freezing is an important production-limiting factor. Pearce et al. (2013) screened ten SHW and identified only one accession having nine-gene deletion at a major cold tolerance locus *Fr-B2*. The deletion is associated with freezing susceptibility and was contributed by the durum parent 'Altar-84' of the SHW. Winter wheat breeding programs interested in using SHW and improving freezing tolerance may benefit from preliminary screening of SHW accessions for those that do not carry deletions at the *Fr-B2* locus.

Alpha-amylase (α-Amylase-amy1) gene plays a key role in seed germination and its activity determines levels of starch degradation, seed germination, and pre-harvest sprouting (PHS), which is a serious problem in wheat production. Yang et al. (2014) isolated and characterized high α-Amylase coding genes from the wheat cultivar Chuanmai32 (PHS susceptible) and the synthetic wheat SHW-L1 (PHS resistant). Expression profiling of amy1indicated that mRNA transcript accumulation began at a late stage of grain development. amy1 transcript accumulation in Chuanmai32 was 4.32- and 18.36-fold higher than observed in SHW-L1 at DPA25 and DPA30, respectively. Two significant expression quantitative trait loci (eQTLs) on chromosome 1BS and one on 3DS were characterized by expression analysis of amy1 transcripts and genetic analysis of SHW-L1/Chuanmai32-derived recombinant inbred lines. The genes that encoded high PI amylase were located near the centromere on chromosomes 6AL/6BL/6DL. These results suggest that these eQTL regions may provide candidate genes that play potential roles in regulating PHS through effects on amy1 expression, and points to the possible use of SHW to improve PHS tolerance. This result is consistent with the finding of Imtiaz et al. (2008) who reported that the enhanced expression of PHS resistance in SBLs led to the development of white PHS-resistant wheat germplasm from red-grained *Ae. tauschii* accession.

Besides their use as sources of genes for improving abiotic and biotic stresses, several wheat cultivars have been released that are derived from SHWs. These include Lalma and KT-2010 in Pakistan (CIMMYT Wheat Atlas), Maravilla in Mexico (CIMMYT Wheat Atlas), Carmona in Spain (van Ginkel and Ogbonnaya 2007) and Chuanmai-42 and its derivatives in China. Recently, Li et al. (2014) reviewed the current status of synthetic-derived wheat cultivars released in Southwestern China. They reported that 16 commercial wheat varieties including Chuanmai 28, 42, 43 and 47 have been released from using SHW. Apart from released cultivars, a significant proportion of international bread wheat screening nurseries by CIMMYT and ICARDA comprises of synthetic-derived germplasm which are distributed on annual basis worldwide.

10.7 International Initiatives

Several large investments are being made by the British (Wheat LOLA project), French (BREEDWHEAT) and Mexican government (SeeD) on the characterisation and utilisation of wild relatives for wheat improvement. A major componet of these initiatives involves the use of synthetics including the development of introgression lines. The SeeD used more than 100 SHW in the development of Linked-Top-Cross (LTP) populations. The TC1F5 (~5000 derived lines) are currently being evaluated at Obregon under drought and heat stresses. In UK, a project was funded to introduce novel genetic variation from wheat progenitor species (*Ae. tauschii*) via SHW. Genetic diversity from 50 SHWs was backcrossed into Paragon and Xi-19 to produce over 5600 BC1-derived lines for field selection. In China, breeders began to cross CIMMYT SHWs with their local varieties in the mid-1990s, and released their first SHW-derived variety in 2003, which yielded over 20 % more than checks in provincial trials. SBLs are now reported to be grown on over five million hectares in China, some 25 % of the wheat acreage. The Synthetic Evaluation Project was funded by the Grain Research and Development Corporation (GRDC), Australia from 2001 to 2007 to improve the productivity and sustainability of rainfed wheat production in the Mediterranean environments of Australia. The project aimed to identify sources of disease resistance and abiotic stress tolerance in SHW and to incorporate them into Australian adapted germplasm.

References

Ali A, Arshad M, Naqvi SMS, Ahmad M, Sher H, Fatima S, Kazi AG, Rasheed A, Mujeeb-Kazi A (2014) Exploitation of synthetic-derived wheats through osmotic stress responses for drought tolerance improvement. Acta Physiol Plant 36:2453–2465

Arraiano LS, Worland AJ, Ellerbrook C, Brown JKM (2001) Chromosomal location of a gene for resistance to *Septoria tritici* blotch (*Mycosphaerella graminicola*) in the hexaploid wheat synthetic 6x. Theor Appl Genet 103:758–764

Bockus WW, Appel JA, Bowden RL, Fritz AK, Gill BS, Martin TJ, Sears RG, Seifers DL, Brown-Guedira G, Eversmeyer MG (2001) Success stories: breeding for wheat disease resistance in Kansas. Plant Dis 85:453–461

Burrows M, Franc G, Rush C, Blunt T, Ito D, Kinzer K, Olson J, O'Mara J, Price J, Tande C, Ziems A, Stack J (2009) Occurrence of viruses in wheat in the great plains region. Plant Health Prog. doi:10.1094/PHP-2009-0706-01-RS

Calderini DF, Reynolds MP (2000) Changes in grain weight as a consequence of de-graining treatments at pre- and post-anthesis in synthetic hexaploid wheats. Aust J Plant Physiol 27:183–191

Cesari S, Bernoux M, Moncuquet P, Kroj T, Dodds PN (2014) A novel conserved mechanism for plant NLR protein pairs: the integrated decoy' hypothesis. Front Plant Sci 5:606. doi:10.3389/fpls.2014.00606

Chapman JA, Mascher M, Buluç A, Barry K, Georganas E, Session A et al (2015) A whole-genome shotgun approach for assembling and anchoring the hexaploid.bread wheat genome. Genome Biol 16:26. doi:10.1186/s13059-015-0582-8

Cooper JK, Ibrahim AMH, Rudd J, Malla S, Hays DB, Baker J (2012) Increasing hard winter wheat yield potential via synthetic wheat: I. Path-coefficient analysis of yield and its components. Crop Sci 52:2014–2022

Cooper JK, Ibrahim AM, Rudd J, Hays D, Malla S, Baker J (2013) Increasing hard winter wheat yield potential via synthetic hexaploid wheat: II. Heritability and combining ability of yield and its components. Crop Sci 53:67–73

Crespo-Herrera LA, Akhunov E, Garkava-Gustavsson L, Jordan KW, Smith CM, Singh RP, Åhman I (2014) Mapping resistance to the bird cherry-oat aphid and the greenbug in wheat using sequence-based genotyping. Theor Appl Genet 127:1963–1973

Dreccer MF, Ogbonnaya FC, Borgognone G, Wilson J (2003) Boron tolerance is present in primary synthetic wheats. In: Pogna N (ed) Proceedings of the 10th international wheat genetics symposium, 1–6 Sept 2003, Paestum, Italy, pp 1130–1132

Dreccer MF, Ogbonnaya FC, Borgognone G (2004) Sodium exclusion in primary synthetic wheats. In: Proceedings of the 11th wheat breeding assembly, 'symposium seeding the future' conference, Canberra, Sept 2004, pp 118–121

Dreccer MF, Borgognone MG, Ogbonnaya FC, Trethowan RM, Winter B (2007) CIMMYT-selected derived synthetic bread wheats for rainfed environments: yield evaluation in Mexico and Australia. Field Crops Res 100:218–228

Dubin HJ, Duveiller E (2000) Helminthosporium leaf blights of wheat: integrated control and prospects for the future. In: Proceedings of the international conference on integrated plant disease management for sustainable agriculture. Indian, New Delhi, India, pp 575–579

Eastwood RF, Lagudah ES, Appels R (1994) A directed search for DNA sequences tightly linked to cereal cyst nematode(CCN) resistance genes in *Triticum tauschii*. Genome 37:311–319

Eig A (1929) Monographisch-kritische Ubersicht der Gattung Aegilops. Repert Spec Nov Regn Veg Beih 55:1–228

El Bouhssini MA, Ogbonnaya FC, Chen M, Lhaloui S, Rihawi F, Dabbous A (2013) Sources of resistance in primary synthetic hexaploid wheat (*Triticum aestivum* L.) to insect pests: Hessian fly, Russian wheat aphid and Sunn pest in the fertile crescent. Genet Resour Crop Evol 60:621–627

Ellis MH, Reetzke GJ, Mago R, Chu P (2003) First report of wheat streak mosaic virus in Australia. Aust Plant Pathol 32:551–553
Emebiri LC, Ogbonnaya FC (2015) Exploring the synthetic hexaploid wheat for novel sources of tolerance to excess boron. Mol Breeding 35:68. doi:10.1007/s11032-015-0273-x
Francis TR, Kannenberg LW (1978) Yield stability studies in short season maize 1. A descriptive method for grouping genotypes. Can J Plant Sci 58:1029–1034
Friesen TL, Xu SS, Harris MO (2008) Stem rust, tan spot, *Stagonospora nodorum* blotch, and hessian fly resistance in Langdon durum- *Aegilops tauschii* synthetic hexaploid wheat lines. Crop Sci 48:1062–1070
Fu YB, Peterson GW, McCallum BD, Huang L (2010) Population-based resequencing analysis of improved wheat germplasm at wheat leaf rust resistance locus Lr21. Theor Appl Genet 121: 271–281
Gatford KT, Hearnden P, Ogbonnaya F, Eastwood RF, Halloran GM (2002) Novel resistance to pre-harvest sprouting in Australian wheat from the wild relative *Triticum tauschii*. Euphytica 126:67–76
Gegas VC, Nazari A, Griffiths S, Simmonds J, Fish L, Orford S, Sayers L (2010) A genetic framework for grain size and shape variation in wheat. Plant Cell 22:1046–1056
Gill BS, Raupp JW (1987) Direct genetic transfers from *Aegilops squarrosa* L. to hexaploid wheat. Crop Sci 27:445–450
Gill BS, Sharma HC, Raupp WJ, Browder LE, Hatchett JH, Harvey TL, Moseman JG, Waines JG (1985) Evaluation of Aegilops species for resistance to wheat powdery mildew, wheat leaf rust, hessian fly, and greenbug. Plant Dis 69:314–316
Gororo NN, Eagles HA, Eastwood RF, Nicolas ME, Flood RG (2002) Use of *Triticum tauschii* to improve yield of wheat in low-yielding environments. Euphytica 123:241–254
Haley SD, Martin TJ, Quick JS, Seifers DL, Stromberger JA, Clayshulte SR, Clifford BL et al (2002) Registration of CO960293-2 wheat germplasm resistant to and Russian wheat aphid. Crop Sci 42:1381–1382
Halloran GM, Ogbonnaya FC, Lagudah ES (2008) Triticum (Aegilops) tauschii in the natural and artificial synthesis of hexaploid wheat. Aust J Agr Res 59:475–490
Hanif U, Rasheed A, Kazi AG, Munir M, Mujeeb-Kazi A (2014) Analysis of genetic diversity in synthetic wheat assemblage (*T. turgidum* × *Aegilops tauschii*; 2n = 6x = 42; AABBDD) for winter wheat breeding. Cytologia 79:485–500
Huang L, Brooks SA, Li W, Fellers JP, Trick HN, Gill BS (2003) Map-based cloning of leaf rust resistance gene *Lr21* from the large and polyploidy genome of bread wheat. Genetics 164:655–664
Huang L, Brooks S, Li W, Fellers J, Nelson JC, Gill B (2009) Evolution of new disease specificity at a simple resistance locus in a crop weed complex: reconstitution of the *Lr21* gene in wheat. Genetics 182:595–602
Iehisa JCM, Shimizu A, Sato K, Nishijima R, Sakaguchi K, Matsuda R, Nasuda S, Takumi S (2014a) Genome-wide marker development for the wheat D genome based on single nucleotide polymorphisms identified from transcripts in the wild wheat progenitor *Aegilops tauschii*. Theor Appl Genet 127:261–271
Iehisa JCM, Matsuura T, Mori IC, Yokota H, Kobayashi F, Takumi S (2014b) Identification of quantitative trait loci for abscisic acid responsiveness in the D-genome of hexaploid wheat. J Plant Physiol 171:830–841
Ilyas M, Mahmood T, Ahmad A, Babar M, Mujeeb-Kazi A, Rasheed A (2015) Characterizing of D genome diversity for tolerance to boron toxicity in synthetic hexaploid wheat and *in silico* analysis of candidate genes. Acta Physiol Plant 37:17. doi:10.1007/s11738-014-1765-8
Imtiaz M, Ogbonnaya FC, Oman J, van Ginkel M (2008) Characterization of QTLs controlling genetic variation for pre-harvest sprouting in synthetic backcross derived wheat lines. Genetics 178:1725–1736
Jones H, Gosman N, Horsnell R, Rose GA, Everest LA, Bentley AR, Tha S, Uauy C, Kowalski A, Novoselovic D, Simek R, Kobiljski B, Kondic-Spika A, Brbaklic L, Mitrofanova O, Chesnokov

Y, Bonnett D, Greenland A (2014) Strategy for exploiting exotic germplasm using genetic, morphological, and environmental diversity: the *Aegilops tauschii* Coss. example. Theor Appl Genet 126:1793–1808

Joukhadar R, El-Bouhssini M, Jighly A, Ogbonnaya FC (2013) Genome-wide association mapping for five major pest resistances in wheat. Mol Breeding 32:943–960

Kajimura T, Murai K, Takumi S (2011) Distinct genetic regulation of flowering time and grain-filling period based on empirical study of D genome diversity in synthetic hexaploid wheat lines. Breed Sci 61:130–141

Kihara H (1944) Discovery of the DD analyser, one of the ancestors of Triticum vulgare. Agric Hortic 19:889–890

Landjeva S, Neumann K, Lohwasser U, Börner A (2008) Molecular mapping of genomic regions associated with wheat seedling growth under osmotic stress. Biol Plant 52:259–266

Law CN, Worland AJ (1973) Aneuploidy in wheat and its uses in genetic analysis. Plant Breed Inst Annu Rep 1972:25–65

Leonova IN, Laikova LI, Popova OM, Unger O, Börner A, Röder MS (2007) Detection of quantitative trait loci for leaf rust resistance in wheat–*T. timopheevii/T. tauschii* introgression lines. Euphytica 155:79–86

Li Y, Zhou R, Wang J, Liao X, Branlard G, Jia J (2012) Novel and favorable allele clusters for end use quality revealed by introgression lines derived from synthetic wheat. Mol Breeding 29:627–643

Li J, Wan HS, Yang WY (2014) Synthetic hexaploid wheat enhances variation and adaptive evolution of bread wheat in breeding processes. J Syst Evol 52:735–742

Lindsell KJ, Rahman MS, Taylor JD, Davey RS, Gogel BJ, Wallwork H, Forrest KL, Hayden MJ, Taylor SP, Oldach KH (2014) QTL for resistance to root lesion nematode (*Pratylenchus thornei*) from a synthetic hexaploid wheat source. Theor Appl Genet 127:1409–1421

Lohwasser U, Röder MS, Börner A (2005) QTL mapping for the domestication traits pre-harvest sprouting and dormancy in wheat (*Triticum aestivum* L.). Euphytica 143:247–249

Ma Z-Q, Gill BS, Sorrells ME, Tanksley SD (1993) RFLP markers linked to two Hessian fly-resistance genes in wheat (*Triticum aestivum* L.) from *Triticum tauschii* (coss.) Schmal. Theor Appl Genet 85:750–754

Mandeep S, Bains NS, Kuldeep S, Sharma SC, Parveen C (2010) Molecular marker analysis of Karnal bunt resistant wheat-*Aegilops tauschii* introgression lines. Plant Dis Res 25:107–112

Martin TJ, Harvey TL, Bender CG, Seifers DL (1984) Control of Wheat streak mosaic virus with vector resistance in wheat. Phytopathology 74:963–964

McFadden ES, Sears ER (1946) The origin of Triticum spelta and its free-threshing hexaploid relatives. J Hered 37:81–89

McFadden ES, Sears ER (1947) The genome approach in radical wheat breeding. J Am Soc Agron 39:1011–1026

McIntyre CL, Rattey A, Kilian A, Dreccer MF, Shorter R (2014) Preferential retention of chromosome regions in derived synthetic wheat lines: a source of novel alleles for wheat improvement. Crop Pasture Sci 65:125–138

Miranda LM, Murphy JP, Marshall D, Leath S (2006) *Pm34*: a new powdery mildew resistance gene transferred from *Aegilops tauschii* Coss. to common wheat (*Triticum aestivum* L.). Theor Appl Genet 113:1497–1504

Mizuno N, Yamasaki M, Matsuoka Y, Kawahara T, Takumi S (2010) Population structure of wild wheat D-genome progenitor *Aegilops tauschii* Coss.: implications for intraspecific lineage diversification and evolution of common wheat. Mol Ecol 19:999–1013

Mujeeb-Kazi A, Cano S, Rosas V, Cortes A, Delgado R (2001) Registration of five synthetic hexaploid wheat and seven bread wheat lines resistant to wheat spot blotch. Crop Sci 41:1653–1654

Mujeeb-Kazi A, Gul A, Farooq M, Rizwan S, Ahmad I (2008) Rebirth of synthetic hexaploids with global implications for wheat improvement. Aust J Agric Res 59:391–398

Mulki MA, Jighly A, Ye G, Emebiri LC, Moody D, Ansari O, Ogbonnaya FC (2013) Association mapping for soilborne pathogen resistance in synthetic hexaploid wheat. Mol Breeding 31:299–311

Munns R, Tester M (2008) Mechanisms of salinity tolerance. Annu Rev Plant Biol 59:651–681

Navakode S, Weidner A, Lohwasser U, Röder MS, Börner A (2009) Molecular mapping of quantitative trait loci (QTLs) controlling aluminium tolerance in bread wheat. Euphytica 166:283–290

Nicol JM, Rivoal R (2008) Global knowledge and its application for the integrated control and management of nematodes on wheat. In: Ciancio A, Mukerji KG (eds) Integrated management and biocontrol of vegetable and grain crops nematodes. Springer, Amsterdam, pp 243–287

Nishijima R, Iehisa JCM, Matsuoka Y, Takumi S (2014) The cuticular wax inhibitor locus *Iw2* in wild diploid wheat *Aegilops tauschii*: phenotypic survey, genetic analysis, and implications for the evolution of common wheat. BMC Plant Biol 14:246

Ogbonnaya FC, Ye G, Trethowan R, Dreccer F, Lush D, Shepperd J, van Ginkel M (2007) Yield of synthetic backcross-derived lines in rainfed environments of Australia. Euphytica 157: 321–336

Ogbonnaya FC, Abdalla O, Mujeeb-Kazi A, Kazi AG, Xu SS, Gosman N, Lagudah ES, Bonnett DG, Sorells ME, Tsujimoto H (2013) Synthetic hexaploids: harnessing species of primary gene pool for wheat improvement. Plant Breed Rev 37:35–122

Okamoto Y, Nguyen AT, Yoshioka M, Iehisa JC, Takumi S (2013) Identification of quantitative trait loci controlling grain size and shape in the D genome of synthetic hexaploid wheat lines. Breed Sci 63:423–429

Olson EL, Rouse MN, Pumphrey MO, Bowden RL, Gill BS, Poland JA (2013) Introgression of stem rust resistance genes *SrTA10187* and *SrTA10171* from *Aegilops tauschii* to wheat. Theor Appl Genet 126:2477–2484

Pallotta M, Schnurbusch T, Hayes J, Hay A, Baumann U, Paull J, Langridge P, Sutton T (2014) Molecular basis of adaptation to high soil boron in wheat landraces and elite cultivars. Nature 514:88–91

Pearce S, Zhu J, Boldizsár A, Vágújfalvi A, Burke A, Garland-Campbell K, Galiba G, Dubcovsky J (2013) Large deletions in the CBF gene cluster at the *Fr-B2* locus are associated with reduced frost tolerance in wheat. Theor Appl Genet 126:2683–2697

Pecenkova T, Hala M, Kulich I, Kocourkova D, Drdova E, Fendrych M, Toupalova H, Zarsky V (2011) The role for the exocyst complex subunits Exo70B2 and Exo70H1 in the plant-pathogen interaction. J Exp Bot 2:2107–2116

Periyannan S, Moore J, Ayliffe M, Bansal U, Wang X, Huang L, Deal K, Luo M, Kong X, Bariana H, Mago R, McIntosh R, Dodds P, Dvorak J, Lagudah E (2013) The gene Sr33, an ortholog of barley Mla genes, encodes resistance to wheat stem rust race Ug99. Science 341:786–788

Periyannan S, Bansal U, Bariana H, Deal K, Luo MC, Dvorak J, Lagudah E (2014) Identification of a robust molecular marker for the detection of the stem rust resistance gene *Sr45* in common wheat. Theor Appl Genet 127:947–955

Pestsova EG, Börner A, Röder MS (2001) Development of a set of *Triticum aestivum-Aegilops tauschii* introgression lines. Hereditas 135:139–143

Pestsova EG, Börner A, Röder MS (2006) Development and QTL-assessment of *Triticum aestivum-Aegilops tauschii* introgression lines. Theor Appl Genet 112:634–647

Rasheed A, Xia XC, Ogbonnaya F, Mahmood T, Zhang Z, Mujeeb-Kazi A, He ZH (2014) Genome-wide association for grain morphology in synthetic hexaploid wheats using digital imaging analysis. BMC Plant Biol 14:128

Rebetzke GJ, Ellis MH, Bonnett DG, Condon AG, Falk D, Richards RA (2011) The Rht13 dwarfing gene reduces peduncle length and plant height to increase grain number and yield of wheat. Field Crops Res 124:323–331

Reif JC, Zhang P, Dreisigacker S, Warburton ML, van Ginkel M, Hoisington D, Bohn M, Melchinger AE (2005) Wheat genetic diversity trends during domestication and breeding. Theor Appl Genet 110:859–864

Röder MS, Korzun V, Wendehake K, Plaschke J, Tixier M-H, Leroy P, Ganal MW (1998) A microsatellite map of wheat. Genetics 149:2007–2023

Rouse MN, Olson EL, Gill BS, Pumphrey MO, Jin Y (2011) Stem rust resistance in Aegilops tauschii germplasm. Crop Sci 51:2074–2078

Rupp JLS, Simon ZG, Gillet-Walker B, Fellers JP (2014) Resistance to wheat streak mosaic virus identified in synethetic wheat lines. Euphytica 198:223–229

Ryan PR, Raman H, Gupta S, Sasaki T, Yamamoto Y, Delhaize E (2010) The multiple origins of aluminium resistance in hexaploid wheat include *Aegilops tauschii* and more recent *cis* mutations to *TaALMT1*. Plant J 64:446–455

Sharma RC, Duveiller E (2006) Spot blotch continues to cause substantial grain yield reductions under resource-limited farming conditions. J Phytopathol 154:482–488

Sharma P, Sareen S, Saini M, Verma A, Tyagi BS, Sharma I (2014) Assessing genetic variation for heat tolerance in synthetic wheat lines using phenotypic data and molecular markers. Aust J Crop Sci 8:512–522

Simón MR, Worland AJ, Cordo CA, Struik PC (2001) Chromosomal location of resistance to *Septoria tritici* in seedlings of a synthetic hexaploid wheat, *Triticum spelta* and two cultivars of *Triticum aestivum*. Euphytica 109:151–155

Simón MR, Perelló A, Cordo CA, Larrán S, van der Putten PEL, Struik PC (2005) Association between Septoria tritici blotch, plant height, and heading date in wheat. Agron J 97:1072–1081

Simón MR, Ayala FM, Cordo CA, Röder MS, Börner A (2007) The exploitation of wheat (*Triticum aestivum*)/*Aegilops tauschii* introgression lines for the detection of gene(s) determining resistance to septoria tritici blotch (*Mycosphaerella graminicola*). Euphytica 154:249–254

Strand E (1965) Studies on seed dormancy in barley. Meld Norg Landbruks Hoegs 44:1–23

Takumi S, Naka Y, Morihiro H, Matsuoka Y (2009) Expression of morphological and flowering time variation through allopolyploidization: an empirical study with 27 wheat synthetics and their parental *Aegilops tauschii* accessions. Plant Breeding 128:585–590

Talbot SJ (2011) Introgression of genetic material from primary synthetic hexaploids into an Australian bread wheat (*Triticum aestivum* L.). Mem Agric Sci. University of Adelaide, Adelaide, SA

Tang Y, Rosewarne GM, Li C, Wu X, Yang W, Wu C (2014) Physiological factors underpinning grain yield improvements of synthetic derived wheat in Southwestern China. Crop Sci. doi:10.2135/cropsci2014.02.0124

Trethowan R (2014) Delivering drought tolerance to those who need it; from genetic resource to cultivar. Crop Pasture Sci 65:645–654

Valluru R, Reynolds MP, Salse J (2014) Genetic and molecular bases of yield-associated traits: a translational biology approach between rice and wheat. Theor Appl Genet 127:1463–1489

van Ginkel M, Ogbonnaya F (2007) Novel genetic diversity from synthetic wheats in breeding cultivars for changing production conditions. Field Crops Res 104:86–94

Vikas VK, Sivasamy M, Kumar J, Jayaprakash P, Kumar S, Parimalan R et al (2014) Stem and leaf rust resistance in wild relatives of wheat with D genome (*Aegilops* spp.). Genet Res Crop Evol 61:861–874

Wang J, Luo M-C, Chen Z, You FM, Wei Y, Zheng Y, Dvorak J (2013) *Aegilops tauschii* single nucleotide polymorphisms shed light on the origins of wheat D-genome genetic diversity and pinpoint the geographic origin of hexaploid wheat. New Phytol 198:925–937

Watanabe N, Fujii Y, Takesada N, Martinek P (2006) Cytological and microsatellite mapping of genes for brittle rachis in a *Triticum aestivum*-*Aegilops tauschii* introgression line. Euphytica 151:63–69

Williams K, Sorrells ME (2014) Three-dimensional seed size and shape QTL in hexaploid wheat (*Triticum aestivum* L.) populations. Crop Sci 54:98–110

Yang J, Liu Y, Pu Z, Zhang L, Yuan Z, Chen G, Wei Y, Zheng Y, Liu D, Wang J (2014) Molecular characterization of high pI α-amylase and its expression QTL analysis in synthetic wheat RILs. Mol Breeding 34:1075–1085

Zegeye H, Rasheed A, Makdis F, Badebo A, Ogbonnaya FC (2014) Genome-wide association mapping for seedling and adult plant resistance to stripe rust in synthetic hexaploid wheat. PLoS One 9, e105593

Zhang L, Zhao YL, Gao LF, Zhao GY, Zhou RH, Zhang BS (2012) *TaCKX6-D1*, the ortholog of rice *OsCKX2*, is associated with grain weight in hexaploid wheat. New Phytol 195:574–584

Zhu Z, Bonnett D, Ellis M, Singh P, Heslot N, Dreisigacker S, Gao C, Mujeeb-Kazi A (2014) Mapping resistance to spot blotch in a CIMMYT synthetic-derived bread wheat. Mol Breeding 34:1215–1228

Chapter 11
Wheat-Perennial Triticeae Introgressions: Major Achievements and Prospects

Carla Ceoloni, Ljiljana Kuzmanovic, Paola Forte, Maria Elena Virili, and Alessandra Bitti

11.1 Introduction

Species belonging to the Triticeae tribe of the grass family (Poaceae) have constantly played a decisive role in development and livelihood of mankind. In addition to about 100 annual such species, including some of the most important domesticated cereal crops (e.g., wheat, barley, and rye), the tribe encompasses around 400 perennials. The latter group comprises several species, commonly referred to as wheatgrasses and wildryes, which, either as natural invaders or purposefully introduced and even selected by man, represent excellent sources of forage and habitat for livestock and wildlife, and also contribute to soil upkeep and to many other aspects of environmental management. In addition to their utility as species per se, many perennial grasses have been successfully hybridized with cultivated, annual cereal crops and notably wheat, for which they have worked as highly valuable sources to improve resistance to biotic and abiotic stresses, as well as quality and yield-related traits, and even to try conferring a perennial habitus to the typically annual wheat.

The relative ease with which hybridization, both natural and man-made, and transfer of desirable attributes has been accomplished from several perennial Triticeae into wheat is due to the existence of sufficient cytogenetic and cytogenomic affinity between the former group of species, belonging to the wheat tertiary gene pool (Harlan and de Wet 1971), and the cultivated forms of *Triticum*. Based on this, not only complete or partial wheat-perennial Triticeae amphiploids, but also addition and substitution lines of single alien chromosomes, and even radiation-induced--> translocation or recombinant lines with segmental introgressions have

C. Ceoloni (✉) • L. Kuzmanovic • P. Forte • M.E. Virili • A. Bitti
Department of Agriculture, Forestry, Nature and Energy (DAFNE), University of Tuscia, Viterbo 01100, Italy
e-mail: ceoloni@unitus.it

M. Molnár-Láng et al. (eds.), *Alien Introgression in Wheat*,
DOI 10.1007/978-3-319-23494-6_11

been developed. In recent years, several extensive reviews have dealt with this ample field of basic and applied research (see, e.g., Li and Wang 2009; Wang 2011; Mujeeb-Kazi et al. 2013; Ceoloni et al. 2014a; Chaudhary et al. 2014). The present contribution will focus on major achievements and latest advances in gene transfer from those perennial Triticeae species that have more significantly contributed, or have the potential, to enhance the breeding performance of cultivated wheats, both bread wheat (*Triticumaestivum* L., $2n=6x=42$) and durum wheat (*Triticum durum* Desf., $2n=4x=28$), in the light of a changing ecological and socioeconomic agricultural perspective.

11.2 A Brief Survey on the Cytogenomic Makeup of Perennial Triticeae

Depending on the taxonomic treatment, between 200 and 250 wheatgrass and wildrye species have been described worldwide, the majority being native to Eurasia and a few to North America, but, as a whole, spread and adapted to the most different environmental conditions of all continents (reviewed in Wang 2011). Not differently from the other representatives of the Triticeae tribe, perennial grasses represent fascinating and puzzling examples of reticulate evolution, in which besides polyploidization, hybridization and interspecific introgression among polyploid lineages and/or their diploid progenitors played a key role in shaping their genomes (e.g. Mahelka et al. 2011; Mason-Gamer 2013; Sun 2014). The resulting genomic heterogeneity has made taxonomic treatment of this ample group of species quite challenging, and often controversial. Moreover, as for other species groups, rather different criteria have animated classification systems through the years (e.g. Barkworth et al. 2009). Even the many genetic and cytogenetic approaches used to investigate intra- and interspecific genome relationships, from classical analyses of chromosome pairing in hybrids, to their "modern" version, with differentially painted genomes, based on genomic in situ hybridization (GISH), up to the ever increasing use of gene/sequence comparisons, and of the many other molecular tools recently available, in several cases still require a thorough interpretation. Thus, although the genome-based classifications proposed by Dewey (1984) and Löve (1984) remain important reference points, subsequent literature revealed them insufficient or incongruent in the face of new findings, and novel interpretations for some controversial biosystematics questions concerning perennial Triticeae species have been recently proposed (see, e.g., Mahelka et al. 2013; Wang and Lu 2014; Wang et al. 2015). As a matter of fact, it is quite obvious that, as more and robust information becomes available, taxonomic treatments change to reflect this information; however, extent and timing of their acceptance and adoption remain subjective. As a consequence, often multiple taxonomic treatments are in use at any given time (Barkworth and von Bothmer 2009). Because taxonomic consideration are not within the scopes of this chapter, in line with what stated by Yen et al. (2005), taxonomy will be used here as "a tool for species recognition," and, perhaps primarily, as "a guide for germplasm utilization, and a common language for communication." In this view, the currently most-used nomenclature will be employed here, both for genera and species names, and for their genomic formulas. Table 11.1 summarizes this information for perennial Triticeae species that are representatives of pivotal genomes and/or of genome combinations. In the following sections of the present chapter, those that, at various levels, have been involved in alien gene transfer into wheat will be recalled. Species of *Hordeum* genus (H genome) will be treated in a different chapter (Chap. 12).

Most recent classifications identify ~10 basic genome types represented among perennial Triticeae species, the majority of which are contained in varying combinations (and most likely, for what above said, in variously rearranged forms) in the numerous polyploid representatives (Table 11.1). Among them, genome P identifies the current *Agropyron* genus, formerly reported to comprise about 20 genera and 400–500 species, and now universally restricted to the "crested wheatgrasses" such as *A. cristatum* (Wang 2011). Diploid, tetraploid, and hexaploid species in this genus form a common gene pool, in which gene flow occurs among cross-pollinating species of the three ploidy levels. Tetraploid *A. cristatum*, the most common of the three crested wheatgrasses, represents a case of segmental autosomy, in that it probably originated from hybridization between diploid *A. cristatum* and *A. mongolicum*,

Table 11.1 Perennial Triticeae species representatives of pivotal genomes or genome combinations

Genus	Basic genome(s)	Ploidy	Representative species	Genomic formula
Agropyron	P	2*x*, $\underline{4x}$[a], 6*x*	*A. desertorum*; *A. cristatum*	P
Australopyrum	W	2*x*	*A. pectinatum*; *A. retrofractum*	W
Dasypyrum	V	2*x*, 4*x*	*D. breviaristatum*	V^b
Elymus	St, H, Y, P, W	4*x*, 6*x*		
		6*x*	*E. rectisetus*	StWY
		6*x*	*E. repens*	StStH
Leymus	Ns, Xm	4*x*	*L. racemosus*; *L. multicaulis*; *L. mollis*	NsXm or Ns_1Ns_2
Psathyrostachys	Ns	$\underline{2x}$, 4*x*	*P. huashanica*	Ns
Pseudoroegneria	St	2*x*, 4*x*	*Ps. spicata*; *Ps. stipifolia*	St; StSt or St_1St_2
Thinopyrum	E (≅J), St or S	2*x*–10*x*		
		2*x*	*Th. bessarabicum*	J or J^b or E^b
		2*x*	*Th. elongatum*	E or E^e or J^e
		4*x*	*Th. curvifolium*	EE or J^eJ^e or J^bJ^e
		6*x*	*Th. junceum*	$E^bE^bE^e$ or JJE
		6*x*	*Th. intermedium*	E^eE^bSt or E_1E_2St or JJ^sS or $J^{vs}J^rSt$
		10*x*	*Th. ponticum*	$E^eE^bE^xStSt$ or JJJ^sJ^s

For comments and references, particularly on alternative genomic formulas of a given species, see text

[a]The most frequently detected ploidy level within a given genus is underlined

both containing the same basic P genome, but distinguished from each other by structural rearrangements (see Han et al. 2014 and references therein; see also Sects. 11.3.2 and 11.3.3).

An important genus is *Pseudoroegneria*, whose St genome (designated S before Wang et al. 1995) characterizes its taxa, with diploid and auto- or near-autopolyploid representatives. *Pseudoroegneria* species have been prolific contributors, most probably as maternal parents (Zhang et al. 2009a; Mahelka et al. 2011), to many allopolyploids of different genera, notably *Thinopyrum* and *Elymus* (Table 11.1), hence St is definitely a core genome of the Triticeae tribe (Wang et al. 2010a, 2015; Mason-Gamer 2013).

Noteworthy is then the E genome of *Thinopyrum* species, whose symbol was differentiated into E^e and E^b to designate the haplomes of two diploid representatives of the genus, i.e., *Th. elongatum* and *Th. bessarabicum*, respectively (Wang et al. 1995). Alternative symbols, namely J, or J^e, and J^b for the respective diploid species, are used, and also included in the genome formulas of *Thinopyrum* polyploids (Table 11.1; see also Chen et al. 1998; Fedak and Han 2005; Chang et al. 2010; Wang 2011). In fact, the genus encompasses a large number of perennial species and a wide range of ploidy levels, from diploidy up to decaploidy. As several examples will illustrate in the following sections, both the diploids and many polyploids, particularly the hexaploid *Th. intermedium* and the decaploid *Th. ponticum*, have been among the most extensively exploited in wheat breeding, not only among perennial Triticeae, but among wheat relatives as a whole (see Sect. 11.3). Consequently, many cytogenetic and cytogenomic aspects of their chromosome makeup have been extensively analysed. Relatively close relationships have been established for the genomes of diploid *Th. elongatum* and *Th. bessarabicum* (reviewed in Wang and Lu 2014), although accompanied by various types of chromosomal rearrangements, which differentiate their karyotypes and reduce interspecific pairing (see, e.g., Jauhar 1990; Wang and Hsiao 1989; Wang 1992). As expected, the level of intricacy of intergenomic relationships increases in polyploid representatives of the genus, making interpretation of their origin and definition of their genomic constitution highly debated (Zhang et al. 1996a, b; Chen et al. 1998; Chen 2005; Wang 2011; Wang and Lu 2014; Wang et al. 2015). A shared conviction is that the St (or S) genome from the *Pseudoroegneria* genus definitely enters in the genomic composition of both *Th. intermedium* and *Th. ponticum*. The St/S genome, in turn, shows close relatedness with the E (= E^e) and J (= E^b) genomes, as proved by extensive autosyndetic pairing (Wang 1989a, b, 1992; Jauhar 1995; Cai and Jones 1997) and cross-hybridization in Southern and GISH experiments (Zhang et al. 1996a; Liu et al. 2007). Thus, a conclusive definition of the genomic composition of *Th. intermedium* and *Th. ponticum* has been difficult to reach. The former has been described with various genome formulas, including E^eE^bSt (Wang and Zhang 1996) and E_1E_2St (Zhang et al. 1996b) or JJ^sS (Chen et al. 1998), while $E^eE^bE^xStSt$ (reviewed in Li and Zhang 2002) or $JJJJ^sJ^s$ (Chen et al. 1998) have been indicated for the latter. The controversy has particularly dealt with distinction between chromosomes considered on one hand of fully St/S-genome derivation (Zhang et al. 1996a), and, on the other, hypothesized to result from intergenomic rearrangements in the course of polyploid evolution, with presence of St/S genomic DNA confined to pericentromeric regions (as in J^s type chromosomes, see Chen et al. 1998). In any case, hybridization to the St/S genomic DNA, whether complete or segmental, represents a distinctive mark of the genomic origin of the *Thinopyrum* chromosome(s) involved. However, the picture has been further complicated by evidence of hybridization of J^s chromosomes with either genomic DNA or a repetitive sequence of the V genome of the genus *Dasypyrum* (see ahead), suggesting the involvement of V genome in the evolution of the J^s genome (Kishii et al. 2005; Mahelka et al. 2011; Deng et al. 2013). Very recently, comprehensive reassessments have been provided for some of the most controversial issues of biosystematics and evolutionary relationships of perennial Triticeae species (Wang and Lu 2014; Wang et al. 2015, and references therein). One of them concerns the origin and genome constitution of *Th. intermedium*, which appears to be now nearly resolved. Presence of the St genome from *Pseudoroegneria* is substantiated by all studies, hence considered unequivocal. Moreover, based on assays with EST-SSR primer sequences derived from the putative diploid progenitor species carrying the St, J^b, and J^egenomes, the St genome in *Th. intermedium* results the least modified from the present-day *Pseudoroegneria* diploids. On the other hand, the same assays showed both J and J^s to differ from present day J^e (*Th. elongatum*) and J^b (*Th. bessarabicum*) genomes, respectively: the former distinction (J vs. J^e) would be based on presence of long-terminal repeat sequences of rye (R genome) origin, the latter (J^s vs. J^b) on presence of repetitive sequences of *Dasypyrum* (V genome) derivation. Taking into account such evidence, a novel designation has been proposed for the *Th. intermedium* genome formula, that is $J^{vs}J^rSt$, with J^{vs} and J^r representing ancestral genomes of J^b and J^e (Wang et al. 2015).

A complex history of genome rearrangements has been also described for species of the genus *Elymus* (Mahelka and Kopecky 2010; Zeng et al. 2013b; Sun 2014; Wang et al. 2014; Wang and Lu 2014), the largest genus in the Triticeae tribe, including, in its broadest sense, around 200 species that are widely distributed all over the world. *Elymus* is an exclusively allopolyploid genus, in which five basic genomes (St, H, Y, P, and W; Table 11.1) have been identified (Wang 2011; Wang and Lu 2014). Among them, the St genome is recognized as common to all *Elymus* species, while Y, another pivotal genome of the genus, is still of debated origin (Sun 2014; Wang and Lu 2014).

Chromosomal rearrangements were also frequently detected in species of the polyploid *Leymus* genus, such as tetraploid *L. racemosus* and *L. multicaulis* (Qi et al. 1997; Jia et al. 2002; Zhang et al. 2010; see Sect. 11.3.2). All *Leymus* species are based on the Ns-genome from *Psathyrostachys* (see ahead) and the Xm-genome of still unknown origin, with genomes P of *Agropyron* and F of *Eremopyrum triticeum* hypothetically considered in its ancestry (reviewed in Wang and Lu 2014). However, a different genomic constitution has been recently proposed (Anamthawat-Jónsson 2014). The study, based on FISH experiments with *Leymus* specific dispersed retroelement-like repeats as probes, showed them to be distributed over all *Leymus* chromosomes, without any differentiation between chromosomes. The same repeats were also abundant in the Ns genome progenitor in *Leymus*, i.e.,

Psathyrostachys. Experiments on *Leymus* chromosomes using *Psathyrostachys* genomic DNA as probes further supported the proposal of NsNs (Ns_1Ns_2) genome constitution for *Leymus*. The possibility that an Xm genome might have been involved in the beginning of the allopolyploidization process was not discarded, but in this case, the Ns genome specific elements would have spread predominantly and rapidly across genomes, leading to genome homogenization.

As mentioned, the Ns genome characterizes the predominantly diploid species of the small *Psathyrostachys* genus, containing species, such as *P. huashanica*, endemic to the Shaanxi Province of China, which has provided a number of desirable genes for wheat improvement (see Sect. 11.3.3.2).

Finally, the V^b genome is included in Table 11.1, which symbolizes the perennial representative of the *Dasypyrum* genus, currently considered to comprise two species only, the other one being the annual *D. villosum*, with a V^v genome designation (Gradzielewska 2006a; De Pace et al. 2011). *D. breviaristatum* is largely tetraploid, while *D. villosum* is strictly diploid. The origin and genomic constitution of 4*x* *D. breviaristatum* is debated. A general consensus exists on its autotetraploid origin. However, a direct derivation from *D. villosum* appears to contrast with the results of various types of investigations (reviewed in De Pace et al. 2011). Thus, the most likely candidate for the diploid species in which the genome duplication event occurred to give rise to the current 4*x* *D. breviaristatum* genome seems to be 2*x* *D. breviaristatum* rather than *D. villosum* (reviewed in De Pace et al. 2011). Nonetheless, several morphological and cytomolecular features are definitely indicative of a common ancestry of the two species, with differentiation between them probably due to adaptability to diverse ecogeographic areas occupied by the common ancestor. In comparison with *D. villosum* (Gradzielewska 2006b; De Pace et al. 2011), research on *D. breviaristatum* and, hence, exploitation of its positive attributes in wheat breeding, is very limited. However, some examples are given in Sect. 11.3.2.

11.3 Exploitation of Useful Traits

As above anticipated, exploitation of the ample variability present in perennial Triticeae germplasm has been accomplished through incorporation into the wheat genome of as much as the entire alien genome, particularly in the form of chromosome-doubled hybrids, i.e., amphiploids, down to a single chromosome or chromosome arm pair (either added or substituted), or just a small chromosomal segment. As expected, the relative degree of success has been strongly correlated with several factors, primarily interspecific and consequent intergenomic relatedness, but also degree of crossability, as well as stability of cross combinations and of their derived lines. For all these aspects, species belonging to the *Thinopyrum* genus turned out to be the most amenable (see, e.g., Jiang et al. 1994; Mujeeb-Kazi and Wang 1995; Wang 2011; Mujeeb-Kazi et al. 2013; Ceoloni et al. 2014a), hence the most extensively used in the production of one or more types of assembly with the wheat genome. Several examples of wheat–alien combinations will be illustrated in the following, both involving *Thinopyrum* spp. and also more distant wheat relatives among perennial Triticeae species, gathering them on the basis of the amount of alien genome(s) contribution.

11.3.1 Hybrids and Amphiploids

A plentiful array of hybrids and complete or partial amphiploids was obtained with representatives of many species and genera of perennial Triticeae (Wang 1989a, 1989b, 1992, 2011; Jiang et al. 1994; Mujeeb-Kazi and Wang 1995; Fedak and Han 2005; Mujeeb-Kazi et al. 2013; Ceoloni et al. 2014a, and references therein). Such hybrid combinations have provided fundamental knowledge of the intergenomic affinities between the donor species and the recipient wheat, and, as in most wide crosses, represented the first step in the course of targeted introgression of desired alien traits, generally associated with alien transfers of limited entity (see ahead, Sects. 11.3.2 and 11.3.3).

Moreover, amphiploids involving *Thinopyrum* species, sometimes referred to as "Tritipyrums" (e.g., Marais et al. 2014) or "Trigopiros" (e.g., Fradkin et al. 2012), obtained from colchicine-induced or even spontaneous doubling of the F_1's chromosome number, probably represent the only other case, besides that of the well-known triticale (× Triticosecale Wittmack; Larter 1976), that may have practical utility. In fact, as an alternative route to direct domestication of some perennial species, such as *Th. intermedium* and *Th. ponticum* (see, e.g., Cox et al. 2010; Bell et al. 2010; DeHaan et al. 2014), derivatives from hybridization of *Thinopyrum* species with either durum or bread wheat have long been looked to as possible gateways to development of a perennial wheat. Unlike conventional wheat, that requires tilling and seeding the soil every growing season, develops shallow roots and grows on soil exposed to wind and water erosion for much of the year, perennial wheat would be planted once and harvested several times, would take greater advantage of precipitation during its longer growing seasons, and, thanks to deeper and larger roots, would also reduce soil erosion, nitrogen losses and salinization, as well as sequester carbon from the atmosphere. It would also require fewer farming operations and less herbicide supply, additional key features for sustainability of cereal cropping in less developed regions and marginal lands. Furthermore, greater complexity of the perennial cereal crown may act as a barrier to soil diseases, and this, coupled with ample resistance to foliar diseases conferred by genes of the perennial donor species, can greatly reduce challenges that perennial wheat production might face in terms of disease control (e.g., Cox et al. 2005a, b; Hayes et al. 2012; Turner et al. 2013). Therefore, breeding programmes aimed at capitalizing on perenniality-associated traits, yet providing agronomically and economically acceptable yields, are being conducted in the United States, Australia and other countries, also pointing at a dual-purpose perennial wheat, able to produce grain and additional forage during summer and autumn, hence representing a sustainable and profitable option for mixed crop–livestock farming systems (Cox et al. 2006, 2010; Bell et al. 2010; Ward et al. 2011; Larkin et al. 2014).

Perennial habit embodies a highly complex suite of traits, expected to be largely quantitative in nature. Therefore, notwithstanding the evidence of a gene or genes on chromosome 4E of diploid *Th. elongatum* determining some ability of post-harvest regrowth (Lammer et al. 2004; see also Sect. 11.3.2), it is not surprising that a common outcome from studies on perennial wheat is that plants derived from intergeneric combinations tend to be perennial only when the proportion of their total genome derived from the perennial parent is conspicuous (Cox et al. 2002; Hayes et al. 2012; Larkin et al. 2014). As a matter of fact, similarly to the regrowth phenotype conferred by chromosome 4E, which results less strong and more environmentally dependent than that observed in the *T. aestivum-Th. elongatum* complete amphiploid (Lammer et al. 2004), in wheatgrass derivatives reasonable capacity to regrow post-harvest and yield grain over successive years are only observed when many chromosomes are added to wheat from the perennial donor species (Hayes et al. 2012; Larkin et al. 2014). On reviewing the genomic composition of the most promising wheat-perennial derivatives, it has been recently suggested that the best prospects for a productive breeding program in the medium term would derive from complete amphiploids between wheat (either tetraploid or hexaploid wheat) and a diploid perennial donor, such as *Th. elongatum*, contributing, as in the triticale case, a whole-genome equivalent (Mujeeb-Kazi et al. 2008; Larkin et al. 2014). Providing the necessary chromosomes for the desired "package" of perenniality traits are present, other possibilities for generating perennial wheat amphiploids are of course possible; in fact, besides *Th. elongatum*, successful donor perennial parents have largely included the polyploid *Th. ponticum* and *Th. intermedium*. Also in these cases, if the wheat parent was a hexaploid, an entry required at least 56 chromosomes to achieve any substantial post-harvest regrowth, and even this was no guarantee of a capacity to survive post-harvest (Cox et al. 2010; Hayes et al. 2012; Larkin and Newell 2014; Larkin et al. 2014). Clearly, it is the presence, not solely numerical, of critical alien chromosomes to assure perenniality traits, as the expression of robust perennial habit in partial amphiploids derived from various hybridization strategies demonstrates. One notable case is that of MT-2 lines, derived from an original *T. durum/Th. intermedium* decaploid amphiploid and, following chromosome loss, averaging $2n = 56$, with around 30 *Thinopyrum* chromosomes (2:1 ratio between E (or J)-genome and St-genome chromosomes, see Table 11.1) and 26 wheat chromosomes (Jones et al. 1999). This genomic constitution contrasts with that of other octoploid partial amphiploids, such as OK-906 and Agrotana, having 40 wheat and only 16 *Thinopyrym* (E/J + St) chromosomes, characterized by an annual habit (Jones et al. 1999). In all cases, including primary types, as well as derivatives from their inter-crossing or even from backcrossing to the perennial parent, several rounds of breeding cycles and heavy selection will likely be required for the novel wheat type to achieve the desired performance and stability of all target traits. Such stability is expected to correspond to achievement of an inter-genomic "equilibrium," following a variety of "revolutionary" changes triggered by newly established allopolyploid conditions (Ma and Gustafson 2008; Feldman and Levy 2012; Wang et al. 2014).

Apart from the "perenniality" suite of traits, the numerous amphiploids, both complete and partial types synthetized through the years, represent a key step to early assess expression into the recipient wheat background of the valuable attributes identified in perennial Triticeae germplasm. As for closer gene pools, mostly stress resistance genes, particularly those against fungal and viral diseases, with a frequently monogenic control, have been more easily detected and, through subsequent steps, handled for targeted transfer. Genes of this kind were shown to provide excellent resistance to leaf, stem and stripe rust, powdery mildew, karnal bunt, spot blotch, *Stagonospora nodorum* blotch, Fusarium head blight (FHB) or scab, tan spot, eyespot, barley yellow dwarf virus (BYDV), wheat streak mosaic virus (WSMV) and its vector, wheat curl mite (WCM), and aphids (see, e.g., Oliver et al. 2006; Li and Wang 2009; Chang et al. 2010; Wang 2011; Zeng et al. 2013a). Remarkable tolerance to abiotic constraints, notably salinity, has also been detected in various amphiploid combinations involving various perennials, such as *Th. elongatum*, *Th. bessarabicum* and *Th. distichum*, and both bread and durum wheat (e.g., Dvorak and Ross 1986; King et al. 1997b; Colmer et al. 2006; Mujeeb-Kazi et al. 2013; Marais et al. 2014). Although major genes have been sometimes identified also for such more complex traits and eventually transferred into wheat (see Sects. 11.3.2 and 11.3.3), it was not infrequent to detect a truly quantitative type of inheritance, with several alien genes scattered on different chromosomes acting in an additive manner (Zhong and Dvorak 1995; Colmer et al. 2006; Mujeeb-Kazi et al. 2013). A similar outcome was observed for FHB resistance conferred by genes on *Th. junceum* chromosomes (McArthur et al. 2012), and for BYDV resistance from *Th. elongatum* (Anderson et al. 2010). In these instances, a "genome-wide" approach of transfer promotion can be profitably applied, by introducing into the hybrid or amphiploid genotype a recessive condition at the main wheat locus normally suppressing homoeologous chromosome pairing, i.e., *Ph1* (reviewed in Ceoloni and Jauhar 2006; Qi et al. 2007; see also Sect. 11.3.3). This strategy is not only effective in capturing the most of the alien donor genetic determinants for a given polygenic trait, but also when multiple genes/quantitative trait loci (QTLs) for various desirable attributes are scattered in the alien genome(s). In fact, this was proved to be the case for a large number of complete or partial amphiploids (see, e.g., Chen 2005; Fedak and Han 2005; Wang 2011; Zeng et al. 2013a), and for some of these (e.g., for a *T. aestivum-Th. bessarabicum* amphiploid, Kazi 2011) extension of pairing and recombination promotion to potentially all wheat-alien homoeologous partners was successfully exploited in parallel with the more frequently pursued strategy of backcrossing the *Ph1* amphiploid to a normal wheat genotype, to "scale down" the alien donor genomic component to single chromosome additions and substitution lines, and leaving the *ph1* mutant effect to be active only on single, specific alien chromosomes (see Sect. 11.3.3).

11.3.2 Chromosome Addition and Substitution Lines

In general, starting from a hybrid or a complete/partial amphiploid combination with one or more appealing attributes for potential enhancement of wheat performance, a step forward for reduction of unwanted alien genetic material is the isolation of single alien chromosome addition and substitution lines into the wheat background. Molecular and phenotypic evaluation of these materials enables chromosomal assignment of gene(s) of interest, besides that genome and homoeologous group attribution of the specific alien chromosome(s). To this respect, sometimes the picture may be complicated by intergenomic rearrangements not seldom occurring in hybrids and amphiploids, as it is in the case of related genomes of polyploid *Thinopyrum* species (reviewed in Fedak and Han 2005), and hence maintained in derived addition and substitution lines. Various examples are illustrative of this phenomenon, including *Th. junceum* chromosomes in AJDAj5 and AJDAj6 addition lines, which, based on EST-SSR (Wang et al. 2010b) and RAPD (Wang et al. 2003) markers, appear as complexly restructured chromosomes, carrying portions with segmental homoeology to groups 1 + 5 and 2 + 5 + 1, respectively. Such complex patterns of homoeology with wheat chromosomes (Moustakas 1991; Wang et al. 2010b; Wang 2011) are likely the result of structural rearrangements differentiating the E^b and E^egenomes which make up the hexaploid *Th. junceum* genome (Table 11.1), and of the complex, reticulate evolution characterizing polyploid lineages of these and all Triticeae species (see Sect. 11.2).

The same reasoning can explain what observed in combinations with wheat of *Thinopyrum* chromosomes belonging to polyploid species, containing, besides E/J--type genomes, one or more St genomes, consistently proved to be closely related to E/J genomes (Zhang et al. 1996a; Fan et al. 2007; Liu et al. 2007; Wang 2011). Thus, the *Th. intermedium* chromosome present in addition line L1, derived from the TAF46 partial amphiploid with *T. aestivum*, was interpreted as a prevailingly St chromosome, with pericentromeric chromatin of E-genome derivation (Wang and Zhang 1996). The BYDV resistance gene present in TAF46 and L1 was allocated to the distal 7St region of the long arm of such a chromosome (Zhang et al. 1996b). Carrier of an St-E translocation was also considered one of the seven *Th. intermedium* chromosomes present in the Zhong 5 partial amphiploid, and from it incorporated into disomic additions Z1, Z2 and Z6, and substitutions Yi 4212 and HG 295 (Tang et al. 2000; Zhang et al. 2001; Ayala-Navarrete et al. 2009). All lines carrying this chromosome, showing group 2 homoeology, showed high resistance to BYDV.

The array of disease resistance phenotypes assigned to given *Thinopyrum* spp. chromosomes via evaluation of addition or substitution lines is indeed plentiful. Among the most relevant for having represented starting materials for subsequent use in breeding of their gene content, are various examples in which *Th. ponticum* chromosomes are involved. One notable case is that of the 6Ag chromosome, containing the durable and wide-spectrum stem rust resistance gene *Sr26*. Remarkably, since its identification in the 1950s (Shebeski and Wu 1952), *Sr26* remains still effective against all known races of the causing agent, including all pathotypes belonging to the Ug99 lineage (FAO 2015). After several backcrosses with bread wheat cv. Thatcher of an original *T. aestivum-Th. ponticum* partial amphiploid with $2n=56$ (Shebeski and Wu 1952), addition and substitution lines were obtained. These, besides proving that the strong resistance was fully associated with the alien 6Ag chromosome, also showed the latter to be homoeologous to wheat group 6, and to compensate well for wheat 6A, consistently replaced by 6Ag in the substitution lines (Knott 1964). Since the 1960s, the long-lasting *Sr26*-based resistance was introduced in Australia in the form of a radiation-induced 6AgL-6AL translocation, derived from a 6Ag addition line (Knott 1961), which has been widely used commercially, in spite of the 6AgL-associated yield penalty (The et al. 1988). Further 6Ag manipulations subsequently undertaken to obviate this defect will be described ahead (Sect. 11.3.3).

Another valuable *T. aestivum-Th. ponticum* pre-breeding material is the 7Ag(7D) substitution line called Agrus. Initially the line was used as a source of the highly effective leaf rust resistance gene *Lr19*, and, as illustrated in the following section (Sect. 11.3.3), through different strategies (Sharma and Knott 1966; Sears 1973, 1978; see also ahead), 7Ag was engineered to give wheat translocation and recombinant lines carrying *Lr19*. In rather close linkage with *Lr19*, along the 7AgL arm, the *Sr25* stem rust resistance gene was also found to be located (McIntosh et al. 1977, and both genes still provide strong resistance to the respective rust disease (e.g., Gennaro et al. 2009; Liu et al. 2010; FAO 2015). Of somewhat lower efficacy in time and space (Friebe et al. 1996; FAO 2015) has been the resistance to leaf and stem rust conferred by the *Lr24* and *Sr24* gene, respectively, located on a *Th. ponticum* 3AgL arm of a 3Ag *Th. ponticum* chromosome, substituted for wheat chromosome 3D in the TAP67 derivative line from the (*T. aestivum* × *Th. ponticum*) × *T. aestivum* cross (Bakshi and Schlehuber 1959). TAP 67, showing normal vigor and fertility and reasonably good yield, was used as donor of the *Lr24* gene to a series of bread wheat recombinant lines mostly involving the homoeologous wheat 3DL arm (Sears 1973, 1978). Similarly to the *Lr19* + *Sr25* case, it was later discovered that *Lr24* was linked to *Sr24* in all recombinant and translocation lines of the same 3Ag chromosome (McIntosh et al. 1977). Based on GISH evidence, the 3Ag chromosome appears to belong to a J^s (= St) genome of the donor species (Li et al. 2003).

As to valuable resistance sources associated to chromosomes of other *Thinopyrum* species, resistance to stem rust, including Ug99 strains (Xu et al. 2009), was found to be conferred by a *Th. junceum* chromosome, largely homoeologous to wheat group 4, present in the addition line HD3505 (Wang et al. 2010b). Besides this, a group 2 *Th. junceum* chromosome in the AJDAj3 addition line contained an effective gene(s) for resistance to FHB (McArthur et al. 2012).

Further, considering wheat cv. Chinese Spring-*Th. elongatum* disomic substitutions, chromosomes 2E and 3E provided excellent resistance to cereal yellow dwarf virus (CYDV), while substitution lines for 1E and 6E were significantly more resistant to *Septoria tritici* blotch compared to Chinese Spring (Anderson et al. 2010). However, neither chromosome by itself conferred resistance as high as that of several wheatgrass accessions; similarly, genes on multiple *Th. elongatum* chromosomes were apparently required for complete resistance to BYDV. On the other

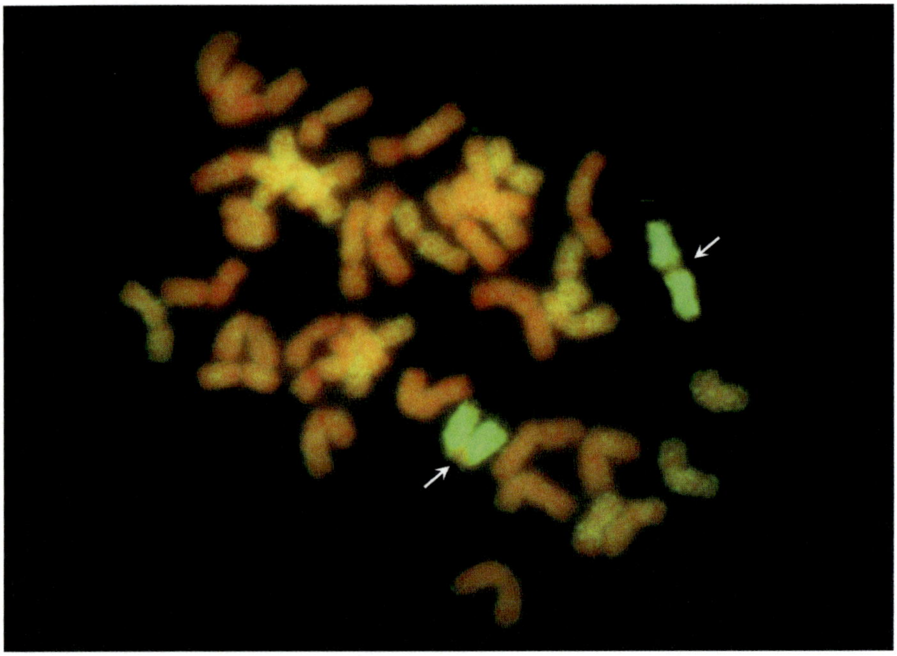

Fig. 11.1 Partial somatic metaphase plate of a bread wheat-*Thinopyrum elongatum* 7E(7D) substitution line, subjected to fluorescence GISH, as described in Forte et al. (2014). Total *Th. elongatum* DNA, used as one probe, is marked by green fluorescence (FITC fluorochrome), which highlights the two 7E chromosomes (*arrowed*). Total wheat DNA, used as the other probe, was labelled by Cy3 (red fluorescence). The more greenish wheat chromosomes belong to the D genome, with which the *Thinopyrum* DNA cross hybridizes more intensely (i.e., shares higher homology) than with those of the A and B genomes

hand, the excellent resistance to FHB observed in 7E(7A), 7E(7B), and 7E(7D) substitution lines (Fig. 11.1), was shown to be due to the presence of a gene(s) for type II resistance on the 7EL arm (Shen et al. 2004; Shen and Ohm 2006). Transfer of this effective gene in both bread and durum wheat has been recently undertaken (Forte et al. 2011 and unpublished).

Molecular and cytogenetic techniques were used to identify a new wheat-*Th. intermedium* ssp. *trichophorum* substitution line from the cross of TE-3 partial amphiploid and the Chinese wheat line ML-13 (Hu et al. 2011). The *Thinopyrum* derived St-chromosome, substituting for wheat chromosome 1D, was proved to contain new gene(s) for stripe rust resistance. Furthermore, wheat-*Th. intermedium* lines carrying a group 4 homoeologous chromosome or its short arm were resistant to eyespot (Li et al. 2005). Eyespot resistance was similarly associated to a group 4 chromosome of *Th. ponticum*, but the same was not true for group 4 chromosomes of the diploid *Th. elongatum* and *Th. bessarabicum* species (Li et al. 2005).

On the other hand, taking into account tolerance to abiotic stresses, chromosomes of diploid *Thinopyrum* species with different homoeology to wheat chromosomes were shown to harbour genes enhancing wheat salt tolerance (Colmer et al. 2006). Dissection of this complex trait by use of addition or substitution lines, revealed *Th. elongatum* chromosome 3E to have a major dominant effect on salt tolerance, with 3E substitution lines showing superior "exclusion" of Na+ and better maintenance of K+ in flag leaves, higher dry mass and grain yields when compared with wheat cv. Chinese Spring (Omielan et al. 1991). It was, instead, the $5E^b$ of *Th. bessarabicum* to confer higher ability to "exclude" Na+ from both mature and newly developed leaves to substitution, and to a lesser extent, addition lines compared to normal wheat (Mahmood and Quarrie 1993). Although other *Th. bessarabicum* chromosomes are likely involved in fuller expression of the trait, attempts to transfer the $5E^b$ gene(s) were undertaken (King et al. 1997a). Partly homoeologous to wheat group 5 is also the *Th. junceum* chromosome ($5E^b$ or 5J) present in the AJDAj5 addition line (see also above), resulting salt tolerant (Wang et al. 2003).

Another interesting trait that could allow wheat to better respond to environmental constraints, namely waterlogged conditions, was identified in diploid *Th. elogatum* chromosomes homoeologous to groups 2 and 4. In fact, presence of 2E and 4E in *T. aestivum* addition lines was associated with a positive effect on root growth and penetration in waterlogged soil (Taeb et al. 1993).

Investigation of addition lines of *Th. bessarabicum*, *Th. elongatum* and *Th. intermedium* chromosomes, mostly homoeologous to wheat group 1, have also provided interesting information as to the presence of several novel alleles for high molecular weight glutenin subunits (HMW-GS), known to play an important role in determining end-use quality in wheat (Niu et al. 2011). Novel HMW-GS have been also identified and fully characterized in *Th. ponticum*: some of them contain extra cysteine residues in their amino acid sequences, a feature that makes them potentially able to exert a positive influence on wheat dough properties (Liu et al. 2008). Similarly, good potential for improvement of wheat end-quality was shown for a HMW-GS gene located on a group 1 St chromosome of *Th. intermedium* ssp. *trichophorum*, present into spontaneously occurred substitution and homoeologous recombinant bread wheat lines (Li et al. 2013). The genomic sequence of the *Thinopyrum*-derived HMW-GS was characterized and designated *Glu-1St#2x*, since it closely resembled x-type wheat glutenins. Phylogenetic analysis revealed that *Glu-1St#2x* subunit clearly clustered with *Glu-R1* from *Secale cereale* and *Glu-E1* from *Th. elongatum*, and evolved earlier than the split of wheat *Glu-1* homoeologous genes.

A peculiar trait that *Thinopyrum* species, besides that other Triticeae, showed to possess is blue aleurone. The blue pigmentation of aleuronic layer is due to the presence of anthocyanins, one of the major groups of flavonoid compounds that differ from those in red or white wheat grains, and play active roles in several human metabolic activities, such as antioxidant activity, anti-inflammatory, anticancer, and hypoglycemic effects (Abdel-Aal et al. 2006). Development of synthetic blue-seeded wheats from intergeneric crosses with *Thinopyrum* species has a long history in North America, Europe, and Asia (Morrison et al. 2004; Zheng et al. 2009). Blue aleurone represents an easily scorable marker, used in various genetic studies and breeding practice of wheat (reviewed in Zheng et al. 2009). Moreover, increased interest in health beneficial flavonoids, has attracted attention to the accumulation of anthocyanin pigments in blue-grained wheat as a new dietary source of these compounds, whose biosynthesis might be upregulated during seed development by blue-aleurone (*Ba*) genes. Analysis of substitution and addition lines of *Thinopyrum* chromosomes into wheat backgrounds showed effective *Ba* genes to be consistently associated with homoeologous group 4. A *Th. ponticum* 4Ag chromosome pair was found to substitute for a 4D pair in the blue-grained Blue Dark wheat developed in Canada and in Blue 58 produced in China (Zheng et al. 2009). Crossing the substitution line Blue 58 with euploid *T. aestivum*, Li et al. (1986) isolated monosomic substitution plants ($2n=41$), in whose selfed progeny distinct seed colors (dark blue, medium-light blue, and white) were observed: dark blue seeds corresponded to disomic substitutions ($2n=42$), seeds of medium-light blue color were of monosomic substitutions, whereas white seeds belonged to nullisomic plants ($2n=40$). Thus, a clear-cut association between the grain color and the dosage of the *Th. ponticum* blue-grained gene (named *Ba1*) on 4Ag was established. A strong dosage effect was similarly observed for a gene, designated *BaThb*, on chromosome 4 J ($=4E^b$) of diploid *Th. bessarabicum* (Shen et al. 2013). By molecular cytogenetic and marker analyses of spontaneous or induced translocation lines, the *Ba* genes of the two *Thinopyrum* species were assigned to non-colinear positions along the long arms of the respective group 4 chromosomes (Zheng et al. 2009; Shen et al. 2013).

Although in much more limited number, and never as complete sets, addition and substitution lines of chromosomes of perennial grasses more distantly related to wheat than the *Thinopyrum* genus have been obtained. Some involve species of the large *Elymus* genus, such the allohexaploid *Elymus rectisetus*, an apomictic species carrying the St, W, and Y genomes (Table 11.1). In backcross progeny from crosses with bread wheat, also attempted to transfer apomixis into wheat (Liu et al. 1994), a disomic addition for a 1Y chromosome was identified, exhibiting moderate resistance to both tan spot and Stagonospora nodorum blotch (Oliver et al. 2008), as well as a 1St addition with a good level of resistance to FHB (Dou et al. 2012; McArthur et al. 2012).

Chromosomes of another polyploid, mostly tetraploid, group of perennials, belonging to the *Leymus* genus (Table 11.1), were also shown to possess valuable traits for wheat improvement. Particularly noteworthy is the biological nitrification inhibition (BNI), i.e., the capacity of root exudates to suppress NO_3^- formation and keep the largest part of soil's inorganic-N in the NH^+_4-form, found to be highly expressed in Volga or mammoth wildrye, *L. racemosus*, while virtually lacking in wheat. Analysis of wheat *L. racemosus* addition lines revealed that such BNI capacity is expressed in the genetic background of wheat cv. Chinese Spring, and is mostly controlled by one *L. racemosus* chromosome, named Lr#n, whose presence increases by about fourfolds that of the Chinese Spring control (Subbarao et al. 2007). However, the same Lr#n chromosome, recently shown to be homoeologous mostly to group 2 wheat chromosomes (Larson et al. 2012), does not provide tolerance to NH^+_4, which appears to be under the control of chromosome 7Lr#1-1, showing group 7 homoeology. Both attributes would be beneficial for a sustainable wheat production, since while NH^+_4 could represent the necessary signal to make the BNI capacity responsive to the environment, the ability to combat nitrification in intensive wheat farming systems has the potential to reduce nitrogen pollution from such systems (Subbarao et al. 2007).

Furthermore, two of the several wheat-*L. racemosus* addition lines developed in China showed high resistance to FHB (Wang and Chen 2008). From pollen irradiation of the MA7Lr monosomic addition line, with the alien chromosome showing homoeology to wheat group 7 chromosomes, a ditelosomic substitution line was isolated, where a pair of 7Lr#1S telocentric chromosomes, to which the FHB resistance gene(s) could to assigned, replaced wheat chromosome 7A.

Another attribute that could profitably be introgressed from *L. racemosus* into wheat is tolerance to Aluminium (Al) toxicity, a key factor limiting its production in acidic soils, which represent 40 % of the world's cultivated land. Recently, two addition lines, for chromosome A (group 2 homoeology) and E (unknown homoeology) were shown to significantly enhance wheat Al tolerance in terms of relative root growth (Mohammed et al. 2013). The markedly increased tolerance conferred by chromosome E was attributed to improved cell membrane integrity. The same study also showed the importance of wheat chromosome 2B in the expression of the Al tolerance of *L. racemosus* chromosome A, not detected in the substitution line lacking this chromosome, and also the negative effect of other *L. racemosus* chromosomes on the same trait, evidently resulting from interaction between wheat and alien genes. Targeted chromosome engineering with the two positively contributing lines is expected to allow attainment of Al-tolerant wheat cultivars.

In the search for sources of tolerance to heat stress, one of the major factors limiting wheat production in tropical and subtropical environments, the same set of *L. racemosus* addition and substitution lines was evaluated under controlled and field stressful conditions (Mohammed et al. 2014). Chromosomes A, 2Lr#1 and 5Lr#1, added to lines TAC1, TAC12 and TAC13, showed early heading and maturity, which enabled these lines to fill their grains normally and escape the late heat stress occurring at the end of the season. In addition to this avoidance mechanism, the most tolerant TAC12 line probably possesses a heat tolerance mechanism correlated with a more efficient mitochondrial electron transport activity, hence cell viability (Mohammed et al. 2014). Higher mitochondrial efficiency under heat stress conditions also appeared to underlie the heat tolerance of TAC6 addition line, harboring a *Leymus* chromosome of homoeologous group 5. Yield-related traits were also observed in the various lines, among which TAC14, carrying the group 7 chromosome 7Lr#1, stood out for its considerable yield potential, resulting from both high tiller number and kernel weight. Interestingly, group 7 chromosomes of wheat (Quarrie et al. 2006) and *Th. ponticum* (Kuzmanovic et al. 2014), also carry genes for yield-contributing traits (see Sect. 11.3.3).

Analysis of addition/translocation lines of another *Leymus* species, i.e., *L. multicaulis*, into Chinese bread wheat cultivars, showed different *Leymus* chromosomes as capable to confer resistance to FHB, CYDV and stem rust (Zhang et al. 2010). Where revealed by use of SSR markers, homoeology of these chromosomes corresponded to wheat groups 1 and 3, to which most of the 24 tested lines appeared to be ascribable (Jia et al. 2002; Zhang et al. 2010). Both RFLP and SSR markers revealed as well rearranged chromosomes, of frequent occurrence in cross-pollinated species like *L. racemosus* (Qi et al. 1997) and *L. multicaulis* (Jia et al. 2002; Zhang et al. 2010; see Sect. 11.2). Curiously, all the added/translocated chromosomes of the 24 lines illustrated above appeared to belong to only one of *Leymus*genomes, namely Xm, of yet unknown origin (Wang 2011). Conversely, of Ns genome likely derivation are the three *L. mollis* chromosomes substituted into the $2n=42$ line selected among F5 progenies from the cross of an octoploid *Tritileymus* amphiploid (*T. aestivum*×*L. mollis*, $2n=56$) with *T. durum* (Zhao et al. 2013). The retained *L. mollis* chromosomes belong to homoeologous groups 1, 5 and 6. The triple alien substitution line, meiotically stable and well compensated, is remarkably resistant to stripe rust and of convenient short stature; thus, it can be employed as a bridge parent in wheat breeding via chromosome engineering.

Desirable genes for wheat improvement have also been identified in species of the genus *Agropyron* (P genome, Table 11.1) (Han et al. 2014 and references therein). A series of disomic addition lines was obtained from the cross of a Chinese accession of tetraploid *A. cristatum* with common wheat cv. Fukuhokomugi (Wu et al. 2006; Han et al. 2014). In all of them, SSR, EST-SSR and STS markers specific to the *Agropyron* chromosome were primarily related to homoeologous group 6; however, the group 6 markers, mainly located in the 6P pericentromeric region, were not completely identical among the different addition lines. Moreover, there were several markers belonging to other homoeologous groups distally located along the various 6Ps. Such rearrangements, probably differentiating the two P genomes of *A. cristatum* (see Sect. 11.2), led to distinguish four 6P types ($6P_I$–$6P_{IV}$) with different genetic make-up. Among them, $6P_I$ was proved to carry a gene(s) conferring high grain number per spikelet and per spike and also gene(s) for resistance to wheat powdery mildew (Han et al. 2014).

Various novel disease resistance genes have been also identified on specific V^b chromosomes of the perennial tetraploid *D. breviaristatum*. Addition and substitution lines were isolated in the progeny of wheat-*D. breviaristatum* amphiploids crossed with cultivated wheat, including different addition lines carrying genes for stripe rust (Yang et al. 2008), as well as stem rust and powdery mildew (Liu et al. 2011) resistance. Marker data indicated that the V^b chromosomes in the latter two addition lines were rearranged with respect to wheat homoeologous groups. On the other hand, various molecular markers confirmed a group 2 homoeology for the V^b chromosome substituted into a Chinese bread wheat in place of chromosome 2D, able to confer stripe rust resistance at the adult plant stage (Li et al. 2014). Interestingly, FISH, C-banding, and PCR-based molecular marker analyses indicated that the $2V^b$ of *D. breviaristatum* was completely different from $2V^v$ of *D. villosum*, in line with the current view about the origin of 4*x* *D. breviaristatum* (see Sect. 11.2).

All the addition and substitution lines described above were obtained in the hexaploid background of *T. aestivum*. Development and maintenance of intra- and interspecific aneuploid types is known to be much more difficult at the tetraploid level (reviewed in Ceoloni and Jauhar 2006). Thus, a very limited number of chromosomes of alien species belonging to the secondary and tertiary gene pools (containing genomes that are nonhomologous to those of wheat) could be stably added to the *T. durum* genome or substituted for its component chromosomes. With the only notable exception of the complete set of D-genome disomic substitution lines, of which a complete set was developed in the variety Langdon, the remaining ones were mostly incomplete and/or of a monosomic type (reviewed in Ceoloni et al. 2005a). However, four out of the seven chromosome pairs of diploid *Th. elongatum* were added to *T. durum* cv. Stewart (Mochizuki 1960, 1962). Isozyme analysis allowed identification of the homoeology of the added chromosomes with those of wheat, showing individual relationship to group 1, 6, 3 and 4 (Ono et al. 1983). More recently, from an initial F_1 hybrid between durum wheat cv. Langdon and a *Th. elogatum* accession tolerant to Fusarium heat blight, subjected to several backcrosses with the durum parent and selfings, a disomic addition line with $2n=30$ was obtained (Jauhar et al. 2009; Jauhar and Peterson 2011). Molecular markers allowed identification of group 1 homoeology of the added chromosome, hence named 1E, and will be useful in the transfer of the FHB resistance into durum wheat in a more stable chromosomal condition.

11.3.3 Segmental Introgression Lines

The excessive amount of alien genetic material makes the type of cytogenetic stocks described above still unsuited for practical breeding use; they represent, however, potent resources from which further chromosome manipulations can give rise to exploitable products where undesired linkage drag is largely minimized.

Although none of the wheat-alien transfers so far produced has probably equalled the worldwide success of the spontaneous 1BL.1RS wheat–rye translocation (Mujeeb-Kazi et al. 2013), for an appreciable number of the beneficial traits originating from perennial Triticeae the transfer into wheat has reached the final step, i.e., that of a segmental introduction(s) of sub-chromosomal entity, well harmonized with the wheat genomic environment. The most significant progress has been registered in the last few years, in coincidence with the great advancements in molecular genetic, cytogenetic, and genomic tools, and consequent ability to precisely monitor the alien introduction process and finely target the desired outcomes.

As with other alien sources not sharing with wheat completely homologous genomes, various methods have been used to induce translocations or recombination events between wheat chromosomes and those of perennial wheatgrasses and wildryes. While in the early attempts of chromosome engineering the use of radiation-induced-->translocations was almost invariably adopted (e.g., Sharma and Knott 1966; Knott 1968), genetic promotion of intergenomic homoeologous pairing and recombination has been later the method of election, particularly after the isolation of mutants at the *Ph1* locus in both bread (Sears 1977) and durum wheat (Giorgi 1983). In a few instances, exchanges associated with cell culture-induced breakage and fusion (Banks et al. 1995), or the ability of *Aegilopsspeltoides* Ph^I gene(s) to partly inhibit the wheat *Ph1* effect (Wang et al. 2003), were the way followed to promote transfer of BYDV resistance and, respectively, salt tolerance from addition lines for a group 7 *Th. intermedium* chromosome (7Ai-1) and for a group 5 *Th. junceum* chromosome (AJDAj5). On the other hand, in a number of cases (see ahead) exchange products of potential breeding value have spontaneously occurred, probably as a result of the action of pairing promoting genes present in perennial Triticeaegenomes, which, to various extent, appear to counteract the suppressing effect of wheat *Ph* genes (Dvorak 1987; Zhang and Dong 1995; Jauhar and Almouslem 1998; Jauhar and Peterson 2000; Kang et al. 2008; Mullan et al. 2009). Whatever the method adopted for promoting exchanges, segmental introductions that involved homoeologous chromosomes and were of relatively more limited length gave rise to the best compensating and hence useful products for breeding exploitation.

11.3.3.1 Transfers Involving *Th. intermedium* and *Th. ponticum*

Favoured by the close relationships of their E/J and St basic genomes (Table 11.1) with those of wheat, particularly D and A, *Th. intermedium* and *Th. ponticum*, and so diploid species of the same genus, result the most valuable species contributing to wheat cultivar development among perennial Triticeae, and probably among wild relatives altogether (e.g., Li and Wang 2009; Wang 2011; Mujeeb-Kazi et al. 2013). As for other alien donors, among the many useful genes that have been stably transferred from *Thinopyrum* species into compensating wheat translocation/recombinant lines, the more easily manageable genes conferring disease resistance prevail. Many of these were described in previous extensive reviews (e.g., McIntosh et al. 1995; Friebe et al. 1996; Li and Wang 2009). In recent years, the list of exploitable resistance sources has widened, with addition of several new genes, and of genes previously included in transfer types unsuitable for breeding use (e.g., associated with excessive alien chromatin amount).

Introgressions from *Th. intermedium*

As to *Th. intermedium*, transfers into wheat include *Pm40* (Luo et al. 2009) and *Pm43* (He et al. 2009) for resistance to wheat powdery mildew, a disease toward which the donor species shows complete immunity (Wang et al. 2000). Both transfers appeared to consist of a spontaneously occurred cryptic translocation (the former on wheat 7BS, the latter on 2DL), giving rise to cytologically and phenotypically suitable bread wheat lines for use in breeding, especially in humid Chinese environments, where the disease is constantly epidemic. From the same wheat-*Th. intermedium* partial amphiploid used as donor material of the *Pm43* gene, an additional, spontaneous translocation product was recently obtained, consisting of a bread wheat line with a pair of chromosomes 6A carrying a *Th. intermedium* segment which occupies most of the short arm, and contains a gene with dominant inheritance, determining an immune reaction to powdery mildew races collected from wheat fields of the Southwestern China (Tang et al. 2014). GISH analyses revealed the alien segment to originate from an St chromosome of the *Thinopyrum* donor species, which likely carries numerous *Pm*-type genes on different chromosomes and genomes. Various genes for resistance to rusts were also incorporated into wheat chromosomes within *Th. intermedium* chromosome segments. One such cases is that of the stem rust resistance gene *Sr44* (initially designated as *SrAgi*, Mclntosh et al. 1995), which confers resistance to the Ug99 race complex (Pretorius et al. 2010), and, starting from addition and substitution lines for the complete 7Ai (or 7J) chromosome, has been incorporated into a compensating 7DL.7JS (=7DL.7Ai-1S) Robertsonian translocation (Liu et al. 2013a), as well as into 7AL.7AS-7Ai-1S *ph1b*-induced homoeologous recombinant lines (Khan 2000). Distally located on the long arm of the same 7Ai chromosome, the *Bdv2* gene conferring BYDV resistance is located; several subchromosomal introgressions containing *Bdv2* were obtained, as either cell culture- or radiation-induced translocations (Banks et al. 1995; Crasta et al. 2000), or *ph1b*-induced homoeologous recombinants (Xin et al. 2001) into wheat 7DL. Lines carrying the shortest alien segment (particularly TC14, with just 20 % distal 7DL replaced by 7AiL, see e.g Ayala-Navarrete et al. 2009) have been deployed in breeding in China and Australia (Zhang et al. 2009b; Ayala-Navarrete et al. 2013). Multi-environment yield trials conducted in Australia showed the impact of TC14 on various genetic backgrounds to be generally benign, except for a frequent delayed maturity, which makes this translocation useful in BYDV-prone areas that experience a less pronounced terminal drought (Rosewarne et al. 2015).

An additional highly effective gene toward this relevant viral disease, named *Bdv4*, is located on a *Th. intermedium* chromosome with group 2 homoeology incorporated into the Zhong 5 partial amphiploid and derived substitution lines (see Sect. 11.3.2). A combination of translocations containing *Bdv2* and *Bdv4* is being endeavoured, which would likely confer to wheat an even stronger and more durable resistance than either gene alone, like that expressed by the wheatgrass donor species (Ayala-Navarrete et al. 2009).

Introgressions from *Th. ponticum*

Perhaps even more impacting on breeding are transfers originating from *Th. ponticum*. Among the many genes introduced from this species into the wheat background (see, e.g., Li and Wang 2009; Ceoloni et al. 2014a), one illustrative case of how chromosome engineering can be the key to effective exploitation of desirable alien variation is that of the stem rust resistance gene *Sr26*. As above anticipated (Sect. 11.3.2), a radiation-induced 6AgL-6AL (=6Ae#1) translocation was introduced in Australia and from that many wheat cultivars carrying *Sr26* have been released. However, although *Sr26* still remains effective against all known races of stem rust, including all currently described pathotypes of the Ug99 lineage (FAO 2015), its resistance has only been used commercially in Australia, where use of 6Ae#1-bearing cultivars has been declining over the past two decades. The reason probably lies in the up to 15 % yield reduction associated with the presence of the 6Ae#1 segment carrying the target gene (Knott 1968; The et al. 1988), a segment that occupies about 90 % of 6AL (Friebe et al. 1996; Dundas et al. 2015). Work was therefore undertaken to reduce the segment size of the 6Ae#1 translocation present in the Australian cv. Eagle by induced recombination with its wheat homoeologues (Dundas and Shepherd 1998; Dundas et al. 2007). By use of biochemical and molecular markers, over 1400 critical individuals were effectively screened, and among them 11 were found to have reduced size of 6Ae#1 chromatin (Dundas et al. 2015). Of the seven proved to carry *Sr26*, five involved chromosome 6A, one 6D and the last an unidentified wheat chromosome. GISH-based physical mapping placed the *Sr26* gene at the distal end of the *Th. ponticum* chromosome arm, and selected recombinant lines with around 30 % of distal 6AgL chromatin have already shown higher grain yields than the recurrent wheat cultivars in preliminary field evaluations (Dundas et al. 2015).

As mentioned before (Sect. 11.3.2), some *Th. ponticum* chromosome regions present the interesting occurrence of more than one useful gene for potential use in wheat. One such case concerns the distal portion of 3AgL arm, harbouring the *Lr24* and *Sr24* genes. Cultivars have been developed mostly from Agent, a spontaneous, compensating translocation carrying both genes in its terminal 3AgL segment, spanning about 30 % of the wheat 3DL arm (Friebe et al. 1996); however, several recombinant lines, involving mostly chromosome 3D but also 3B, were also obtained by genetically induced homoeologous recombination (Sears 1973, 1978). One of the 3BS.3BL-3AgL recombinant lines, carrying about 20 % of distal 3AgL, has been employed in recent years to introduce the two genes on a 3B chromosome of durum wheat (Ceoloni, unpublished). In contrast with the short-term protection provided in North America and South Africa, breakdown of leaf rust resistance conferred by *Lr24* was only detected in Australia after almost 20 years of its extensive exploitation (Park et al. 2002), and the gene can still be useful in resistance gene combinations (Bariana et al. 2007). On the other hand, *Lr24* confers complete immunity to all leaf rust pathotypes spread in China, where was recently tagged by a STS marker for MAS breeding (Zhang et al. 2011b), and in India (Kumar et al. 2010). As to *Sr24*, the gene maintains its efficacy in Australia and in many wheat producing regions worldwide, although some Ugandan "Ug99" pathotypes mutated in recent years to acquire virulence toward *Sr24* (FAO 2015).

Transfers Involving *Th. ponticum* Group 7 Chromosomes

Amongst *Sr* genes effective against all Ug99 races emerged so far, besides *Sr44* and *Sr26* already described, *Sr25* and *Sr43*, both from *Th. ponticum*, have to be recalled. The latter was originally introduced into common wheat in the form of chromosome substitution and translocation lines involving the alien group 7 chromosome, designated $7el_2$, and wheat chromosome 7D (Knott et al. 1977; Kibiridge-Sebunya and Knott 1983). However, even the best of the translocation lines (e.g., KS24-1 or KS24-2 in Kim et al. 1993), turned out to be $7DS.7el_2L$ Robertsonian translocations, with the complete $7el_2L$ arm determining undesirable linkage drag. Therefore, to eliminate unwanted alien genes, such as one enhancing yellow flour pigmentation (*Yp*) of the recipient bread wheat, a *ph1b*-mediated chromosome engineering strategy was applied to the 7DS-$7el_2S$.$7el_2L$ KS10-2 initial translocation line (Niu et al. 2014). To identify new wheat lines carrying *Sr43* on shortened alien segments, several stem rust resistant plants were screened for dissociation of *Sr43* from one or more of the six codominant SSR markers located on 7DL, and two recombinants with *Th. ponticum* segments of limited size, interstitially located in the subterminal region of 7DL, were eventually isolated. GISH revealed the $7el_2L$ portions to be inferior to 20 % of the recombinant 7DL; however, in both lines the yellow pigmentation was still higher than in wheat control lines, though inferior to the KS10-2 initial translocation line. Moreover, the *Sr43* gene they contain could have a limited deployment because of its temperature-sensitive expression, making the gene largely ineffective at 26 °C (Niu et al. 2014).

No doubt, the most extensively targeted group 7 *Thinopyrum* chromosome is the one originally named 7Ag (Sears 1973) but also $7el_1$ (Sharma and Knott 1966; Knott et al. 1977), because of its nearly complete homology with $7el_2$, as probably deriving from a different accession of the same species. In fact, $7el_1$ and $7el_2$ not only show almost complete pairing (Dvorak 1975; Forte et al. 2014), but also exhibit considerable correspondence in gene content, particularly at the L arm level. Both arms possess genes controlling resistance to leaf rust (*Lr19* on $7el_1L$ and a weaker, unknown *Lr* gene on $7el_2L$), and to stem rust (*Sr25* and *Sr43*, respectively), as well as genes determining yellow flour pigmentation (*Yp*) and Segregation distortion (*Sd*) (reviewed in Ceoloni et al. 2014a). Contrasting phenotypes for reaction to *Fusarium* head blight (FHB) differentiate the two 7el sources, $7el_1$ being susceptible and $7el_2$ bearing a major QTL in the distal end of its long arm (Shen and Ohm 2007; Forte et al. 2014).

The extensive work addressed to traits of $7el_1$ derivation, of both theoretical and practical value, has been enabled by availability of a wide array of translocation and recombinant lines involving this chromosome, produced both in bread and in durum wheat backgrounds. The consistent interest for $7el_1$ transfers was primarily addressed to the *Lr19* gene, conferring a largely effective resistance to wheat leaf rust across time and space (Gennaro et al. 2009 and references therein). Using the $7el_1$(7D) substitution line, named Agrus, as starting material (Sect. 10.3.2), *Lr19* was incorporated into bread wheat cultivars through both irradiation (Sharma and Knott 1966; Knott 1968) and induced homoeologous recombination (Sears 1973, 1978). Among the radiation-induced translocations, the one named T4 (= Agatha), consisting of a 70 % long $7el_1L$ segment inserted onto the wheat 7DL arm (Fig. 11.2a; Dvorak and Knott 1977; Friebe et al. 1996), proved to have a good compensating ability (Friebe et al. 1994).

An additional resistance gene, namely *Sr25*, conferring resistance to several races of wheat stem rust (McIntosh et al. 1976; Knott 1989a), and recently shown to be effective even against Ug99 (Li and Wang 2009; Liu et al. 2010), enhanced the validity of T4. As such, this sizable translocation has been incorporated into several bread wheat varieties, including the CIMMYT cultivar Oasis 86 and various

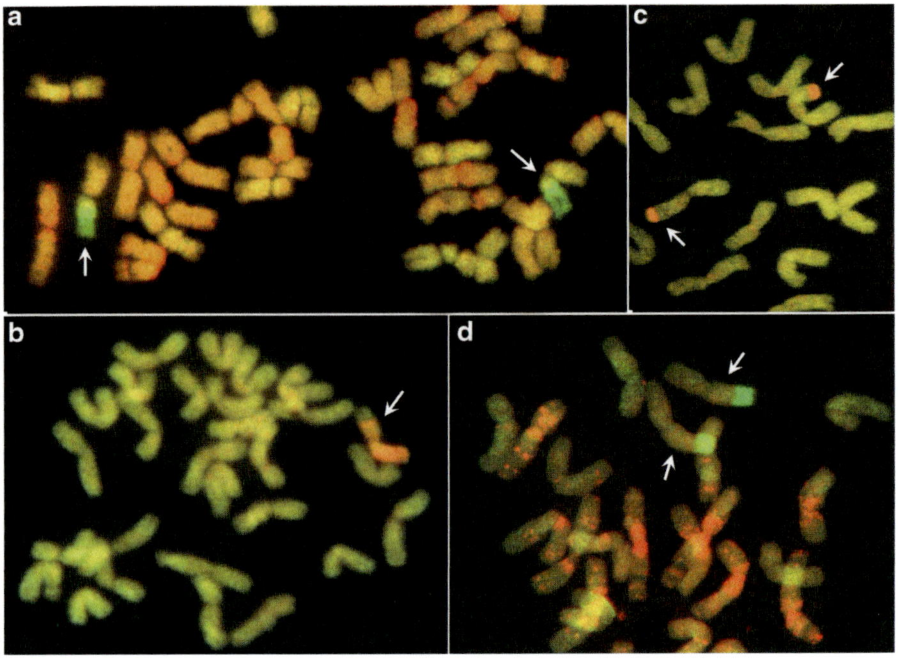

Fig. 11.2 Somatic chromosomes of wheat translocation and *ph1*-induced recombinant lines, carrying different segments of *Thinopyrum ponticum* chromosome 7Ag, subjected to fluorescence GISH, as described in Forte et al. (2014). Total DNAs of wheat and *Th. ponticum* were labelled with different fluorochromes in separate experiments; as a result, *Th. ponticum* DNA is marked by green fluorescence in **a** and **d**, and by red fluorescence in **b** and **c**. In all *plates*, the wheat-*Th. ponticum* chromosomes are *arrowed*. (**a**) Partial metaphase *plate* of the bread wheat T4 translocation line, carrying a pair of 7D chromosomes with 70 % of 7AgL on their long arms; (**b**) Complete metaphase *plate* of a primary durum wheat recombinant, hemizygous for a 7A chromosome with 7Ag chromatin encompassing the entire long arm and about half of the short arm; (**c**–**d**) partial metaphase *plates* of secondary 7A-7Ag durum wheat recombinants, with shortened amounts of 7Ag chromatin compared to the primary recombinant type: R5-2-10, with 23 % 7AgL on distal 7AL (**c**), and R23-1, with 40 % 7AgL on its 7AL arms (**d**)

derivatives still bred in several world Countries endangered by Ug99 (e.g., DRRW 2011), and the Sweden cultivar Sunnan (McIntosh et al. 1995). In India, the second largest wheat producer in the world after China, with about 12 % share in total world wheat production, cultivars with *Lr19*+*Sr25* have become popular, and besides the rust protection, this alien transfer seems to be associated with slow leaf senescence and increased yield (Sivasamy et al. 2010).

As to the *Yp* gene(s), also associated with *Lr19* and *Sr25*, as it is one of its most likely candidates, i.e., *Psy1* (coding for phytoene synthase, required for endosperm carotenoid accumulation, see Gallagher et al. 2004; Pozniak et al. 2007), the current perception by breeders and consumers seems to be changing as compared to the past. In fact, higher carotenoid content in the wheat grain is becoming a desirable trait for the multiple beneficial health effects determined by these compounds, leaving the aesthetic aspect of the increased flour yellowness of relatively minor importance (Ravel et al. 2013). However, reduction of pigmentation was attempted by various means in past years. By subjecting Agatha (=T4) to EMS treatment, Knott (1980, 1984) produced mutant lines with reduced flour pigment levels but poor agronomic performance (Knott 1984, 1989b): one (Agatha 28-4) was found to carry a point mutation for the *Psy1*-7el$_1$ gene (Zhang and Dubcovsky 2008), the other (Agatha 235-6) has apparently lost a terminal portion of the original 7el$_1$L segment, also including *Sr25* (Friebe et al. 1994), but retains an intact *Psy1*-7el$_1$ gene. A second *Yp* gene, more proximally located along the T4 7el$_1$L segment, is possibly mutated in Agatha 235-6 (Zhang and Dubcovsky 2008). In a recent study (Rosewarne et al. 2015), Australian adapted genetic backgrounds carrying the original T4 translocation or each of Knott's two mutant T4's, were trialled over a wide range of Australian environments and growth conditions. The effects of T4 and mutant 28-4 on yield in the different sites were similar to those of previous reports by CIMMYT, indicating they provide a yield advantage in high yielding, particularly non-moisture-stressed, environments. For both lines the percentage effect of the translocation in comparison to non-T4 lines increased as the site yield increased. By contrast, the mutant 235-6 translocation was not found to provide a yield increase, but rather to cause depression of several yield-related traits even under water-favourable conditions, which suggests the loss of yield contributing gene(s), besides that yellow pigmentation and *Sr25* resistance, in its rearranged 7el$_1$L segment.

Additional white-endosperm, T4 derivatives were also obtained by gamma-ray irradiation (Marais 1992a), and via *ph1b*-induced recombination (Marais 1992b; Prins et al. 1997; Marais et al. 2001; Somo et al. 2014). Through the latter strategy, repeated rounds of homoeologous recombination eventually resulted in production of recombinants with the desired endosperm phenotype, but with their differently sized alien segments relocated from the wheat 7DL arm to 7BL. This event was accompanied by structural aberrations in the region of exchange, which, even in the absence of the *Sd* gene (named *Sd1*, see ahead), might be partly responsible for some detrimental effects on the breeding performance of the carrier lines. However, one recombinant line showed to have the aberrant region replaced by normal wheat chromatin, thus being potentially suitable for breeding exploitation (reviewed in Somo et al. 2014). In all such lines, a second *Sd* gene (named *Sd2*, see Groenewald et al. 2005) was hypothesized to be retained and to cause, similarly to *Sd1*, aberrant segregation of the recombinant chromosomes, particularly through the male germline, as well as negative effects on fertility and seed set. These undesirable *Sd*-associated effects were, however, manifested in heterozygotes only.

In a separate attempt (Zhang et al. 2005), Sears' Transfer#1 primary recombinant line (Sears 1973, 1978) instead of the T4 translocated chromosome was fractionated into various 7D/7el$_1$ secondary recombinant chromosomes. One of the selected recombinants was proved to have lost the very distal *Yp* gene, yet retaining *Lr19* and the proximally associated *Sd1* gene.

Genetic and physical mapping of all the breakpoints present in the several 7D/7el$_1$ translocation/recombinant lines, by means of various molecular markers and genes/phenotypes assigned to deletion bins or chromosome segments revealed by GISH (e.g., Groenewald et al. 2005), showed substantial synteny and colinearity between the *Th. ponticum* and wheat group 7 chromosomes, and the critical alien genes, with a centromere-*Sd1-Lr19- Yp/Sr25* order, to reside in the distal half of the arm, with the *Lr19*-to-telomere portion spanning about a quarter of it. This type of information is also instrumental to selection of the most suited lines for breeding applications, i.e., in principle, those that possess the minimum amount of alien chromatin exceeding the gene(s) of interest. Such a selection criterion becomes particularly stringent when the recipient species is durum wheat, markedly less tolerant to genic and chromosomal imbalances than hexaploid bread wheat (reviewed in Ceoloni et al. 2005a). For the same reason, chromosome engineering in this *Triticum*species poses more difficulties and has provided, as a whole, a relatively limited number of transfer lines of breeding value (Ceoloni et al. 2014b).

Nonetheless, the *Lr19* + *Sr25* + *Yp* association looked very appealing for a multitargeted improvement of the durum wheat crop. Therefore, a chromosome engineering strategy, based on the use of the *ph1c* mutant of durum cv. Creso (Giorgi 1983), was specifically targeted to the concomitant incorporation of the three genes into durum wheat (Ceoloni et al. 2005b). Through this approach, assisted by proper selection tools (molecular markers, FISH/GISH), the excessive length of $7el_1$ chromatin present on chromosome 7A of a primary recombinant line (Fig. 11.2b) was sufficiently shortened to give secondary and tertiary recombinant types still carrying all target genes and being well tolerated by the recipient durum genome (Fig. 11.2c, d; Ceoloni et al. 2005b, 2014a). From one of them, named R5-2-10 (Fig. 11.2c), with its 23 % of distal $7el_1L$ including *Lr19* + *Sr25* + *Yp* but no *Sd* gene, exhibiting very good agronomic and quality performance (Gennaro et al. 2003, 2007), a variety was released in Italy in 2010 with the name of Cincinnato.

The 7AL-$7el_1L$ durum wheat recombinants, for several of which near-isogenic lines (NILs) have been obtained, represented a highly valuable tool to carry out a variety of studies, from integrated genetic and physical mapping of the 7L critical arms, with assignment of numerous markers and genes to several 7L sub-regions, to the analysis of some structural and functional characteristics associated with defined $7el_1L$ portions. Among the most meaningful *Th. ponticum* traits that could be detected and precisely evaluated are traits with a great potential to positively impact on yield. Effects on yield parameters started to be associated to $7el_1L$ introgression following field trials of NILs carrying the T4 translocation introduced by CIMMYT into the cultivar Oasis 86 and various other bread wheat cultivars (Singh et al. 1998; Reynolds et al. 2001; Monneveux et al. 2003). In all backgrounds, 10–15 % increase in biomass, grain yield and grain number/spike was observed compared to controls lacking the translocation. Through CIMMYT lines, such material has been extensively used in many breeding programs worldwide. However, no precise notion was available on the location along the sizable $7el_1L$ segment of the multiple genes/QTL likely underlying the yield-contributing traits. NILs of some of the durum wheat-Th. *ponticum* recombinants developed in the background of the well adapted but rust-susceptible cv. Simeto, helped in gaining this knowledge. NILs of recombinant lines R5-2-10, R112-4 and R23-1, carrying 23, 28 and 40 % distal $7el_1L$, respectively, on the corresponding 7AL arms (Fig. 11.2c, d), have been field-tested in a

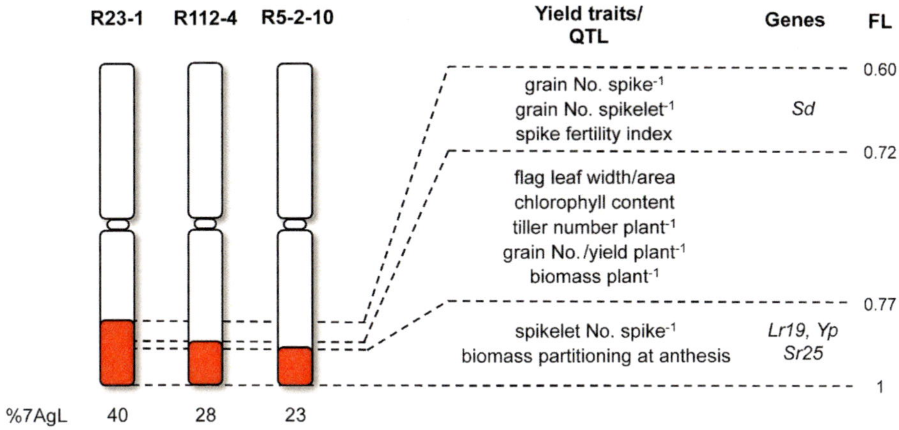

Fig. 11.3 Structural–functional dissection of 7A-7Ag chromosomes representing three durum wheat-*Th. ponticum* recombinant lines. Within the 7AgL regions defined by the different 7AL-7AgL breakpoints in the three chromosome types, known genes (*Lr19*, *Sr25*, *Yp*, *Sd*), and putative genes/QTL for yield-contributing traits could be assigned (adapted from Kuzmanovic et al. 2014 and unpublished). *White background*=7A, *red background*=7AgL. *FL* fractional arm length of the breakpoint to centromere distance, from 0=centromere to 1=telomere

central Italy environment where durum wheat is typically grown. Significant and differential impacts on relevant yield traits could be detected in homozygous carriers (HOM+) vs. non-carriers (HOM−) of the different $7el_1L$ segments (Ceoloni et al. 2014b; Kuzmanovic et al. 2014). $7el_1L$ genes/QTL determining positive effects on grain and tiller number, grain and biomass yield, flag leaf dimensions and chlorophyll content, were found to be all located within the distal 40 % of the alien arm. Most of them turned out to reside within the 28 %-long $7el_1L$ segment of line R112-4, except for the locus/i controlling grain number per spike and per spikelet, present in the most proximal $7el_1L$ portion of the total 40 % assayed, specific to R23-1 (Fig. 11.3).

Many of the valuable loci detected and confirmed across years appear to be readily exploitable in durum wheat breeding, as they are included in the R112-4 $7el_1L$ segment that has no undesirable linkage drag for the recipient species. Less straightforward may turn out to be the harnessing of the grain number per spike/spikelet locus, in most instances found to be associated with considerably decreased kernel weight and, at times, also depending on the wheat genotype, with other abnormal plant phenotypes. These have been tentatively attributed to the presence, in the same R23-1-specific $7el_1L$ segment, of a *Sd* gene (probably *Sd1*, see also above), determining in durum wheat backgrounds a more or less severe segregation distortion of such a recombinant chromosome, always in the direction of self-elimination through the male germline (Ceoloni et al. 2014a).

Taking advantage of the $7el_1$-$7el_2$ homologous relationships, via homologous recombination in hybrids between translocation/recombinant lines carrying the two long arms or portions of them, useful genes originally present in either one have been recently assembled in novel combinations in both bread wheat (Shen and Ohm 2007; Zhang et al. 2011a; Forte et al. 2014) and durum wheat backgrounds (Forte et al. 2014). Thus, both species have been equipped not only with *Lr19* + *Yp* + *Sr25* and yield-contributing QTLs of $7el_1L$ origin, but also with a $7el_2L$-derived effective FHB resistance gene/QTL (named *Fhb-7el*$_2$ in Forte et al. 2014). The latter can considerably reinforce resistance to a disease which, favoured by climate changes, has recently become a threat also in unusual environments and toward largely susceptible species, such as durum wheat. In both transfer schemes, the bread wheat $7DS.7el_2L$ centric translocation line (KS24-1 genotype, see above) was employed as donor of *Fhb-7el*$_2$, while, depending on the target wheat species, the bread wheat T4 translocation onto 7D or durum wheat recombinant lines onto 7A were used as source of the $7el_1L$ genes. In contrast to the nearly complete pairing and recombination between the $7el_1L$ and $7el_2L$ portions shared by the two cross parents of the bread wheat transfer, in *T. aestivum* × *T. durum* pentaploid F_1s considerable reduction in pairing and recombination frequency was observed, mostly ascribable to the location of the parental 7elL portions on otherwise homologous 7D chromosomes vs. 7D and 7A homoeologs in the two F_1 types. Nevertheless, pyramiding of target genes/QTL from the two *Th. ponticum* accessions into both wheat species, greatly assisted by use of advanced selection (MAS) and characterization (GISH) tools, was successfully achieved. Such materials, already exhibiting highly satisfactory agronomic performance (Forte et al. 2014), are expected to lead in the short-term run to durum and also bread wheat cultivars with potential to greatly contribute to enhanced security and safety of the wheat crop.

A highly significant example of multiple *Th. ponticum* gene introgressions with proved positive impact on breeding is that of the Chinese cultivar Xiaoyan 6, in which at least two wheat chromosomes (2A and 7D) carry chromosomal segments from *Th. ponticum* with genes contributing durable resistance to stripe rust and *Septoria tritici* blotch, wide environmental tolerance to high temperature, strong light, and hot-dry winds during grain-filling stages (Li et al. 2008 and references therein). Because of its superior performance under both drought and irrigated conditions, its strong yield stability under diverse environments and good bread-making quality, Xiaoyan 6 was cultivated as the main variety in Shaanxi Province for 16 years (1980–1995). Moreover, Xiaoyan 6 has been used as a core parent for wheat breeding in China in the past 20 years, with more than 50 wheat varieties apparently involving it in their pedigree. These varieties have been grown in more than 20 million hectares, and increased wheat grain production by 7.5 billion kg in total. Thus, Xiaoyan 6 provides an unparalleled example of successful wheat breeding through segmental alien introgression.

11.3.3.2 Additional Introgressions with Actual or Potential Impact on Wheat Improvement

Additional wheat-*Thinopyrum* derivatives have been used in breeding programs to combat major wheat production constraints. One such case concerned the Mayoor and Chirya lines, selected from the intergeneric combination between *T. aestivum* and *Th. curvifolium* (the latter containing two modified E or J^e/J^bgenomes, see Table 11.1; see also Liu and Wang 1993; Wang 2011), used to develop wheat varieties resistant to the spot blotch disease, caused by *Bipolaris sorokiniana*, in various parts of the world, particularly Central-Southern America and South Asia (Duveiller et al. 1998; Mujeeb-Kazi et al. 2013). Another interesting case is that of a wheat-*Th. disticum* derivative (the latter with a similar genome to that of *Th. curvifolium*, see Liu and Wang 1993), able to confer remarkable salt tolerance to Pakistani and Mexican varieties (see Mujeeb-Kazi et al. 2013).

A potentially appealing resource for wheat breeders has been recently produced which combines genes from two *Thinopyrum* species, namely the already mentioned *Lr19* + *Sr25* from $7el_1L$ of *Th. ponticum*, and the *Bdv2* gene from the closely homoeologous arm 7AiL of *Th. intermedium* (Ayala-Navarrete et al. 2013). Donors of the respective genes to the recombinant translocations were T4 mutant 28-4 (lacking the *Yp* gene, see above), with 70 % of wheat 7DL replaced by *Th. ponticum* chromatin, and TC14, with 20 % distal 7DL replaced by 7AiL (Ayala-Navarrete et al. 2009). Thus, from recombination in a *ph1b* background, desired recombinants were selected by MAS, containing all genes within a single block of alien chromatin as short as the distal 20 % of 7DL chromosome arm (Ayala-Navarrete et al. 2013).

Of similarly limited size was found to be the *Th. elongatum* 3EL segment replacing the most distal 3AL portion in the best of a series of recombinant lines (named 524–568) developed to improve bread wheat salt tolerance (Mullan et al. 2009). Thanks to high-resolution genetic and physical mapping of wheat-alien chromosome breakpoints, coupled with phenotypic analysis of the 3D-3E and 3A-3E recombinant set, the gene(s) responsible for the Na^+ “exclusion” mechanism, previously ascribed to the alien 3E chromosome, turned out to be confined to the distal end of 3EL. This makes line 524–568 a good candidate to obviate or at least alleviate yield penalties caused by high accumulation of Na^+ that modern commercial wheat varieties experience under salinity stress. Interestingly, a comparable level of 3E-wheat group 3 homoeologous recombination was detected in the absence of the wheat *Ph1* gene and under the pairing promoting effect of gene(s) on *Th. elongatum* chromosome 3E, combined with absence of other *Ph* wheat genes on chromosomes 3A and 3D, as in 3E(3A) and 3E(3D) substitution lines (Mullan et al. 2009).

Also in a normal *Ph1* background of the cross progeny to wheat of a wheat-*Th. bessarabicum* amphiploid, a T2JS-2BS·2BL translocation line was recovered (Qi et al. 2010). This, when compared to normal wheat lines, showed enhancement of many yield-related traits, namely more fertile spikes per plant, longer spikes, more grains per spike, and higher yield per plant. At the same time, the quite syntenic replacement included the wheat dominant allele for the *Ppd-B1*, controlling photoperiod response and determining early heading date (Snape et al. 2001). Thus, the T2JS-2BS.2BL translocation line headed considerably later than control lines. However, the size of the 2JS segments could be further engineered to eliminate the undesirable trait(s), and only exploit the positive yield attributes.

Transfers onto wheat of subchromosomal segments from perennial Triticeaegenomes more distantly related to the former than those of *Thinopyrum* spp. have been documented in recent years. One interesting example concerns tetraploid crested wheatgrass, *Agropyroncristatum* (Table 11.1), which, besides genes for disease resistance and stress tolerance, was shown to harbour genes/QTL able to enhance wheat yield performance. Already in addition lines (Wu et al. 2006; Han et al. 2014; see Sect. 11.3.2), and, in translocations lines later obtained (Luan et al. 2010), several useful genes were allocated to *A. cristatum* chromosome 6P; Song et al. 2013; Ye et al. 2015). In particular, genes/QTL enhancing relevant traits such as number of fertile tillers and of grains per spike could be ascribed to specific 6P subregions, and were expressed in translocation lines with minimal amount of 6P chromatin (Luan et al. 2010; Ye et al. 2015). Precise characterization of these lines, including identification of 6P-specific markers, provides a good basis for the utilization of multiple *A. cristatum* genes in wheat improvement.

The same applies to a highly effective stripe rust resistance gene present on chromosome arm 3NsS of *Psathyrostachys huashanica* (Table 11.1), an endemic and endangered wild species in China, stably transferred onto wheat arm 3BS (Kang et al. 2011). The 3BL.3BS-3NsS translocation had a spontaneous occurrence in the selfed progeny of a 3Ns monosomic addition line into wheat. Of spontaneous occurrence were also numerous translocations detected in advanced backcross/selfed generations from a *T. aestivum* x *Elymusrepens* cross, several of which fully expressed the high FHB resistance of the wild parent (Zeng et al. 2013b).

11.4 Concluding Remarks and Future Prospects

The examples above illustrated provide ample evidence of the richness of perennial Triticeae gene pools in beneficial traits for wheat improvement. Recent progress has also been described for many such traits in the path of their stable incorporation into the wheat genome, and, hence, of their well advanced "state of art" in view of practical exploitation as novel breeding materials. Although the history of distant hybridization involving this large group of wheat relatives goes back to the 1930s (Tsitsin 1960), the major progress, as, indeed, for wheat relatives in general, is concentrated in the very last years, in parallel with remarkable advancements in the amount of information and tools for Triticeae genomics and related fields of knowledge. Not only these extraordinary developments have speeded up and made much more effective characterization and selection procedures, but also have considerably increased the ability to capture novel variability for much more complex traits, e.g., yield-related traits, than those almost exclusively targeted in past years. Thus, the answer is now ready to Cox's question of some years ago (Cox 1997), concerning the perspectives for deepening the wheat primary gene pool. He, in fact, while appreciating the fact that "humans have resorted to interspecific crossing to improve wheat's pest resistance," at the same time wondered "...why should wheat's progenitors not be regarded as sources of useful genetic variation for all economic traits?" The rather obvious answer is in the above statements, that is to say we now can manipulate genomes with knowledge and tools virtually inconceivable just a few years ago. As a result, also the overall success of cytogenetic strategies in finely engineering the genomes of cultivated wheat species has been greatly enhanced. Thus, although the route from the trait identification stage to the output of practical products, ending with varietal release, may well remain relatively long and complex, a holistic and highly effective approach is nowadays applicable; no doubt, this will help to better cope with current and future environmental and social changes, and to meet the requirements for food security and safety sustainably.

In a broader and longer-run perspective, other approaches are expected to complement and even succeed in effecting finer manipulations than those currently achievable by chromosome engineering. In this context, disregarding a number of limitations, the transgenic approach, or perhaps better, where applicable, its cisgenic "version" (Holme et al. 2013), has the advantage of avoiding "linkage drag" potentially associated with cytogenetics-based manipulations. A cisgenesis-mediated transfer of a gene coding for a MYB transcription factor isolated from *Th. intermedium* and conferring to the transformed wheat enhanced resistance to the take-all disease was recently described (Liu et al. 2013b). The number of "wild" genes available for this kind of transfer is expected to rapidly increase in a short-term run, as whole genome sequences of the wild donors of the A and D genomes to polyploidy wheats are already available (Jia et al. 2013; Ling et al. 2013). Further, to preferentially target genic regions and greatly reduce the repetitive sequence content of large Triticeaegenomes, reduced representation-based approaches are available (Hirsch et al. 2014), and, among them, exome capture has been recently demonstrated to be a powerful approach for variant discovery in cultivated barley and related species (Mascher et al. 2013), as well as in rice and in polyploid wheat species (Saintenac et al. 2011; Winfield et al. 2012; Henry et al. 2014). For species with large genome size, such as cultivated wheats and the many polyploid perennial taxa discussed in the present context, an exome-wide basis can be cost-effective and high-throughput. In fact, even if a reference sequence is not available, a transcriptome assembly can be performed at relatively low cost and can serve as a starting point for the design of capture targets for any species (Henry et al. 2014). Furthermore, progress in chromosome genomics, i.e., genome analysis using chromosome-based approaches, allowed by high-throughput flow sorting technology and recently empowered by the high discriminatory capabilities of FISH labeling (Giorgi et al. 2013), enables isolation of pure chromosome fractions from virtually any genome, including the complex ones of polyploid wheats and of related Triticeae species (reviewed in Doležel et al. 2014; see also Chap. 13 in this book). Thus, specific alien chromosomes, or even recombinant/translocated wheat-alien chromosomes as those described in Sect. 11.3.3, can be the "substrate" for a highly focused exome capture or other gene-targeted approaches. Desirable variant alleles from a given donor species thus identified, can, on one hand, be made available for cisgenic manipulations, although these will retain their inherent drawback of random insertion and unpredictable expression of the delivered gene in the recipient genome. On the other hand, they can also serve as externally supplied DNA templates for targeted gene/allele correction or replacement via recent "genome editing" approaches, holding good promise also in a breeding perspective (reviewed in Podevin et al. 2013).

In conclusion, there is ample consensus on the absolute need that intensification of agricultural production necessary to "close the yield gap," particularly crucial in the wheat case, has to imply a package of measures that only a changing landscape of breeding strategies can successfully harness. Technical or other types of limitations will have to be overcome before the latest ground-breaking strategies will be in place. Fortunately, however, several novel wheat materials, many reinforced with valuable genes from perennial Triticeae, have been developed in which yet untapped potential for breeding gains has been effectively unlocked, and many more are expected to be readily available in the near future. This effort, coupled with the ever paramount contribution of traditional breeding to incorporate the beneficial transfers into elite and adapted cultivars, will hopefully soften the highly complex scenario threatening mankind's staple wheat crop.

References

Abdel-Aal EM, Young JC, Rabalski I (2006) Anthocyanin composition in black, blue, pink, purple, and red cereal grains. J Agric Food Chem 54:4696–4704

Anamthawat-Jónsson K (2014) Molecular cytogenetics of *Leymus*: mapping the Ns genome specific repetitive sequences. J Syst Evol 52:716–721

Anderson JM, Bucholtz DL, Sardesai N, Santini JB, Gyulai G, Williams CE, Stephen B, Goodwin SB (2010) Potential new genes for resistance to *Mycosphaerella graminicola* identified in Triticum aestivum-*Lophopyrum elongatum* disomic substitution lines. Euphytica 172:251–262

Ayala-Navarrete L, Tourton E, Mechanicos AA, Larkin PJ (2009) Comparison of *Thinopyrum intermedium* derivatives carrying barley yellow dwarf virus resistance in wheat. Genome 52:537–546

Ayala-Navarrete L, Mechanicos AA, Gibson JM, Singh D, Bariana HS, Fletcher J, Shorter S, Larkin PJ (2013) The *Pontin* series of recombinant alien translocations in bread wheat: single translocations integrating combinations of *Bdv2*, *Lr19* and *Sr25* disease-resistance genes from *Thinopyrum intermedium* and *Th. ponticum*. Theor Appl Genet 126:2467–2475

Bakshi JS, Schlehuber AM (1959) Identification of a substituted chromosome pair in a *Triticum-Agropyron* line. Proc Okla Acad Sci 1958:16–21

Banks PM, Larkin PJ, Bariana HS, Lagudah ES, Appels R, Waterhouse PM, Brettell RIS, Chen X, Xu HJ, Xin ZY, Qian YT, Zhou XM, Cheng ZM, Zhou GH (1995) The use of cell culture for subchromosomal introgressions of barley yellow dwarf virus resistance from *Thinopyrum intermedium* to wheat. Genome 38:395–405

Bariana HS, Brown GN, Bansal UK, Miah H, Standen GE, Lu M (2007) Breeding triple rust resistant wheat cultivars for Australia using conventional and marker-assisted selection technologies. Aust J Agric Res 58:576–587

Barkworth ME, von Bothmer R (2009) Scientific names in the Triticeae. In: Feuillet C, Muehlbauer GJ (eds) Genetics and genomics of the Triticeae. Part I – genetics of the Triticeae. Springer, New York, NY, pp 3–30. doi:10.1007/978-0-387-77489-3

Barkworth ME, Cutler DR, Rollo JS, Jacobs SWL, Rashid A (2009) Morphological identification of genomic genera in the Triticeae. Breed Sci 59:561–570. doi:10.1270/jsbbs.59.561

Bell LW, Wade LJ, Ewing MA (2010) Perennial wheat: a review of environmental and agronomic prospects for development in Australia. Crop Pasture Sci 61:679–690

Cai X, Jones S (1997) Direct evidence for high level of autosyndetic pairing in hybrids of *Thinopyrum intermedium* and *Th. ponticum* with *Triticum aestivum*. Theor Appl Genet 95:568–572

Ceoloni C, Jauhar PP (2006) Chromosome engineering of the durum wheat genome: strategies and applications of potential breeding value. In: Singh RJ, Jauhar PP (eds) Genetic resources, chromosome engineering, and crop improvement: cereals. CRC Press, Boca Raton, FL, pp 27–59

Ceoloni C, Pasquini M, Simeone R (2005a) The cytogenetic contribution to the analysis and manipulation of the durum wheat genome. In: Royo C, Nachit MN, Di Fonzo N, Araus JL, Pfeiffer WH, Slafer GA (eds) Durum wheat breeding: current approaches and future strategies. Haworth Press, New York, NY, pp 165–196

Ceoloni C, Forte P, Gennaro A, Micali S, Carozza R, Bitti A (2005b) Recent developments in durum wheat chromosome engineering. Cytogenet Genome Res 109:328–344

Ceoloni C, Kuzmanović L, Gennaro A, Forte P, Giorgi D, Grossi MR, Bitti A (2014a) Genomes, chromosomes and genes of perennial Triticeae of the genus *Thinopyrum*: the value of their transfer into wheat for gains in cytogenomic knowledge and 'precision' breeding. In: Tuberosa R, Graner A, Frison E (eds) Advances in genomics of plant genetic resources. Springer Science, Dordrecht, pp 333–358. doi:10.1007/978-94-007-7575-6_14

Ceoloni C, Kuzmanović L, Forte P, Gennaro A, Bitti A (2014b) Targeted exploitation of gene pools of alien Triticeae species for sustainable and multi-faceted improvement of the durum wheat crop. Crop Pasture Sci 65:96–111

Chang ZJ, Zhang XJ, Yang ZJ, Zhan HX, Li X, Liu C, Zhang CZ (2010) Characterization of a partial wheat–*Thinopyrum intermedium* amphiploid and its reaction to fungal diseases of wheat. Hereditas 147:304–312

Chaudhary HK, Kaila V, Rather SA, Badiyal A, Hussain W, Jamwal NS, Mahato A (2014) Wheat. In: Pratap A, Kumar J (eds) Alien gene transfer in crop plants, vol 2: achievements and impacts. Springer, New York, NY. doi:10.1007/978-1-4614-9572-7_1

Chen Q (2005) Detection of alien chromatin introgression from *Thinopyrum* into wheat using S genomic DNA as a probe – a landmark approach for *Thinopyrum* genome research. Cytogenet Genome Res 109:350–359

Chen Q, Conner RL, Laroche A, Thomas JB (1998) Genome analysis of *Thinopyrum intermedium* and *Thinopyrum ponticum* using genomic in situ hybridization. Genome 41:580–586

Colmer TD, Flowers TJ, Munns R (2006) Use of wild relatives to improve salt tolerance in wheat. J Exp Bot 57:1059–1078

Cox TS (1997) Deepening the wheat gene pool. J Crop Prod 1:1–25

Cox TS, Bender M, Picone C, Van Tassel DL, Holland JB, Brummer CE, Zoeller BE, Paterson AH, Jackson W (2002) Breeding perennial grain crops. Crit Rev Plant Sci 21:59–91

Cox CM, Garrett KA, Bockus WW (2005a) Meeting the challenge of disease management in perennial grain cropping systems. Renew Agr Food Syst 20:15–24

Cox CM, Garrett KA, Cox TS, Bockus WW, Peters T (2005b) Reactions of perennial grain accessions to four major cereal pathogens of the great plains. Plant Dis 89:1235–1240

Cox TS, Glover JD, Van Tassel DL, Cox CM, DeHaan LR (2006) Prospects for developing perennial-grain crops. Bioscience 56:649–659

Cox TS, Van Tassel DL, Cox CM, DeHaan LR (2010) Progress in breeding perennial grains. Crop Pasture Sci 61:513–521

Crasta OR, Francki MG, Bucholtz DB, Sharma HC, Zhang J, Wang RC, Ohm HW, Anderson JM (2000) Identification and characterization of wheat–wheatgrass translocation lines and localization of barley yellow dwarf virus resistance. Genome 43:698–706

De Pace C, Vaccino P, Cionini PG, Pasquini M, Bizzarri M (2011) *Dasypyrum*. In: Kole C (ed) Wild crop relatives: genomic and breeding resources, cereals. Springer, Berlin, pp 185–292

DeHaan LR, Wang S, Larson SR, Cattani DJ, Zhang X, Kantarski T (2014) In: Batello C, Wade L, Cox S, Pogna N, Bozzini A, Choptiany J (eds) Current efforts to develop perennial wheat and domesticate *Thinopyrum intermedium* as a perennial grain. Proceedings of the FAO expert workshop, Rome, Italy, 28–30 Aug 2013, pp 72–89

Deng CL, Bai LL, Fu SL, Yin WB, Zhang YX, Chen YH, Wang RRC, Zhang XQ, Han FP, Hu ZM (2013) Microdissection and chromosome painting of the alien chromosome in an addtition line of wheat-*Thinopyrum intermedium*. PLoS One 8, e72564. doi:10.1371/journal.pone.0072564

Dewey DR (1984) The genomic system of classification as a guide to intergeneric hybridization with the perennial Triticeae. In: Gustafson JP (ed) Gene manipulation in plant improvement. Plenum Publishing Corporation, New York, NY, pp 209–279

Doležel J, Vrána J, Cápal P, Kubaláková M, Burešová V, Šimková H (2014) Advances in plant chromosome genomics. Biotechnol Adv 32:122–136

Dou QW, Lei YT, Li XM, Mott IW, Wang RRC (2012) Characterization of alien grass chromosomes in backcross derivatives of *Triticum aestivum-Elymus rectisetus* hybrids by using molecular markers and multi-color FISH/GISH. Genome 55:337–347

DRRW—Durable Rust Resistance in Wheat. http://wheatrust.cornell.edu, Project impacts 2011: Appendix F—Project-Wide Impact Metrics and Varieties Released November 2011.

Dundas IS, Shepherd KW (1998) Shortening the Agropyron chromosome segment carrying gene Sr26 utilising chromosome engineering and molecular markers. In: Slinkard AE (ed) Proceedings of the 9th international wheat genetics symposium, vol 2. University Extension Press, University of Saskatchewan, Saskatoon, pp 35–37.

Dundas IS, Anugrahwati DR, Verlin DC, Park RF, Bariana HS, Mago R, Islam AKMR (2007) New sources of rust resistance from alien species: meliorating linked defects and discovery. Aust J Agric Res 58:545–549

Dundas I, Zhang P, Verlin D, Graner A, Shepherd K (2015) Chromosome engineering and physical mapping of the *Thinopyrum ponticum* translocation in wheat carrying the rust resistance gene *Sr26*. Crop Sci 55:648–657

Duveiller E, Dubin HJ, Reeves J, McNab A (eds) (1998) Helminthosporium blights of wheat: spot blotch and tan spot. CIMMYT, Mexico, DF. ISBN 970-648-001-3

Dvorak J (1975) Meiotic pairing between single chromosomes of diploid *Agropyron elongatum* and decaploid *A. elongatum* in *Triticum aestivum*. Can J Genet Cytol 17:329–336

Dvorak J (1987) Chromosomal distribution of genes in diploid *Elytrigia elongata* that promote or suppress pairing of wheat homoeologous chromosomes. Genome 29:34–40

Dvorak J, Knott DR (1977) Homoeologous chromatin exchange in a radiation-induced gene transfer. Can J Genet Cytol 19:125–131

Dvorak J, Ross K (1986) Expression of tolerance of Na+, K+, Mg2+, Cl–, and SO2-4 ions and sea water in the amphiploid of *Triticum aestivum* x *Elytrigia elongata*. Crop Sci 26:658–660

Fan X, Zhang HQ, Sha LN, Zhang L, Yang RW, Ding CB, Zhou YH (2007) Phylogenetic analysis among *Hystrix, Leymus* and its affinitive genera (Poaceae: Triticeae) based on the sequences of a gene encoding plastid acetyl-CoA carboxylase. Plant Sci 172:701–707

FAO (2015) RUST SPORE – a global wheat rust monitoring system: rust report – stem (black) rust. http://www.fao.org/agriculture/crops/rust/stem/stem-pathotypetracker/stem-effectivesrgenes/en/. Accessed 4 Mar 2015

Fedak G, Han F (2005) Characterization of derivatives from wheat-*Thinopyrum* wide crosses. Cytogenet Genome Res 109:360–367

Feldman M, Levy AA (2012) Genome evolution due to allopolyploidization in wheat. Genetics 192:763–774

Forte P, Kuzmanovic L, Gennaro A, Bitti A, Ceoloni C (2011) Using wild species of *Thinopyrum* genus in breeding wheat resistant to Fusarium head blight. Proceedings of the joint meeting AGI-SIBV-SIGA, Assisi, Italy, abstract 6A.21. ISBN 978-88-904570-2-9

Forte P, Virili ME, Kuzmanović L, Moscetti I, Andrea Gennaro A, D'Ovidio R, Ceoloni C (2014) A novel assembly of Thinopyrum ponticum genes into the durum wheat genome: pyramiding Fusarium head blight resistance onto recombinant lines previously engineered for other beneficial traits from the same alien species. Mol Breeding 34:1701–1716

Fradkin M, Ferrari MR, Ferreira V, Grassi E, Greizerstein EJ, Poggio L (2012) Chromosome and genome composition of a *Triticum* x *Thinopyrum* hybrid by classical and molecular cytogenetic techniques. Genet Resour Crop Evol 59:231–237

Friebe B, Jiang J, Knott DR, Gill BS (1994) Compensation indices of radiation-induced wheat-Agropyronelongatum translocations conferring resistance to leaf rust and stem rust. Crop Sci 34:400–404

Friebe B, Jiang J, Raupp WJ, McIntosh RA, Gill BS (1996) Characterization of wheat-alien translocations conferring resistance to diseases and pests: current status. Euphytica 91:59–87
Gallagher CE, Matthews PD, Li F, Wurtzel ET (2004) Gene duplication in the carotenoid biosynthetic pathway preceded evolution of the grasses. Plant Physiol 135:1776–1783
Gennaro A, Borrelli GM, D'Egidio MG, De Vita P, Ravaglia S, Ceoloni C (2003) A chromosomally engineered durum wheat-Thinopyrum ponticum recombinant line with novel and promising attributes for varietal development. In: Pogna NE, Romanò M, Pogna EA, Galterio G (eds) Proceeding of the 10th international wheat genetics symposium. S.I.M.I, Rome, Italy, pp 881–883
Gennaro A, Forte P, Carozza R, Savo Sardaro ML, Ferri D, Bitti A, Borrelli GM, D'Egidio MG, Ceoloni C (2007) Pyramiding different alien chromosome segments in durum wheat: feasibility and breeding potential. Isr J Plant Sci 55:267–276
Gennaro A, Koebner RMD, Ceoloni C (2009) A candidate for *Lr19*, an exotic gene conditioning leaf rust resistance in wheat. Funct Integr Genomics 9:325–334
Giorgi B (1983) Origin, behaviour and utilization of a *Ph1* mutant of durum wheat, *Triticum turgidum* (L.) var. *durum*. In: Sakamoto S (ed) Proceedings of the 6th international wheat genetics symposium, Kyoto. Plant Germplasm Institute, Kyoto University, Kyoto, pp 1033–1040
Giorgi D, Farina A, Grosso V, Gennaro A, Ceoloni C, Lucretti S (2013) FISHIS: fluorescence in situ hybridization in suspension and chromosome flow sorting made easy. PLoS One 8, e57994. doi:10.1371/journal.pone.0057994
Gradzielewska A (2006a) The genus *Dasypyrum* – part 1. The taxonomy and relationships within Dasypyrum and with Triticeae species. Euphytica 152:429–440
Gradzielewska A (2006b) The genus *Dasypyrum* – part 2. A wild species used in wheat improvement. Euphytica 152:441–454
Groenewald JZ, Fourie M, Marais AS, Marais GF (2005) Extention and use of a physical map of the *Thinopyrum*-derived *Lr19* translocation. Theor Appl Genet 112:131–138
Han H, Bai L, Su J, Zhang J, Song L, Gao A, Yang X, Li X, Liu W, Li L (2014) Genetic rearrangements of six wheat–*Agropyron cristatum* 6P addition lines revealed by molecular markers. PLoS One 9, e91066
Harlan J, de Wet JMJ (1971) Towards a rationale classification of cultivated plants. Taxon 20:509–517
Hayes RC, Newell MT, DeHaan LR, Murphy KM, Crane S, Norton MR, Wade LJ, Newberry M, Fahim M, Jones SS, Cox TS, Larkin PJ (2012) Perennial cereal crops: an initial evaluation of wheat derivatives. Field Crops Res 133:68–89
He RL, Chang ZJ, Yang ZJ, Yuan ZY, Zhan HX, Zhang XJ, Liu JX (2009) Inheritance and mapping of powdery mildew resistance gene *Pm43* introgressed from *Thinopyrum intermedium* into wheat. Theor Appl Genet 118:1173–1180
Henry IM, Nagalakshmi U, Lieberman MC, Ngo KJ, Krasileva KV et al (2014) Efficient genome-wide detection and cataloguing of EMS-induced mutations using exome capture and next-generation sequencing. Plant Cell 26:1382–1397
Hirsch CD, Evans J, Buell CR, Hirsch CN (2014) Reduced representation approaches to interrogate genome diversity in large repetitive plant genomes. Brief Funct Genomics 13:257–67
Holme IB, Wendt T, Holm PB (2013) Intragenesis and cisgenesis as alternatives to transgenic crop development. Plant Biotechnol J 11:395–407
Hu LJ, Li GR, Zeng ZX, Chang ZJ, Liu C, Zhou JP, Yang ZJ (2011) Molecular cytogenetic identification of a new wheat-*Thinopyrum* substitution line with stripe rust resistance. Euphytica 177:169–177
DRRW – Durable Rust Resistance in Wheat, http://wheatrust.cornell.edu, Project impacts 2011: appendix F – project-wide impact metrics and varieties released, Nov 2011
Jauhar PP (1990) Multidisciplinary approach to genome analysis in the diploid species, *Thinopyrum bessarabicum* and *Th. elongatum* (*Lophopyrum elongatum*), of the Triticeae. Theor Appl Genet 80:523–536

Jauhar PP (1995) Meiosis and fertility of F_1 hybrids between hexaploid bread wheat and decaploid tall wheatgrass (*Thinopyrum ponticum*). Theor Appl Genet 90:865–871

Jauhar PP, Almouslem AB (1998) Production and meiotic analyses of intergeneric hybrids between durum wheat and *Thinopyrum* species. In: Jaradat AA (ed) Proceedings of the 3rd international Triticeae symposium. Scientific Publishers, Enfield, NH, pp 119–126

Jauhar PP, Peterson TS (2000). Toward transferring scab resistance from a diploid wild grass, *Lophopyrum elongatum*, into durum wheat. In: Proceedings of the 2000 National Fusarium Head Blight Forum, Cincinnati, 10–21 Dec 2000, pp 201–204

Jauhar PP, Peterson TS (2011) Cytological and molecular characterization of homoeologous group-1 chromosomes in hybrid derivatives of a durum disomic alien addition line. Plant Genome 4:102–109

Jauhar PP, Peterson TS, Xu SS (2009) Cytogenetic and molecular characterization of a durum alien disomic addition line with enhanced tolerance to Fusarium head blight. Genome 52:467–483

Jia J, Zhou R, Li P, Zhao M, Dong Y (2002) Identifying the alien chromosomes in wheat-*Leymus multicaulis* derivatives using GISH and RFLP techniques. Euphytica 127:201–207

Jia J, Zhao S, Kong X, Li Y, Zhao G et al (2013) Aegilops tauschii draft genome sequence reveals a gene repertoire for wheat adaptation. Nature 496:91–95

Jiang J, Friebe B, Gill BS (1994) Recent advances in alien gene transfer in wheat. Euphytica 73:199–212

Jones TA, Zhang XY, Wang RRC (1999) Genome characterization of MT-2 perennial and OK-906 annual wheat x intermediate wheatgrass hybrids. Crop Sci 39:1041–1043

Kang HY, Zhang HQ, Wang Y, Jiang Y, Yuan HJ, Zhou YH (2008) Comparative analysis of the homoeologous pairing effects of *phKL* gene in common wheat x *Psathyrostachys huashanica* Keng ex Kuo. Cereal Res Commun 36:429–440

Kang H, Wang Y, Fedak G, Cao W, Zhang H, Fan X, Sha L, Xu L, Zheng Y, Zhou Y (2011) Introgression of chromosome 3Ns from *Psathyrostachys huashanica* into wheat specifying resistance to stripe rust. PLoS One 6, e21802. doi:10.1371/journal.pone.0021802

Kazi AG (2011) Utilization of Triticeae gene pool diversity for wheat improvement. Ph.D. dissertation, Quaid-i-Azam University, Islamabad, Pakistan, pp 1–240

Khan IA (2000) Molecular and agronomic characterization of wheat-*Agropyron intermedium* recombinant chromosomes. Plant Breeding 119:25–29

Kibiridge-Sebunya I, Knott DR (1983) Transfer of stem rust resistance to wheat from an *Agropyron* chromosome having a gametocidal effect. Can J Genet Cytol 25:215–221

Kim NS, Armstrong K, Knott DR (1993) Molecular detection of *Lophopyrum* chromatin in wheat–*Lophopyrum* recombinants and their use in the physical mapping of chromosome 7D. Theor Appl Genet 85:561–567

King IP, Forster BP, Law CC, Cant KA, Orford SE, Gorham J, Reader S, Miller TE (1997a) Introgression of salt-tolerance genes from *Thinopyrum bessarabicum* into wheat. New Phytol 137:75–81

King IP, Law CN, Cant KA, Orford SE, Reader SM, Miller TE (1997b) Tritipyrum, a potential new salt-tolerant cereal. Plant Breeding 116:127–132

Kishii M, Wang RRC, Tsujimoto H (2005) (GISH analysis revealed new aspect of genomic constitution of *Thinopyrum intermedium*. Czec J Genet Plant Breed 41:92–95

Knott DR (1961) The inheritance of rust resistance. VI. The transfer of stem rust resistance from *Agropyron elongatum* to common wheat. Can J Plant Sci 41:109–123

Knott DR (1964) The effect on wheat of an *Agropyron* chromosome carrying rust resistance. Can J Genet Cytol 6:500–507

Knott DR (1968) Translocations involving *Triticum* chromosomes and *Agropyron* chromosomes carrying rust resistance. Can J Genet Cytol 10:695–696

Knott DR (1980) Mutation of a gene for yellow pigment linked to *Lr19* in wheat. Can J Genet Cytol 22:651–654

Knott DR (1984) The genetic nature of mutations of a gene for yellow pigment linked to *Lr19* in 'Agatha' wheat. Can J Genet Cytol 26:392–393

Knott DR (1989a) Genetic analysis of resistance. In: The wheat rusts—Breeding for resistance. Springer, Berlin, pp 58–83.

Knott DR (1989b) The effect of transfers of alien genes for leaf rust resistance on the agronomic and quality characteristics of wheat. Euphytica 44:65–72

Knott DR, Dvorak J, Nanda JS (1977) The transfer to wheat and homoeology of an *Agropyron elongatum* chromosome carrying resistance to stem rust. Can J Genet Cytol 19:75–79

Kumar J, Sivasamy M, Nisha R (2010) Performance of brown and black rust resistance genes in some wheat cultivars of central, peninsular, and south India. Ann Wheat Newsl 56:80–82

Kuzmanovic L, Gennaro A, Benedettelli S, Dodd IC, Quarrie SA, Ceoloni C (2014) Structural-functional dissection and characterization of yield-contributing traits originating from a group 7 chromosome of the wheatgrass species *Thinopyrum ponticum* after transfer into durum wheat. J Exp Bot 65:509–525

Lammer D, Cai X, Arterburn M, Chatelain J, Murray T, Jones S (2004) A single chromosome addition from Thinopyrum elongatum confers a polycarpic, perennial habit to annual wheat. J Exp Bot 55:1715–1720

Larkin PJ, Newell MT (2014) Perennial wheat breeding: current germplasm and a way forward for breeding and global cooperation. In: Batello C, Wade L, Cox S, Pogna N, Bozzini A, Choptiany J (eds) Perennial crops for food security – proceedings of the FAO expert workshop. Rome, Italy, 28–30 Aug 2013, pp 39–53

Larkin PJ, Newell MT, Hayes RC, Aktar J, Norton MR, Moroni SJ, Wade LJ (2014) Progress in developing perennial wheats for grain and grazing. Crop Pasture Sci 65:1147–1164

Larson SR, Kishii M, Tsujimoto H, Qi LL, Chen PD, Lazo GR, Jensen KB, Wang RRC (2012) *Leymus* EST linkage maps identify 4NsL–5NsL reciprocal translocation, wheat-*Leymus* chromosome introgressions, and functionally important gene loci. Theor Appl Genet 124:189–206

Larter EN (1976) Triticale. In: Simmonds NW (ed) Evolution of crop plants. Longman Group, London, pp 117–120

Li HJ, Wang XM (2009) *Thinopyrum ponticum* and *Th. intermedium*: the promising source of resistance to fungal and viral diseases of wheat. J Genet Genomics 36:557–565

Li D, Zhang XY (2002) Physical localization of the 18S-5.8S-26S rDNA sequence analysis of ITS regions in *Thinopyrum ponticum* (Poaceae: Triticeae): implications for concerted evolution. Ann Bot 90:445–452

Li ZS, Mu SM, Zhou HP, Wu J (1986) The establishment and application of blue-grained monosomics in wheat chromosome engineering. Cereal Res Commun 14:133–137

Li H, Chen Q, Conner RL, Guo B, Zhang Y, Graf RJ, Laroche A, Jia X, Liu G, Chu C (2003) Molecular characterization of a wheat–*Thinopyrum ponticum* partial amphiploid and its derivatives for resistance to leaf rust. Genome 46:906–913

Li HJ, Arterburn M, Jones SS, Murray TD (2005) Resistance to eyespot of wheat, caused by *Tapesia yallundae*, derived from *Thinopyrum intermedium* homoeologous group 4 chromosome. Theor Appl Genet 111:932–940

Li Z, Li B, Tong Y (2008) The contribution of distant hybridization with decaploid *Agropyron elongatum* to wheat improvement in China. J Genet Genomics 35:451–456

Li GR, Liu C, Li CH, Zhao JM, Zhou L, Dai G, Yang EN, Yang ZJ (2013) Introgression of a novel *Thinopyrum intermedium* St-chromosome-specific HMW-GS gene into wheat. Mol Breeding 31:843–853

Li GR, Zhao JM, Li DH, Yang EN, Huang YF, Liu C, Yang ZJ (2014) A novel wheat-*Dasypyrum breviaristatum* substitution line with stripe rust resistance. Cytogenet Genome Res 143:280–287

Ling HQ, Zhao S, Liu D, Wang J, Sun H et al (2013) Draft genome of the wheat A-genome progenitor Triticum urartu. Nature 496:87–90

Liu ZW, Wang RRC (1993) Genome constitution of *Thinopyrum curvifolium*, *T. scirpeum*, *T. distichum*, and *T. junceum* (Triticeae: Gramineae). Genome 36:641–651

Liu ZW, Wang RRC, Carman JG (1994) Hybrids and backcross progenies between wheat (*Triticum aestivum* L.) and apomictic Australian wheatgrass [*Elymus rectisetus* (Nees in Lehm.) A. Löve and Connor]: karyotypic and genomic analyses. Theor Appl Genet 89:599–605
Liu Z, Li DY, Zhang XY (2007) Genetic relationships among five basic genomes St, E, A, B and D in Triticeae revealed by genomic southern and in situ hybridization. J Integr Plant Biol 49:1080–1086
Liu S, Gao X, Guangmin X (2008) Characterizing HMW-GS alleles of decaploid Agropyron elongatum in relation to evolution and wheat breeding. Theor Appl Genet 116:325–334
Liu S, Yu L-X, Singh RP, Jin Y, Sorrells ME, Anderson JA (2010) Diagnostic and co-dominant PCR markers for wheat stem rust resistance genes *Sr25* and *Sr26*. Theor Appl Genet 120:691–697
Liu C, Li G, Yan H, Zhou J, Hu L, Lei M, Ran L, Yang Z (2011) Molecular and cytogenetic identification of new wheat-*Dasypyrum breviaristatum* additions conferring resistance to stem rust and powdery mildew. Breed Sci 61:366–372
Liu W, Danilova T, Rouse M, Bowden R, Friebe B, Gill BS, Pumphrey M (2013a) Development and characterization of a compensating wheat-*Thinopyrum intermedium* Robertsonian translocation with *Sr44* resistance to stem rust (Ug99). Theor Appl Genet 126:1167–1177
Liu X, Yang L, Zhou X, Zhou M, Lu Y, Ma L, Maand H, Zhang Z (2013b) Transgenic wheat expressing *Thinopyrum intermedium* MYB transcription factor TiMYB2R-1 shows enhanced resistance to the take-all disease. J Exp Bot 64:2243–2253. doi:10.1093/jxb/ert084
Löve A (1984) Conspectus of the Triticeae. Feddes Repert 95:425–521
Luan Y, Wang XG, Liu WH, Li CY, Zhang JP, Gao AN, Wang YM, Yang XM, Li LH (2010) Production and identification of wheat-*Agropyron cristatum* 6P translocation lines. Planta 232:501–510
Luo PG, Luo HY, Chang ZJ, Zhang HY, Zhang M, Ren ZL (2009) Characterization and chromosomal location of *Pm40* in common wheat: a new gene for resistance to powdery mildew derived from *Elytrigia intermedium*. Theor Appl Genet 118:1059–1064
Ma XF, Gustafson JP (2008) Allopolyploidization-accommodated genomic sequence changes in triticale. Ann Bot 101:825–832
Mahelka V, Kopecky D (2010) Gene capture from across the grass family in the allohexaploid *Elymus repens* (L.) Gould (Poaceae, Triticeae) as evidenced by ITS, GBSSI, and molecular cytogenetics. Mol Biol Evol 27:1370–1390
Mahelka V, Kopecky D, Pastova L (2011) On the genome constitution and evolution of intermediate wheatgrass (*Thinopyrum intermedium*: Poaceae, Triticeae). Mol Biol Evol 30:2065–2086
Mahelka V, Kopecky D, Baum BR (2013) Contrasting patterns of evolution of 45S and 5S rDNA families uncover new aspects in the genome constitution of the agronomically important grass *Thinopyrum intermedium* (Triticeae). BMC Evol Biol 11:127
Mahmood A, Quarrie SA (1993) Effects of salinity on growth, ionic relations and physiological traits of wheat, disomic addition lines from *Thinopyrum bessarabicum*, and two amphiploids. Plant Breeding 110:265–276
Marais GF (1992a) Gamma-irradiation induced deletions in an alien chromosome segment of the wheat 'Indis' and their use in gene mapping. Genome 35:225–229
Marais GF (1992b) The modification of a common wheat-*Thinopyrum distichum* translocated chromosome with a locus homoeoallelic to *Lr19*. Theor Appl Genet 85:73–78
Marais GF, Marais AS, Groenwald JZ (2001) Evaluation and reduction of *Lr19*-149, a recombined form of the *Lr19* translocation of wheat. Euphytica 121:289–295
Marais F, Simsek S, Poudel RS, Cookman D, Mohamed Somo M (2014) Development of hexaploid (AABBJJ) Tritipyrums with rearranged *Thinopyrum distichum* (Thunb.) Á Löve-derived genomes. Crop Sci 54:2619–2630
Mascher M, Richmond TA, Gerhardt DJ, Himmelbach A, Clissoldet L et al (2013) Barley whole exome capture: a tool for genomic research in the genus *Hordeum* and beyond. Plant J 76:494–505
Mason-Gamer RJ (2013) Phylogeny of a genomically diverse group of *Elymus* (Poaceae) allopolyploids reveals multiple levels of reticulation. PLoS One 8, e78449. doi:10.1371/journal.pone.0078449

McArthur RI, Zhu X, Oliver RE, Klindworth DL, Xu SS, Stack RW, Wang RRC, Cai X (2012) Homoeology of *Thinopyrum junceum* and *Elymus rectisetus* chromosomes to wheat and disease resistance conferred by the *Thinopyrum* and *Elymus* chromosomes in wheat. Chromosome Res 20:699–715

McIntosh RA, Dyck PL, Green GJ (1976) Inheritance of leaf rust and stem rust resistances in wheat cultivars Agent and Agatha. Aust J Agric Res 28:37–45

McIntosh RA, Dyck PL, Green GJ (1977) Inheritance of leaf rust and stem rust resistance in wheat cultivars Agent and Agatha. Aust J Agric Res 28:37–45

McIntosh RA, Wellings CR, Park RE (1995) Wheat rusts: an atlas of resistance genes. CSIRO, Canberra, ACT

Mochizuki A (1960) Addition of individual chromosomes of *Agropyron* to *durum* wheat. Wheat Inf Serv 11:22–23

Mochizuki A (1962) *Agropyron* addition lines of *durum* wheat. Seiken Ziho 13:133–138

Mohammed YSA, Eltayeb AE, Tsujimoto H (2013) Enhancement of aluminum tolerance in wheat by addition of chromosomes from the wild relative *Leymus racemosus*. Breed Sci 63:407–416

Mohammed YSA, Tahir ISA, Kamal NM, Eltayeb AE, Ali AM, Tsujimoto H (2014) Impact of wheat-*Leymus racemosus* added chromosomes on wheat adaptation and tolerance to heat stress. Breed Sci 63:450–460

Monneveaux P, Reynolds MP, Gonzalez Aguilar J, Singh RP (2003) Effects of the 7DL.7Ag translocation from Lophopyrum elongatum on wheat yield and related morphophysiological traits under different environments. Plant Breed 122:379–384

Morrison LA, Metzger RJ, Lukaszewski AJ (2004) Origin of the Blue-Aleurone Gene in Sebesta blue wheat genetic stocks and a protocol for its use in apomixis screening. Crop Sci 44:2063–2067

Moustakas M (1991) Further evidence of the genome relationships between *Thinopyrum bessarabicum* and *T. elongatum*. Cytobios 68:197–206

Mujeeb-Kazi A, Wang RRC (1995) Perennial and annual wheat relatives in the Triticeae. In: Mujeeb-Kazi A, Hettel GP (eds) Utilizing wild grass biodiversity in wheat improvement: 15 years of wide cross research at CIMMYT, vol 2, CIMMYT Research Report No. CIMMYT, Mexico, DF, pp 5–13

Mujeeb-Kazi A, Cortes A, Gul A, Farooq M, Majeed F, Ahmad I, Bux H, William M, Rosas V, Delgad R (2008) Production and cytogenetics of a new *Thinopyrum elongatum/Triticum aestivum* hybrid, its amphiploid and backcross derivatives. Pakistan J Bot 40:565–579

Mujeeb-Kazi A, Kazi AG, Dundas I, Rasheed A, Ogbonnaya F, Kishii M, Bonnett D, Wang RRC, Xu S, Chen P, Mahmood T, Bux H, Farrakh S (2013) Genetic diversity for wheat improvement as a conduit to food security. In: Sparks DL (ed) Adv agron, vol 122. Academic, Burlington, MA, pp 179–257

Mullan DJ, Mirzaghaderi G, Walker E, Colmer TD, Francki MG (2009) Development of wheat–*Lophopyrum elongatum* recombinant lines for enhanced sodium ‘exclusion’ during salinity stress. Theor Appl Genet 119:1313–1323

Niu ZX, Klindworth R, Wang RRC, Jauhar PP, Larkin PJ, Xu SS (2011) Characterization of HMW glutenin subunits in *Thinopyrum intermedium*, *Th. bessarabicum*, *Lophopyrum elongatum*, *Aegilops markgrafii*, and their addition lines in wheat. Crop Sci 51:667–677

Niu Z, Klindworth DL, Friesen TL, Chao S, Ohm JB, Xu SS (2014) Development and characterization of wheat lines carrying stem rust resistance gene *Sr43* derived from *Thinopyrum ponticum*. Theor Appl Genet 127:969–980

Oliver RE, Xu SS, Stack RW, Friesen TL, Jin Y, Cai X (2006) Molecular cytogenetic characterization of four partial wheat-*Thinopyrum ponticum* amphiploids and their reactions to Fusarium head blight, tan spot and *Stagonospora nodorum* blotch. Theor Appl Genet 112:1473–1479

Oliver RE, Cai X, Wang RC, Xu SS, Friesen TL (2008) Resistance to tan spot and Stagonospora nodorum blotch in wheat-alien species derivatives. Plant Dis 92:150–157

Omielan JA, Epstein E, Dvorak J (1991) Salt tolerance and ionic relations of wheat as affected by individual chromosomes of salt-tolerant *Lophopyrum elongatum*. Genome 34:961–974

Ono H, Nagayoshi T, Nakamura C, Shinotani K (1983) *Triticum durum–Elytrigia elongata* addition line – citological, morphological and biochemical characteristics. In: Sakamoto S (ed) Proceedings of the sixth international wheat genetics symposium, Kyoto. Plant Germplasm Institute, Kyoto University, Kyoto, pp 1049–1053
Park RF, Bariana HS, Wallwork H, Wellings CR (2002) Detection of virulence for leaf rust resistance gene *Lr24* in Australia. Aust J Agr Res 53:1069–1076
Podevin N, Davies HV, Hartung F, Nogué F, Casacuberta JM (2013) Site-directed nucleases: a paradigm shift in predictable, knowledge-based plant breeding. Trends Biotechnol 31:375–383
Pozniak C, Knox R, Clarke F, Clarke J (2007) Identification of QTL and association of a phytoene synthase gene with endosperm colour in durum wheat. Theor Appl Genet 114:525–537
Pretorius ZA, Bender CM, Visser B, Terefe T (2010) First report of a *Puccinia graminis* f. sp. *tritici* race virulent to the Sr24 and Sr31 wheat stem rust resistance genes in South Africa. Plant Dis 94:784
Prins R, Marais GF, Pretorius ZA, Janse BJH, Marais AS (1997) A study of modified forms of the *Lr19* translocation of common wheat. Theor Appl Genet 95:424–430
Qi LL, Wang SL, Chen PD, Liu DJ, Friebe B, Gill BS (1997) Molecular cytogenetic analysis of *Leymus racemosus* chromosomes added to wheat. Theor Appl Genet 95:1084–1091
Qi L, Friebe B, Zhang P, Gill BS (2007) Homoeologous recombination, chromosome engineering and crop improvement. Chromosome Res 15:3–19
Qi ZJ, Chen PD, Qian B, Zhuang L, Chen H, Chen T, Shen J, Guo J, Feng Y, Pei Z (2010) Characterization of a wheat–*Thinopyrum bessarabicum* (T2JS-2BS.2BL) translocation line. Theor Appl Genet 121:589–597
Quarrie SA, Pekic-Quarrie S, Radosevic R, Rancic D, Kaminska A, Barnes JD, Leverington M, Ceoloni C, Dodig D (2006) Dissecting a wheat QTL for yield present in a range of envi the QTL to candidate genes. J Exp Bot 57:2627–2637
Ravel C, Dardevet M, Leenhardt F, Bordes J, Joseph JL, Perretant MR, Exbrayat F, Poncet C, Balfourier F, Chanliaud E, Charmet G (2013) Improving the yellow pigment content of bread wheat flour by selecting the three homoeologous copies of *Psy1*. Mol Breeding 31:87–99
Redinbaugh MG, Jones TA, Zhang Y (2000) Ubiquity of the St chloroplast genome in St containing Triticeae polyploids. Genome 43:846–852
Reynolds MP, Calderini DF, Condon AG, Rajaram S (2001) Physiological basis of yield gains in wheat associated with the Lr19 translocation from Agropyron elongatum. Euphytica 119:137–141
Rosewarne G, Bonnett D, Rebetzke G, Lonergan P, Larkin PJ (2015) The potential of *Lr19* and *Bdv2* translocations to improve yield and disease resistance in the high rainfall wheat zones of Australia. Agronomy 5:55–70
Saintenac C, Jiang D, Akhunov ED (2011) Targeted analysis of nucleotide and copy number variation by exon capture in allotetraploid wheat genome. Genome Biol 12:R88
Sears ER (1973) *Agropyron*-wheat transfers induced by homoeologous pairing. In: Sears ER, Sears LMS (eds) Proceeding of the 4th international wheat genetics symposium. University of Missouri, Columbia, MO, pp 191–199
Sears ER (1977) An induced mutant with homoeologous pairing in common wheat. Can J Genet Cytol 19:585–593
Sears ER (1978) Analysis of wheat-*Agropyron* recombinant chromosomes. In: Sanchez-Monge E, Garcia-Olmedo F (eds) Interspecific hybridization in plant breeding. Proceedings of the 8th EUCARPIA congress. Ciudad Universitaria, Madrid, pp 63–72, Escuela Técnica Superior de Ingenieros Agrónomos
Sharma D, Knott DR (1966) The transfer of leaf rust resistance from *Agropyron* to *Triticum* by irradiation. Can J Genet Cytol 8:137–143
Shebeski LH, Wu YS (1952) Inheritance in wheat of stem rust resistance derived from *Agropyron elongatum*. Sci Agric 32:26–32
Shen Y, Shen J, Dawadondup, Zhuang L, Wang Y, Pu J, Feng Y, Chu C, Wang X, Qi Z (2013) Physical localization of a novel blue-grained gene derived from *Thinopyrum bessarabicum*. Mol Breeding 31:195–204

Shen X, Ohm H (2006) Fusarium head blight resistance derived from *Lophopyrum elongatum* chromosome 7E and its augmentation with *Fhb1* in wheat. Plant Breeding 125:424–429
Shen X, Ohm H (2007) Molecular mapping of *Thinopyrum*-derived Fusarium head blight resistance in common wheat. Mol Breeding 20:131–140
Shen X, Kong L, Ohm H (2004) Fusarium head blight resistance in hexaploid wheat (Triticum aestivum)-*Lophopyrum* genetic lines and tagging of the alien chromatin by PCR markers. Theor Appl Genet 108:808–813
Singh RP, Huerta-Espino J, Rajaram S, Crossa J (1998) Agronomic effects from chromosome translocations 7DL.7Ag and 1BL.1RS in spring wheat. Crop Sci 38:27–33
Sivasamy M, Kumar J, Menon MK, Tomar SMS (2010) Developing elite, durable disease resistant wheat cultivars combining high grain yield and end-use quality by introgressing effective genes employing conventional and modern breeding approaches. Ann Wheat Newsl 56:87–94
Snape JW, Butterworth K, Whitechurch E, Worland AJ (2001) Waiting for fine times: genetics of flowering time in wheat. Euphytica 119:185–190
Somo M, Chao S, Acevedo M, Zurn J, Cai X, Marais F (2014) A genomic comparison of homoeologous recombinants of the *Lr19* (T4) translocation in wheat. Crop Sci 54:565–575
Song L, Jiang L, Han H, Gao A, Yang X, Li L, Liu W (2013) Efficient induction of wheat–*Agropyron cristatum* 6P translocation lines and GISH detection. PLoS One 8, e69501. doi:10.1371/journal. pone.0069501
Subbarao GV, Ban T, Kishii M, Ito O, Samejima H, Wang HY, Pearse SJ, Gopalakrishinan S, Nakahara K, Zakir Hossain AKM, Tsujimoto H, Berry WL (2007) Can biological nitrification inhibition (BNI) genes from perennial *Leymus racemosus* (Triticeae) combat nitrification in wheat farming? Plant Soil 299:55–64
Sun G (2014) Molecular phylogeny revealed complex evolutionary process in *Elymus* species. J Syst Evol 52:706–711
Taeb M, Koebner RMD, Forster BP (1993) Genetic variation for waterlogging tolerance in the Triticeae and the chromosomal location of genes conferring waterlogging tolerance in *Thinopyrum elongatum*. Genome 36:825–830
Tang S, Li Z, Jia X, Larkin PJ (2000) Genomic in situ hybridization (GISH) analyses of *Thinopyrum intermedium*, its partial amphiploid Zhong 5, and disease-resistant derivatives in wheat. Theor Appl Genet 100:344–352
Tang X, Shi D, Xu J, Li Y, Li W, Ren Z, Fu T (2014) Molecular cytogenetic characteristics of a translocation line between common wheat and *Thinopyrum intermedium* with resistance to powdery mildew. Euphytica 197:201–210
The TT, Latter BDH, McIntosh RA, Ellison FW, Brennan PS, Fisher J, Hollamby GJ, Rathjen AJ, Wilson RE (1988) Grain yields of near isogenic lines with added genes for stem rust resistance. In: Miller TE, Koebner RMD (eds) Proceedings of the seventh international wheat genetics symposium, Cambridge, UK, pp 901–906
Tsitsin NV (1960) Wide hybridization in plants. In: Tsitsin NV (ed) The significance of wide hybridization in the evolution and production of new species and forms of plant and animals. Israel Program for Scientific Translations, Jerusalem, pp 2–30
Turner MK, DeHaan LR, Jin Y, Anderson JA (2013) Wheatgrass–wheat partial amphiploids as a novel source of stem rust and Fusarium head blight resistance. Crop Sci 53:1994–2005
Wang RR-C (1989a) Intergeneric hybrids involving perennial Triticeae. Genetics 8:57–64
Wang RR-C (1989b) An assessment of genome analysis based on chromosome pairing in hybrids of perennial Triticeae. Genome 32:179–189
Wang RR-C (1992) Genome relationships in the perennial Triticeae based on diploid hybrids and beyond. Hereditas 116:133–136
Wang RR-C (2011) *Agropyron* and *Psathyrostachys*. In: Kole C (ed) Wild crop relatives: genomic and breeding resources, cereals. Springer, Berlin, pp 77–108
Wang LS, Chen PD (2008) Development of *Triticum aestivum-Leymus racemosus* ditelosomic substitution line 7Lr#1S(7A) with resistance to wheat scab and its meiotic behavior analysis. Chinese Sci Bull 53:3522–3529

Wang RR-C, Hsiao CT (1989) Genome relationship between *Thinopyrum bessarabicum* and *T. elongatum*: revisited. Genome 32:802–809
Wang RR-C, Lu B (2014) Biosystematics and evolutionary relationships of perennial Triticeae species revealed by genomic analyses. J Syst Evol 52:697–705
Wang RR-C, Zhang XY (1996) Characterization of the translocated chromosome using fluorescence in situ hybridization and random amplified polymorphic DNA on two *Triticum aestivum-Thinopyrum intermedium* translocation lines resistant to wheat streak mosaic or barley yellow dwarf virus. Chromosome Res 4:583–587
Wang RR-C, von Bothmer R, Dvorak J, Fedak G, Linde-Laursen I, Muramatsu M (1995) Genome symbols in the Triticeae (Poaceae). In: Wang RRC, Jensen KB, Jaussi C (eds) Proceedings of 2nd international Triticeae symposium, 20–24 June 1994. Utah State University Publications on Design and Production, Logan, UT, pp 29–34
Wang HG, Liu SB, Qi ZJ, Kong FJ, Gao JR (2000) Application studies of *Elytrigia intermedium* in hereditary improvement of wheat. J Shandong Agric Univ (Nat Sci) 31:333–336
Wang RR-C, Li XM, Hu ZM, Zhang JY, Larson SR, Zhang XY, Grieve CM, Shannon MC (2003) Development of salinity-tolerant wheat recombinant lines from a wheat disomic addition line carrying a *Thinopyrum junceum* chromosome. Int J Plant Sci 164:25–33
Wang Q, Xiang J, Gao A, Yang X, Liu W, Li X, Li L (2010a) Analysis of chromosomal structural polymorphisms in the St, P and Y genomes of Triticeae (Poaceae). Genome 53:241–249
Wang RR-C, Larson SR, Jensen KB (2010b) Analyses of *Thinopyrum bessarabicum*, *T. elongatum*, and *T. junceum* chromosomes using EST-SSR markers. Genome 53:1083–1089
Wang QX, Gao AN, Yang XM, Li LH (2014) Chromosome changes after polyploidization in Triticeae. J Syst Evol 52:790–793
Wang RR-C, Larson SR, Jensen KB, Bushman S, DeHaan L, Wang S, Yan X (2015) Genome evolution of intermediate wheatgrass as revealed by EST-SSR markers developed from its three progenitor diploid species. Genome 58:63–70
Ward PR, Palta JA, Waddell HA (2011) Root and shoot growth by seedlings of annual and perennial medic, and annual and perennial wheat. Crop Pasture Sci 62:367–373
Winfield MO, Wilkinson PA, Allen AM, Barker GLA, Coghill JA et al (2012) Targeted re-sequencing of the allohexaploid wheat exome. Plant Biotechnol J 10:733–742
Wu J, Yang X, Wang H, Li H, Li L, Li X, Liu W (2006) The introgression of chromosome 6P specifying for increased numbers of florets and kernels from *Agropyron cristatum* into wheat. Theor Appl Genet 114:13–20
Xin ZY, Zhang ZY, Chen X, Lin ZS, Ma YZ, Xu HJ, Banks PM, Larkin PJ (2001) Development and characterization of common wheat-*Thinopyrum intermedium* translocation lines with resistance to barley yellow dwarf virus. Euphytica 119:161–165
Xu SS, Jin Y, Klindworth DL, Wang RR-C, Cai X (2009) Evaluation and characterization of seedling resistance to stem rust Ug99 races in wheat-alien species derivatives. Crop Sci 49:2167–2175
Yang ZJ, Zhang T, Liu C, Li GR, Zhou JP, Zhang Y, Ren ZL (2008) Identification of wheat Dasypyrum breviaristatum addition lines with stripe rust resistance using C-banding and genomic in situ hybridization. In: Appels R, Eastwood R, Lagudah E, Langridge P, Mackay M, McIntyre L, Sharp P (eds) Proceedings of the 11th international wheat genetics symposium. University Press, Sydney, Australia, http://hdl.handle.net/2123/3272
Ye X, Lu Y, Liu W, Chen G, Han H, Zhang J, Yang X, Li X, Gao A, Li L (2015) The effects of chromosome 6P on fertile tiller number of wheat as revealed in wheat-*Agropyron cristatum* chromosome 5A/6P translocation lines. Theor Appl Genet 128:797–811
Yen C, Yang JL, Yen Y (2005) Hitoshi Kihara, Áskell Löve and the modern genetic concept of the genera in the tribe Triticeae (Poaceae). Acta Phytotaxon Sin 43:82–93
Zeng J, Cao W, Fedak G, Sun S, McCallum B, Fetch T, Xue A, Zhou Y (2013a) Molecular cytological characterization of two novel durum-*Thinopyrum intermedium* partial amphiploids with resistance to leaf rust, stem rust and Fusarium head blight. Hereditas 150:10–16
Zeng J, Cao W, Hucl P, Yang Y, Xue A, Chi D, Fedak G (2013b) Molecular cytogenetic analysis of wheat – *Elymus repens* introgression lines with resistance to Fusarium head blight. Genome 56:75–82

Zhang W, Dubcovsky J (2008) Association between allelic variation at the Phytoene synthase 1 gene and yellow pigment content in the wheat grain. Theor Appl Genet 116:635–645
Zhang XY, Dong YS (1995) Cytogenetic research on hybrids of *Triticum* with both *Th. ponticum* and *Th. intermedium* as well as their derivatives. III. Primary detection of a genetic base for useful gene introgression from the two alien species into wheat. Acta Genet Sin 22:217–222
Zhang XY, Dong YS, Wang RRC (1996a) Characterization of genomes and chromosomes in partial amphiploids of the hybrid *Triticum aestivum* x *Thinopyrum ponticum* by in situ hybridization, isozyme analysis, and RAPD. Genome 39:1062–1071
Zhang XY, Koul A, Petroski R, Ouellet T, Fedak G, Dong YS, Wang RR-C (1996b) Molecular verification and characterization of BYDV-resistant germ plasms derived from hybrids of wheat with *Thinopyrum ponticum* and *Th. intermedium*. Theor Appl Genet 93:1033–1039
Zhang ZY, Xin ZY, Larkin PJ (2001) Molecular characterization of a *Thinopyrum intermedium* group 2 chromosome (2Ai-2) conferring resistance to barley yellow dwarf virus. Genome 44:1129–1135
Zhang W, Lukaszewski AJ, Kolmer J, Soria MA, Goyal S, Dubcovsky J (2005) Molecular characterization of durum and common wheat recombinant lines carrying leaf rust resistance (*Lr19*) and yellow pigment (*Y*) genes from *Lophopyrum ponticum*. Theor Appl Genet 111:573–582
Zhang C, Fan X, Yu HQ, Zhang L, Wang XL, Zhou YH (2009a) Different maternal genome donor to *Kengyilia* species inferred from chloroplast trnL-F sequences. Biol Plant 53:759–763
Zhang ZY, Lin ZS, Xin ZY (2009b) Research progress in BYDV resistance genes derived from wheat and its wild relatives. J Genet Genomics 36:567–573
Zhang X, Cai J, Anderson JM, Zhao M, Ohm HW, Kong L (2010) Identification of disease resistances in wheat-*Leymus multicaulis* derivatives and characterization of *L. multicaulis* chromatin using microsatellite DNA markers. Front Agric China 4:394–405
Zhang XL, Shen XR, Hao YF, Cai JJ, Ohm HW, Kong L (2011a) A genetic map of *Lophopyrum ponticum* chromosome 7E, harboring resistance genes to Fusarium head blight and leaf rust. Theor Appl Genet 122:263–270
Zhang N, Yang WX, Liu DQ (2011b) Identification and molecular tagging of leaf rust resistance gene (*Lr24*) in wheat. Agric Sci China 10:1898–1905
Zhao JX, Du WL, Wu J, Cheng XN, Gao Y, Pang YH, Chen XH, Liu SH, Yang QH, Fu J (2013) Development and identification of a wheat-*Leymus mollis* multiple alien substitution line. Euphytica 190:45–52
Zheng Q, Li B, Li HW, Li ZS (2009) Utilization of blue-grained character in wheat breeding derived from *Thinopyrum ponticum*. J Genet Genomics 36:575–580
Zhong GY, Dvorak J (1995) Chromosomal control of the tolerance of gradually and suddenly imposed salt stress in the *Lophopyrum elongatum* and wheat, *Triticum aestivum* L., genomes. Theor Appl Genet 90:229–236

Chapter 12
Wheat–Barley Hybrids and Introgression Lines

Márta Molnár-Láng and Gabriella Linc

12.1 Wheat (*T. aestivum*) × Barley (*H. vulgare*) Hybridization

Wheat (*Triticum aestivum* L.) and barley (*Hordeum vulgare* L.) are two of the major crops cultivated in the temperate zones. Besides its agronomic importance, barley has been an important material for genetic and genomic studies on the *Triticeae*, a tribe of the grass family. The advantages of barley as an experimental material reside basically in its diploid nature. Wheat/barley intergeneric hybridization could make it possible to incorporate the major agronomical traits of barley (earliness, β-glucan content, favourable amino acid composition, salt and drought tolerance, good tillering ability, etc.) into the bread wheat genome (Molnár-Láng et al. 2014). Although experiments had already begun in the early twentieth century, the first demonstrably successful cross between the two species was reported by the Danish scientist Kruse in 1973. Encouraged by this success, attempts were made worldwide, and at first hybrids were produced with relatively greater frequency when barley was used as the female parent (Islam and Shepherd 1990). Barley × wheat hybrids were developed in numerous combinations by several scientists: Islam et al. (1975), Fedak (1977), Thomas et al. (1977), Mujeeb-Kazi (1981), Clauss (1980), Shumny et al. (1981), Wojciechowska (1985) and Molnár-Láng et al. (1985). In crosses between a total of 18 barley varieties and 15 wheat varieties, the highest seed set was achieved when the wheat variety Chinese Spring (CS) was hybridized with the barley varieties Betzes and Ketch. A seed set of 15.4 % was reported by Islam et al. (1975), while Fedak (1980) achieved 49 % seed set, though only 2 % developed into plants. Fedak and Jui (1982) used the CS-Hope substitution line series to determine the chromosomal location of genes in CS that permit

M. Molnár-Láng (✉) • G. Linc
Agricultural Institute, Centre for Agricultural Research, Hungarian Academy of Sciences, Brunszvik 2, Martonvásár 2462, Hungary
e-mail: molnar.marta@agrar.mta.hu

M. Molnár-Láng et al. (eds.), *Alien Introgression in Wheat*,
DOI 10.1007/978-3-319-23494-6_12

crossability with Betzes. No progeny were obtained from substitution lines 5A, 5B, 5D, indicating that the homoeologous group 5 chromosomes of CS are the chromosomes chiefly responsible for permitting crossability with Betzes. All the hybrid plants obtained were raised in embryo culture, since hybrid seeds have no endosperm (Islam and Shepherd 1990). The barley × wheat hybrids exhibited complete male sterility, but when backcrossed with wheat it proved possible to produce BC_1 and BC_2 plants. The seed set in the first backcross was extremely low (0.5–1.2 seeds/spike) (Islam and Shepherd 1990). Due to the pistilloidy observed in the BC_1 and BC_2 plants the progeny remained sterile despite several backcrosses, making it impossible to develop fertile addition lines. In order to eliminate pistilloidy, attempts were made to make reciprocal crosses, using wheat as the female parent. Far fewer laboratories were able to report successful crosses and these involved a much smaller number of combinations (Islam and Shepherd 1990; Fedak 1980; Wojciechowska and Pudelska 1993; Molnár-Láng and Sutka 1994; Molnár-Láng et al. 2000b; Jauhar 1995; Taketa et al. 1998). Altogether about 100 combinations between 29 wheat cultivars and 55 barley accessions were tested, of which about 45 combinations produced 1 % of hybrid embryos. It was found by Islam et al. (1978, 1981) that crosses between CS and Betzes gave the highest seed set, but this was only 1.3 %. In experiments carried out under controlled conditions in the Martonvásár phytotron, 3.3 % seed set was achieved with the same combination (Molnár-Láng and Sutka 1994). A relatively short time after the development of the first wheat × barley hybrids, addition lines (2H, 3H, 4H, 5H, 6H, 7H) were also produced for the first time between CS wheat and the spring barley Betzes (Islam et al. 1978, 1981) (Table 12.1). By crossing this addition series with the relevant monosomic lines, substitution lines were developed for all the chromosomes except 1H and 5H (Islam and Shepherd 1992b, 1995; Ya-Ping et al. 2003). Despite many attempts, it proved extremely difficult to expand the number of genotypes that could be successfully crossed, and very few hybrids were developed from genotypes with satisfactory agronomic traits (Wojciechowska and Pudelska 1993; Jauhar 1995; Taketa et al. 1998). It proved impossible to develop BC_1 seed on a substantial proportion of the new hybrids, so no fertile progeny could be obtained from the new combinations (Wojciechowska and Pudelska 1993; Jauhar 1995). The efficiency of hybrid development was greatly improved by Koba et al. (1991), who used the 2,4-D treatment that had been successfully applied in wheat × maize crosses and also proved that F_1 hybrids could be produced from most embryos through embryo culture. A number of Japanese wheat varieties were included in the crosses, among which Norin 12, Norin 61 and Shinchunaga gave better seed set than CS × Betzes crosses. The highest seed set (8.25 %) was obtained from the Norin 12 × Betzes combination. Addition lines containing the barley chromosomes 5H and 6H were developed from a cross between Shinchunaga and Nyugoruden. Translocation lines were produced containing the 5HS.5BL translocation chromosome pair in addition to 42 wheat chromosomes (Koba et al. 1997). However, it has recently been reported that, despite resulting in several new combinations, the application of 2,4-D treatment also led to a significantly higher frequency of maternal haploids (Polgári et al. 2014). Backcross progenies (BC_1 and BC_2) were developed from the Shinchunaga × barley line

Table 12.1 Wheat (*Triticum aestivum*)—barley (*Hordeum*) disomic addition lines

Triticum aestivum genotype	*Hordeum* genotype	Added barley chromosome	References
Chinese Spring	*Hordeum vulgare* cv. Betzes (2-row, spring, German)	2H, 3H, 4H, 5H, 6H, 7H, 1H/1HS + 6H 1HS, 2HS, 2HL, 3HS, 3HL, 4HS, 4HL,5HS, 5HL, 6HS, 6HL, 7HS, 7HL	Islam et al. (1978, 1981) Islam and Shepherd (2000) Islam (1983), Islam and Shepherd (1990)
Shinchunaga (Japanese, spring)	*H. vulgare* cv. Nyugoruden (New Golden) (2-row, spring, Japanese)	6H, 7H	Koba et al (1997)
Martonvásári 9 kr_1 (Mv9kr1) (Hungarian, winter)	*H. vulgare* cv Igri (2-row, winter, German)	2H, 3H, 4H, 7H, 1HS, 6HS,	Molnár-Láng et al. (2007) Szakács and Molnár-Láng (2007; 2010)
Asakaze komugi (Japanese, facultative)	*H. vulgare* cv. Manas (6-row, winter, Ukrainian)	2H, 3H, 4H, 6H, 7H 2HS, 2HL, 3HS, 3HL, 5HS, 5HL, 6HS, 6HL, 7HS, 7HL	Molnár-Láng et al. (2012) Türkösi et al. (2014b)
Chinese Spring	*Hordeum chilense*	$1H^{ch}$, $4H^{ch}$, $5H^{ch}$, $7H^{ch}$	Miller et al. (1981)
Shinchunaga (Japanese, spring)	*Hordeum vulgare* ssp. *spontaneum* OUH602	2H, 3H, 4H, 5H, 6H, 7H, 1HS, 5HS, 6HS, 6HL,	Taketa and Takeda (2001)
Chinese Spring	*Hordeum marinum*	$1\ H^{m}$, $2H^{m}$, $4H^{m}$, $5H^{m}$, $6H^{m}$, $7H^{m}$	Islam and Colmer (2008)

T3-7aai combination by Malysheva et al. (2003). The genome composition of the backcross progenies was analysed using genomic in situ hybridization (GISH) and microsatellite markers. Some of the barley chromosomes (2H, 4H) were entirely eliminated from the BC_2 plants, and the presence of 1H caused sterility, while chromosome segments from other barley chromosomes (3H, 5H, 6H, 7H) were detected in some BC_2 plants. The development of disomic addition lines from this combination was not reported.

Barley has great genetic diversity for many agronomically important traits (spring or winter habit, two-rowed or six-rowed, tolerance to abiotic stresses, yield ability, earliness, quality, adaptation, etc.). In order to utilize these traits, it would be worth producing wheat/barley addition and introgression lines with agronomically adaptable winter barley cultivars. Two new additions were reported from the wheat × barley hybrids produced using winter barley cultivars in Martonvásár (Molnár-Láng et al. 2000b) (Fig 12.1). Backcross progenies were developed on the hybrids at very low frequency (Molnár-Láng et al. 2000b, 2005). Wheat–barley disomic addition lines (2H, 3H, 4H, 6HS, 7H, 1HS isochromosome) were pro-

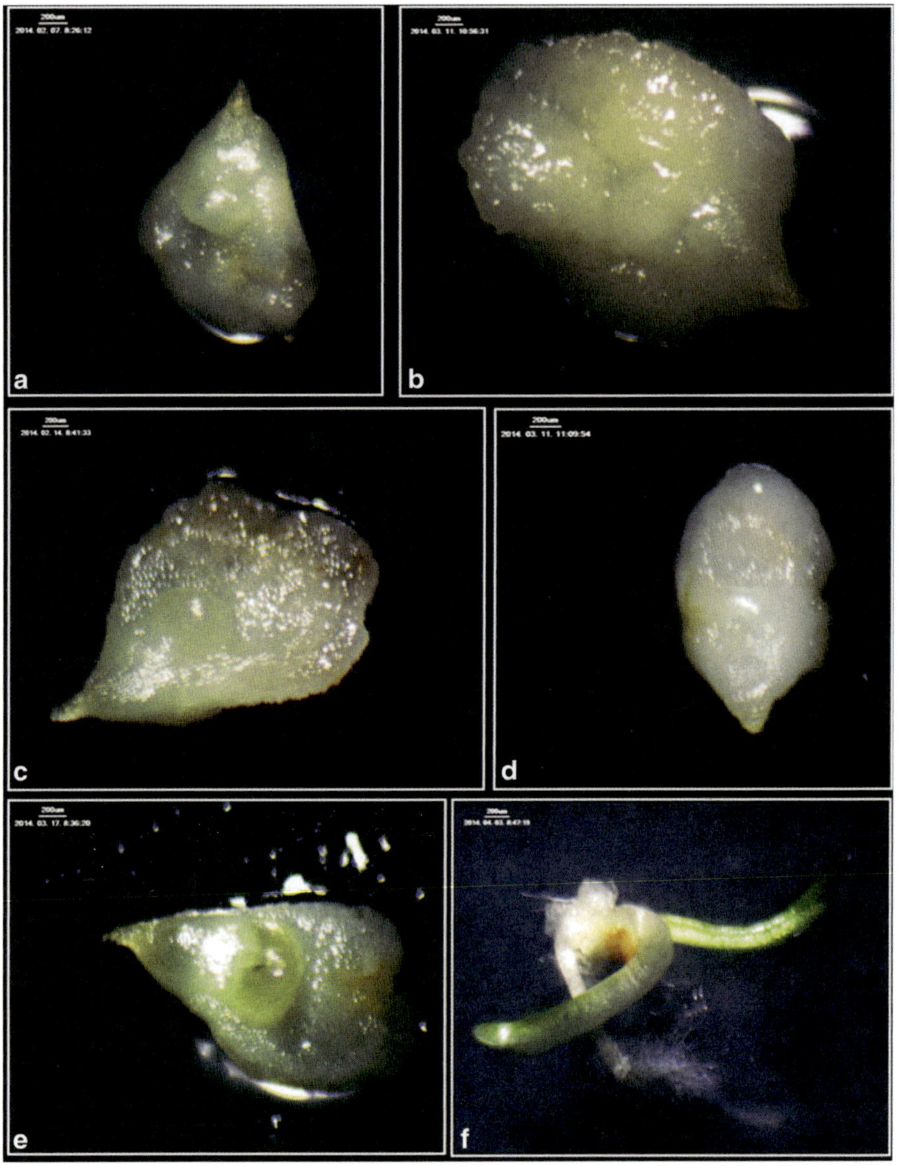

Fig. 12.1 (**a**–**e**) Wheat × barley hybrid embryos excised 3 weeks after pollination from developing seeds with no endosperm (each bar = 200 μm). (**f**) Developing plantlet, maintained in vitro

duced from hybrids between winter wheat line Mv9kr1 and the German two-rowed winter barley Igri (Molnár-Láng et al. 2007; Szakács and Molnár-Láng 2007; Szakács and Molnár-Láng 2010) and were identified using molecular cytogenetic methods. In order to increase the allelic variation in wheat/barley introgressions, new wheat/barley disomic addition lines were developed containing the 2H, 3H,

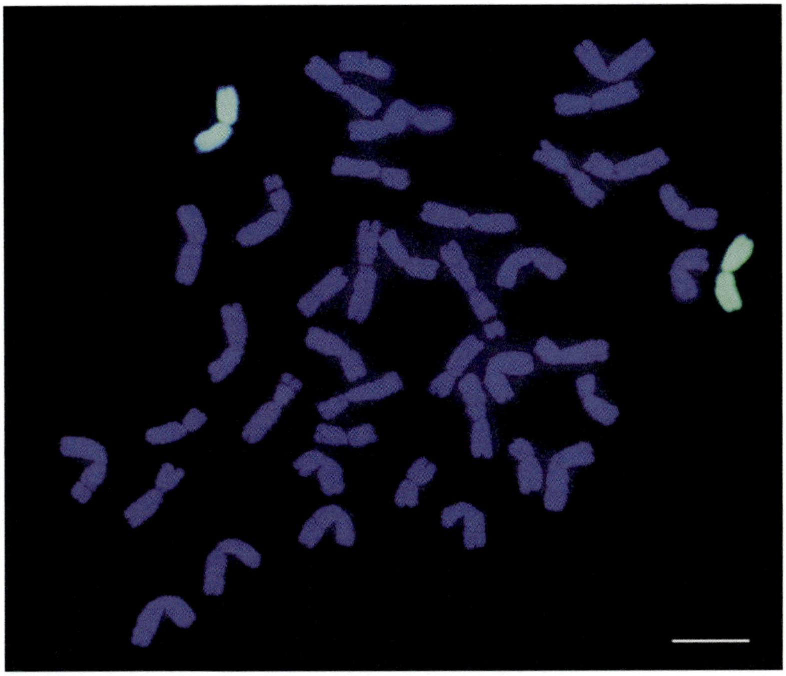

Fig. 12.2 Mitotic chromosomes of the 2H disomic addition line after GISH. Total genomic *H. vulgare* DNA was labelled with biotin-11-dUTP and detected with streptavidin-FITC (*green*), while wheat chromosomes were counterstained with DAPI (*blue*)

4H, 6H and 7H chromosomes of the six-rowed Ukrainian winter barley cultivar Manas (Molnár-Láng et al. 2012) (Figs. 12.2 and 12.3). This cultivar is agronomically much better adapted to Central European environmental conditions than the two-rowed spring barley cultivar Betzes previously used.

12.2 Production of Wheat × Barley Hybrids Using Alternative *Hordeum* Species

In addition to hexaploid wheat (*T. aestivum* L.) and diploid barley (*Hordeum vulgare* L.), hybrids have also been developed between other *Triticum*and *Hordeum* species. The most successful of these is hexaploid tritordeum, which arose from a cross between *Triticum turgidum* L. ssp. durum (Desf.) Husn. (synonym: *T. durum*) and *Hordeum chilense* Roem. et Schulz. (Martín and Sanchez-Monge Laguna 1980, 1982) (Table 12.2). *Hordeum chilense* was first pollinated with hexaploid *T. aestivum* to produce an F_1 hybrid (Martín and Chapman 1977), and a partially fertile amphidiploid was produced by means of colchicine treatment (Chapman and Miller 1978). The amphidiploid was then backcrossed to wheat to develop wheat/*H. chilense*

Fig. 12.3 Spike morphology of the Asakaze/Manas disomic addition lines compared with that of the parental Asakaze (wheat) and Manas (barley) genotypes

Table 12.2 Wheat (*Triticum*) – barley (*Hordeum*) amphidiploids

Wheat genotype	Hordeum species	Amphidiploid	References
Triticum timopheevii	*Hordeum bogdanii*	$2n=42$	Kimber and Sallee (1976)
Triticum aestivum	*Hordeum chilense* (maternal plant)	$2n=56$ designated as octoploid *Tritordeum*	Martín and Chapman (1977) Martin et al. (1987)
Triticumturgidum	*H. chilense* (maternal plant)	$2n=42$ designated as hexaploid *Tritordeum*	Martín and Sanchez-Monge Laguna (1982) Martín and Cubero (1981)
T.aestivum	*H. californicum* (maternal plant)	$2n=56$	Gupta and Fedak (1985)
T.aestivum	*Hordeum marinum* (maternal plant)	$2n=56$	Islam and Colmer (2008)
T. aestivum	*Hordeum vulgare*	–	Not reported

addition lines (Miller et al. 1981). Later, *H. chilense* was pollinated with *T. durum*, after which the hybrid was treated with colchicine to develop fertile amphidiploids with 42 chromosomes (Martín and Sanchez-Monge Laguna 1980, 1982). As this new amphidiploid exhibited fewer meiotic chromosome pairing anomalies and fertility problems than the primary triticale, it was assumed that, like triticale, it could be cultivated as a new plant species, and was named tritordeum. Hexaploid tritordeum was found to have a protein content of 19–24 % (Martín and Cubero 1981), so numerous analyses were made to provide a detailed description of the quality parameters. After 6–7 years of self-fertilization in field experiments, it was established that the new species yielded only 20–40 % as much as cultivated wheat, but had a protein content amounting to 17.6–25.2 % of the dry matter (Cubero et al. 1986). Its other quality parameters (fibre, lignin, cellulose and hemicellulose contents, and amino acid composition) were similar to those of cultivated wheat. A multi-location study under diverse growth conditions revealed important information about the effect of water availability on the yield of tritordeum. In the lowest yielding environments tritordeum and triticale had similar yields (Villegas et al. 2010). However, under better growth conditions tritordeum was found to yield less than wheat and triticale. It is suggested that tritordeum could be a new option for cultivation in very dry environments. In the course of cytological analyses, the chromosome number and chromosome pairing of the new species were first monitored using the Feulgen method, which revealed a high level of chromosome stability (Martín and Cubero 1981). Later the *H. chilense* chromosomes were studied by means of C-banding (Fernandez et al. 1985) and fluorescence in situ hybridization (FISH) using repetitive DNA sequences (Cabrera et al. 1995). Hybridization with the pAs1 DNA clone (isolated from *Aegilopstauschii* Coss.) gave a hybridization pattern similar to that of the D genome chromosomes of wheat for the *H. chilense* chromosomes, with strong hybridization signals on the telomeres. This DNA probe gives diffuse signals on the *H. vulgare* chromosomes and cannot be identified. The hybridization pattern obtained with C-banding bore more resemblance to that of the wheat chromosomes and strong telomeric bands were observed on the *H. chilense* chromosomes, while in the case of *H. vulgare* chromosomes, C-banding revealed interstitial bands near the centromere (Cabrera et al. 1995). Molecular cytogenetic analysis showed that *H. chilense* was genetically distant from cultivated barley. Numerous papers have been published on the taxonomical classification of *Hordeum* species (Löve 1982, 1984; Dewey 1984), which were first classified on the basis of morphological observations, then on the basis of chromosome pairing in interspecific hybrids, and later in terms of the conclusions drawn from molecular genetic analysis. Bothmer et al. (1986, 1987) used the data of chromosome pairing analysis to divide the *Hordeum* species into four basic genomes (I, Y, X and H). Molecular genetic analysis later confirmed this classification (Svitashev et al. 1994), showing that *H. vulgare* and *H. bulbosum* contained genome I and *H. murinum* genome Y, while *H. chilense* was one of the species carrying the H genome, and *H. marinum* Huds. had an X genome. This classification confirmed the relatively distant relationship between *H. vulgare* and *H. chilense.*

In an effort to improve the agronomic traits of tritordeum, further crosses were made, primarily with triticale. The progeny were then analysed using various cytogenetic methods (Fernandez-Escobar and Martín 1985; Lima-Brito et al. 1996). The chromosomes of both *H. chilense* and rye could be identified by means of FISH (Lima-Brito et al. 1996). The hexaploid tritordeum was also crossed with *H. vulgare*, but the amphidiploid developed by treating the F_1 hybrid with colchicine proved to be sterile (Martín et al. 1995). Transgenic lines were developed by transforming tritordeum

(Barcelo et al. 1994) and these were later studied in nursery experiments (Hernandez et al. 2001). A new cytoplasmic male sterility (CMS) source designated msH1 was reported in bread wheat by Martín et al. (2009). This system uses the cytoplasm of *H. chilense*. The male sterility of alloplasmic wheat containing *H. chilense* cytoplasm is stable under various environmental conditions, and the plants exhibit no developmental or floral abnormalities, except for slightly reduced height and some delay in heading. There is thus real potential for the development of a viable technology for hybrid wheat production. The addition of chromosome $6H^{ch}S$ from *H. chilense* accession H1 was able to restore the pollen fertility of the CMS phenotype induced by the presence of *H. chilense* cytoplasm in wheat. An optimal combination for fertility restoration was the translocation $T6H^{ch}.6DL$, developed by Martín et al. (2009). In addition to *H. chilense*, the following *Hordeum* species have been hybridized with wheat:

- *H. spontaneum* [syn.: *H. vulgare* ssp. *spontaneum* (C. Koch) Thell] (Islam and Shepherd 1990; Taketa et al. 1995).
- *H. bulbosum* L. (Barclay 1975; Blanco et al. 1986).
- *H. bogdanii* Wil. (Kimber and Sallee 1976).
- *H. pussillum* Nutt. (Finch and Bennett 1980).
- *H. geniculatum* All. (Clauss 1983; Pershina et al. 1988).
- *H. pubiflorum* Hook. f. (Fedak 1983).
- *H. californicum* Covas & Stebbins [syn.: *H. brachyantherum* Nevski ssp. *californicum* (Covas & Stebbins)] (Gupta and Fedak 1985).
- *H. marinum* Huds. (Jiang and Dajun 1987; Islam et al. 2007; Pershina et al. 2009).
- *H. depressum* (Scribn. & Smith) Rydb. (Jiang and Dajun 1987).

A complete set of wheat–wild barley (*Hordeum vulgare* ssp. *spontaneum)* chromosome addition lines was developed by Taketa and Takeda (2001). The chromosome constitution of the addition lines was confirmed by C-banding and GISH hybridization. Addition lines for the entire 1H chromosome and its long arm are only available as monosomic and monotelosomic additions, respectively, because of sterility. Disomic additions involving individual chromosomes of sea barleygrass (*Hordeum marinum* Huds.) in CS were obtained by Islam and Colmer (2008). The salt tolerance of the wheat–*H. marinum* amphiploid was intermediate to that of its parents (Islam et al. 2007). Alloplasmic wheat–barley substitution and addition lines were produced by Pershina et al. (2009) from *H. marinum* ssp. *gussoneanum* Huds. × *T. aestivum* hybrids.

12.3 Maintenance of Wheat × Barley Hybrids in Tissue Culture

Sterile interspecific and intergeneric hybrids can be maintained and multiplied in a vegetative manner through callus formation in tissue culture (in vitro) (Fedak 1985). Interspecific and intergeneric hybrids produced in wide crosses are often not only male sterile, but also have such a low level of female fertility that seeds are only set at extremely low frequency even when pollinated with one of the parents. Tissue culture makes it possible to multiply hybrids and produce enough progeny for backcrossing. The in vitro multiplication of interspecific and intergeneric hybrids developed between cereal species has been reported for various combinations: barley × rye (Shumny and Pershina 1979); wheat × rye (Armstrong et al. 1983; Doré et al. 1988); *Aegilops crassa* (*Triticum crassum*) × *Hordeum vulgare* (Nakamura et al. 1981); wheat × *Agropyron* hybrids (Sharma et al. 1984; Bai and Knott 1993); *Elymuscanadensis* L. × *Psathyrostachys juncea* (Fisch.) Nevski (Park et al. 1990). When the progeny were subjected to cytological analysis, deviations were observed in all cases compared with the initial hybrids. It was established that the somaclonal variability (SV) observed during the in vitro multiplication of plants (Larkin and Scowcroft 1981) could lead to useful rearrangements during the maintenance of interspecificand intergeneric hybrids in tissue culture (Fedak 1985). Amphidiploids with a doubled chromosome number have been successfully produced from F_1 hybrids (Doré et al. 1988; Ter Kuile et al. 1988), translocations have been observed (Sharma et al. 1984), and in some cases the regenerants have been found to have increased fertility (Sharma et al. 1984; Fedak and Grainger 1986; Molnár-Láng et al. 1991). Wheat–barley hybrids were multiplied in tissue culture by Pershina and Shumny (1981), Chu et al. (1984), Junming et al. (1985), Galiba et al. (1986), Surikov and Kissel (1988) and Koba et al. (1988). Detailed cytological analyses on the regenerants were published by Junming et al. (1985), Fedak and Grainger (1986), Shimada et al. (1987), Fedak et al. (1987) and Molnár-Láng et al. (1991). Chromosome numbers differing from that of the initial hybrid (28) were observed by Junming et al. (1985) and Koba et al. (1988) in some regenerants (21–27) and all the authors recorded the occurrence of amphidiploid cells. The appearance of telocentric chromosomes in the regenerants was observed in several experiments (Junming et al. 1985; Koba et al. 1988; Molnár-Láng et al. 1991). A detailed analysis was made of the meiosis of regenerant hybrids by Molnár-Láng et al. (1991), who found an increase in the rate of homoeologous chromosome pairing. A similar conclusion was drawn by Dahleen (1999) when investigating the progeny regenerated from barley × wild rye hybrids in tissue culture. As no backcross seeds were obtained from the initial hybrid of facultative wheat cv. Asakaze × winter barley cv. Manas, young inflorescences of the hybrids were used for in vitro multiplication in three consecutive cycles until a backcross progeny was developed. The chromosome constitution of the regenerated hybrids was analysed using GISH after each in vitro multiplication cycle (Molnár-Láng et al. 2005). The seven barley chromosomes were present even after the third cycle, but abnormalities were observed. Due to chromosome breakages, the number of barley telocentrics became significantly higher after the third cycle and amphidiploid cells with 56 chromosomes were counted. The number of wheat–barley chromosome arm associations, i.e. the homoeologous pairing frequency, increased after in vitro multiplication (Molnár-Láng et al. 2005).

12.4 Meiotic Pairing Behaviour of Wheat × Barley Hybrids

At first, the Feulgen technique was used to analyse the meiotic pairing behaviour of wheat × barley hybrids. In most cases Islam and Shepherd (1980) observed 28 univalent chromosomes when analysing pollen mother cells, though chromosome pairing could be seen in a few cells, with an average of 0.7 bivalents per pollen mother cell. A higher rate of chromosome pairing was recorded by Fedak (1977), resulting in a chiasma frequency of 1.82 per pollen mother cell. This was higher than the rate reported earlier in wheat haploids (Riley and Law 1965), suggesting that pairing also took place between barley and wheat chromosomes. Fedak (1977) drew attention to the phenomenon of homoeologous pairing between the chromosomes of two distantly related genera, and suggested that this should be confirmed with the Giemsa technique, the best method available at the time. Later Jauhar (1995) demonstrated a chiasma frequency of 2.16–6.72 per pollen mother cell in wheat × barley hybrids developed using the barley variety Luther. These data pointed to pairing between wheat and barley chromosomes, but as the chromosomes were analysed in meiosis using the Feulgen method, it was not possible to identify the individual chromosomes. An average of 5.03–6.63 bivalents per pollen mother cell could be observed in wheat × barley hybrids produced using the *Ph* mutant of CS, together with a small number of trivalents and quadrivalents (Sethi et al. 1986), but pairing between wheat and barley could not be demonstrated with the Feulgen method. Wojciechowska (1985) performed detailed meiosis analysis in several barley × wheat hybrid combinations and found a chiasma frequency of 1.17–1.98 per pollen mother cell in hybrid cells containing 28 chromosomes. Islam and Shepherd (1988) elaborated a method for the detection of pairing between wheat and barley chromosomes. They crossed ditelosomic wheat/barley addition lines with a high-pairing strain of an *Ae. speltoides* genotype carrying the *Ph* suppressor gene. F_1 hybrids possessing 28 + 1 telocentric somatic chromosomes (21 wheat + 7 *Ae. speltoides* + 1 barley telocentric) were grown. Pairing between telocentric and non-telocentric chromosomes was observed in 1.2–4.5 % of the pollen mother cells. Triple monosomic addition lines were developed in a wheat monosomic background, one of which contained 19 pairs of wheat chromosomes together with one 5B *Ph* mutant chromosome, one 3HL barley chromosome arm and a 3A wheat chromosome (Islam and Shepherd 1992a). In another line the 5B *Ph* mutant was accompanied by one 6HS and one 6B chromosome. In the triple monosomic addition lines, plants carrying the 3HL and 6HS barley chromosome arms only exhibited pairing in 0.3–0.7 % of the cells. These experiments proved that chromosomes of the distantly related species wheat and barley are capable of pairing with each other, thus allowing recombinations to occur. The meiotic instability of wheat × barley hybrids was noted by a number of authors, who found many cells with hypo- or hyperploid chromosome numbers in addition to cells with 28 chromosomes (Fedak 1980; Mujeeb-Kazi and Rodriguez 1983; Islam and Shepherd 1980; Wojciechowska 1985). Islam and Shepherd (1980) observed that the chromosome number became doubled in some cells during meiosis (restitution nuclei). In these hybrids the univalent chromosomes assembled in the equatorial plate during meiotic metaphase I, but in many cells, instead of migrating to the two poles in anaphase I, the chromosomes remained together, forming a chromatin mass, thus leading to the formation of cells with a doubled chromosome number. In these cells, however, it was often observed that the second phase of meiosis did not take place, preventing the development of microspores with the full chromosome complement, which would restore the fertility of the hybrids (Islam and Shepherd 1980). It can be assumed that the egg-cells, which became fertilized and set seed when the hybrids were backcrossed arose from megaspores with a doubled chromosome number. Wheat–barley chromosome pairing was first detected using GISH by Molnár-Láng et al. (2000b). Meiotic analysis of the wheat × barley hybrid Mv9 kr1 × Igri revealed 1.59 chromosome arm associations per cell using the Feulgen method (Molnár-Láng et al. 2000b). The number of chromosome arm associations increased to 4.72 after in vitro culture. According to GISH analysis, wheat–barley chromosome arm associations made up 3.6 % of the total in the initial hybrid and 16.5 % of the total in progenies of the Mv9 kr1 × Igri hybrids regenerated in vitro. The meiotic pairing behaviour of a wheat–winter barley hybrid (Asakaze × Manas) was analysed using GISH after long-term maintenance in tissue culture (Molnár-Láng et al. 2005) (Fig. 12.4a). As no backcross seeds were obtained

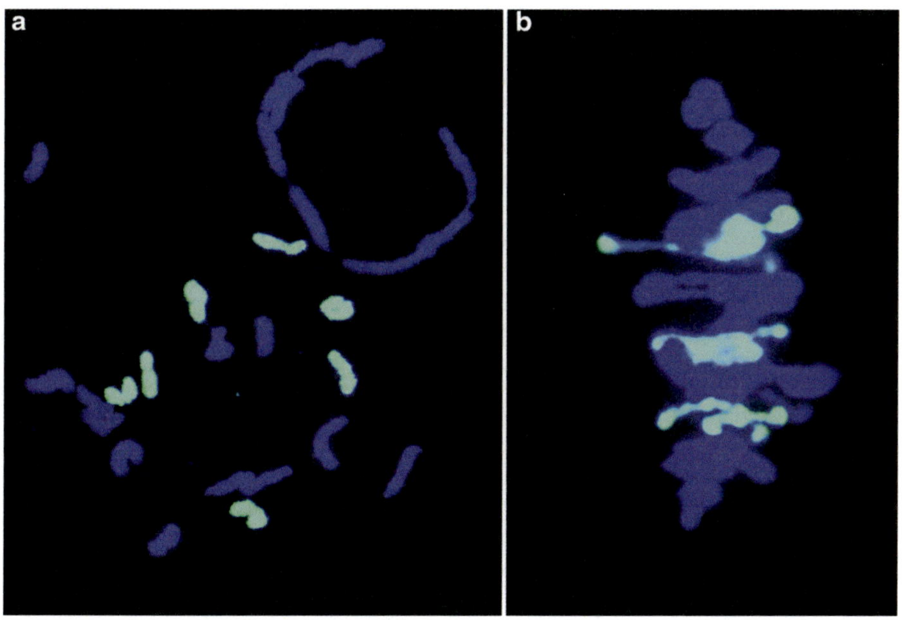

Fig. 12.4 GISH on meiotic chromosomes from the wheat × barley (Asakaze × Manas) hybrid multiplied in vitro. Total barley genomic DNA was labelled with Fluorogreen and used as probe. Barley chromosomes are *green*, and wheat chromosomes are *blue* as a result of counterstaining with DAPI. (**a**) Seven barley univalents + 14 wheat univalents + 2 wheat rod bivalents and 1 wheat trivalent. (**b**) An amphidiploid cell. Five barley rod bivalents + 2 barley ring bivalents + 4 wheat rod bivalents + 17 wheat ring bivalents

from the initial hybrid, young inflorescences were used for in vitro multiplication in three consecutive cycles until a backcross progeny was developed. The chromosome constitution of the regenerated hybrids was analysed using GISH after each in vitro multiplication cycle. The number of wheat–barley chromosome arm associations increased after the second and third cycles. Amphidiploid cells containing seven barley bivalents were counted after the third cycle (Fig. 12.4b). The use of the GISH technique to demonstrate wheat–barley chromosome pairing in the hybrids, and especially in their in vitro-regenerated progenies, proved the possibility of producing recombinants between these two genera, and thus of transferring useful characters from barley into wheat (Molnár-Láng et al. 2000b, 2005). In some regenerants in vitro conditions caused an increase in chromosome arm association frequency and in fertility.

12.5 Wheat–Barley Introgression Lines

Recombinant lines are the vehicles for transferring barley chromatin and its characters into the wheat genome. Islam and Shepherd (1992a) were the first to develop recombinations from wheat × barley crosses. Triple monosomics were developed from crosses between wheat/barley ditelosomic substitution lines and the *Ph* mutant of CS wheat. In addition to 19 wheat chromosome pairs, the triple monosomic additions contained one barley telocentric chromosome, the homoeologous wheat chromosome and one 5B *Ph* mutant chromosome. With this method six wheat/barley recombinations involving 6HL and 3HL chromosome segments were detected among the progeny. The presence of the recombinations was proved by isoenzyme analysis: the progeny were found to contain isoenzymes located either on the 6A and 6H or on the 3A and 3H chromosomes. Sherman et al. (2001) also utilized the effect of the *Ph* mutant gene to develop recombinations from the 4H and 5H wheat/barley addition lines produced by Islam et al. (1981). The presence of the recombinations was confirmed with PCR-based molecular markers. The use of GISH to detect the occurrence of wheat–barley translocations was first reported by Schwarzacher et al. (1992). The translocation line was developed by Islam and Shepherd (unpublished data) and isoenzyme analysis proved that at least the segment of the 4HL barley chromosome arm carrying the gene coding for the barley β-amylase isoenzyme had been incorporated into this line. GISH analysis then demonstrated that the whole of the 4HL chromosome arm was present in the translocation line, i.e. a centric fusion had occurred between wheat and barley. The occurrence of spontaneous translocations was observed by Koba et al. (1997) in the progeny of a cross between Shinchunaga wheat and Nyugoruden barley. The translocation chromosome was identified with C-banding and, using an earlier nomenclature, it was found that it involved the short arm of barley chromosome 7H and the long arm of wheat chromosome 5B. When the names of the barley chromosomes were later revized, the old chromosome 7H was renamed 5H (Linde-Laursen et al. 1997) and it became clear that a homoeologous translocation had indeed taken place.

Various methods are available for producing translocations, including irradiation (Sears 1956; Szakács et al. 2010) or the induction of homoeologous pairing (Riley and Chapman 1958; Sears 1972; Griffiths et al. 2006). A number of genes from common wheat promote chromosome pairing and several act as inhibitors (Sears 1976). The pairing homoeologous gene, *Ph1*, on the long arm of chromosome 5B, has the most decisive effect. In its presence, pairing is restricted to homologues; in its absence, homoeologues also pair, albeit less frequently than homologues. The simple deletion of *Ph1*, or the counteraction of its effect by high-pairing types of *Ae. speltoides* or *Ae. mutica*, can induce many *Triticinae* chromosomes to pair with their wheat homoeologues. Such induced homoeologous pairing is usually the method of choice for transferring genes from alienchromosomes to those of wheat. The "*Ph* system" was used by Islam and Shepherd (1992a) and by Sherman et al. (2001) to produce wheat–barley translocations. A unique genetic system exists in common wheat, which induces frequent chromosomal structural rearrangements. The gametocidal (Gc) system involves alien chromosomes called Gc chromosomes, which were introduced into common wheat from certain wild species belonging to the genus*Aegilops* (Endo 2007). This system proved to be effective in inducing structural rearrangements in the barley chromosomes added to common wheat, as well as in common wheat chromosomes (Endo 2009).

The rearranged chromosomes thus induced include deletions of barley chromosomes and translocations between the barley and wheat chromosomes. Lines carrying rearranged barley chromosomes are designated as "dissection lines" (Endo 2009). Schubert et al. (1998) developed wheat–barley translocations from wheat/barley disomic addition lines by exploiting the gametocidal effect of the 2C chromosome of *Aegilops cylindrica*. The 7H wheat/barley addition line was crossed with the 2C wheat/*Ae. cylindrica* addition line and the resulting line, containing two different alien chromosomes, was self-fertilized. Lines carrying barley deletions and wheat–barley translocations were selected from the progeny. More than ten translocation lines carrying segments of the 7H barley chromosome were produced. The incorporation of the barley chromosome segments was detected by means of GISH, and with FISH using the repetitive probe HvT01. These 7H deletion and translocation lines were then used for the physical mapping of the 7H barley chromosome (Serizawa et al. 2001).

The Gc system was used for the dissection of several barley chromosomes using CS/Betzes disomic addition lines. When the barley chromosome 5H added to common wheat was dissected, chromosomes with structural chromosomal changes involving 5H were selected. Barley-specific EST markers were screened and the authors proved the usefulness of the 5H dissection line for the physical mapping of DNA markers (Ashida et al. 2007). A chromosome 3H addition was used to establish 50 common wheat lines carrying single rearranged (or dissected) 3H chromosomes of independent origin. The dissected 3H chromosomes were either deletions or translocations with wheat chromosomes. These lines were used to map EST markers, after which then polymorphic markers were selected to construct a 3H genetic map (Sakai et al. 2009). The Gc chromosome induced chromosomal structural rearrangements in the progeny of the 4H addition line of common wheat, and the rearranged chromosomes were characterized by sequential C-banding and in situ hybridization. 4H chromosome-specific EST markers were used for cytological mapping (Sakata et al. 2010), while Nasuda et al. (2005) performed chromosomal assignment and deletion mapping of barley EST markers. EST markers were demonstrated to be amplified differently on wheat and barley was assigned to all seven barley chromosomes. By using a set of Betzes ditelosomic additions of CS, the chromosome arm location of 90 % of the EST markers assigned to each barley chromosome was determined. Barley chromosomes 1H and 6H were dissected by thegametocidal system and structural changes were identified by means of GISH and FISH (Ishihara et al 2014). Five aberrations of chromosome 1H were found and 33 dissection lines carrying single aberrant 6H chromosomes were established. PCR analysis of the aberrant barley chromosomes was conducted using 75 and 81 EST markers specific to chromosomes 1H and 6H, respectively. A cytological map of chromosome 6H was compared to the previously reported genetic and physical map. The cytological map had better resolution in the proximal region than the corresponding genetic map (Ishihara et al 2014). The agronomical value of the dissection lines have not been analysed in the above mentioned reports. The barley dissection lines were produced from CS-Betzes addition lines, so they all carried chromosome segments from Betzes barley.

Molnár-Láng et al. (2000a) developed translocations from wheat–barley hybrids multiplied in tissue culture, using GISH for confirmation (Fig. 12.5). The origin of the barley chromosome segments involved in the selected homozygous translocation lines was determined using molecular markers (Nagy et al. 2002). Segments of various sizes from the 1H, 3H, 4H and 5H chromosomes were found to have been incorporated in the translocation lines. These lines were then used for the physical mapping of microsatellite markers previously located on the barley chromosomes. Sepsi et al. (2006) produced wheat/barley translocations as the result of induced homoeologous chromosome pairing in a 4H(4D) wheat–barley substitution line by crossing with the line CO4-1, which carries the *Ph* suppressor gene from *Aegilopsspeltoides*. Kruppa et al. (2013) reported the development of a 4HL.5DL Robertsonian translocation line after crossing the 4H(4D) wheat–barley substitution line with the CS*ph1b* mutant. The rearrangement was confirmed with sequential GISH, FISH and SSR markers. A spontaneous wheat–barley translocation was identified using sequential GISH, FISH and SSR markers by Cseh et al. (2011) in the progenies of the Asakaze × Manas hybrid. This translocation line carries a 4BS wheat chromosome arm and a 7HL chromosome arm from the Ukrainian six-rowed winter barley. Another spontaneous wheat/barley translocation line, identified as 5HS-7DS.7DL, was detected among the progenies of the Mv9kr1 × Igri wheat–barley hybrid (Kruppa et al. 2013) (Fig. 12.6). Despite the non-compensating nature of the translocation, the plants showed good viability. Of the 45 microsatellite markers analysed, ten failed to amplify any 7DS-specific fragments, signalling the elimination of a short chromosome segment in the telomeric region. The breakpoint of the 5HS-7DS.7DL translocation appeared to be more distal than that of reported deletion lines, thus providing a new physical landmark for future deletion mapping studies.

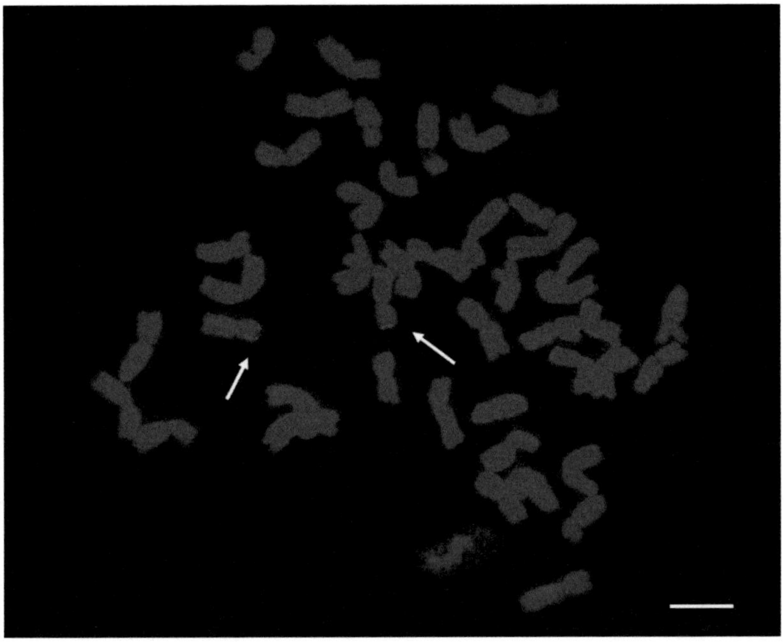

Fig. 12.5 Detection of barley chromosome segments in the 6BS.6BL-4HL translocation (arrows) using GISH. Total barley DNA was labelled with digoxigenin and detected with anti-DIG-Rhodamine (*red*). Wheat chromosomes were counterstained with DAPI (*blue*)

12.6 Morphological and Agronomic Traits of Wheat–Barley Introgressions

Alien additions are primarily produced to add specific desirable genes to a crop species (Gale and Miller 1987), but addition lines can be used for many other purposes, such as mapping genes and markers on introgressed alien chromosomes, examining alien gene regulation, understanding meiotic pairing behaviour and chromosome structure, and isolating individual chromosomes and genes of interest (Chang and de Jong 2005; Cho et al. 2006). The wheat–barley addition lines produced in various cultivar combinations (CS × Betzes, Mv9 kr1 × Igri, Asakaze × Manas) had several morphological traits in common (Molnár-Láng et al. 2012). The 4H additions had the best fertility and 7H the lowest in all three combinations. The 2H addition line had a lax spike structure each of the cultivar combinations. The 3H addition had the shortest, most compact spike of all the addition lines in the Mv9kr1-Igri and CS-Betzes sets (Szakács and Molnár-Láng 2007; Aranyi et al. 2014a). The 3H Asakaze–Manas addition also had a short spike, but it was not as dense as in the other two combinations. This addition line showed a high level of genetic instability, which cannot yet be explained. The 4H addition line had the tallest plants and

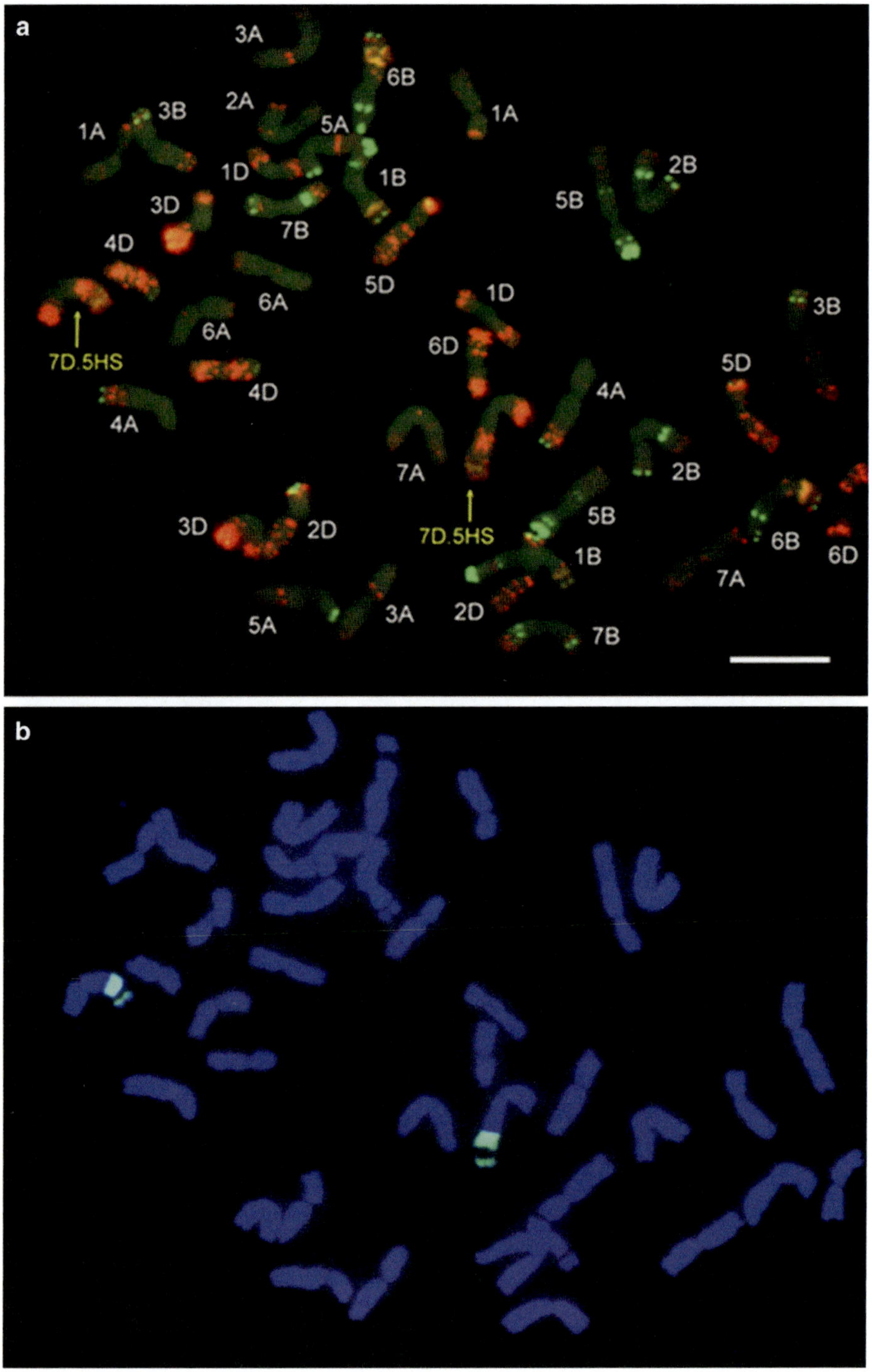

Fig. 12.6 Sequential GISH and FISH on mitotic chromosomes from 7DL.7DS-5HS wheat–barley disomic translocation line. (**a**) Identification of the chromosomes using DNA probes pSc119.2 (*green*), Afa family (*red*) and pTa71 (*yellow*). (**b**) Barley total genomic DNA was labelled with biotin and detected with streptavidin-FITC (*green*). Wheat chromosomes were counterstained with DAPI (*blue*)

3H the shortest. The 6H and 7H additions were shorter than 4H, as also observed in the Mv9kr1-Igri addition lines. Unfortunately no 5H additions could be selected from the Mv9 kr1 × Igri and Asakaze × Manas combinations, as this chromosome was eliminated most frequently from the backcross progenies (Molnár-Láng et al. 2005). Barley chromosome 1H caused sterility even in the presence of other barley chromosomes such as 2H, 3HS, 4H, 5HS and 7H. A fertile addition line involving the entire barley chromosome 1H could not be produced by Islam et al. (1978), because a gene on the long arm of this chromosome caused sterility when present in a wheat background. The double monosomic 1H and 6H addition became partly female fertile and a few backcross seeds were produced after pollinating them with normal wheat (Islam and Shepherd 1990). Partially self-fertile plants disomic for 6H and monosomic for 1H were developed (Islam and Shepherd 2000). Unfortunately none of the BC_2 plants from the Asakaze × Manas and Mv9 kr1 × Igri combinations carried the barley chromosomes 6H and 1H together. Hart et al. (1980) used differences between wheat and barley isozymes to determine the chromosomal locations of barley structural genes for these isozymes. Genes controlling more than 58 isozymes have been allocated to specific barley chromosomes or to the arms thereof using wheat–barley addition lines (Islam and Shepherd 1990). The effect of the added barley chromosomes on heading characters was studied by Murai et al. (1997) using the CS-Betzes addition lines produced by Islam et al. (1978) together with the 5H and 6H Shinchunaga/New Golden additions produced by Koba et al. (1997). The earliest flowering was observed for the CS-Betzes 5H addition line and for the Shinchunaga/New Golden 5H addition lines. Murai et al. (1997) demonstrated that the heading characters of wheat may be altered by barley genes. The Mv9kr-Igri and Asakaze-Manas wheat/winter barley addition lines made it possible to study the effects of chromosomes from winter barley cultivars on flowering time in the wheat genetic background under various environmental conditions (Farkas et al. 2014) (Table 12.3). The winter barley chromosome additions significantly influenced the flowering time of wheat both in a controlled environment test and under field-sown conditions. Unfortunately the 5H addition was missing from both combinations, because 5H was the first chromosome to be eliminated from the backcross derivatives (Molnár-Láng et al. 2005, 2012; Szakács and Molnár-Láng 2010). Of all the barley addition lines, the effect of the 4H and 7H additions was the most characteristic. The 7H addition lines were the earliest in both cultivar combinations in each treatment (Farkas et al. 2014). In the Mv9kr1-Igri combination the 4H addition was the latest under all the environmental conditions (Aranyi et al. 2014a). In the Asakaze–Manas combination the 4H addition was the latest under short-day and long-day illumination in the phytotron, but the 6H addition was the latest without vernalization and in the field in several years (2012, 2013, 2014). There was 12 and 11 days' difference between the flowering times of the 7H and 4H Mv9kr1-Igri and Asakaze-Manas addition lines in the field in 2012, which increased to 52 and 44 days under short-day illumination in the phytotron (Farkas et al. 2014). Only two days' difference was observed between the CS-Betzes 7H and 4H addition lines by Murai et al. (1997) under short-day illumination in the phytotron, which could be

Table 12.3 Flowering time of wheat (*T. aestivum*) -barley (*H. vulgare*) addition lines (Mv9 kr1/Igri, Asakaze/Manas) in the field in Martonvásár in three consecutive years (2012, 2013, 2014)

Genotype	2012 May	2013 May	2014 May
Mv9 kr1	11th	13th	8th
Igri	6th	5th	1st
2H Mv9kr1/Igri	20th	20th	18th
3H Mv9kr1/Igri	18th	24th	16th
4H Mv9kr1/Igri	22nd	21st	20th
6HS Mv9kr1/Igri	16th	14th	9th
7H Mv9kr1/Igri	10th	12th	11th
Asakaze	8th	9th	4th
Manas	4th	6th	1st
2H Asakaze/Manas	13th	10th	7th
3H Asakaze/Manas	16th	10th	8th
4H Asakaze/Manas	17th	14th	12th
6H Asakaze/Manas	22nd	24th	20th
7H Asakaze/Manas	6th	8th	2nd

primarily due to the fact that Betzes, being a spring barley, did not carry the *ZCCT-H* genes at the *Vrn-H2* locus, giving further indirect proof that the effect of *Vrn-H2* was detected in these addition lines.

The dietary fibre (1,3;1,4)- β-D-glucan (β-glucan), is a major quality parameter of cereals. Barley β-glucans are beneficial to human health, as they are a major source of soluble dietary fibre and have been recognized both as potential cholesterol-lowering polysaccharides (Kerckhoffs et al. 2003) and as non-specific immune-activators (Allendorf et al. 2005). The grain of barley is one of the most important β-glucan sources, having a β-glucan content ten times higher than that of wheat. The cellulose synthase-like F6 (*CslF6*) gene, encoding a putative β-glucan synthase, has been assigned to the 7H chromosome (Burton et al. 2008). The presence of the *HvCslF6* gene, responsible for β-glucan production, was revealed in the centromeric region of 7HL using the 4BS.7HL Asakaze-Manas translocation line (Cseh et al. 2011). An increased β-glucan level was also detected in the translocation line, demonstrating that the *HvCslF6* gene is of potential relevance for the manipulation of wheat β-glucan levels. The Mv9kr1-Igri 1HS ditelosomic and Mv9kr1-Igri 7H disomic wheat/barley addition lines carrying the *HvCslF9* and *HvCslF6* barley genes, respectively, were used to investigate the additive effect of barley cellulose synthase-like genes on the wheat β-glucan content (Cseh et al. 2013). A significantly higher β-glucan level was detected in the leaves and grains of the wheat/barley 1HS and 7H addition lines compared to the control wheat line. The expression of the *HvCslF9* and *HvCslF6* genes in the genetic background of wheat was also determined by quantitative RT-PCR, and the *HvCslF9* gene was mapped to the short arm of the 1H chromosome (Cseh et al. 2013). The *HvGlb1* barley gene, encoding β-glucan endohydrolase isoenzyme EI, is possibly involved in the regulation of the β-glucan level during grain development. Previously this was also mapped to the barley 1H chromosome, and this study made it clear that it was located on the 1HL chromosome arm. Zou et al. (2012) recently identified wheat–barley 2HL chromosometranslocation lines derived from crosses between CS-Betzes 2H disomic substitution lines and Chinese wheat varieties. These translocations carry the *Isa* gene encoding the barley bifunctional α-amylase/subtilisin inhibitor (BASI). Because BASI is more effective in inhibiting wheat AMY2 than the α-amylase inhibitors of other cereals (Henry et al. 1992), the introduction of the barley *Isa* gene into wheat may regulate endogenous α-amylase activity during starch granule synthesis in the developing grain and reduce the level of preharvest sprouting damage.

The drought tolerance of a spontaneous 4H(4D) substitution line was studied under laboratory and field conditions (Molnár et al. 2007; Hoffmann et al. 2009) to investigate the ability of the barley 4H chromosome to compensate for wheat 4D in response to mild drought stress (15 % PEG) in barley and in the 4H(4D) substitution line. Mild osmotic stress induced intensive stomatal closure, resulting in reduced water loss through transpiration and unchanged relative water content in the leaves. The water use efficiency under mild osmotic stress increased greatly in these lines (Molnár et al. 2007). The drought tolerance of the 4H(4D) substitution line and of wheat/barley addition and translocation lines was studied under a rain shelter in the field in Keszthely. The difference in water supply between the control and stress treatment was 180 mm (Hoffmann et al. 2009). The largest root/shoot ratio was observed in the 4H(4D) substitution line. Large root biomass could contribute to better drought tolerance. The grain yield of the genotypes was also analysed and no yield loss was observed in the 4H(4D) substitution line during drought treatment. This was confirmed by observations in the next vegetative season (Hoffmann et al. 2010). However, the grain yield of the 4H(4D) substitution was much lower than that of the 3HS.3BL or the 5HS-7DS.7DL translocation lines.

The aluminium tolerance of wheat/barley disomic addition, substitution and translocation lines carry chromosomes from three different barley cultivars (Manas, Igri, Betzes) was evaluated by comparing the root growth in a solution containing 75 μM $AlCl_3$ at pH 4.0 to that of known Al-tolerant and sensitive wheat genotypes (Darkó et al. 2012). The wheat Asakaze komugi, the barley Manas cultivar and their hybrid derivatives were found to have high levels of Al tolerance, while the wheat line Mv9kr1, the barley cultivar Igri and their hybrid progenies were sensitive to Al. In most cases, the Al tolerance of the wheat/barley introgression lines derived from Al-sensitive wheat Mv9kr1 and barley Betzes, which has moderate Al tolerance, was similar to that of the wheat parents, but the 2DS.2DL-1HS translocation line of Mv9kr1/Betzes exhibited more intensive root growth, while accumulating less Al than the parental lines. This indicates that either the lack of the distal part of chromosome 2DL or the presence of the distal part of 1HS improved the Al tolerance level (Darkó et al. 2012).

Salt responses were studied by Darkó et al. (2015) during germination and in young plants of the wheat–barley disomic addition lines 2H, 3H, 4H, 6H and 7H, in ditelosomic addition lines 3HS and 7HL, developed from the Asakaze × Manas hybrid (Molnár-Láng et al. 2000b, 2012), and in the parental genotypes. Two other wheat genotypes, namely Mv9kr1, a winter wheat line originating from Martonvásár and used as a parental genotype in other addition lines and Chinese Spring (CS), the wheat parental genotype of addition lines previously tested for salt tolerance (Colmer et al. 2006) were also used for comparison. Asakaze possesses relatively high salt tolerance, as indicated by the less pronounced reduction in germination % and in root and shooth growth and the retention of high leaf water content and photosynthetic activity, as compared to CS and Mv9kr1. The barley cv Manas showed better salt tolerance than wheat cv Asakaze, although Manas accumulated more Na in the root, its transport to the shoots was restricted. Among the addition lines tested, the disomic addition line 7H and the ditelosomic addition line 7HL exhibited higher salt tolerance both during germination and in the early developmental stages than the wheat parent, which may be related to the elevated osmotic adjustment capacity of these addition lines, similar to that found for barley cv Manas (Darkó et al. 2015).

A spontaneous 3HS.3BL Robertsonian translocation was obtained from the progenies of a Chinese Spring × Betzes wheat–barley hybrid produced in Martonvásár (Molnár-Láng and Sutka 1994). The hybrid was backcrossed with wheat line Mv9kr1, after which it was transferred into the modern Martonvásár wheat variety Mv Bodri. The translocation was detected by means of GISH (Molnár-Láng et al. 2000a). Fluorescence in situ hybridization (FISH) using the barley telomere- and centromere-specific repetitive DNA probes (HvT01 and (AGGGAG)n confirmed that the complete barley chromosome arm was involved in the Robertsonian translocation. Wheat-specific repetitive DNA probes identified the presence of the whole wheat genome, except for the short arm of the 3B chromosome (Fig. 12.7). Genotypes homozygous for the 3HS.3BL translocation were selected, after which morphological analysis was performed on the plants and the yield components were measured in the field during two consecutive vegetative seasons. The introgression of the 3HS.3BL translocation into the modern wheat cultivar Mv Bodri significantly reduced the plant height of the initial translocation line due to the incorporation of the dwarfing allele *RhtD1b* from Mv Bodri and increased its productivity (seeds/spike). The presence of the 3HS.3BL translocation in the Mv9kr1 and MvBodri wheat background improved tillering and seeds/plant productivity in field experiments carried out in Martonvásár and Keszthely, Hungary (Türkösi et al. 2014a, b).

Infection with fungal biotrophic pathogens causing powdery mildew diseases was studied on wheat–barley addition and translocation lines by Aranyi et al. (2014b). Most powdery mildew species are strictly host-specific, colonizing only a narrow range of species or one particular host species. *Blumeria graminis* f. sp. *tritici* isolate 14 (HM484334) was identified on the wheat parent and all the wheat/barley inrogression lines, while *B.graminis f.* sp. *hordei* isolate MUMH1723 (AB 273556) was identified on the barley parent, so the added barley chromosomes did not result in host range expansion for barley powdery mildew.

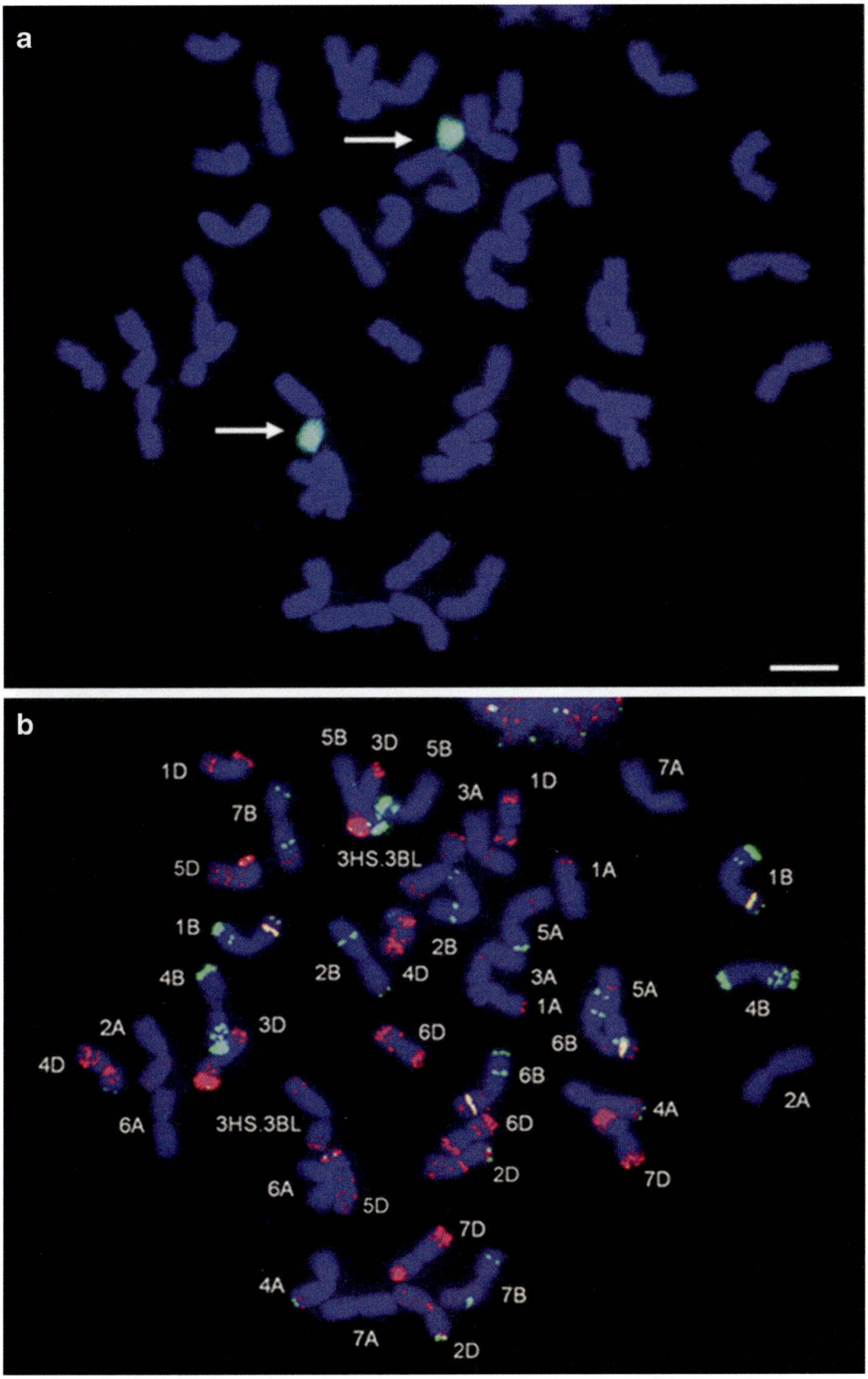

Fig. 12.7 Sequential GISH-FISH analysis on mitotic chromosomes of 3HS.3BL wheat–barley Robertsonian translocation line. (**a**) Labelled barley genomic DNA was used as probe and barley chromosome arm 3HS is highlighted in *green*. Wheat chromosomes were counterstained with DAPI. (**b**) Chromosomes were identified by means of fluorescence in situ hybridization using DNA repetitive probes: Afa family (*red*), pSc119.2 (*green*), and pTa71 (*yellow*)

12.7 Molecular Genetic Studies on Wheat–Barley Introgression Lines

Wheat–barley chromosome addition lines are useful genetic resources for studying the transcript accumulation patterns of barley in a wheat genetic background and for the large-scale physical mapping of genes. In a study performed by Cho et al. (2006), CS-Betzes addition lines were examined with the Barley1 Affymetrix GeneChip probe array and a total of 1787 barley transcripts were detected and physically mapped to barley chromosomes and to the long and short arms of chromosome 6H. The same method and plant materials were used to physically map barley genes to their respective chromosome arm locations by Bilgic et al. (2007), who mapped 1257 barley genes to chromosome arms 1HS, 2HS, 2HL, 3HS, 3HL, 4HS, 4HL, 5HS, 5HL, 7HS and 7HL. The number of genes assigned to individual chromosome arms ranged from 24 to 197. Flow sorting can be effective for isolating large samples of alien chromosomes from metaphase suspensions if the flow karyograms of sorted additions demonstrate distinct peaks not present in those of the parental species (Doležel et al. 2005). The telocentric chromosomes of Betzes barley were isolated from CS-Betzes ditelosomic addition lines, thus allowing the barley genome to be dissected into fractions each representing only about 6–12 % of the total genome (Suchánková et al. 2006). The DNA of flow-sorted chromosomes can be used for the isolation of molecular markers, for physical mapping using PCR and FISH, for the integration of genetic and physical maps and for the construction of chromosome-specific DNA libraries, including sequences cloned in bacterial artificial chromosome vectors. The first barley chromosome to be isolated by flow sorting and shotgun sequencing was 1H (Mayer et al. 2009). As there is no significant difference in the size of the barley chromosomes, the other six chromosomes could only be sorted from wheat/barley ditelosomic addition lines, although some barley chromosomes are identifiable based on morphology. Twelve barley chromosome arms (2HS to 7HL) were purified separately by flow cytometry (Suchánková et al. 2006), after which the DNA was amplified by multiple displacement amplification (MDA) and then shotgun sequenced (Mayer et al. 2011). Using this procedure, between 2261 and 3616 genes were tentatively positioned along each of the individual barley chromosomes, representing a cumulative set of 21,766 genes across the entire barley genome. An additional set of 5815 genes could not be integrated into the genome zippers based on conserved synteny models, but were associated with individual chromosomes/chromosome arms. Overall, it was possible to tentatively position 27,581 barley genes, or 86 % of the estimated 32,000 gene repertoire of the barley genome, into chromosomal regions (Mayer et al. 2011). Among the 21,766 genes anchored to the genome zipper, 3125 genes (14 %) were allocated to the genetic centromeres. Based on the 454 sequence and array-based gene assignments to chromosome arms, all but nine of these 3125 genes were distributed to specific arms of chromosomes 1H to 7H. Not much later, using the whole genome approach, a deep physical map of 4.98 Gb was developed and more than 3.90 Gb was anchored to a high-resolution genetic map (The International Barley Genome Sequencing Consortium 2012). By combining the physical map with a complementary short-read whole-genome assembly and with high-coverage RNAseq data, approximately 80 % of the barley genome could be delineated, including more than 90 % of the expressed genes. These chromosome- or whole-genome-derived genomic resources provide an essential platform to advance gene discovery and genome-assisted crop improvement.

12.8 Conclusions

From the practical point of view new wheat–barley hybrids need to be produced using a wider range of barley genotypes carrying genes responsible for useful agronomic traits (e.g. drought tolerance, high β-glucan content, salt tolerance, earliness). The major limitation for successful gene transfer from barley into wheat is the low crossability between these species. The efficiency with which wheat × barley hybrids can be produced should be increased by hormone treatment at pollination, and by improving the yield of embryo culture. Among the various methods available for producing translocations from wheat × barley hybrids and additions, the gametocidal system is currently the most promising. Although Tritordeum already exists as the product of the wheat × *H. chilense* combination, and fertile amphiploids have been produced with *H. marinum* and *H. californicum*, fertile *T. aestivum* × *H. vulgare* amphiploids have not yet been developed. Unfortunately, chromosome 1H of *H. vulgare* carries a gene (*Shw*) that causes sterility in the wheat background (Taketa et al. 2002), thus preventing the production of a fertile amphiploid. The development of small barley introgressions in the wheat genetic background has been started, but a much larger number of translocations carrying genes responsible for useful traits is needed. Wheat/barley translocations are ideal material for the physical mapping of wheat and barley chromosomes, as the two genomes can be clearly identified using GISH, and the physical landmarks can be used in genome mapping. The efficient manipulation of alien chromatin and the selection of proper recipient genotypes play a central role in the success of alien introgression. The physical, genetic and functional assembly of the barley genome (Mayer et al. 2011; The International Barley Genome Sequencing Consortium 2012) has made an important contribution to the targeted introgression of barleygenes into the wheat genome and the exact identification of wheat–barley introgression lines.

Acknowledgements This work was supported by the Hungarian National Scientific Research Fund (OTKA K 104382 and K 108555) and by the "Wheat and Barley Legacy for Breeding Improvement"—WHEALBI EU FP7 project. Thanks are due to Barbara Hooper for revising the manuscript linguistically.

References

Allendorf DJ, Yan J, Ross GD, Hansen RD, Baran JT, Subbarao K, Wang L, Haribabu B (2005) C5a-mediated leukotriene B4-amplified neutrophil chemotaxis is essential in tumor immunotherapy facilitated by anti-tumor monoclonal antibody and beta-glucan. J Immunol 174:7050–7056

Aranyi NR, Molnár-Láng M, Hoffmann S, Hoffmann B (2014a) Phenotypic characterization of wheat-barley 'Mv9 kr1'/'Igri' introgression lines in the field. Plant Breeding 133(6):718–721. doi:10.1111/pbr.12213

Aranyi NR, Varga I, Poczai P, Cernák I, Vida G, Molnár-Láng M, Hoffmann B (2014b) What types of powdery mildew can infect wheat-barley introgression lines? Eur J Plant Pathol 139:19–25. doi:10.1007/s10658-014-0382-0

Armstrong KC, Nakamura C, Keller WA (1983) Karyotype instability in tissue culture regenerants of Triticale (× *Triticosecale* Wittmack) cv. 'Welsh' from 6-month-old callus cultures. Z Pflanzenzüch 91:233–245

Ashida T, Nasuda S, Sato K, Endo TR (2007) Dissection of barley chromosome 5H in common wheat. Genes Genet Syst 82:123–133. doi:10.1266/ggs.82.123

Bai D, Knott DR (1993) The effects of level of 2,4-D and time in culture on regeneration rate and chromosome numbers of regenerants from calli of the hybrid *Triticum aestivum* cv. Chinese Spring *ph1b*×*Thinopyrum ponticum* (2n=10x=70). Genome 36:166–172. doi:10.1139/g93-022

Barcelo P, Hagel C, Becker D, Martín A, Lörz H (1994) Transgenic cereal (tritordeum) plants obtained at high efficiency by microprojectile bombardment of inflorescence tissue. Plant J 5:583–592. doi:10.1007/s11627-997-0003-0

Barclay IR (1975) High frequencies of haploid production in wheat (*Triticum aestivum*) by chromosome elimination. Nature (London) 256:410–411. doi:10.1038/256410a0

Bilgic H, Cho S, Garvin DF, Muehlbauer GJ (2007) Mapping barley genes to chromosome arms by transcript profiling of wheat-barley ditelosomic chromosome addition lines. Genome 50:898–906. doi:10.1139/G07-059

Blanco A, Fracchiolla GV, Greco B (1986) Intergeneric wheat×barley hybrid. J Hered 77: 98–100

Bothmer R, von Flink J, Landström T (1986) Meiosis in interspecific *Hordeum* hybrids. I. Diploid combinations. Can J Genet Cytol 28:525–535. doi:10.1139/g86-077

Bothmer R, von Flink J, Landström T (1987) Meiosis in interspecific *Hordeum* hybrids. II. Triploid hybrids. Evol Trends Plants 1:41–50

Burton RA, Jobling SA, Harvey AJ, Shirley NJ, Mather DE, Bacic A, Fincher GB (2008) The genetics and transcriptional profiles of the cellulose synthase-like *HvCSLF* gene family in barley. Plant Physiol 146:1821–1833. doi:10.1104/pp. 107.114694

Cabrera A, Friebe B, Jiang J, Gill BS (1995) Characterization of *Hordeum chilense* chromosomes by C-banding and in situ hybridization using highly repeated DNA probes. Genome 38:435–442. doi:10.1139/g95-057

Chang SB, de Jong H (2005) Production of alien chromosome additions and their utility in plant genetics. Cytogenet Genome Res 109:335–343. doi:10.1159/000082417

Chapman V, Miller TE (1978) The amphiploid of *Hordeum chilense*×*Triticum aestivum*. Cereal Res Commun 6:351–352

Cho S, Garvin DF, Muehlbauer GJ (2006) Transcriptome analysis and physical mappping of barley genes in wheat-barley addition lines. Genetics 172:1277–1285. doi:10.1534/genetics.105.049908

Chu CC, Sun CS, Chen X, Zhang WX, Du ZH (1984) Somatic embryogenesis and plant regeneration in callus from inflorescences of *Hordeum vulgare*×*Triticum aestivum* hybrids. Theor Appl Genet 68:375–379. doi:10.1007/BF00267892

Clauss E (1980) Trigeneric hybrids between barley, wheat and rye. Cereal Res Commun 2:341–347

Clauss E (1983) Bastarde aus *Hordeum geniculatum* all Und Triticum aestivum L. Arch Züchtungf 13:413–418

Colmer TD, Flowers TJ, Munns R (2006) Use of wild relatives to improve salt tolerance in wheat. J Exp Bot 57:1059–1078

Cseh A, Kruppa K, Molnár I, Rakszegi M, Doležel J, Molnár-Láng M (2011) Characterization of a new 4BS.7HL wheat-barley translocation line using GISH, FISH, and SSR markers and its effect on the β-glucan content of wheat. Genome 54:795–804. doi:10.1139/g11-044

Cseh A, Soós V, Rakszegi M, Türkösi E, Balázs E, Molnár-Láng M (2013) Expression of *HvCslF9* and *HvCslF6* barley genes in the genetic background of wheat and their influence on the wheat β-glucan content. Ann Appl Biol 163:142–150

Cubero JI, Martín A, Millan T, Gomez-Cabrera A, De Haro A (1986) Tritordeum: a new alloploid of potential importance as a protein source crop. Crop Sci 26:1186–1190

Dahleen LS (1999) Tissue culture increases meiotic pairing of regenerants from barley × Canada wild rye hybrids. J Hered 90:265–269. doi:10.1093/jhered/90.2.265

Darkó É, Barnabás B, Molnár-Láng M (2012) Characterization of newly developed wheat/barley introgression lines in respect of aluminium tolerance. Am J Plant Sci 3:1462–1469. doi:10.4236/ajps.2012.310176

Darkó É, Janda T, Majláth I, Szopkó D, Dulai S, Molnár I, Türkösi E, Molnár-Láng M (2015) Salt stress response of wheat-barley addition lines carrying chromosomes from the winter barley "Manas". Euphytica 203:491–504 doi:10.1007/s10681-014-1245-7

Dewey DR (1984) The genomic system of classification as a guide to intergeneric hybridization with the perennial *Triticeae*. In: Gustafson JP (ed) Gene manipulation in plant improvement. Plenum, New York, NY, pp 209–279

Doležel J, Kubaláková M, Bartoš J, Macas J (2005) Chromosome flow sorting and physical mapping. In: Khalid M, Günter K (eds) The handbook of plant genome mapping: genetic and physical mapping. Wiley-VCH, Berlin, pp 151–171. doi:10.1002/3527603514.ch7

Doré C, Cauderon Y, Chueca MC (1988) Inflorescence culture of *Triticum aestivum* × *Secale cereale* hybrids: production, characterization and cytogenetic analysis of regenerated plants. Genome 30:511–518. doi:10.1139/g88-086

Endo TR (2007) The gametocidal chromosome as a tool for chromosome manipulation in wheat. Chromosome Res 15:67–75. doi:10.1007/s10577-006-1100-3

Endo TR (2009) Cytological dissection of barley genome by the gametocidal system. Breed Sci 59:481–486. doi:10.1270/jsbbs.59.481

Farkas A, Molnár I, Kiss T, Karsai I, Molnár-Láng M (2014) Effect of added barley chromosomes on the flowering time of new wheat/winter barley addition lines in various environments. Euphytica 195:45–55. doi:10.1007/s10681-013-0970-7

Fedak G (1977) Increased homoeologous chromosome pairing in *Hordeum vulgare* × *Triticum aestivum* hybrids. Nature 266:529–530. doi:10.1038/266529a0

Fedak G (1980) Production, morphology and meiosis of reciprocal barley-wheat hybrids. Can J Genet Cytol 22:117–123. doi:10.1139/g80-014

Fedak G (1983) Hybrids between *Hordeum pubiflorum* and *Triticum aestivum*. Barley Genet Newsl 13:59

Fedak G (1985) Propagation of intergeneric hybrids of Triticeae through callus culture of immature inflorescence. Z Pflanzenzüch 94:1–7

Fedak G, Grainger J (1986) Chromosome instability in somaclones of a *Triticum aestivum* × *Hordeum vulgare* hybrid. Can J Genet Cytol 28:618–623. doi:10.1139/g86-090

Fedak G, Jui PY (1982) Chromosomes of Chinese Spring wheat carrying genes for crossability with Betzes barley. Can J Genet Cytol 24:227–233

Fedak G, Armstrong KC, Handyside RJ (1987) Chromosome instability in wheat plants regenerated from suspension culture. Genome 29:627–629. doi:10.1139/g87-104

Fernandez JA, Gonzalez JM, Jouve N (1985) Meiotic pairing of the amphiploid *Hordeum chilense* × *Triticum turgidum* conv. durum studied by means of Giemsa C-banding technique. Theor Appl Genet 70:85–91. doi:10.1007/BF00264487

Fernandez-Escobar J, Martín A (1985) Morphology, cytology and fertility of a trigeneric hybrid from *Triticale* × *Tritordeum*. Z Pflanzenzüch 95:311–318
Finch RA, Bennett MD (1980) Meiotic and mitotic chromosome behaviour in new hybrids of *Hordeum* with *Triticum* and *Secale*. Heredity 44:201–209. doi:10.1038/hdy.1980.17
Gale MD, Miller TE (1987) The introduction of alien genetic variation into wheat. In: Lupton FGH (ed) Wheat breeding: its scientific basis. Chapman and Hall, London, pp 173–210
Galiba G, Molnár-Láng M, Sutka J (1986) In vitro multiplication of a barley (*Hordeum vulgare* L.) × wheat (*Triticum aestivum* L.) hybrid. Növénytermelés 35:481–485
Griffiths S, Sharp R, Foote TN, Bertin I, Wanous M, Reader S, Colas I, Moore G (2006) Molecular characterization of *Ph1* as a major chromosome pairing locus in polyploid wheat. Nature 439:749–752. doi:10.1038/nature04434
Gupta PK, Fedak G (1985) Intergeneric hybrids between *Hordeum californicum* and *Triticum aestivum*. J Hered 76:365–368
Hart GE, Islam AKMR, Shepherd KW (1980) Use of isozymes as chromosome markers in the isolation and characterization of wheat-barley chromosome addition lines. Genet Res Camb 36:311–325
Henry RJ, Battershell VG, Brennan PS, Onon K (1992) Control of wheat α-amylase using inhibitors from cereals. J Sci Food Agric 58:281–284. doi:10.1002/jsfa.2740580218
Hernandez P, Barcelo P, Lazzeri PA, Lörz H, Martín A, Bohanec B (2001) Agronomic performance of transgenic Tritordeum. Biotechnological approaches for utilisation of gametic cells. COST 824 final meeting, Bled, Slovenia, 1–5 July 2000, pp 199–204
Hoffmann B, Aranyi NR, Hoffmann S, Molnár-Láng M (2009) Possibilities to increase stress tolerance of wheat. Cereal Res Commun 37(Suppl):93–96
Hoffmann B, Aranyi NR, Molnár-Láng M (2010) Characterization of wheat/barley introgression lines for drought tolerance. Acta Agron Hung 58:211–218
Ishihara A, Mizuno N, Islam AKMR, Doležel J, Endo TR, Nasuda S (2014) Dissection of barley chromosomes 1H and 6H by the gametocidal system. Genes Genet Syst 89:203–214
Islam AKMR (1983) Ditelosomic additions of barley chromosomes to wheat. In: Sakamoto S (ed) Proceedings of the 6th International Wheat Genet Symposium, Kyoto, Japan, pp 233–238
Islam AKMR, Colmer TD (2008) Attempts to transfer salt- and waterlogging tolerances from sea barleygrass (*Hordeum marinum* Huds.) to wheat. In: Appels R, Eastwood R, Lagudah E, Langridge P, Mackay M, McIntyre L, Sharp P (eds) Proceedings of the 11th international wheat genetics symposium, Brisbane, Australia. Sydney University Press, Brisbane, QLD, pp 336–338
Islam AKMR, Shepherd KW (1980) Meiotic restitution in wheat-barley hybrids. Chromosoma 79:363–372. doi:10.1007/BF00327326
Islam AKMR, Shepherd KW (1988) Induced pairing between wheat and barley chromosomes. In: Miller TE, Koebner RMD (eds) Proceedings of the 7th International Wheat Genetics Symposium, Cambridge, England, pp 309–314
Islam AKMR, Shepherd KW (1990) Incorporation of barley chromosomes into wheat. In: Bajaj YPS (ed) Biotechnology in agriculture and forestry. Volume 13, wheat, vol 13. Springer, Berlin, pp 128–151
Islam AKMR, Shepherd KW (1992a) Production of wheat-barley recombinant chromosomes through induced homoeologous pairing. 1. Isolation of recombinants involving barley arms 3HL and 6HL. Theor Appl Genet 83:489–494. doi:10.1007/BF00226538
Islam AKMR, Shepherd KW (1992b) Substituting ability of individual barley chromosomes for wheat chromosomes. 1. Substitutions involving barley chromosomes 1, 3 and 6. Plant Breeding 109:141–150. doi:10.1111/j.1439-0523.1992.tb00164.x
Islam AKMR, Shepherd KW (1995) Substitution of barley chromosome 4 for group 4 homoelogues of wheat. In: Li ZS, Xin ZY (eds) Proceedings of the 8th International Wheat Genetics Symposium, Beijing, China. China Agricultural Scientech Press, Beijing, pp 141–144, 20–25 July 1993

Islam AKMR, Shepherd KW (2000) Isolation of a fertile wheat-barley addition line carrying the entire barley chromosome 1H. Euphytica 111:145–149. doi:10.1023/A:1003822719317

Islam AKMR, Shepherd KW, Sparrow DHB (1975) Addition of individual barley chromosomes to wheat. Proceedings of the 3rd International Barley Genetics Symposium, Garching, W. Germany, pp 260–270

Islam AKMR, Shepherd KW, Sparrow DHB (1978) Production and characterization of wheat-barley addition lines. In: Ramanujam S (ed) Proceedings of the 11th International Wheat Genetics Symposium, New Delhi, India, pp 356–371

Islam AKMR, Shepherd KW, Sparrow DHB (1981) Isolation and characterization of euplasmic wheat-barley chromosome addition lines. Heredity 46:160–174. doi:10.1038/hdy.1981.24

Islam S, Malik AI, Islam AKMR, Colmer TD (2007) Salt tolerance in a *Hordeum marinum-Triticum aestivum* amphiploid and its parents. J Exp Bot 58:1219–1229. doi:10.1093/jxb/erl293

Jauhar PP (1995) Morphological and cytological characteristics of some wheat × barley hybrids. Theor Appl Genet 90:872–877. doi:10.1007/BF00222025

Jiang J, Dajun L (1987) New *Hordeum – Triticum* hybrids. Cereal Res Commun 15:95–99

Junming L, Mei Z, Tishu C (1985) Somaclonal variation in intergeneric hybrids of *Hordeum vulgare* × *Triticum aestivum*. Acta Genet Sin 12:434–439

Kerckhoffs DA, Hornstra G, Mensink RP (2003) Cholesterol-lowering effect of beta-glucan from oat bran in mildly hypercholesterolemic subjects may decrease when beta-glucan is incorporated into bread and cookies. Am J Clin Nutr 78:221–227

Kimber G, Sallee PJ (1976) A hybrid between *Triticum timopheevii* and *Hordeum bogdanii*. Cereal Res Commun 4:33–37

Koba T, Shimada T, Otani M, Niizeki H (1988) Chromosomal and morphological variation in plants regenerated from calli of immature embryos and inflorescences of a barley-wheat hybrid. In: Miller TE, Koebner RMD (eds) Proceedings of the 7th International Wheat Genetics Symposium, Cambridge, UK, pp 757–762

Koba T, Handa T, Shimada T (1991) Efficient production of wheat-barley hybrids and preferential elimination of barley chromosomes. Theor Appl Genet 81:285–292. doi:10.1007/BF00228665

Koba T, Takumi S, Shimada T (1997) Isolation, identification and characterization of disomic and translocated barley chromosome addition lines of common wheat. Euphytica 96:289–296. doi:10.1023/A:1003081619338

Kruppa K, Sepsi A, Szakács É, Röder MS, Molnár-Láng M (2013) Characterization of a 5HS-7DS.7DL wheat-barley translocation line and physical mapping of the 7D chromosome using SSR markers. J Appl Genet. doi:10.1007/s13353-013-0152-2

Kruse A (1973) *Hordeum* × *Triticum* hybrids. Hereditas 73:157–161. doi:10.1111/j.1601-5223.1973.tb01078.x

Larkin PJ, Scowcroft WR (1981) Somaclonal variation – a novel source of variability from cell cultures for plant improvement. Theor Appl Genet 60:197–204. doi:10.1007/BF02342540

Lima-Brito J, Guedes Pinto H, Harrison GE, Heslop-Harrison JS (1996) Chromosome identification and nuclear architecture in triticale × tritordeum F_1 hybrids. J Exp Bot 47:583–588. doi:10.1093/jxb/47.4.583

Linde-Laursen I, Heslop-Harrison JS, Shepherd KW, Taketa S (1997) The barley genome and its relationship with the wheat genomes. A survey with internationally agreed recommendation for barley chromosome nomenclature. Hereditas 126:1–16. doi:10.1111/j.1601-5223.1997.00001.x

Löve A (1982) Generic evolution of the wheatgrasses. Biol Zentralbl 101:199–212

Löve A (1984) Conspectus of the *Triticeae*. Feddes Repert 95:425–521

Malysheva L, Sjakste T, Matzk F, Röder M, Ganal M (2003) Molecular cytogenetic analysis of wheat-barley hybrids using genomic in situ hybridization and barley microsatellite markers. Genome 46:314–322. doi:10.1139/g02-117

Martín A, Chapman V (1977) A hybrid between *Hordeum chilense* and *Triticum aestivum*. Cereal Res Commun 5–4:365–366

Martín A, Cubero JI (1981) The use of *Hordeum chilense* in cereal breeding. Cereal Res Commun 9:317–323
Martín A, Sanchez-Monge Laguna E (1980) A hybrid between *Hordeum chilense* and *Triticum turgidum*. Cereal Res Commun 8:349–353
Martín A, Sanchez-Monge Laguna E (1982) Cytology and morphology of the amphiploid *Hordeum chilense* × *Triticum turgidum* conv. durum. Euphytica 31:261–267. doi:10.1007/BF00028329
Martin A, Padilla JA, Fernandez-Escobar J (1987) The amphiploid *Hordeum chilense* × *Triticum aestivum* ssp. *sphaerococcum*. Variability in Octoploid Tritordeum. Plant Breeding 99: 336–339
Martín A, Rubiales D, Rubio JM, Cabrera A (1995) Hybrids between *Hordeum vulgare* and tetra-hexa-, and octoploid tritordeums (amphiploid *H. chilense* × *Triticum* ssp.). Hereditas (Lund) 123:175–182. doi:10.1111/j.1601-5223.1995.00175.x
Martín AC, Atienza SG, Ramirez MC, Barro F, Martín A (2009) Chromosome engineering in wheat to restore male fertility in the msH1 CMS system. Mol Breeding 24:397–408. doi:10.1007/s11032-009-9301-z
Mayer KFX, Taudien S, Martis M, Simková H, Suchánková P, Gundlach H, Wicker T, Petzold A, Felder M, Steuernagel B, Scholz U, Graner A, Platzer M, Doležel J, Stein N (2009) Gene content and virtual gene order of barley chromosome 1H. Plant Physiol 151:496–505. doi:10.1104/pp. 109.142612
Mayer KF, Martis M, Hedley PE, Simková H, Liu H, Morris JA, Steuernagel B, Taudien S, Roessner S, Gundlach H, Kubaláková M, Suchánková P, Murat F, Felder M, Nussbaumer T, Graner A, Salse J, Endo T, Sakai H, Tanaka T, Itoh T, Sato K, Platzer M, Matsumoto T, Scholz U, Doležel J, Waugh R, Stein N (2011) Unlocking the barley genome by chromosomal and comparative genomics. Plant Cell 23:1249–1263. doi:10.1105/tpc.110.082537
Miller TE, Reader SM, Chapman V (1981) The addition of *Hordeum chilense* chromosomes to wheat. Induced variability in plant breeding. International Symposium Eucarpia Pudoc, Wageningen, pp 79–81
Molnár I, Linc G, Dulai S, Nagy ED, Molnár-Láng M (2007) Ability of chromosome 4H to compensate for 4D in response to drought stress in a newly developed and identified wheat–barley 4H(4D) disomic substitution line. Plant Breeding 126:369–374. doi:10.1111/j.1439-0523.2007.01300.x
Molnár-Láng M, Sutka J (1994) The effect of temperature on seed set and embryo development in reciprocal crosses of wheat and barley. Euphytica 78:53–58. doi:10.1007/BF00021397
Molnár-Láng M, Sutka J, Barnabás B, Sági L, Belea A (1985) Production of barley (*Hordeum vulgare* L.) × wheat (*Triticum aestivum* L.) hybrids. Növénytermelés 34:257–262
Molnár-Láng M, Galiba G, Kovács G, Sutka J (1991) Changes in the fertility and meiotic behaviour of barley (*Hordeum vulgare*) × wheat (*Triticum aestivum*) hybrids regenerated from tissue cultures. Genome 34:261–266. doi:10.1139/g91-041
Molnár-Láng M, Linc G, Friebe BR, Sutka J (2000a) Detection of wheat-barley translocations by genomic in situ hybridization in derivatives of hybrids multiplied in vitro. Euphytica 112:117–123. doi:10.1023/A:1003840200744
Molnár-Láng M, Linc G, Logojan A, Sutka J (2000b) Production and meiotic pairing behaviour of new hybrids of winter wheat (*Triticum aestivum*) x winter barley (*Hordeum vulgare*). Genome 43:1045–1054. doi:10.1139/g00-079
Molnár-Láng M, Novotny C, Linc G, Nagy ED (2005) Changes in the meiotic pairing behaviour of a winter wheat-winter barley hybrid maintained for a long term in tissue culture, and tracing the barley chromatin in the progenies using GISH and SSR markers. Plant Breeding 124:247–252. doi:10.1111/j.1439-0523.2005.01097.x
Molnár-Láng M, Szakács É, Linc G, Nagy ED (2007) Development and molecular cytogenetic identification of new winter wheat/winter barley disomic addition lines. In: Buck HT, Nisi JE, Salomón N (eds) Wheat production in stressed environments, vol 12, Developments in plant breeding. Springer, Dordrecht, pp 707–713

Molnár-Láng M, Kruppa K, Cseh A, Bucsi J, Linc G (2012) Identification and phenotypic description of new wheat – six-rowed winter barley disomic additions. Genome 55:302–311. doi:10.1139/g2012-013

Molnár-Láng M, Linc G, Szakács É (2014) Wheat-barley hybridization: the last 40 years. Euphytica 195:315–329. doi:10.1007/s10681-013-1009-9

Mujeeb-Kazi A (1981) Apomictic progeny derived from intergeneric *Hordeum-Triticum* hybrids. J Hered 72:284–285

Mujeeb-Kazi A, Rodriguez R (1983) Meiotic instability in *Hordeum vulgare* × *Triticum aestivum* hybrids. J Hered 74:292–296

Murai K, Koba T, Shimada T (1997) Effects of barley chromosome on heading characters in wheat-barley chromosome addition lines. Euphytica 96:281–287. doi:10.1023/A:1003025501591

Nagy ED, Molnár-Láng M, Linc G, Láng L (2002) Identification of wheat-barley translocations by sequential GISH and two-colour FISH in combination with the use of genetically mapped barley SSR markers. Genome 45:1238–1247. doi:10.1139/g02-068

Nakamura C, Keller WA, Fedak G (1981) In vitro propagation and chromosome doubling of a *Triticum crassum* × *Hordeum vulgare* intergeneric hybrid. Theor Appl Genet 60:89–96. doi:10.1007/BF00282423

Nasuda S, Kikkawa Y, Ashida T, Islam AKMR, Sato K, Endo TR (2005) Chromosomal assignment and deletion mapping of barley EST markers. Genes Genet Syst 80:357–366. doi:10.1266/ggs.80.357

Park CH, Kumar PS, Walton PD (1990) Plant regeneration and chromosomal stability in tissue cultures of the hybrids of *Elymus canadensis* with *Psathyrostachys juncea* and *Secale cereale*. Plant Breeding 104:184–189. doi:10.1111/j.1439-0523.1990.tb00421.x

Pershina LA, Shumny VK (1981) A characterization of clonal propagation of barley × rye and barley × wheat hybrids by means of tissue cultures. Cereal Res Commun 9:273–279

Pershina LA, Numerova OM, Belova LI, Devyatkina EP, Shumny VK (1988) Fertility in barley × wheat hybrids *H. geniculatum* All. × *T. aestivum* L., their regenerants and hybrid progeny of backcrosses to *T. aestivum* L. Cereal Res Commun 16:157–163

Pershina LA, Deviatkina EP, Belova LI, Trubacheeva NV, Arbuzova VS, Kravtsova LA (2009) Features of alloplasmic wheat-barley substitution and addition lines (*Hordeum marinum* subsp. *gussoneanum* – *Triticum aestivum*). Genetika 45:1386–1392. doi:10.1134/S102279540910010X

Polgári D, Cseh A, Szakács E, Jager K, Molnár-Láng M, Sági L (2014) High-frequency generation and characterization of intergeneric hybrids and haploids from new wheat-barley crosses. Plant Cell Rep 33(8):1323–1331

Riley R, Chapman V (1958) Genetic control of cytologically diploid behaviour of hexaploid wheat. Nature 182:713–715. doi:10.1038/182713a0

Riley R, Law CN (1965) Genetic variation in chromosome pairing. Adv Genet 13:57–114. doi:10.1016/S0065-2660(08)60047-4

Sakai K, Nasuda S, Sato K, Endo TR (2009) Dissection of barley chromosome 3H in common wheat and a comparison of physical and genetic maps. Genes Genet Syst 84:25–34. doi:10.1266/ggs.84.25

Sakata M, Nasuda S, Endo TR (2010) Dissection of barley chromosome 4H in common wheat by the gametocidal system and cytological mapping of chromosome 4H with EST markers. Genes Genet Syst 85:19–29

Schubert I, Shi F, Fuchs J, Endo TR (1998) An efficient screening for terminal deletions and translocations of barley chromosomes added to common wheat. Plant J 14:489–495. doi:10.1046/j.1365-313X.1998.00125.x

Schwarzacher T, Anamthawat-Jónsson K, Harrison GE, Islam AKMR, Jia JZ, King IP, Leitch AR, Miller TE, Reader SM, Rogers WJ, Shi M, Heslop-Harrison JS (1992) Genomic in situ hybridization to identify alien chromosomes and chromosome segments in wheat. Theor Appl Genet 84:778–786. doi:10.1007/BF00227384

Sears ER (1956) The transfer of leaf rust resistance from *Aegilops umbellulata* to wheat. Brookhaven Symp Biol 9:1–22

Sears ER (1972) Chromosome engineering in wheat. Stadler Symp 4:23–38
Sears ER (1976) Genetic control of chromosome pairing in wheat. Ann Rev Genet 10:31–51. doi:10.1146/annurev.ge.10.120176.000335
Sepsi A, Németh K, Molnár I, Szakács É, Molnár-Láng M (2006) Induction of chromosome rearrangements in a 4H(4D) wheat-barley substitution using a wheat line containing a *ph* suppressor gene. Cereal Res Commun 34:1215–1222. doi:10.1556/CRC.34.2006.4.261
Serizawa N, Nasuda S, Shi F, Endo TR, Prodanovic S, Schubert I, Künzel G (2001) Deletion-based physical mapping of barley chromosome 7H. Theor Appl Genet 103:827–834. doi:10.1007/s001220100703
Sethi GS, Finch RA, Miller TE (1986) A bread wheat (*Triticum aestivum*) × cultivated barley (*Hordeum vulgare*) hybrid with homoeologous chromosome pairing. Can J Genet Cytol 28:777–782. doi:10.1139/g86-109
Sharma HC, Gill BS, Sears RG (1984) Inflorescence culture of wheat-*Agropyron* hybrids: callus induction, plant regeneration, and potential in overcoming sterility barriers. Plant Cell Tissue Org 3:247–255. doi:10.1007/BF00040344
Sherman JD, Smith LY, Blake TK, Talbert LE (2001) Identification of barley genome segments introgressed into wheat using PCR markers. Genome 44:38–44. doi:10.1139/g00-092
Shimada T, Koba T, Otani M, Niizeki H (1987) Morphology, meiosis and in vitro propagation of barley-wheat hybrids. Proceedings of the 5th International Barley Genetics Symposium, Okayama, pp 343–350
Shumny VK, Pershina LA (1979) Production of barley-rye hybrids and their clonal propagation by the method of isolated tissue cultivation. Dokl Acad Nauk SSSR 249:218–220
Shumny VK, Pershina LA, Belova LI (1981) Production of barley x rye and barley x wheat hybrids. Cereal Res Commun 4:265–272
Suchánková P, Kubaláková M, Kovářová P, Bartoš J, Číhalíková J, Molnár-Láng M, Endo TR, Doležel J (2006) Dissection of the nuclear genome of barley by chromosome flow sorting. Theor Appl Genet 113:651–659. doi:10.1007/s00122-006-0329-8
Surikov JM, Kissel NI (1988) Mikroklonirovanie yachmenno-pshenichnikh gibridov. Dokl Vaskhnil 1:2–4
Svitashev S, Bryngelsson T, Vershinin A, Pedersen C, Säll T, von Bothmer R (1994) Phylogenetic analysis of the genus *Hordeum* using repetitive DNA sequences. Theor Appl Genet 89:801–810. doi:10.1007/BF00224500
Szakács É, Molnár-Láng M (2007) Development and molecular cytogenetic identification of new winter wheat/winter barley (Martonvásári 9 kr1/Igri) disomic addition lines. Genome 50:43–50. doi:10.1139/g06-134
Szakács É, Molnár-Láng M (2010) Identification of new winter wheat – winter barley addition lines (6HS and 7H) using fluorescence in situ hybridization and stability of the whole 'Martonvásári 9 kr1'–'Igri' addition set. Genome 53:35–44. doi:10.1139/G09-085
Szakács É, Kruppa K, Molnár I, Molnár-Láng M (2010) Induction of wheat/barley translocations by irradiation and their detection by in situ hybridization. Acta Agron Hung 58:203–209. doi:10.1556/AAgr.58.2010.3.2
Taketa S, Takeda K (2001) Production and characterization of a complete set of wheat-wild barley (*Hordeum vulgare* ssp. *spontaneum*) chromosome addition lines. Breed Sci 51:199–206. doi:10.1270/jsbbs.51.199
Taketa S, Kato J, Takeda K (1995) High crossability of wild barley (*Hordeum spontaneum* C. Koch) with bread wheat and the differential elimination of barley chromosomes in the hybrids. Theor Appl Genet 91:1203–1209. doi:10.1007/BF00220930
Taketa S, Takahashi H, Takeda K (1998) Genetic variation in barley of crossability with wheat and its quantitative trait loci analysis. Euphytica 103:187–193. doi:10.1023/A:1018344119747
Taketa S, Choda M, Ohashi R, Ichii M, Takeda K (2002) Molecular and physical mapping of a barley gene on chromosome arm 1HL that causes sterility in hybrids with wheat. Genome 45:617–625. doi:10.1139/g02-024

Ter Kuile N, Nabors M, Mujeeb-Kazi A (1988) Callus culture induced amphiploids of *Triticum aestivum* and *T. turgidum* × *Aegilops variabilis* F_1 hybrids: production, cytogenetics and practical significance. In: 80th Annual Meetings of the American Society of Agron Abstract 98

The International Barley Genome Sequencing Consortium (2012) A physical, genetic and functional sequence assembly of the barley genome. Nature 491:711–717

Thomas JB, Mujeeb-Kazi A, Rodriguez R, Bates LS (1977) Barley × wheat hybrids. Cereal Res Commun 5:181–188

Türkösi E, Farkas A, Aranyi NR, Hoffmann B, Tóth V, Molnár-Láng M (2014a) Improvement of the agronomic traits of a wheat/barley centric fusion by introgressing the 3HS.3BL translocation into a modern wheat cultivar. Genome 57:601–607. doi:10.1139/gen-2014-0187

Türkösi E, Cseh A, Molnár-Láng M (2014b) Development and identification of new wheat-barley ditelosomic addition lines using fluorescence in situ hybridization and molecular markers. In Kőszegi I (ed) Advances in plant breeding & biotechnology techniques. Book of abstracts. Pannonian Plant Biotechnology Association, The Faculty of Agricultural and Food Science University of West Hungary, Mosonmagyaróvár, Hungary, 27–29 Apr 2014, pp 62–63

Villegas D, Casadesús J, Atienza SG, Martos V, Maalouf F, Karam F, Aranjuelo I, Nogués S (2010) Tritordeum, wheat and triticale yield components under multi-local Mediterranean drought conditions. Field Crop Res 116:68–74. doi:10.1016/j.fcr.2009.11.012

Wojciechowska B (1985) Hybrids between *Hordeum vulgare* L. and *Triticum aestivum* L. Genet Pol 26:457–462

Wojciechowska B, Pudelska H (1993) Hybrids from reciprocal barley-wheat crosses. Genet Pol 34:1–13

Ya-Ping Y, Xiao C, Si-He X, Islam AKMR, Zhi-Yong X (2003) Identification of wheat-barley 2H alien substitution lines. Acta Bot Sin 45:1096–1102

Zou H, Wu Y, Liu H, Lin Z, Ye X, Chen X, Yuan Y (2012) Development and identification of wheat–barley 2H chromosome translocation lines carrying the *Isa* gene. Plant Breeding 131:69–74. doi:10.1111/j.1439-0523.2011.01910.x

Chapter 13
Genomics of Wild Relatives and Alien Introgressions

Elodie Rey, István Molnár, and Jaroslav Doležel

13.1 Introduction

As one of the most important staple food crops, bread wheat (*Triticum aestivum*, L.) continues to play a major role in ensuring global food security. The growing human population is estimated to reach nine billion by 2050, and in order to meet the expected demand, the annual yield increase of wheat should reach 2 %. This is a great challenge, as climate change and land degradation act against this endeavor. Apart from improved agronomic practice and reduction of postharvest losses, the key elements will be new varieties with increased resistance to diseases and pests, adverse environmental conditions, and with improved yield.

According to the most widely accepted scenario, bread wheat ($2n=6x=42$, BBAADD genome) arose about 8000 years ago when a cultivated form of tetraploid *Triticum turgidum* ($2n=4x=28$, BBAA genome) migrated to south of the Caspian Sea and in the area of Fertile Crescent crossed with a wild diploid grass *Aegilops tauschii* Coss. ($2n=2x=14$, DD genome). The union of unreduced gametes, or somatic chromosome doubling in the hybrid (Feuillet et al. 2008), resulted in a new allohexaploid species. The genetic diversity of bread wheat was restricted at the onset of its origin by the limited diversity of parental populations and was eroded subsequently during domestication and thousands years of cultivation and breeding.

One option to recover the useful variation that was lost and to acquire new and valuable genes and alleles is to utilize wild relatives of wheat, which were not

E. Rey • J. Doležel (✉)
Institute of Experimental Botany, Centre of the Region Haná for Biotechnological and Agricultural Research, Šlechtitelů 31, Olomouc 78371, Czech Republic
e-mail: dolezel@ueb.cas.cz

I. Molnár
Agricultural Institute, Centre for Agricultural Research, Hungarian Academy of Sciences, Brunszvik u. 2, Martonvásár 2462, Hungary

M. Molnár-Láng et al. (eds.), *Alien Introgression in Wheat*,
DOI 10.1007/978-3-319-23494-6_13

subjected to human selection, and thus represent a rich source of diversity. The tribe Triticeae comprises wild annual and perennial species related to wheat, facilitating the production of interspecific hybrids. The efforts to use this approach date back 140 years, and the first experiments at the end of nineteenth century and beginning of twentieth century involved hybridization between wheat and rye (Wilson 1876), wheat and barley (Farrer 1904), and between wheat and *Aegilops* (Kihara 1937). However, larger-scale production of interspecific hybrids was delayed until the introduction of colchicine treatment in 1930s (Blakeslee 1937), allowing the production of fertile amphiploids by doubling chromosome number in otherwise sterile hybrids. Among other, this provided a way to develop triticale as a new cereal crop (Meurant 1982). With the advances in hybridization techniques (Kruse 1973) and establishment of *in vitro* embryo rescue methodology (Murashige and Skoog 1962), wide hybridization became more accessible, and the experiments involved a larger group of wild and cultivated wheat relatives (Mujeeb-Kazi 1995).

An extensively used approach to utilize wild germplasm in wheat breeding has been the production of synthetic hexaploid wheat by hybridizing tetraploid durum wheat (*T. turgidum* ssp. *durum* (Desf.) Husn.) ($2n=4x=28$; BBAA genome) with *Ae. tauschii*. Both synthetic hexaploid and bread wheat have the same genomic constitution and therefore can be readily hybridized to transfer novel alleles and genes from different accessions of the D-genome progenitor. This strategy has been employed at CIMMYT where more than 1000 synthetic wheats were created (del Blanco et al. 2001; Warburton et al. 2006; van Ginkel and Ogbonnaya 2008; Li et al. 2014).

Genetic diversity suitable for wheat improvement is not limited to *Ae. tauschii*, and over the years, a range of interspecific hybrids, chromosome addition and translocation lines were obtained between perennial and annual Triticeae species and bread wheat (Mujeeb-Kazi 1995; Friebe et al. 1996; Schneider et al. 2008; Molnár-Láng et al. 2014). Probably the best example of a successful wheat–alien introgression has been the spontaneous 1BL.1RS chromosome translocation (Mujeeb-Kazi 1995). It was estimated that between 1991 and 1995, 45 % of 505 commercial cultivars of bread wheat in 17 countries carried 1BL.1RS translocation, which confers increased grain yield by providing race-specific disease resistance to major rust diseases (including *Lr29/Yr26* leaf and yellow rust resistance genes), improved adaptation and stress tolerance, superior aerial biomass, and higher kernel weight (Rabinovich 1998; Feuillet et al. 2008; Zarco-Hernandez et al. 2005). However, too few other alien introgressions into wheat made their way to agricultural practice.

This chapter reviews the progress in characterizing nuclear genomes of wild relatives of wheat and wheat–alien introgression lines at chromosomal and DNA levels, and the potential of these approaches to support wheat–alien introgression breeding. After introducing the diversity of wild relatives of wheat and the difficulties of the introgression breeding, methods of cytogenetics and genomics are outlined and examples of their uses are given. The need for better understanding the mechanisms controlling chromosome behavior and for better knowledge of genome structure of wild relatives is explained. The last part of the chapter is devoted to the interaction of the introgressed chromatin with the host wheat genome. This research area has been poorly developed so far, and the lack of information may hamper the attempts to develop improved cultivars of wheat with alien introgressions.

13.2 Wild Relatives of Wheat and Difficulties with Alien Introgression

The tribe Triticeae comprises a group of species belonging to the Poaceae grass family commonly named Gramineae. In addition to economically important bread wheat (*T. aestivum* L.), durum wheat (*Triticum turgidum* L. subsp. *durum (Desf.) Husn.*), barley (*Hordeum vulgare* L.), and rye (*Secale cereale* L.), the tribe comprises over 500 wild and cultivated species of genera *Aegilops*, *Agropyron*, *Ambylopyrum*, *Anthosachne*, *Campeiostachys*, *Dasypyrum*, *Elymus*, *Hordeum*, *Leymus*, *Lophopyrum*, *Psathyrostachys*, *Pseudoroegneria*, *Secale*, *Thinopyrum*, and *Triticum.*

The Triticeae species exhibit a large diversity in terms of geographical distribution, environmental requirements, and agronomically interesting traits. The latter includes increased yield (Reynolds et al. 2001), resistance to pests and diseases (Friebe et al. 1996), early maturity (Koba et al. 1997), drought tolerance (Fatih 1983; Molnár et al. 2004; Dulai et al. 2014), salt tolerance (Fatih 1983; Dulai et al. 2010; Darkó et al. 2015), micronutrient content and efficiency (Schlegel et al. 1998, Farkas et al. 2014), lodging resistance (Chen et al. 2012), heat tolerance (Pradhan and Prasad 2015), high dietary fibre content (Cseh et al. 2011), and high protein content (Pace et al. 2001). Donors for these traits have been identified and some of the traits have been transferred to wheat (Gill et al. 2011). Some of the genes responsible for the traits have been tagged, and a few of them were even cloned (Feuillet et al. 2008; Hajjar and Hodgkin 2007; Jiang et al. 1993). However, the degree of genetic and genomic characterization of wild Triticeae species is highly variable and uneven.

Although the potential of wild relatives for wheat improvement has been recognized since a long time, the available genetic diversity remains largely underexploited. In order to utilize its full potential, it is important to understand genome organization in wild wheat relatives, increase the number of genome-specific molecular tools and identify loci underlying traits of interest (Hajjar and Hodgkin 2007). The poor knowledge on genome structure of Triticeae species and the lack of high resolution genetic maps hampers identification of genes underlying important traits, identification of unwanted sequences and their elimination using suitable large-scale screening platforms.

Elimination of unwanted alleles may be challenging due to low level of recombination between chromosomes of wild relatives and wheat. Two principal approaches have been developed to overcome this hindrance. The first is based on decreasing the effect of *Ph1* locus by the use of wheat genotypes *ph1b* or Ph^I (Riley and Chapman 1958; Griffiths et al. 2006), which promotes recombination between homoeologous wheat and alien chromosomes. The second approach involves induction of donor chromosome breakage by ionizing irradiation, or gametocidal chromosomes (Jiang et al. 1993) to stimulate insertion of alien chromosome fragments into wheat chromosomes.

Evolutionary chromosome rearrangements broke down the collinearity between the homoeologous wheat and alien chromosomes (Devos and Gale 1993). As a consequence, genes on alien chromosome segments may not compensate for the loss of wheat genes. This may negatively affect agricultural performance of the wheat–alien introgression lines and represents another obstacle in using wheat–alien translocations in breeding. Little is known about different levels of interaction between the host genome and the alien chromatin, which may lead to unexpected and even undesirable effects. Insertion of alien chromosome segment may interfere with functionality of the host genome at genomic, epigenomic, transcriptomic and proteomic levels, and may explain the failure of some introgressed genes to function in the host background, although their sequences remained intact after the introgression.

13.3 Tools to Support Alien Introgression in Wheat

13.3.1 Cytogenetics Techniques

The development of alien chromosome addition and translocation lines and their characterization greatly profits from the ability to identify chromosomes involved. Originally, the repertoire of selection methods was limited to cytological techniques that visualize mitotic and meiotic chromosomes. When Sears (1956) transferred leaf rust resistance from *Ae. umbellulata* to wheat, cytological characterization of the wheat—*Ae. umbellulata* addition line was limited to microscopic observation of mitotic chromosomes in root tips, and the translocation event was identified based on the leaf rust-resistance phenotype (Sears 1956). The advent of chromosome banding techniques such as Giemsa C-banding (Gill and Kimber 1974), permitted description of genomic constitution in interspecific hybrids, identification of alien chromosomes and characterization of translocations at subchromosomal level. C-banding was particularly effective in characterizing wheat–rye translocations because of diagnostic terminal bands of rye chromosomes (Lukaszewski and Gustafson 1983; Lapitan et al. 1984; Friebe and Larter 1988). However, it has been less useful if chromosomal segments of interest lacked diagnostic bands.

Introduction of techniques for *in situ* hybridization further stimulated the development and characterization of alien introgression lines. Following the pioneering work of Rayburn and Gill (1985), fluorescence *in situ* hybridization (FISH) was developed in wheat (Yamamoto and Mukai 1989). The potential of FISH to identify chromosomes and their segments depends on the availability of suitable probes. The most popular probes included the pAs1 repeat (Rayburn and Gill 1985; Nagaki et al. 1995), which permits identification of D-genome chromosomes, the rye subtelomeric repeat pSc119.2 (Bedbrook et al. 1980), which is useful to identify B-genome chromosomes, and pTa71 DNA clone (Gerlach and Bedbrook 1979), which identifies nucleolus organizing regions on satellite chromosomes. FISH with these probes discriminates the whole set of D- and B-genome chromosomes and, depending on the quality of hybridization, partially or completely the A-genome chromosomes of bread wheat. The same set of DNA probes has been applied to examine genetic diversity and construct karyotypes of wild species in *Aegilops* (Badaeva et al. 1996a, 1996b), *Agropyron* (Linc et al. 2012), and *Hordeum* (de Bustos et al. 1996; Szakács et al. 2013;), and to identify their chromosomes introgressed into wheat (Molnár et al. 2009; Sepsi et al. 2008; Nagy et al. 2002, Molnár-Láng et al. 2012) (see Figs. 13.1 and 13.2)

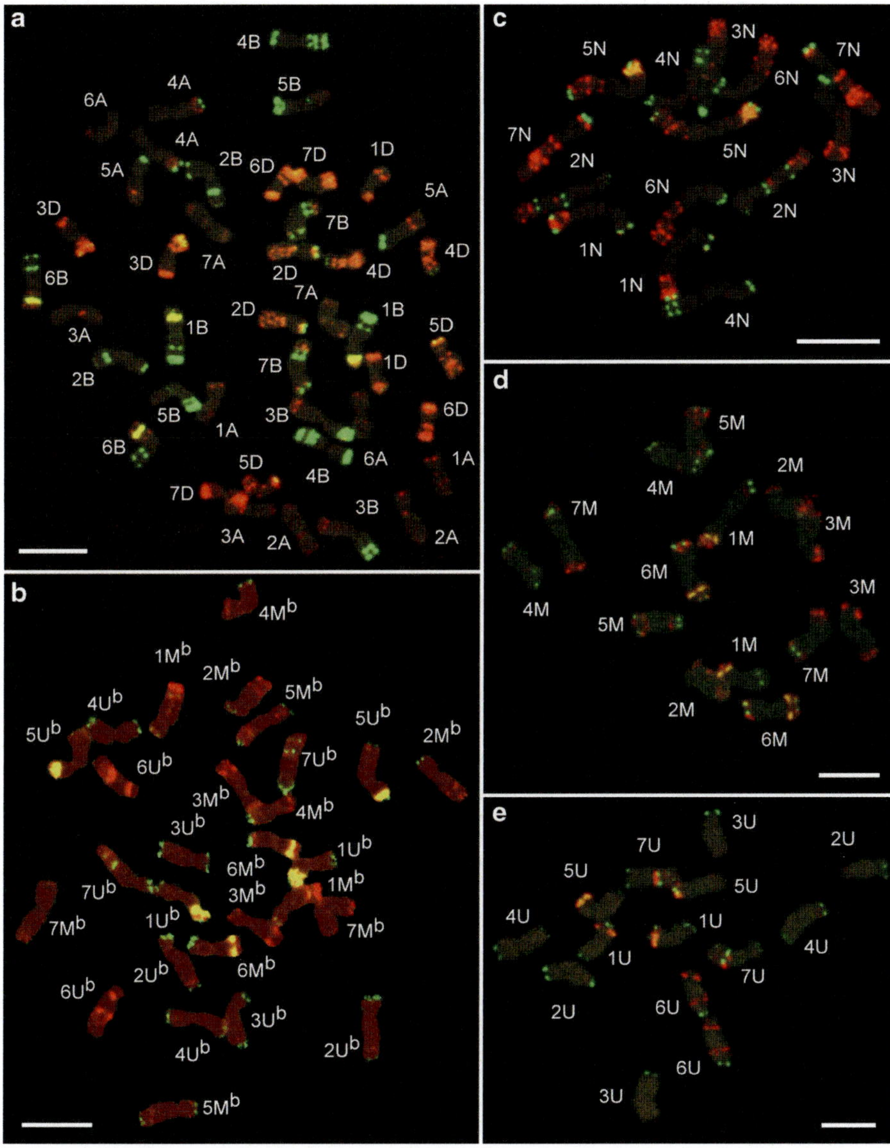

Fig. 13.1 Molecular cytogenetic identification of mitotic metaphase chromosomes in (**a**) *T. aestivum* cv. Chinese Spring ($2n = 6x = 42$; BBAADD); (**b**) *Ae. biuncialis* MvGB382 ($2n = 4x = 28$; $U^bU^bM^bM^b$); (**c**) *Ae. uniaristata* JIC2120001 ($2n = 2x = 14$; NN); (**d**) *Ae. comosa* MvGB1039 ($2n = 2x = 14$; MM); and (**e**) *Ae. umbellulata* AE740/03 ($2n = 2x = 14$; UU). Fluorescence *in situ* hybridization (FISH) was done using repetitive DNA probes for Afa family repeat (red), pSc119.2 repeat (green) and pTa71 repeat (yellow) and allowed identification of all chromosomes in the karyotypes. Scale bar = 10 μm

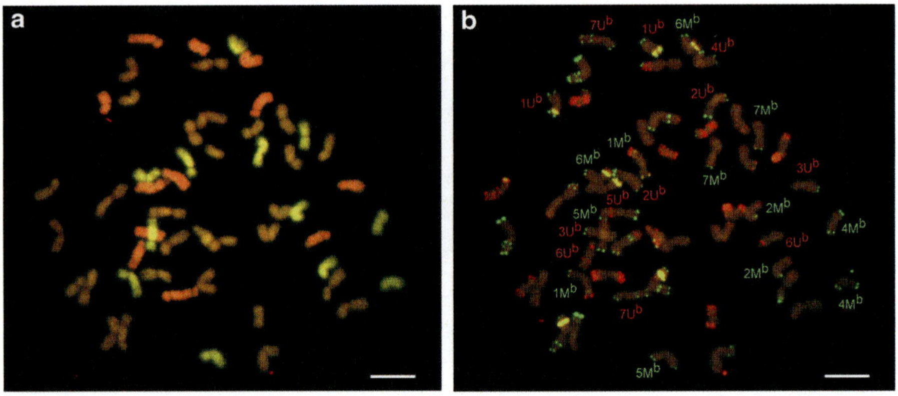

Fig. 13.2 Multicolor genomic *in situ* hybridization (mcGISH) using U- and M-genomic probes (**a**) and FISH with probes for DNA repeats (**b**) on mitotic metaphase chromosomes of a partial meristem root tip cell of wheat-*Ae. biuncialis* amphiploid plant. (**a**) mcGISH allows discrimination of U^b genome (red color), M^b genome (green color), and wheat (brown color) chromosomes. (**b**) FISH with probes for pSc119.2 repeat (green color), Afa family repeat (red color), and pTa71 repeat (yellow color) enables identification of all alien chromosomes in the wheat background. Scale bar = 10 μm

Characteristic FISH labeling patterns of HvT01 tandem repeat (Schubert et al. 1998), and the Triticeae-specific AT-rich tandem repeat pHvMWG2315 (Busch et al. 1995), permitted identification of all chromosomes in barley. In wheat genetic background, barley chromosomes could be discriminated with various combinations of repetitive DNA probes (Szakács and Molnár-Láng 2007). In rye, FISH with the 120-bp repeat family pSc119.2 together with pTa71 or AAC repeats identifies the whole chromosome complement (McIntyre et al. 1990; Szakács and Molnár-Láng 2008). In order to enrich chromosomes with diagnostic landmarks, microsatellite trinucleotide repeats (GAA, AAC, ACG) were found useful in wheat, barley, and rye (Cuadrado et al. 2008) as well as in *Aegilops* (Molnár et al. 2011a) and *Dasypyrum* (Grosso et al. 2012).

Inserts from DNA libraries cloned in a BAC (Bacterial Artificial Chromosome) vector were also tested to identify new repetitive sequences (both dispersed and tandem types), and to develop locus-specific cytogenetic markers (Zhang et al. 2004a). FISH with BAC clones (BAC FISH) was shown useful to discriminate the three subgenomes in hexaploid wheat (Zhang et al. 2004b), and for physical mapping of a powdery mildew-resistance gene (Yang et al. 2013). Unfortunately, BAC FISH suffers from the presence of dispersed repetitive DNA sequences in BAC clones, which often prevent localization of BAC clones to single loci. A possible solution is to use short single-copy probes free of repeats (Karafiátová et al. 2013).

Danilova et al. (2014) used wheat cDNAs as probes for FISH to develop cytogenetic markers specific for single-copy genic loci in wheat. They localized several cDNA markers on each of the 14 homoeologous chromosome arms and studied chromosome structure and homoeology in wild Triticeae species. The work revealed

1U-6U chromosome translocation in *Ae. umbellulata*, showed collinearity between the chromosome A of *Ae. caudata* and group-1 wheat chromosomes, and between chromosome arm 7S#3L of *Thinopyrum intermedium* and the long arm of the group-7 wheat chromosomes. A limitation inherent to performing FISH on condensed mitotic and meiotic chromosomes is the low spatial resolution. This can be improved by performing FISH on stretched mitotic chromosomes (Valárik et al. 2004), on extended DNA fibers (Fiber-FISH) (Jackson et al. 1998; Ersfeld 2004), and on hyper-expanded chromosomes obtained by flow cytometry (Endo et al. 2014).

Genomic *in situ* hybridization (GISH) uses genomic DNA as a probe (Schwarzacher et al. 1989) and permits determination of genomic constitution of allopolyploid Triticeae, and to detect alien chromatin introgressed into wheat. Combined with chromosome banding and/or FISH, the method allows location and identification of wheat–alien translocation breakpoints (Friebe et al. 1992, 1993; Jiang et al. 1993; Molnár-Láng et al. 2000, 2005; Liu et al. 2010; Kruppa et al. 2013). While cytogenetic methods are irreplaceable to verify genomic constitution in interspecific hybrids, the limited sensitivity and spatial resolution, and especially their laborious and time consuming nature seriously limit their suitability for large scale selection of wheat–alien introgressions. High-resolution and high-throughput methods are needed to increase the screening capacity and to identify micro-introgressions and chromosome breakpoints. These include the use of DNA markers and, more recently, DNA sequencing.

13.3.2 Molecular Markers

Morphological, isozyme, and seed storage protein markers were the first markers used in wheat–alien introgression breeding to identify and characterize alien chromosome addition lines (Guadagnuolo et al. 2001; Hart et al. 1980; Tang and Hart 1975). Because of their limited number, they were not suitable to reveal chromosomal rearrangements.

The restriction fragment length polymorphisms (RFLPs), random amplified polymorphic DNA (RAPD) (Williams et al. 1990), and amplified fragment length polymorphism (AFLP) (Vos et al. 1995), were the first DNA markers used to characterize wheat–alien introgression lines (Fedak 1999), since they do not require prior sequence information. They were used in a number of studies to identify chromosome/chromosome-arm addition and substitution lines (Devos and Gale 1993; King et al. 1993; Hernández et al. 1996; Qi et al. 1996; Peil et al. 1998; Wang et al. 1995; Francki et al. 1997; Qi et al. 1997). Despite their temporal popularity, the markers suffered from some drawbacks. Their application was time-consuming, often labor-intensive and expensive, and they were not appropriate for high-throughput genotyping. Moreover, the low level of polymorphism revealed by RAPD markers, and low transferability/conversion of AFLP markers into STS markers, prevented the extensive use of these markers in wheat breeding (Gupta et al. 1999).

RFLPs became the molecular markers of choice for some time due to their codominance and locus specificity (Qi et al. 2007). Wheat RFLPs were used to develop high-resolution genetic and physical maps (Qi et al. 2004; Qi et al. 2003), characterize homoeology of alien chromosomes, and reveal their rearrangements relative to wheat (Devos et al. 1993; Devos and Gale 1993; Zhang et al. 1998; McArthur et al. 2012). RFLP markers identified cryptic alien introgressions where cytogenetic techniques failed (Yingshan et al. 2004), such as the T5DL.5DS-5M^{g}S wheat-*Ae. geniculata* translocation conferring resistance to leaf rust and stripe rust (Kuraparthy et al. 2007). With the advances in molecular biology, informative but cumbersome to use RFLP markers were converted to PCR-based markers such as the sequence-tagged site (STS) markers, which were more suitable for tagging interesting genes (Cenci et al. 1999; Seyfarth et al. 1999; Langridge et al. 2001).

Transposable elements, randomly distributed in nuclear genomes have also been used as molecular markers (Queen et al. 2003; Nagy and Lelley 2003). The sequence-specific amplified polymorphism (S-SAP) technology (Waugh et al. 1997) amplifies regions representing flanking genomic sequences of individual retrotransposons. The advantages of S-SAP for studying genetic diversity are higher amount of accessible polymorphism (Waugh et al. 1997), the markers are more evenly distributed throughout the genome (Nagy and Lelley 2003), and the estimated genetic distances are more consistent with physical mapping (Ellis et al. 1998). Nagy et al. (2006) used the short interspersed nuclear element (SINE) Au identified in *Ae. umbellulata* (Yasui et al. 2001) to develop S-SAP markers specific for U- and M-genome chromosomes of *Aegilops* (Nagy et al. 2006).

Simple Sequence Repeat (SSR) markers (Tautz 1989), or microsatellite markers, were the next generation of molecular markers employed in wheat–alien introgression breeding (Mohan et al. 2007; Bandopadhyay et al. 2004; Yu et al. 2004; Gupta et al. 2003). Efficient development of SSRs requires genomic sequence information, and thus they were developed concomitantly with expressed sequence tags (ESTs), cDNA and BAC libraries. A list of genomic resources currently available for Triticeae is given in Table 13.1.

Together with cDNA libraries and draft genome sequences of barley, bread wheat, *Ae. tauschii* and *T. urartu* (Table 13.2), ESTs are currently the most abundant type of sequence information available for not less than 25 species from 15 Triticeae genera. The release of 16,000 EST loci mapped to chromosome deletion bins (Qi et al. 2004) provided excellent resource for development of markers from specific chromosome regions and helped designing locus-specific markers. Because of the genic and thus conserved nature of ESTs, EST-derived SSR markers are transferable between Triticeae species (Gupta et al. 2008). As ESTs and cDNA resources are much less abundant in other Triticeae, e.g., *Elymus*, *Aegilops* and *Leymus*, numerous studies profited from the high transferability of wheat EST-derived SSR markers across distantly related species for comparative mapping, trait-marker associations and to carry out evolutionary studies to establish the phylogenetic relationships among the wild relatives of wheat and between them and bread wheat (Adonina et al. 2005; Jing et al. 2007; Kroupin et al. 2012).

The conserved orthologous set (COS) markers allowed identification of orthologous regions between wild species and wheat in order to facilitate alien gene-transfer

Table 13.1 Genomic resources available for Triticeae species

Genus (no. of taxonomy entries in NCBI)	Bio Project[a]	Number of genes	Number of ESTs	BAC libraries	cDNA clones	Probe[b]	Map data[c]	SRA[d]	GSS[e]	Genome[f]
Aegilops (42)	35	1172	4546	8	2303	787	4	161	5172	1
Agropyron (16)	0	4	17	–	–	0	–	1	0	1
Amblyopyrum (3)	–	–	–	–	–	–	–	–	–	–
Anthosachne (10)	–	–	–	–	–	–	–	–	–	–
Australopyrum (6)	–	–	–	–	–	–	–	–	–	–
Avena (35)	11	28	79,657	–	–	11,542	24	–	3063	–
Campeiostachys (11)	–	–	–	–	–	–	–	–	–	–
Connorochloa (2)	–	–	–	–	–	–	–	–	–	–
Critesion (4)	–	–	–	–	–	–	–	–	–	–
Crithopsis (2)	–	–	–	–	–	–	–	–	–	–
Dasypyrum (3)	–	–	–	–	–	–	–	–	14	–
Douglasdeweya (3)	–	–	–	–	–	–	–	–	–	–
Elymus (116)	1	–	45,580	–	–	–	–	–	–	1
Eremopyrum (5)	-	–	–	–	–	–	–	–	–	–
Festucopsis (3)	–	–	–	–	–	–	–	–	–	–
Haynaldia (2)	–	–	10	–	–	–	–	–	13	–
Henrardia (5)	–	–	–	–	–	–	–	–	–	–
Heteranthelium (4)	–	–	–	–	–	–	–	–	–	–
Hordelymus (2)	–	–	–	–	–	–	–	–	–	–
Hordeum (62)	148	717	840,120	2	89,452	11,196	76	1894	574,028	4
Hystrix (5)	–	147	–	–	–	–	–	50	–	–
Kengyilia (22)	–	–	–	–	–	–	–	2	–	–
Leymus (50)	4	–	30,749	–	–	1853	3	6	13	–
Lophopyrum (5)	2	–	2	–	–	56	–	1	–	–

(continued)

Table 13.1 (continued)

Genus (no. of taxonomy entries in NCBI)	Bio Project[a]	Number of genes	Number of ESTs	BAC libraries	cDNA clones	Probe[b]	Map data[c]	SRA[d]	GSS[e]	Genome[f]
Pascopyrum (2)	–	–	–	–	–	–	–	–	–	–
Peridictyon (2)	–	–	1	–	–	–	–	–	–	–
Psammopyrum (2)	–	–	–	–	–	–	–	–	–	–
Psathyrostachys (16)	1	–	–	–	–	–	–	1	44	–
Pseudoroegneria (9)	–	–	–	–	–	–	–	–	–	–
<u>*Secale* (16)</u>	21	113	15,903	2	6617	1091	12	36	2956	–
Stenostachys (4)	–	–	–	–	–	–	–	–	–	–
Taeniatherum (6)	–	–	2	–	–	–	–	–	–	–
Thinopyrum (12)	4	–	2385	–	–	–	–	3	7	–
<u>*Triticum* (84)</u>	239	3170	1,358,421	16	10,527	21,164	69	2558	72,374	4
× *Aegilotriticum* (14)	1	–	–	–	–	–	–	–	–	–
× *Triticosecale* (10)	3	–	11	–	–	–	–	–	8	–
× *Tritordeum* (6)	–	–	4	–	–	57	–	–	11	–

The information in this table was collected from NCBI taxonomy (http://www.ncbi.nlm.nih.gov/taxonomy) and GrainGene (http://wheat.pw.usda.gov/GG3/) databases in May 2015. Triticeae genera comprising cultivated species are underlined

[a]Projects initiated in the fields of genomics, functional genomics and genetic studies (NCBI)

[b]Public registry of nucleic acid reagents designed for use in a wide variety of biomedical research applications (NCBI)

[c]Genetic and physical maps available for Triticeae (GrainGene database)

[d]Sequence Read Archive (NCBI) stores sequencing data

[e]Genome Survey Sequences (NCBI) is a collection of unannotated short single-read primarily genomic sequences from GenBank including random survey sequences, clone-end sequences and exon-trapped sequences

[f]Genome (NCBI) reference whole genomes sequencing information, both completely sequenced organisms and those for which sequencing is in progress

Table 13.2 Whole genome sequencing projects in cereals

Species/cultivar	Genome size (1C)	Sequence description	Consortium/team
Oryza sativa ssp. *japonica*	400–430 Mbp	Pseudomolecule	International Rice Genome Sequencing Project (2005)
Zea maize cv. B73	2.4 Gbp	Pseudomolecule	Schnable et al. (2009)
Sorghum bicolor cv. Moench	750 Mbp	Whole-genome draft assembly	Paterson et al. (2009)
Brachypodium distachyon inbread line Bd21	~355 Mbp	Pseudomolecule	The International Brachypodium Initiative (2010)
Hordeum vulgare cv. Morex	~5.3 Gbp	Whole-genome draft assembly	The International Barley Genome Sequencing Consortium (2012)
Aegilops tauschii ssp. *strangulata* accession AL8/78	4.02 Gbp	Whole-genome draft assembly	Luo et al. (2013)
Triticum urartu accession G1812	4.94 Gbp	Whole-genome draft assembly	Ling et al. (2013)
Triticum aestivum cv. Chinese spring (CS) 3B chromosome of *Triticum aestivum* cv. CS	~16 Gbp ~16 Gbp (~1 Gbp)	5× whole-genome draft assembly Chromosome-based draft genome assembly Reference sequence assembly of chromosome 3B	Brenchley et al. (2012) IWGSC (2014) Choulet et al. (2014)

through a better characterization of the potentially recombining regions (Molnár et al. 2013). As the COS markers are PCR based and span exon–intron junctions, they are conserved enough to be transferrable across genera, while the intron sequences provide relatively high polymorphism that allows variants of genes to be discriminated (e.g., between species). Although these markers present interesting tools to support alien-wheat gene transfer, they remain underexploited in this area.

13.3.3 High-Throughput Genotyping

Diversity Arrays Technology (DArT) markers were initially developed as micro-array hybridization-based sequence-independent marker system, and allowed screening thousands of polymorphic loci in a single assay at low cost per data point (Jaccoud et al. 2001). Among other, DArT markers were used to develop high-density genetic map of wheat × wild emmer (Peleg et al. 2008). A new version of DArT marker technology (DArT-seq) is based on next-generation sequencing where the polymorphisms are genotyped by sequencing. Because of its advantages,

DArT has been employed extensively in genetic mapping, genotyping, and diversity assessment in wheat (Cabral et al. 2014; Jighly et al. 2015; Bentley et al. 2014; Yu et al. 2014; Colasuonno et al. 2013; Iehisa et al. 2014), and more recently in its wild and cultivated relatives (Montilla-Bascón et al. 2015; Kalih et al. 2015; Castillo et al. 2014; Bolibok-Brągoszewska et al. 2014; Alheit et al. 2014; Yabe et al. 2014; Cabral et al. 2014; Jing et al. 2009).

The advent of the next generation sequencing technologies changed the paradigm of wheat genetics and genomic and led to the development of Single Nucleotide Polymorphism (SNP) markers. Various platforms have been developed for wheat genotyping such as the 9K and 90K Illumina iSelect® platforms with 9000 and 90,000 SNP markers, respectively (Cavanagh et al. 2013; Wang et al. 2014), the Illumina Infinium® platform (up to 1,000,000 SNP markers), as well as the Axiom® 820K and 35K arrays (with up to 820,000 and 35,000 features) (http://www.cerealsdb.uk.net/cerealgenomics/CerealsDB/axiom_download.php). These platforms provide tools to obtain detailed information on germplasm diversity and characterize allelic variation. However, low representation of wild wheat relatives in the SNP design process may limit the utility of the platforms in wheat alien introgression breeding (Wulff and Moscou 2014). Consequently, a few studies made use of SNP molecular markers to support alien gene transfer in wheat (Tiwari et al. 2014) and very few SNPs derived from wild species are available.

Due to the low cost per data point and ease of development, Kompetitive Allele Specific PCR (KASP) SNP markers (He et al. 2014), a genotyping technology based on allele-specific oligo extension and fluorescence resonance energy transfer for signal generation, are becoming popular and are used in large-scale projects (Petersen et al. 2015). KASP markers can genotype SNP polymorphism, deletions and insertions variations, and have been used in screening wheat–alien hybrids and their back-crossed derivatives to detect recombinants and isolate desired introgressions (King et al. 2013). In order to promote the use of KASP markers, it is important to generate new genomic sequences from wild relatives of wheat.

13.3.4 Genome Sequencing

13.3.4.1 Whole Genome Approaches

Despite the importance of Triticeae species for the humankind (Feuillet et al. 2008), attempts to sequence their genomes were delayed due to the size and complexity. The nuclear genome of bread wheat comprises three structurally similar (homoelogous) subgenomes A, B, and D, and with the size of about 17 Gb/1C, it is 40 times bigger than rice (0.43 Gb) and 126 times bigger than *Arabidopsis thaliana* (0.135 Gb). As the other Triticeae genomes, it is highly redundant and composed mostly from various classes of repetitive DNA sequences (IWGSC 2014).

High throughput of the next generation sequencing technologies makes it possible to sequence even the biggest genomes. However, the problem is to assemble and order the short reads thus obtained (IWGSC 2014). Due to large genome complexity and sequence redundancy, high-quality reference genome assemblies are not yet available for any of the Triticeae species. To date, only draft genome sequences are available for barley (The International Barley Genome Sequencing Consortium 2012), *T. urartu* (Ling et al. 2013)—a progenitor of the A genome of bread wheat, *Ae. tauschii* (Luo et al. 2013)—a D genome progenitor of bread wheat, as well as the whole genome shotgun assembly of hexaploid bread wheat (Brenchley et al. 2012) (see Table 13.2).

Due to their nature, draft sequence assemblies are only partial representations of the genomes, often accounting for less than 50 % of their estimated size. A significant part of expressed genes may be absent, which may compromise efforts with gene discovery and cloning, while the fragmentation of genome sequence and large numbers of unanchored contigs hamper comparative genome analyses.

Despite their preliminary nature, draft genome sequences provided useful insights into the Triticeae genome organization, evolution, and function. They were useful to develop protein-coding gene models, analyze genome organization, assess recombination rates along chromosomes, and characterize synteny and collinearity with other species (Ling et al. 2013; Luo et al. 2013; The International Barley Genome Sequencing Consortium 2012). They served as templates to characterize agronomically important genes and develop genome-specific molecular markers for plant breeding (Ling et al. 2013). The utility and extensive use of whole genome sequences from the main Triticeae crops confirm the need for such resources in wild wheat relatives. Although it may not be possible to sequence genomes of all wild species employed in wheat alien introgression breeding, efforts should be made to obtain as much information on their genomes as possible in order to understand better the genome relationships among Triticeae.

13.3.4.2 Reduced-Complexity Sequencing

One approach to facilitate sequencing and assembly of the huge Triticeae genomes is to reduce sample complexity prior to sequencing. Various strategies have been developed to achieve this, and can be classified into two groups: (1) Transcriptome sequencing and sequence capture approaches, which sequence only certain parts of genomes, and (2) the chromosome-centric approaches, which reduce the complexity in a lossless way by dissecting genomes to small parts (chromosomes and chromosome arms) that are sequenced and assembled separately.

Sequencing conserved genic portions of genomes enables development of cross-species transferable tools, and facilitates functional understanding of important traits. Haseneyer et al. (2011) sequenced transcriptome in five winter rye inbred-lines and identified over 5000 SNPs between the transcriptomes that were subsequently used for genotyping 54 inbred lines using SNP genotyping array. This analysis does not require prior knowledge of genome sequence and allows large-scale molecular marker development for high-throughput genotyping. A recent analysis of *Agropyron cristatum* transcriptome permitted identification of 6172 unigenes specific to *A. cristatum*, including many stress-resistant genes and alleles potentially useful in wheat improvement (Zhang et al. 2015).

Another option to reduce sequencing efforts are sequence-capture approaches, which are used to enrich samples for sequences of interest before carrying out NGS. They are based on hybridization of target sequences to bait probes in solution, or on solid support. This approach usually necessitates preliminary sequence information. However, since it allows high level of mismatches, it permits capturing diverged sequences. Known sequences from more characterized species such as wheat, barley, *Brachypodium*, and rice can be employed to discover uncharacterized sequences from related species and varieties. Accordingly, Jupe et al. (2013) developed an exome capture for nucleotide-binding leucine-rich repeat (NB-LRR) domain for the so-called Resistance gene enrichment Sequencing (RenSeq) in potato. Their work resulted in discovery of 317 previously unannotated NB-LRRs and the method could aid in discovery of new resistance genes in wild relatives of wheat (Wulff and Moscou 2014).

Alternative approach to reduce complexity of large and polyploid genomes is to isolate chromosomes by flow cytometric sorting and sequence them individually (Fig. 13.3). This strategy is called chromosome genomics (Doležel et al. 2007, 2014) and has been adopted by the IWGSC for the bread wheat genome sequencing (IWGSC 2014). The method, originally developed in *Vicia faba* (Doležel et al. 1992), relies on cell cycle synchronization of meristem root tip cells of young seedlings and their accumulation at mitotic metaphase. After mild formaldehyde fixation, intact chromosomes are released into a buffer solution by mechanical homogenization of root tips. Chromosome samples are stained by a DNA fluorochrome DAPI and classified at rates of several thousand per second according to their relative DNA content using flow cytometry. Chromosomes that differ in DNA content from other chromosomes form distinct peaks on histograms of DNA content (flow karyotypes). Such chromosomes, can be sorted individually at rates of about 20 s^{-1}, and several hundred thousand chromosomes of the same type can be collected in one day.

In a majority of species, chromosomes have similar DNA content and cannot be discriminated after DAPI staining alone. The most frequent approach to overcome this difficulty has been the use of cytogenetic stocks in which the size of one or more chromosomes has been changed so that the chromosome of interest can be discriminated and sorted. The stocks included chromosome translocations (Kubaláková et al. 2002), deletions (Kubaláková et al. 2005), alien chromosome addition (Kubaláková et al. 2003) and alien chromosome arm additions (Suchánková et al. 2006). As such stocks are not available for many species, it is important that Giorgi et al. (2013) developed a protocol termed FISHIS, to fluorescently label repetitive DNA on chromosomes prior to flow cytometric analysis. This approach permits discrimination of chromosomes, which have the same or very similar relative DNA content (Fig. 13.3), and has been used successfully to sort chromosomes in *Ae. umbellulata*, *Ae. comosa*, *Ae. speltoides*, and *Ae. markgrafii* (in preparation).

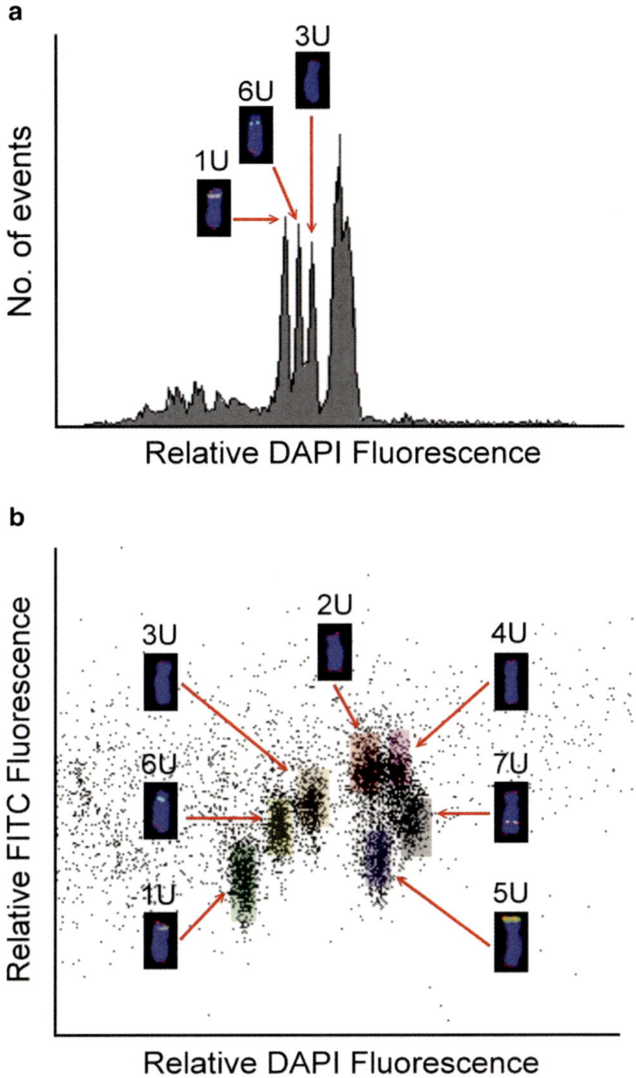

Fig. 13.3 Mono- (**a**) and biparametric (**b**) flow cytometric analysis and sorting of mitotic metaphase chromosomes from *Ae. umbellulata* ($2n=2x=14$; UU). (**a**) Monoparametric analysis of chromosomes stained by DAPI results in a histogram of relative fluorescence intensity (flow karyotype) in which three peaks representing chromosomes 1U, 6U and 3U are discriminated. The remaining four chromosomes form a composite peak and cannot be sorted individually. Biparametric analysis of chromosomes stained by DAPI and with GAA repeats labeled by FITC results in a bivariate flow karyotype on which all seven chromosomes (colored regions) can be discriminated and flow-sorted at a purity of 90–99 %

Table 13.3 List of Triticeae species in which flow cytometric chromosome sorting has been reported (adapted from Doležel et al. (2014))

Genus	Species	Common name	*n*	Reference[a]
Aegilops	*biuncialis*	Goatgrass	14	Molnár et al. (2011b)
	comosa		7	Molnár et al. (2011b)
	cylindrica		14	Molnár et al. (2015)
	geniculata		14	Molnár et al. (2011b); Tiwari et al. (2014)
	markgrafii		7	Molnár et al. (2015)
	speltoides		14	Molnár et al. (2014)
	triuncialis		14	Molnár et al. (2015)
	umbellulata		7	Molnár et al. (2011b)
Avena	*sativa*	Oat	21	Li et al. (2001)
Dasypyrum	*villosum*	Mosquito Grass	7	Grosso et al. (2012); Giorgi et al. (2013)
Hordeum	*vulgare*	Barley	7	Lysák et al. (1999); Lee et al. (2000); Suchánková et al. (2006); **Mayer et al. (2009, 2011)**
Secale	*cereale*	Rye	7	Kubaláková et al. (2003); **Bartoš et al.** (2008); **Martis et al.** (2013)
Triticum	*aestivum*	Bread wheat	21	Wang et al. (1992); Schwarzacher et al. (1997); Lee et al. (1997); Gill et al. (1999); Vrána et al. (2000); Kubaláková et al. (2002); Giorgi et al. (2013); **Hernandez et al.** (2012); **IWGSC** (2014); **Helguera et al.** (2015); Tanaka et al. (2014); Sergeeva et al. (2014); **Lucas et al.** (2014); **Berkman et al.** (2011)
	durum	Durum wheat	14	Kubaláková et al. (2005); Giorgi et al. (2013)
	urartu		7	Molnár et al. (2014)

[a]Reports on chromosome sequencing are printed in bold

To date, chromosome flow-sorting has been reported in at least 29 plant species, including 15 Triticeae (Doležel et al. 2014; Table 13.3). High purity in the sorted fractions and high molecular weight DNA of flow-sorted chromosomes makes them ideal substrate for downstream applications such as PCR-based analysis, development of markers, BAC-vector cloning and construction of optical maps (for review see (Doležel et al. 2014)). Chromosomal DNA can be sequenced or used for other applications either directly, if a sufficient number of chromosomes is sorted, or after representative amplification (Šimková et al. 2008). It is now even possible to sequence a single flow-sorted chromosome (Petr Cápal pers. comm.). The latter is particularly important in cases when the chromosome of interest cannot be discriminated from other chromosomes in karyotype, or if the focus is on the analysis of structural chromosome heterozygosity and allele phasing.

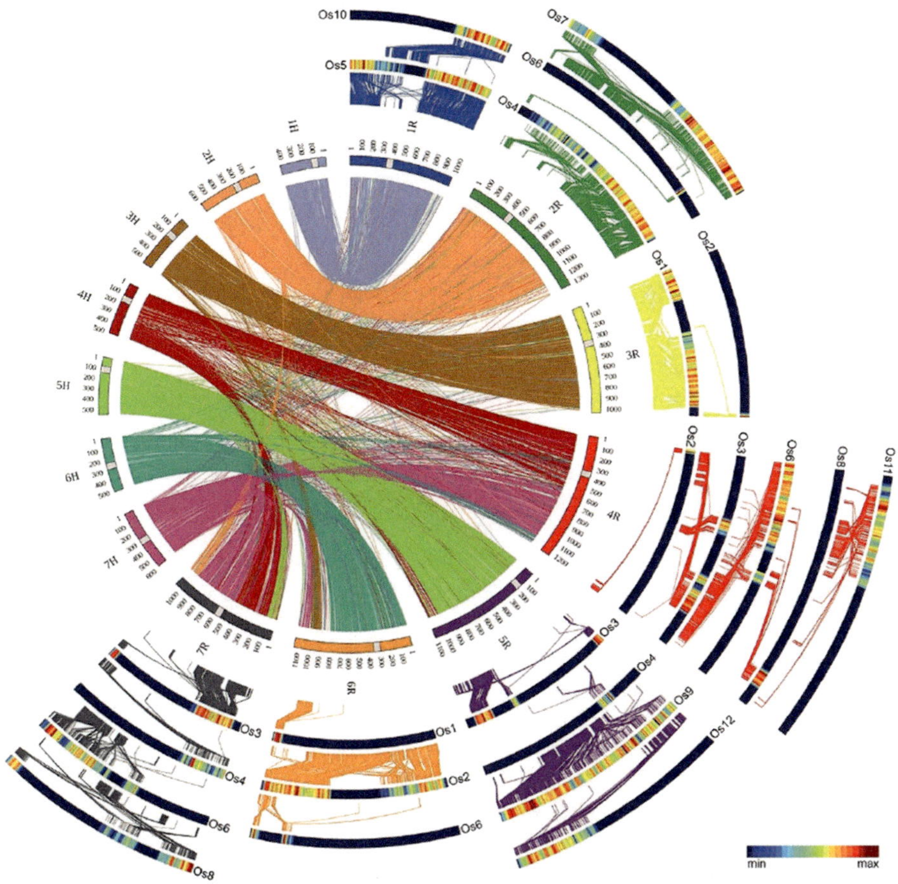

Fig. 13.4 Next-generation sequencing of flow-sorted rye chromosomes allowed characterization of synteny between rye, barley, and rice genomes. Collinearity of the rye and barley genomes is depicted by the inner circle of the diagram. Rye (1R–7R) and barley (1H–7H) chromosomes were scaled according to the rye genetic and barley physical map, respectively. Lines (colored according to barley chromosomes) within the inner circle connect putatively orthologous rye and barley genes. The outer partial circles of heat map colored bars illustrate the density of rice genes hit by chromosome sequencing reads of the corresponding rye chromosomes. Conserved syntenic blocks are highlighted by yellow-red-colored regions of the heat maps. Putatively orthologous genes between rye and rice are connected with lines (colored according to rye chromosomes) and centromere positions are highlighted by grey rectangles. Martis et al., Plant Cell 25: 3685–3698, 2013. www.plantcell.org Copyright American Society of Plant Biologists. Reproduced with permission

For example, BAC-end sequences obtained using 1RS-specific BAC library were used to develop Insertion Site-Based Polymorphism markers (ISBP) specific for 1RS and to identify loci carrying microsatellites suitable for the development of 1RS-specific SSR markers (Bartoš et al. 2008). Next-generation sequencing flow-sorted chromosomes of rye enabled establishing linear gene order model comprising over 22 thousand genes, i.e. 72 % of the detected set of 31,000 rye genes. Chromosome sequencing together with transcript mapping and integration of conserved synteny information of Brachypodium, rice and sorghum enabled a genome-wide high-density comparative analysis of grass genome synteny (Fig. 13.4).

The chromosome genomics approach has been particularly fruitful in genomics of wheat. The chromosome-based draft sequence of bread wheat was obtained by sequencing flow-sorted chromosome arms (except of chromosome 3B), each of them representing only 1.3–3.3 % of the genome. Chromosome arms were sequenced with Illumina technology and the reads were assembled to contigs representing 10.2 Gb (61 %) of the genome with a L50 of repeat-masked assemblies ranging from 1.7 to 8.9 kb. A total of 133,090 loci homologous to related grass genes were classified as high-confidence gene calls. Out of them, 93.3 % were annotated on individual chromosome arm sequences, and 53.2 % were located on syntenic chromosomes compared to brachypodium, rice and sorghum. In total, 81 % raw reads and 76.6 % assembled sequences contained repeats, explaining the difficulty of assembling such genomes from short sequence reads. As demonstrated in chickpea, chromosome genomics can be coupled with whole genome next-generation sequencing to validate whole genome assemblies (Ruperao et al. 2014). This powerful combination could speed up production of good quality whole genome assemblies in wild wheat relatives.

Chromosome genomics was also shown useful to characterize chromosome segments of alien origin, develop markers from these regions, and support cloning alien genes of interest. In a pioneering study, Tiwari et al. (2014) sequenced DNA from flow-sorted short arm of chromosome $5M^g$ of *Ae. geniculata* to develop genome-specific SNP markers. The markers allowed development of two SNP markers identifying introgression of a segment of $5M^g$ to wheat chromosome 5D carrying resistance to leaf rust (*Lr57*) and stripe rust (*Yr40*) (Fig. 13.5). In order to simplify the identification of alien chromatin introgressed into wheat, Abrouk (pers. comm.) developed a method based on comparative analysis. Briefly, using the linear gene order map of a recipient wheat chromosome (IWGSC 2014) and the sequence of flow-sorted chromosome carrying alien introgression, the density of orthologs is calculated along the wheat chromosome. The variation in density makes it possible to detect the alien segment. This approach has been validated recently in wheat *T. aestivum* cv. *Tahti—T. militinae* introgression line 8.1 (Jakobson et al. 2006, 2012), which carries a major QTL for powdery mildew resistance on the distal part of the long arm of chromosome 4A (Michael Abrouk pers. comm.)

13.4 Functional Aspects of Alien Gene Transfer

When introducing alien genes to wheat, the function of introgressed chromosomes or chromosome segments and their interaction with the host genome needs to be considered. It may occur at different levels and concern chromosome behavior during meiosis, changes in chromosomes structure and genome organization, as well as gene expression. Understanding the interaction between the host and alien genomes,

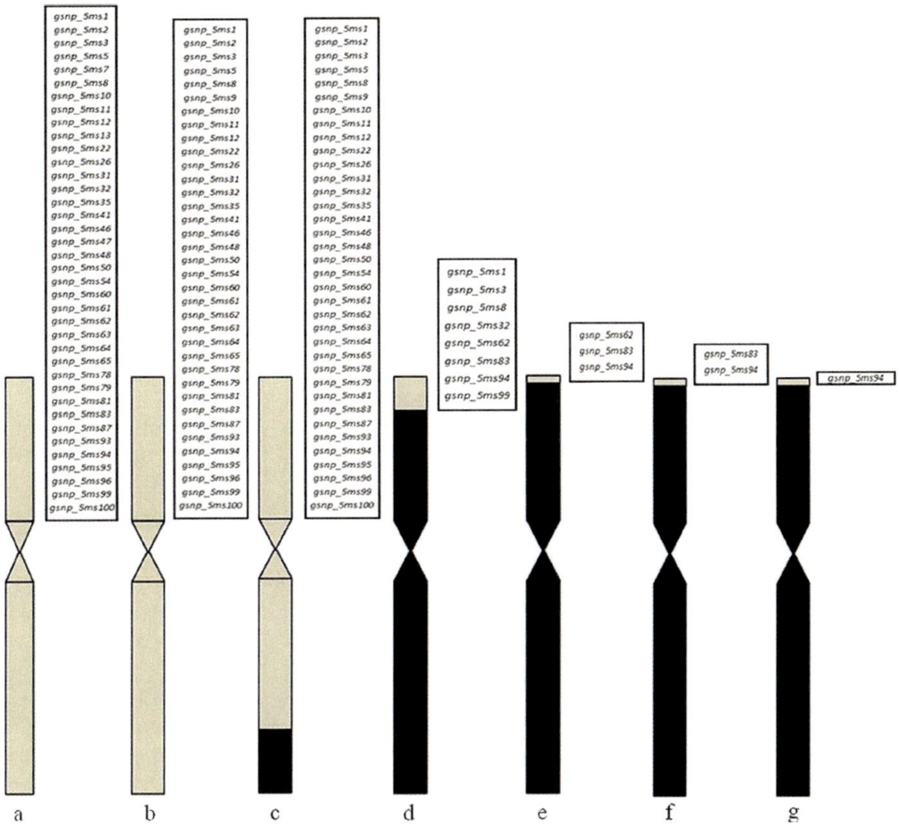

Fig. 13.5 Distribution of validated $5M^gS$-specific SNPs developed from flow-sorted ditelosomic $5M^g$ in different alien introgression-based addition, translocation, and released wheat lines. (**a**) disomic addition line TA7657, (**b**) disomic substitution line TA6675, (**c**) translocation line TA5599, (**d**) terminal translocation line TA5602, (**e**) TA5602 (with very small $5M^g$ segment), (**f**) SNPs validated in germplasm KS11WGGRC53-J and (**g**) SNP validated in germplasm KS11WGGRC53-O. Tiwari et al., BMC Genomics 15: 273, 2014. http://www.biomedcentral.com/bmcgenomics BioMed Central Ltd. Reproduced with permission

the evolution of this relationship from the moment of F1 hybrid formation to a stabilized wheat–alien introgression line, and the way the final equilibrium impacts the performance of the introgression line may contribute to the success of alien gene transfer in wheat improvement.

13.4.1 Interaction Between Host and Donor Genomes

Alien gene transfer involves hybridization and creation of interspecific hybrids, followed by genome duplication to establish fertile amphiploids. A consequence is a shock for both genomes, which may result in activation of mobile genetic elements, various structural changes and lead to changes in epigenetic status of chromatin and novel patterns of gene expression (Comai 2000).

Elimination of specific sequences is commonly reported as rapid genomic rearrangement accompanying allopolyploidization in wheat. The changes include elimination of noncoding and low-copy DNA sequences, and gain of novel fragments (Feldman et al. 1997; Liu et al. 1998). Elimination of rye-specific fragments often representing transposable elements (TEs) and their derivatives was observed in allopolyploid triticales (Ma and Gustafson 2006, 2008; Bento et al. 2008). The analysis of a newly synthesized triticale (Bento et al. 2008; Han et al. 2003) revealed rapid changes in coding sequences upon the induction of allopolyploidy, but the changes did not extend to alterations discernible at cytological level. The molecular mechanisms underlying genome reorganization are not yet fully understood (Tayalé and Parisod 2013). 'Genomic stress' due to polyploidization may activate TEs and promote their proliferation and mobility. At the same time, massive elimination in a TE family-specific manner may be observed (Comai et al. 2003; Parisod and Senerchia 2012). It seems that the degree of TE sequence divergence between progenitors correlates with the degree of restructuring in polyploid TE fractions (Senerchia et al. 2014).

A general observation made in newly created polyploids and synthetic allotetraploids, including wheat, is a change in gene expression immediately after polyploidization (Kashkush et al. 2002; Levy and Feldman 2004). Both genetic and epigenetic mechanisms may alter gene expression (Lynch and Conery 2000; Lee and Chen 2001; Osborn et al. 2003; Soltis et al. 2004). The analysis of cytosine methylation in *Aegilops–Triticum* F1 hybrids and their derivative allotetraploids revealed 13 % of the loci with altered patterns of methylation affecting both repetitive DNA and low-copy DNA (Xiong et al. 1999; Shaked et al. 2001). In leaves of *Arabidopsis* autopolyploids and allotetraploids and their progenitors, Ng et al. (2012) could associate rapid changes in gene expression with quantitative proteomic changes, suggesting rapid changes in post-transcriptional regulation and translational modifications of proteins as a consequence of polyploidization.

Epigenomic rearrangements after allopolyploidization seem to be involved in the processes of uniparental chromosome elimination, a phenomenon observed frequently in interspecific hybrids between *T. aestivum* and *H. bulbosum* (Bennett et al. 1976), *H. vulgare* (Islam et al. 1981) and *Zea mays* (Laurie and Bennett 1986). The loss of centromere-specific histone H3 (CENH3) caused centromere inactivation and triggered mitosis-dependent uniparental chromosome elimination in unstable *H. vulgare* × *H. bulbosum* hybrids (Sanei et al. 2011). Bento et al. (2010), found that chromosome structural rearrangements were more drastic in wheat–rye disomic addition lines than in triticale, indicating that the lesser the amount of rye genome introgressed into wheat, the higher the likelihood of wheat chromosome breakage, chromosome elimination, and chromosome structural rearrangement, including sequence-specific elimination, translocations and TE movement (Fu et al. 2013).

13.4.2 Alien Gene Expression

Various studies indicate complex relationships between the alien and host genes (Pumphrey et al. 2009; Jeffrey Chen and Ni 2006; Bougas et al. 2013; Wu et al. 2015; Yoo et al. 2013; Wulff and Moscou 2014) and, as a result, in some cases alien genes may not function as expected. For example, weaker effect in the wheat background as compared to the wild species was observed in studies involving resistance gene transfer (Wulff and Moscou 2014; Chen et al. 2005; Riley and Chapman 1958; Riley and Macer 1966). One explanation may be that the introgressed genes are involved in polygenic resistance together with other loci, which are not introgressed simultaneously. However, in some cases, resistance genes had no effect at all, as was the case of resistance to wheat leaf rust (*Puccinia triticina* Erikss.) introduced to wheat from rye (Riley and Macer 1966). It seems that the polyploid status of wheat itself may impact alien gene expression. When Kerber and Dyck (1973) transferred stem rust resistance from diploid einkorn wheat (*T. monococcum* L.) to tetraploid durum and hexaploid bread wheat, a progressive loss of the resistance with increasing ploidy from diploid to hexaploid was observed. Chen et al. (2005) described different levels of scab resistance in progenies that involved the same wheat-*Leymus racemosus* alien chromosome translocation, or the same alien chromosome addition, possibly related to other components of resistance in the genetic background.

Suppression of resistance due to negative interaction of homoeologous and non-homoeologous loci between genomes is another effect observed in hexaploid wheat, and the examples include a conserved gene on chromosome arm 7DL that suppresses stem rust resistance, and suppression of powdery mildew locus *Pm8* by *Pm3* locus (Kerber and Aung 1999; Wulff and Moscou 2014). The suppression of introgressed *Pm8* resistance gene by its *Pm3* host ortholog in some wheat–rye 1BL.1RS translocation lines was not due to gene loss, mutation or gene silencing (Hurni et al. 2014). A coexpression analysis of *Pm8* and *Pm3* genes in *Nicotiana benthamiana* leaves followed by co-immunoprecipitation analysis showed that the two proteins interact and form a heteromeric complex, which might result in inefficient or absent signal transmission for the defense reaction. Stirnweis et al. (2014) suggested that the frequently observed failure of resistance genes introduced from the secondary gene pool into polyploid crops could be the result of the expression of closely related NB-LRR-resistance genes or alleles in the host genome, leading to dominant-negative interactions through a posttranslational mechanism involving LRR domains. A recent study showed that genes with low similarity between rye sequences and their closest matches in the *Triticum* genome have a higher probability to be repressed or deleted in the allopolyploid genome (Khalil et al. 2015).

13.4.3 *Spatial Genome Organization and Function*

Little is known how alien chromosome(s) and/or translocated alien chromosome segments influence behavior and position of wheat chromosomes within the 3D space of interphase nucleus, how the position and behavior of alien chromosome differs from that in the nucleus of donor wild relative, and how changes in chromosome position influence gene expression of wheat and alien genes. Numerous studies in human and mouse indicate that chromosome territories are not randomly positioned in the nucleus (Gibcus and Dekker 2013). Small and gene-rich chromosomes localize near the center of nucleus, whereas larger and less-gene-rich chromosomes are more frequently located near the nuclear periphery. In plants, however, 3D-nuclear genome organization has been studied only in a few cases and mostly in *Arabidopsis* (Schubert et al. 2014; Grob et al. 2014) and rice (Mukhopadhyay et al. 2013) with small genomes, whose interphase organization may differ from that of large genomes. The results obtained in rice (Mukhopadhyay et al. 2013) correlated transcriptional regulation with alteration in nucleosome positioning, histone modifications and gene looping, but not DNA methylation. A recent observation using 3D-FISH in wheat–rye chromosome arm introgression lines indicated that the rye alien chromosomes were positioned at the periphery of nuclei (Veronika Burešová, pers. comm.). These preliminary results are consistent with the general observation of negative regulation of the expression of the alien genes introgressed in wheat.

13.5 Concluding Remarks

During more than one century of wheat–alien introgression breeding, a significant progress has been made in developing strategies to produce hybrids of wheat with distant relatives, in devising chromosome engineering techniques to integrate alien-chromosome segments into wheat genome, in the improvement of cytogenetic techniques to identify and characterize introgressed chromatin, and in phenotypical characterization of new introgression lines. These advances led to development of a formidable panel of introgression lines of various types and from a number of wild wheat relatives, carrying important traits. Nevertheless, only a small number of commercially successful wheat cultivars benefitted from these advances, and the potential of alien introgression breeding remains underused.

In order to fully explore it and benefit from the extant genetic diversity of wild wheat relatives, implementation of improved and novel approaches and tools is needed. It is fortunate that new methods of cytogenetics, genomics and phenomics are becoming available for better and, in case of genomics and phenomics, high-throughput characterization of genetic diversity, and identification of donors of important traits. On the other hand, improvement of chromosome engineering methods and better knowledge of molecular mechanisms controlling meiotic recombination are needed to facilitate successful introgression of alien chromatin. This will require a better knowledge of genome structure of wild relatives to assess chances for chromosome recombination and predict its outcomes, in order to decide the best experimental approach to be applied.

The advances in DNA sequencing and DNA marker technologies make it possible to compare genomic organization of wheat and wild relatives, and judge the degree of collinearity. In order to cope with the huge and complex genomes of Triticeae, strategies have been developed to reduce genome complexity prior to sequencing and mapping, such as exome capture and chromosome genomics. The advances in DNA sequencing technologies make it possible to develop powerful and high-throughput DNA marker technologies such as SNP, DArT and KASPAR, which are suitable for development of markers linked tightly to traits of interest, large-scale screening of progenies of wild hybrids and support production of lines with the introgressed genes of interest and minimum of unwanted chromatin.

Altogether these advances provide a toolbox to develop wheat lines enriched for gene(s) of interest with the smallest amount of undesired alien chromatin. At the same time, it is obvious that we are still at the beginning of what one day may become a routine transfer of alien genes to wheat by interspecific hybridization. In fact, there is another potential obstacle, which so far has received little attention, and that is the genome biology. Almost nothing is known on the behavior of introgressed chromosomes, chromosome segments and/or minute amounts of alien chromatin introgressed into the wheat genome. It is not clear how the wheat genome interacts with introgressed genes and how it influences their function. At the same time, it is important to understand if and how the alien DNA affects the function of the recipient wheat genome. There is an urgent need to clarify the interaction between the host and alien genomes to avoid failed attempts. Luckily, the recent advances in genomics, transcriptomics, epigenomics, proteomics, as well as in cytogenetics, and the analysis of 3D organization of interphase nuclei in particular, are promising to deliver the much needed insights.

Acknowledgments We thank our colleagues Michael Abrouk, Veronika Burešová, Petr Cápal and Gabriella Linc for useful comments and sharing their unpublished results. This work has been supported by the National Program of Sustainability (award no. LO 2014), the Czech Science Foundation (award no. P501-12-G090), the Hungarian National Research Fund (K112226 and K116277), János Bólyai Research Scholarship from the Hungarian Academy of Sciences, and an OECD fellowship (TAD/CRP JA00079297).

References

Adonina IG, Salina EA, Pestsova EG, Röder MS (2005) Transferability of wheat microsatellites to diploid *Aegilops* species and determination of chromosomal localizations of microsatellites in the S genome. Genome 48:959–970

Alheit KV, Busemeyer L, Liu W, Maurer HP, Gowda M, Hahn V, Weissmann S, Ruckelshausen A, Reif JC, Würschum T (2014) Multiple-line cross QTL mapping for biomass yield and plant height in triticale (× *Triticosecale* Wittmack). Theor Appl Genet 127:251–260

Badaeva ED, Friebe B, Gill BS (1996a) Genome differentiation in *Aegilops* 1 Distribution of highly repetitive DNA sequences on chromosomes of diploid species. Genome 39:293–306

Badaeva ED, Friebe B, Gill BS (1996b) Genome differentiation in *Aegilops* 2 Physical mapping of 5S and 18S–26S ribosomal RNA gene families in diploid species. Genome 39:1150–1158

Bandopadhyay R, Sharma S, Rustgi S, Singh R, Kumar A, Balyan HS, Gupta PK (2004) DNA polymorphism among 18 species of *Triticum–Aegilops* complex using wheat EST–SSRs. Plant Sci 166:349–356

Bartoš J, Paux E, Kofler R, Havránková M, Kopecký D, Suchánková P, Šafář J, Šimková H, Town CD, Lelley T, Feuillet C, Doležel J (2008) A first survey of the rye (*Secale cereale*) genome composition through BAC end sequencing of the short arm of chromosome 1R. BMC Plant Biol 8:95

Bedbrook JR, Jones J, O'Dell M, Thompson RD, Flavell RB (1980) A molecular description of telomeric heterochromatin in *Secale* species. Cell 19:545–560

Bennett MD, Finch RA, Barclay IR (1976) The time rate and mechanism of chromosome elimination in *Hordeum* hybrids. Chromosoma 54:175–200

Bentley AR, Scutari M, Gosman N, Faure S, Bedford F, Howell P, Cockram J, Rose GA, Barber T, Irigoyen J, Horsnell R, Pumfrey C, Winnie E, Schacht J, Beauchêne K, Praud S, Greenland A, Balding D, Mackay IJ (2014) Applying association mapping and genomic selection to the dissection of key traits in elite European wheat. Theor Appl Genet 127:2619–2633

Bento M, Gustafson P, Viegas W, Silva M (2010) Genome merger: from sequence rearrangements in triticale to their elimination in wheat–rye addition lines. Theor Appl Genet 121:489–497

Bento M, Pereira HS, Rocheta M, Gustafson P, Viegas W, Silva M (2008) Polyploidization as a retraction force in plant genome evolution: sequence rearrangements in triticale. PLoS One 3, e1402

Berkman PJ, Skarshewski A, Manoli S, Lorenc MT, Stiller J, Smits L, Lai K, Campbell E, Kubaláková M, Šimková H, Batley J, Doležel J, Hernandez P, Edwards D (2011) Sequencing wheat chromosome arm 7BS delimits the 7BS/4AL translocation and reveals homoeologous gene conservation. Theor Appl Genet 124:423–432

Blakeslee A (1937) Redoublement du nombre de chromosomes chez les plantes par traitement chimique. Compt Rend Acad Sci Paris 205:476–479

Del Blanco IA, Rajaram S, Kronstad WE (2001) Agronomic potential of synthetic hexaploid wheat-derived populations. Crop Sci 41:670

Bolibok-Brągoszewska H, Targońska M, Bolibok L, Kilian A, Rakoczy-Trojanowska M (2014) Genome-wide characterization of genetic diversity and population structure in *Secale*. BMC Plant Biol 14:184

Bougas B, Normandeau E, Audet C, Bernatchez L (2013) Linking transcriptomic and genomic variation to growth in brook charr hybrids (*Salvelinus fontinalis* Mitchill.). Heredity 110:492–500

Brenchley R, Spannagl M, Pfeifer M, Barker GLA, D'Amore R, Allen AM, McKenzie N, Kramer M, Kerhornou A, Bolser D, Kay S, Waite D, Trick M, Bancroft I, Gu Y, Huo N, Luo M-C, Sehgal S, Gill B, Kianian S, Anderson O, Kersey P, Dvorak J, McCombie WR, Hall A, Mayer KFX, Edwards KJ, Bevan MW, Hall N (2012) Analysis of the bread wheat genome using whole-genome shotgun sequencing. Nature 491:705–710

Busch W, Martin R, Herrmann RG, Hohmann U (1995) Repeated DNA sequences isolated by microdissection. I. Karyotyping of barley (*Hordeum vulgare* L). Genome 38:1082–1090

De Bustos A, Cuadrado A, Soler C, Jouve N (1996) Physical mapping of repetitive DNA sequences and 5S and 18S-26S rDNA in five wild species of the genus *Hordeum*. Chromosome Res 4:491–499

Cabral AL, Jordan MC, McCartney CA, You FM, Humphreys DG, MacLachlan R, Pozniak CJ (2014) Identification of candidate genes, regions and markers for pre-harvest sprouting resistance in wheat (*Triticum aestivum* L). BMC Plant Biol 14:340

Castillo A, Atienza SG, Martín AC (2014) Fertility of CMS wheat is restored by two *Rf* loci located on a recombined acrocentric chromosome. J Exp Bot 65:6667–6677

Cavanagh CR, Chao S, Wang S, Huang BE, Stephen S, Kiani S, Forrest K, Saintenac C, Brown-Guedira GL, Akhunova A, See D, Bai G, Pumphrey M, Tomar L, Wong D, Kong S, Reynolds M, da Silva ML, Bockelman H, Talbert L, Anderson JA, Dreisigacker S, Baenziger S, Carter A, Korzun V, Morrell PL, Dubcovsky J, Morell MK, Sorrells ME, Hayden MJ, Akhunov E (2013) Genome-wide comparative diversity uncovers multiple targets of selection for improvement in hexaploid wheat landraces and cultivars. Proc Natl Acad Sci 110:8057–8062

Cenci A, D'Ovidio R, Tanzarella OA, Ceoloni C, Porceddu E (1999) Identification of molecular markers linked to *Pm13*, an *Aegilops longissima* gene conferring resistance to powdery mildew in wheat. Theor Appl Genet 98:448–454

Chen G, Zheng Q, Bao Y, Liu S, Wang H, Li X (2012) Molecular cytogenetic identification of a novel dwarf wheat line with introgressed *Thinopyrum ponticum* chromatin. J Biosci 37:149–155

Chen P, Liu W, Yuan J, Wang X, Zhou B, Wang S, Zhang S, Feng Y, Yang B, Liu G, Liu D, Qi L, Zhang P, Friebe B, Gill BS (2005) Development and characterization of wheat-*Leymus racemosus* translocation lines with resistance to Fusarium Head Blight. Theor Appl Genet 111:941–948

Choulet F, Alberti A, Theil S, Glover N, Barbe V, Daron J, Pingault L, Sourdille P, Couloux A, Paux E, Leroy P, Mangenot S, Guilhot N, Gouis JL, Balfourier F, Alaux M, Jamilloux V, Poulain J, Durand C, Bellec A, Gaspin C, Safar J, Dolezel J, Rogers J, Vandepoele K, Aury J-M, Mayer K, Berges H, Quesneville H, Wincker P, Feuillet C (2014) Structural and functional partitioning of bread wheat chromosome 3B. Science 345:1249721

Colasuonno P, Maria MA, Blanco A, Gadaleta A (2013) Description of durum wheat linkage map and comparative sequence analysis of wheat mapped DArT markers with rice and *Brachypodium* genomes. BMC Genet 14:114

Comai L (2000) Genetic and epigenetic interactions in allopolyploid plants. Plant Mol Biol 43:387–399

Comai L, Madlung A, Josefsson C, Tyagi A (2003) Do the different parental "heteromes" cause genomic shock in newly formed allopolyploids? Philos Trans R Soc Lond B Biol Sci 358:1149–1155

Cseh A, Kruppa K, Molnár I, Rakszegi M, Doležel J, Molnár-Láng M (2011) Characterization of a new 4BS.7HL wheat/barley translocation line using GISH, FISH and SSR markers and its effect on the β-glucan content of wheat. Genome 54:795–804

Cuadrado A, Cardoso M, Jouve N (2008) Physical organisation of simple sequence repeats (SSRs) in Triticeae: structural, functional and evolutionary implications. Cytogenet Genome Res 120:210–219

Danilova TV, Friebe B, Gill BS (2014) Development of a wheat single gene FISH map for analyzing homoeologous relationship and chromosomal rearrangements within the Triticeae. Theor Appl Genet 127:715–730

Darkó É, Janda T, Majláth I, Szopkó D, Dulai S, Molnár I, Molnár-Láng M (2015) Salt stress response of wheat/barley addition lines carrying chromosomes from the winter barley "Manas". Euphytica 203:491–504

Devos KM, Atkinson MD, Chinoy CN, Francis HA, Harcourt RL, Koebner RMD, Liu CJ, Masojć P, Xie DX, Gale MD (1993) Chromosomal rearrangements in the rye genome relative to that of wheat. Theor Appl Genet 85:673–680

Devos KM, Gale MD (1993) Extended genetic maps of the homoeologous group 3 chromosomes of wheat, rye and barley. Theor Appl Genet 85:649–652

Doležel J, Číhalíková J, Lucretti S (1992) A high-yield procedure for isolation of metaphase chromosomes from root tips of *Vicia faba* L. Planta 188:93–98

Doležel J, Kubaláková M, Paux E, Bartoš J, Feuillet C (2007) Chromosome-based genomics in the cereals. Chromosome Res 15:51–66

Doležel J, Vrána J, Cápal P, Kubaláková M, Burešová V, Šimková H (2014) Advances in plant chromosome genomics. Biotechnol Adv 32:122–136

Dulai S, Molnár I, Haló B, Molnár-Láng M (2010) Photosynthesis in the 7H Asakaze komugi/Manas wheat/barley addition line during salt stress. Acta Agron Hung 58:367–376

Dulai S, Molnár I, Szopkó D, Darkó É, Vojtkó A, Sass-Gyarmati A, Molnár-Láng M (2014) Wheat-*Aegilops biuncialis* amphiploids have efficient photosynthesis and biomass production during osmotic stress. J Plant Physiol 171:509–517

Ellis THN, Poyser SJ, Knox MR, Vershinin AV, Ambrose MJ (1998) Polymorphism of insertion sites of *Ty1*-copia class retrotransposons and its use for linkage and diversity analysis in pea. Mol Gen Genet 260:9–19

Endo TR, Kubaláková M, Vrána J, Doležel J (2014) Hyperexpansion of wheat chromosomes sorted by flow cytometry. Genes Genet Syst 89:181–185

Ersfeld K (2004) Fiber-FISH: fluorescence *in situ* hybridization on stretched DNA. Methods Mol Biol 270:395–402

Farkas A, Molnár I, Dulai S, Rapi S, Oldal V, Cseh A, Kruppa K, Molnár-Láng M (2014) Increased micronutrient content (Zn, Mn) in the $3M^b$(4B) wheat-*Aegilops biuncialis* substitution and $3M^b$.4BS translocation identified by GISH and FISH. Genome 57:61–67

Farrer W (1904) Some notes on the wheat "Bobs"; its peculiarities, economic value and origin. Agric Gaz NSW 15:849–854

Fatih AM (1983) Analysis of the breeding potential of wheat-*Agropyron* and wheat-*Elymus* derivatives. I. Agronomic and quality characteristics. Hereditas 98:287–295

Fedak G (1999) Molecular aids for integration of alien chromatin through wide crosses. Genome 42:584–591

Feldman M, Liu B, Segal G, Abbo S, Levy AA, Vega JM (1997) Rapid elimination of low-copy DNA sequences in polyploid wheat: a possible mechanism for differentiation of homoeologous chromosomes. Genetics 147:1381–1387

Feuillet C, Langridge P, Waugh R (2008) Cereal breeding takes a walk on the wild side. Trends Genet 24:24–32

Francki MG, Crasta OR, Sharma HC, Ohm HW, Anderson JM (1997) Structural organization of an alien *Thinopyrum intermedium* group 7 chromosome in US soft red winter wheat (*Triticum aestivum* L). Genome 40:716–722

Friebe B, Jiang J, Raupp WJ, McIntosh RA, Gill BS (1996) Characterization of wheat-alien translocations conferring resistance to diseases and pests: current status. Euphytica 91:59–87

Friebe B, Larter EN (1988) Identification of a complete set of isogenic wheat/rye D-genome substitution lines by means of Giemsa C-banding. Theor Appl Genet 76:473–479

Friebe B, Zeller FJ, Mukai Y, Forster BP, Bartos P, McIntosh RA (1992) Characterization of rust-resistant wheat-*Agropyron intermedium* derivatives by C-banding, *in situ* hybridization and isozyme analysis. Theor Appl Genet 83:775–782

Fu S, Yang M, Fei Y, Tan F, Ren Z, Yan B, Zhang H, Tang Z (2013) Alterations and abnormal mitosis of wheat chromosomes induced by wheat-rye monosomic addition lines. PLoS One 8, e70483

Gerlach WL, Bedbrook JR (1979) Cloning and characterization of ribosomal RNA genes from wheat and barley. Nucleic Acids Res 7:1869–1885

Gibcus JH, Dekker J (2013) The hierarchy of the 3D genome. Mol Cell 49:773–782

Gill BS, Friebe BR, White FF (2011) Alien introgressions represent a rich source of genes for crop improvement. Proc Natl Acad Sci 108:7657–7658

Gill BS, Kimber G (1974) The giemsa C-banded karyotype of rye. Proc Natl Acad Sci U S A 71:1247–1249

Gill KS, Arumuganathan K, Lee J-H (1999) Isolating individual wheat (*Triticum aestivum*) chromosome arms by flow cytometric analysis of ditelosomic lines. Theor Appl Genet 98:1248–1252

Van Ginkel M, Ogbonnaya F (2008) Using synthetic wheats to breed cultivars better adapted to changing production conditions. Field Crop Res 104:86–94

Giorgi D, Farina A, Grosso V, Gennaro A, Ceoloni C, Lucretti S (2013) FISHIS: fluorescence *in situ* hybridization in suspension and chromosome flow sorting made easy. PLoS One 8, e57994

Griffiths S, Sharp R, Foote TN, Bertin I, Wanous M, Reader S, Colas I, Moore G (2006) Molecular characterization of *Ph1* as a major chromosome pairing locus in polyploid wheat. Nature 439:749–752

Grob S, Schmid MW, Grossniklaus U (2014) Hi-C analysis in *Arabidopsis* identifies the KNOT, a structure with similarities to the flamenco locus of Drosophila. Mol Cell 55:678–693
Grosso V, Farina A, Gennaro A, Giorgi D, Lucretti S (2012) Flow sorting and molecular cytogenetic identification of individual chromosomes of *Dasypyrum villosum* L. (*H. villosa*) by a single DNA probe. PLoS One 7, e50151
Guadagnuolo R, Bianchi DS, Felber F (2001) Specific genetic markers for wheat, spelt, and four wild relatives: comparison of isozymes, RAPDs, and wheat microsatellites. Genome 44:610–621
Gupta PK, Mir RR, Mohan A, Kumar J (2008) Wheat genomics: present status and future prospects. Int J Plant Genomics 2008:896451
Gupta PK, Rustgi S, Sharma S, Singh R, Kumar N, Balyan HS (2003) Transferable EST-SSR markers for the study of polymorphism and genetic diversity in bread wheat. Mol Genet Genomics 270:315–323
Gupta PK, Varshney RK, Sharma PC, Ramesh B (1999) Molecular markers and their applications in wheat breeding. Plant Breeding 118:369–390
Hajjar R, Hodgkin T (2007) The use of wild relatives in crop improvement: a survey of developments over the last 20 years. Euphytica 156:1–13
Han FP, Fedak G, Ouellet T, Liu B (2003) Rapid genomic changes in interspecific and intergeneric hybrids and allopolyploids of Triticeae. Genome 46:716–723
Hart GE, Islam AKMR, Shepherd KW (1980) Use of isozymes as chromosome markers in the isolation and characterization of wheat-barley chromosome addition lines. Genet Res 36:311–325
Haseneyer G, Schmutzer T, Seidel M, Zhou R, Mascher M, Schön C-C, Taudien S, Scholz U, Stein N, Mayer KFX, Bauer E (2011) From RNA-seq to large-scale genotyping - genomics resources for rye (*Secale cereale* L). BMC Plant Biol 11:131
He C, Holme J, Anthony J (2014) SNP genotyping: the KASP assay. Crop Breed (Springer, New York) 2014:75–86
Helguera M, Rivarola M, Clavijo B, Martis MM, Vanzetti LS, González S, Garbus I, Leroy P, Šimková H, Valárik M, Caccamo M, Doležel J, Mayer KFX, Feuillet C, Tranquilli G, Paniego N, Echenique V (2015) New insights into the wheat chromosome 4D structure and virtual gene order, revealed by survey pyrosequencing. Plant Sci 233:200–212
Hernandez P, Martis M, Dorado G, Gálvez S, Pfeifer M, Schaaf S, Jouve N, Šimková H, Valárik M, Doležel J, Mayer KFX (2012) Next-generation sequencing and syntenic integration of flow-sorted arms of wheat chromosome 4A exposes the chromosome structure and gene content. Plant J 69:377–386
Hernández P, Rubio MJ, Martin A (1996) Development of RAPD markers in tritordeum and addition lines of *Hordeum chilense* in *Triticum aestivum*. Plant Breeding 115:52–56
Hurni S, Brunner S, Stirnweis D, Herren G, Peditto D, McIntosh RA, Keller B (2014) The powdery mildew resistance gene *Pm8* derived from rye is suppressed by its wheat ortholog *Pm3*. Plant J 79:904–913
Iehisa JCM, Ohno R, Kimura T, Enoki H, Nishimura S, Okamoto Y, Nasuda S, Takumi S (2014) A high-density genetic map with array-based markers facilitates structural and quantitative trait locus analyses of the common wheat genome. DNA Res 21:555–567
International Rice Genome Sequencing Project (2005) The map-based sequence of the rice genome. Nature 436:793–800
Islam AKMR, Shepherd KW, Sparrow DHB (1981) Isolation and characterization of euplasmic wheat-barley chromosome addition lines. Heredity 46:161–174
The International Wheat Genome Sequencing Consortium (IWGSC) (2014) A chromosome-based draft sequence of the hexaploid bread wheat (*Triticum aestivum*) genome. Science 345:1251788
Jaccoud D, Peng K, Feinstein D, Kilian A (2001) Diversity Arrays: a solid state technology for sequence information independent genotyping. Nucleic Acids Res 29:e25
Jackson SA, Wang ML, Goodman HM, Jiang J (1998) Application of fiber-FISH in physical mapping of *Arabidopsis thaliana*. Genome 41:566–572

Jakobson I, Peusha H, Timofejeva L, Järve K (2006) Adult plant and seedling resistance to powdery mildew in a *Triticum aestivum* x *Triticum militinae* hybrid line. Theor Appl Genet 112:760–769

Jakobson I, Reis D, Tiidema A, Peusha H, Timofejeva L, Valárik M, Kladivová M, Šimková H, Doležel J, Järve K (2012) Fine mapping, phenotypic characterization and validation of non-race-specific resistance to powdery mildew in a wheat–*Triticum militinae* introgression line. Theor Appl Genet 125:609–623

Jeffrey Chen Z, Ni Z (2006) Mechanisms of genomic rearrangements and gene expression changes in plant polyploids. Bioessays 28:240–252

Jiang J, Friebe B, Gill B (1993) Recent advances in alien gene transfer in wheat. Euphytica 73:199–212

Jighly A, Oyiga BC, Makdis F, Nazari K, Youssef O, Tadesse W, Abdalla O, Ogbonnaya FC (2015) Genome-wide DArT and SNP scan for QTL associated with resistance to stripe rust (*Puccinia striiformis* f. sp. *tritici*) in elite ICARDA wheat (*Triticum aestivum* L.) germplasm. Theor Appl Genet 128:1277–1295

Jing H-C, Bayon C, Kanyuka K, Berry S, Wenzl P, Huttner E, Kilian A, Hammond-Kosack KE (2009) DArT markers: diversity analyses, genomes comparison, mapping and integration with SSR markers in *Triticum monococcum*. BMC Genomics 10:458

Jing HC, Kornyukhin D, Kanyuka K, Orford S, Zlatska A, Mitrofanova OP, Koebner R, Hammond-Kosack K (2007) Identification of variation in adaptively important traits and genome-wide analysis of trait-marker associations in *Triticum monococcum*. J Exp Bot 58:3749–3764

Jupe J, Stam R, Howden AJ, Morris JA, Zhang R, Hedley PE, Huitema E (2013) *Phytophthora capsici*-tomato interaction features dramatic shifts in gene expression associated with a hemi-biotrophic lifestyle. Genome Biol 14:R63

Kalih R, Maurer HP, Miedaner T (2015) Genetic architecture of fusarium head blight resistance in four winter triticale populations. Phytopathology 105:334–341

Karafiátová M, Bartoš J, Kopecký D, Ma L, Sato K, Houben A, Stein N, Doležel J (2013) Mapping non-recombining regions in barley using multicolor FISH. Chromosome Res 21:739–751

Kashkush K, Feldman M, Levy AA (2002) Gene loss, silencing and activation in a newly synthesized wheat allotetraploid. Genetics 160:1651–1659

Kerber ER, Aung T (1999) Leaf rust resistance gene *Lr34* associated with nonsuppression of stem rust resistance in the wheat cultivar Canthatch. Phytopathology 89:518–521

Kerber ER, Dyck PL (1973) Inheritance of stem rust resistance transferred from diploid wheat (*Triticum monococcum*) to tetraploid and hexaploid wheat and chromosome location of the gene involved. Can J Genet Cytol 15:397–409

Khalil HB, Ehdaeivand M-R, Xu Y, Laroche A, Gulick PJ (2015) Identification and characterization of rye genes not expressed in allohexaploid triticale. BMC Genomics 16:281

Kihara H (1937) Genomanalyse bei *Triticum* und *Aegilops*. Kurze übersicht über die Ergebnisse der Jahre 1934–36. Mem Coll Agr Kyoto Imp Univ 41:1–61

King J, Armstead I, Harper J, Ramsey L, Snape J, Waugh R, James C, Thomas A, Gasior D, Kelly R, Roberts L, Gustafson P, King I (2013) Exploitation of interspecific diversity for monocot crop improvement. Heredity 110:475–483

King IP, Purdie KA, Rezanoor HN, Koebner RMD, Miller TE, Reader SM, Nicholson P (1993) Characterization of *Thinopyrum bessarabicum* chromosome segments in wheat using random amplified polymorphic DNAs (RAPDs) and genomic in situ hybridization. Theor Appl Genet 86:895–900

Koba T, Takumi S, Shimada T (1997) Isolation, identification and characterization of disomic and translocated barley chromosome addition lines of common wheat. Euphytica 96:289–296

Kroupin PY, Divashuk MG, Fesenko IA, Karlov GI (2012) Evaluating wheat microsatellite markers for the use in genetic analysis of *Thinopyrum*, *Dasypyrum*, and *Pseudoroegneria* species. Dataset Paper Sci 2013, e949637

Kruppa K, Sepsi A, Szakács É, Röder MS, Molnár-Láng M (2013) Characterization of a 5HS-7DS.7DL wheat-barley translocation line and physical mapping of the 7D chromosome using SSR markers. J Appl Genet 54:251–258

Kruse A (1973) *Hordeum* × *Triticum* hybrids. Hereditas 73:157–161
Kubaláková M, Kovářová P, Suchánková P, Číhalíková J, Bartoš J, Lucretti S, Watanabe N, Kianian SF, Doležel J (2005) Chromosome sorting in tetraploid wheat and its potential for genome analysis. Genetics 170:823–829
Kubaláková M, Valárik M, Barto J, Vrána J, Cíhalíková J, Molnár-Láng M, Dolezel J (2003) Analysis and sorting of rye (*Secale cereale* L) chromosomes using flow cytometry. Genome 46:893–905
Kubaláková M, Vrána J, Číhalíková J, Šimková H, Doležel J (2002) Flow karyotyping and chromosome sorting in bread wheat (*Triticum aestivum* L). Theor Appl Genet 104:1362–1372
Kuraparthy V, Chhuneja P, Dhaliwal HS, Kaur S, Bowden RL, Gill BS (2007) Characterization and mapping of cryptic alien introgression from *Aegilops geniculata* with new leaf rust and stripe rust resistance genes *Lr57* and *Yr40* in wheat. Theor Appl Genet 114:1379–1389
Langridge P, Lagudah ES, Holton TA, Appels R, Sharp PJ, Chalmers KJ (2001) Trends in genetic and genome analyses in wheat: a review. Aust J Agric Res 52:1043–1077
Lapitan NLV, Sears RG, Gill BS (1984) Translocations and other karyotypic structural changes in wheat x rye hybrids regenerated from tissue culture. Theor Appl Genet 68:547–554
Laurie DA, Bennett MD (1986) Wheat × maize hybridization. Can J Genet Cytol 28:313–316
Lee H-S, Chen ZJ (2001) Protein-coding genes are epigenetically regulated in *Arabidopsis* polyploids. Proc Natl Acad Sci U S A 98:6753–6758
Lee J-H, Arumuganathan K, Chung Y-S, Kim K-Y, Chung W-B, Bae K-S, Kim D-H, Chung D-S, Kwon O-C (2000) Flow cytometric analysis and chromosome sorting of barley (*Hordeum vulgare* L). Mol Cells 10:619–625
Lee J-H, Arumuganathan K, Yen Y, Kaeppler S, Kaeppler H, Baenziger PS (1997) Root tip cell cycle synchronization and metaphase-chromosome isolation suitable for flow sorting in common wheat (*Triticum aestivum* L). Genome 40:633–638
Levy AA, Feldman M (2004) Genetic and epigenetic reprogramming of the wheat genome upon allopolyploidization. Biol J Linn Soc 82:607–613
Li J, Wan H-S, Yang W-Y (2014) Synthetic hexaploid wheat enhances variation and adaptive evolution of bread wheat in breeding processes. J Syst Evol 52:735–742
Li LJ, Arumuganathan K, Rines HW, Phillips RL, Riera-Lizarazu O, Sandhu D, Zhou Y, Gill KS (2001) Flow cytometric sorting of maize chromosome 9 from an oat-maize chromosome addition line. Theor Appl Genet 102:658–663
Linc G, Sepsi A, Molnár-Láng M (2012) A FISH karyotype to study chromosome polymorphisms for the *Elytrigia elongata* E genome. Cytogenet Genome Res 136:138–144
Ling H-Q, Zhao S, Liu D, Wang J, Sun H, Zhang C, Fan H, Li D, Dong L, Tao Y, Gao C, Wu H, Li Y, Cui Y, Guo X, Zheng S, Wang B, Yu K, Liang Q, Yang W, Lou X, Chen J, Feng M, Jian J, Zhang X, Luo G, Jiang Y, Liu J, Wang Z, Sha Y, Zhang B, Wu H, Tang D, Shen Q, Xue P, Zou S, Wang X, Liu X, Wang F, Yang Y, An X, Dong Z, Zhang K, Zhang X, Luo M-C, Dvorak J, Tong Y, Wang J, Yang H, Li Z, Wang D, Zhang A, Wang J (2013) Draft genome of the wheat A-genome progenitor *Triticum urartu*. Nature 496:87–90
Liu B, Vega JM, Segal G, Abbo S, Rodova M, Feldman M (1998) Rapid genomic changes in newly synthesized amphiploids of *Triticum* and *Aegilops*. I. Changes in low-copy noncoding DNA sequences. Genome 41:272–277
Liu W-H, Luan Y, Wang J-C, Wang X-G, Su J-J, Zhang J-P, Yang X-M, Gao A-N, Li L-H (2010) Production and identification of wheat – *Agropyron cristatum* (1.4P) alien translocation lines. Genome 53:472–481
Lucas SJ, Akpınar BA, Šimková H, Kubaláková M, Doležel J, Budak H (2014) Next-generation sequencing of flow-sorted wheat chromosome 5D reveals lineage-specific translocations and widespread gene duplications. BMC Genomics 15:1080
Lukaszewski AJ, Gustafson JP (1983) Translocations and modifications of chromosomes in triticale × wheat hybrids. Theor Appl Genet 64:239–248
Luo M-C, Gu YQ, You FM, Deal KR, Ma Y, Hu Y, Huo N, Wang Y, Wang J, Chen S, Jorgensen CM, Zhang Y, McGuire PE, Pasternak S, Stein JC, Ware D, Kramer M, McCombie WR,

Kianian SF, Martis MM, Mayer KFX, Sehgal SK, Li W, Gill BS, Bevan MW, Šimková H, Doležel J, Weining S, Lazo GR, Anderson OD, Dvorak J (2013) A 4-gigabase physical map unlocks the structure and evolution of the complex genome of *Aegilops tauschii*, the wheat D-genome progenitor. Proc Natl Acad Sci U S A 110:7940–7945

Lynch M, Conery JS (2000) The evolutionary fate and consequences of duplicate genes. Science 290:1151–1155

Lysák MA, Cíhalíková J, Kubaláková M, Simková H, Künzel G, Dolezel J (1999) Flow karyotyping and sorting of mitotic chromosomes of barley (*Hordeum vulgare* L). Chromosome Res 7:431–444

Martis MM, Zhou R, Haseneyer G, Schmutzer T, Vrána J, Kubaláková M, König S, Kugler KG, Scholz U, Hackauf B, Korzun V, Schön C-C, Doležel J, Bauer E, Mayer KFX, Stein N (2013) Reticulate evolution of the rye genome. Plant Cell 25:3685–3698

Ma X-F, Gustafson JP (2008) Allopolyploidization-accommodated genomic sequence changes in triticale. Ann Bot 101:825–832

Ma X-F, Gustafson JP (2006) Timing and rate of genome variation in triticale following allopolyploidization. Genome 49:950–958

Mayer KFX, Martis M, Hedley PE, Šimková H, Liu H, Morris JA, Steuernagel B, Taudien S, Roessner S, Gundlach H, Kubaláková M, Suchánková P, Murat F, Felder M, Nussbaumer T, Graner A, Salse J, Endo T, Sakai H, Tanaka T, Itoh T, Sato K, Platzer M, Matsumoto T, Scholz U, Dolezel J, Waugh R, Stein N (2011) Unlocking the barley genome by chromosomal and comparative genomics. Plant Cell 23:1249–1263

Mayer KFX, Taudien S, Martis M, Šimková H, Suchánková P, Gundlach H, Wicker T, Petzold A, Felder M, Steuernagel B, Scholz U, Graner A, Platzer M, Doležel J, Stein N (2009) Gene content and virtual gene order of barley chromosome 1H. Plant Physiol 151:496–505

McArthur RI, Zhu X, Oliver RE, Klindworth DL, Xu SS, Stack RW, Wang RR-C, Cai X (2012) Homoeology of *Thinopyrum junceum* and *Elymus rectisetus* chromosomes to wheat and disease resistance conferred by the *Thinopyrum* and *Elymus* chromosomes in wheat. Chromosome Res 20:699–715

McIntyre CL, Pereira S, Moran LB, Appels R (1990) New *Secale cereale* (rye) DNA derivatives for the detection of rye chromosome segments in wheat. Genome 33:635–640

Meurant G (1982) Advances in genetics. Academic, New York, NY

Mohan A, Goyal A, Singh R, Balyan HS, Gupta PK (2007) Physical mapping of wheat and rye expressed sequence tag–simple sequence repeats on wheat chromosomes. Crop Sci 47:3

Molnár I, Gáspár L, Sárvári É, Dulai S, Hoffmann B, Molnár-Láng M, Galiba G (2004) Physiological and morphological responses to water stress in *Aegilops biuncialis* and *Triticum aestivum* genotypes with differing tolerance to drought. Funct Plant Biol 31:1149–1159

Molnár I, Benavente E, Molnár-Láng M (2009) Detection of intergenomic chromosome rearrangements in irradiated *Triticum aestivum-Aegilops biuncialis* amphiploids by multicolour genomic *in situ* hybridization. Genome 52:156–165

Molnár I, Cifuentes M, Schneider A, Benavente E, Molnár-Láng M (2011a) Association between simple sequence repeat-rich chromosome regions and intergenomic translocation breakpoints in natural populations of allopolyploid wild wheats. Ann Bot 107:65–76

Molnár I, Kubaláková M, Šimková H, Cseh A, Molnár-Láng M, Doležel J (2011b) Chromosome isolation by flow sorting in *Aegilops umbellulata* and *Ae. comosa* and their allotetraploid hybrids *Ae. biuncialis* and *Ae. geniculata*. PLoS One 6, e27708

Molnár I, Kubaláková M, Šimková H, Farkas A, Cseh A, Megyeri M, Vrána J, Molnár-Láng M, Doležel J (2014) Flow cytometric chromosome sorting from diploid progenitors of bread wheat, *T. urartu*, *Ae. speltoides* and *Ae. tauschii*. Theor Appl Genet 127:1091–1104

Molnár I, Šimková H, Leverington-Waite M, Goram R, Cseh A, Vrána J, Farkas A, Doležel J, Molnár-Láng M, Griffiths S (2013) Syntenic relationships between the U and M genomes of *Aegilops*, wheat and the model species *Brachypodium* and rice as revealed by COS markers. PLoS One 8, e70844

Molnár I, Vrána J, Farkas A, Kubaláková M, Cseh A, Molnár-Láng M, Doležel J (2015) Flow sorting of C-genome chromosomes from wild relatives of wheat *Aegilops markgrafii*, *Ae triuncia-*

lis and *Ae cylindrica*, and their molecular organization. Ann Bot 116:189–200. doi:10.1093/aob/mcv073

Molnár-Láng M, Kruppa K, Cseh A, Bucsi J, Linc G (2012) Identification and phenotypic description of new wheat: six-rowed winter barley disomic additions. Genome 55:302–311

Molnár-Láng M, Linc G, Friebe BR, Sutka J (2000) Detection of wheat-barley translocations by genomic *in situ* hybridization in derivatives of hybrids multiplied *in vitro*. Euphytica 112:117–123

Molnár-Láng M, Novotny C, Linc G, Nagy ED (2005) Changes in the meiotic pairing behaviour of a winter wheat-winter barley hybrid maintained for a long term in tissue culture, and tracing the barley chromatin in the progeny using GISH and SSR markers. Plant Breeding 124:247–252

Molnár-Láng M, Molnár I, Szakács É, Linc G, Bedö Z (2014) Production and molecular cytogenetic identification of wheat-alien hybrids and introgression lines. In: Tuberosa R, Graner A, Frison E (eds) Genomics of plant genetic resources. Springer, Dordrecht, pp 255–283, doi:10.1007/978-94-007-7572-5

Montilla-Bascón G, Rispail N, Sánchez-Martín J, Rubiales D, Mur LAJ, Langdon T, Howarth CJ, Prats E (2015) Genome-wide association study for crown rust (*Puccinia coronata* f.sp. *avenae*) and powdery mildew (*Blumeria graminis* f.sp. *avenae*) resistance in an oat (*Avena sativa*) collection of commercial varieties and landraces. Front. Plant Sci 6:103

Mujeeb-Kazi A (1995) Intergeneric crosses: hybrid production and utilization. In: Mujeeb-Kazi A, Hettel GP (eds) Utilizing wild grass biodiversity in wheat improvement: 15 years of wide cross research at CIMMYT. CIMMYT, Mexico, p 140

Mukhopadhyay P, Singla-Pareek SL, Reddy MK, Sopory SK (2013) Stress-mediated alterations in chromatin architecture correlate with down-regulation of a gene encoding 60S rpL32 in rice. Plant Cell Physiol 54(4):528–540

Murashige T, Skoog F (1962) A revised medium for rapid growth and bio assays with tobacco tissue cultures. Physiol Plant 15:473–497

Nagaki K, Tsujimoto H, Isono K, Sasakuma T (1995) Molecular characterization of a tandem repeat, *Afa* family, and its distribution among Triticeae. Genome 38:479–486

Nagy ED, Lelley T (2003) Genetic and physical mapping of sequence-specific amplified polymorphic (SSAP) markers on the 1RS chromosome arm of rye in a wheat background. Theor Appl Genet 107:1271–1277

Nagy ED, Molnár I, Schneider A, Kovács G, Molnár-Láng M (2006) Characterization of chromosome-specific S-SAP markers and their use in studying genetic diversity in *Aegilops* species. Genome 49:289–296

Nagy ED, Molnár-Láng M, Linc G, Láng L (2002) Identification of wheat-barley translocations by sequential GISH and two-colour FISH in combination with the use of genetically mapped barley SSR markers. Genome 45:1238–1247

Ng DW-K, Zhang C, Miller M, Shen Z, Briggs SP, Chen ZJ (2012) Proteomic divergence in *Arabidopsis* autopolyploids and allopolyploids and their progenitors. Heredity 108:419–430

Osborn TC, Chris Pires J, Birchler JA, Auger DL, Jeffery Chen Z, Lee H-S, Comai L, Madlung A, Doerge RW, Colot V, Martienssen RA (2003) Understanding mechanisms of novel gene expression in polyploids. Trends Genet 19:141–147

Pace CD, Snidaro D, Ciaffi M, Vittori D, Ciofo A, Cenci A, Tanzarella OA, Qualset CO, Mugnozza GTS (2001) Introgression of *Dasypyrum villosum* chromatin into common wheat improves grain protein quality. Euphytica 117:67–75

Parisod C, Senerchia N (2012) Responses of transposable elements to polyploidy In plant transposable elements. In: Grandbastien M-A, Casacuberta JM (eds) Topics in current genetics. Springer, Berlin, pp 147–168

Paterson AH, Bowers JE, Bruggmann R, Dubchak I, Grimwood J, Gundlach H, Haberer G, Hellsten U, Mitros T, Poliakov A, Schmutz J, Spannagl M, Tang H, Wang X, Wicker T, Bharti AK, Chapman J, Feltus FA, Gowik U, Grigoriev IV, Lyons E, Maher CA, Martis M, Narechania A, Otillar RP, Penning BW, Salamov AA, Wang Y, Zhang L, Carpita NC, Freeling M, Gingle AR, Hash CT, Keller B, Klein P, Kresovich S, McCann MC, Ming R, Peterson DG, Mehboob-

ur-Rahman, Ware D, Westhoff P, Mayer KFX, Messing J, Rokhsar DS (2009) The *Sorghum bicolor* genome and the diversification of grasses. Nature 457:551–556
Peil A, Korzun V, Schubert V, Schumann E, Weber WE, Röder MS (1998) The application of wheat microsatellites to identify disomic *Triticum aestivum-Aegilops markgrafii* addition lines. Theor Appl Genet 96:138–146
Peleg Z, Saranga Y, Suprunova T, Ronin Y, Röder MS, Kilian A, Korol AB, Fahima T (2008) High-density genetic map of durum wheat x wild emmer wheat based on SSR and DArT markers. Theor Appl Genet 117:103–115
Petersen S, Lyerly JH, Worthington ML, Parks WR, Cowger C, Marshall DS, Brown-Guedira G, Murphy JP (2015) Mapping of powdery mildew resistance gene *Pm53* introgressed from *Aegilops speltoides* into soft red winter wheat. Theor Appl Genet 128:303–312
Pradhan GP, Prasad PVV (2015) Evaluation of wheat chromosome translocation lines for high temperature stress tolerance at grain filling stage. PLoS One 10
Pumphrey M, Bai J, Laudencia-Chingcuanco D, Anderson O, Gill BS (2009) Nonadditive expression of homoeologous genes is established upon polyploidization in hexaploid wheat. Genetics 181:1147–1157
Qi L, Cao M, Chen P, Li W, Liu D (1996) Identification, mapping, and application of polymorphic DNA associated with resistance gene *Pm21* of wheat. Genome 39:191–197
Qi L, Echalier B, Friebe B, Gill BS (2003) Molecular characterization of a set of wheat deletion stocks for use in chromosome bin mapping of ESTs. Funct Integr Genomics 3:39–55
Qi L, Friebe B, Zhang P, Gill BS (2007) Homoeologous recombination, chromosome engineering and crop improvement. Chromosome Res 15:3–19
Qi LL, Echalier B, Chao S, Lazo GR, Butler GE, Anderson OD, Akhunov ED, Dvorák J, Linkiewicz AM, Ratnasiri A, Dubcovsky J, Bermudez-Kandianis CE, Greene RA, Kantety R, La Rota CM, Munkvold JD, Sorrells SF, Sorrells ME, Dilbirligi M, Sidhu D, Erayman M, Randhawa HS, Sandhu D, Bondareva SN, Gill KS, Mahmoud AA, Ma XF, Miftahudin, Gustafson JP, Conley EJ, Nduati V, Gonzalez-Hernandez JL, Anderson JA, Peng JH, Lapitan NL, Hossain KG, Kalavacharla V, Kianian SF, Pathan MS, Zhang DS, Nguyen HT, Choi DW, Fenton RD, Close TJ, McGuire PE, Qualset CO, Gill BS (2004) A chromosome bin map of 16,000 expressed sequence tag loci and distribution of genes among the three genomes of polyploid wheat. Genetics 168:701–712
Qi LL, Wang SL, Chen PD, Liu DJ, Friebe B, Gill BS (1997) Molecular cytogenetic analysis of *Leymus racemosus* chromosomes added to wheat. Theor Appl Genet 95:1084–1091
Queen RA, Gribbon BM, James C, Jack P, Flavell AJ (2003) Retrotransposon-based molecular markers for linkage and genetic diversity analysis in wheat. Mol Genet Genomics 271:91–97
Rabinovich SV (1998) Importance of wheat-rye translocations for breeding modern cultivar of *Triticum aestivum* L. Euphytica 100:323–340
Rayburn AL, Gill BS (1985) Use of biotin-labeled probes to map specific DNA sequences on wheat chromosomes. J Hered 76:78–81
Reynolds MP, Calderini DF, Condon AG, Rajaram S (2001) Physiological basis of yield gains in wheat associated with the *Lr19* translocation from *Agropyron elongatum* Wheat in a global environment. In: Bedö Z, Láng L (eds) Developments in plant breeding. Springer, Amsterdam, pp 345–351
Riley R, Chapman V (1958) Genetic control of the cytologically diploid behaviour of hexaploid wheat. Nature 182:713–715
Riley R, Macer RCF (1966) The chromosomal distribution of the genetic resistance of rye to wheat pathogens. Can J Genet Cytol 8:640–653
Ruperao P, Chan C-KK, Azam S, Karafiátová M, Hayashi S, Čížková J, Saxena RK, Šimková H, Song C, Vrána J, Chitikineni A, Visendi P, Gaur PM, Millán T, Singh KB, Taran B, Wang J, Batley J, Doležel J, Varshney RK, Edwards D (2014) A chromosomal genomics approach to assess and validate the desi and kabuli draft chickpea genome assemblies. Plant Biotechnol J 12:778–786
Sanei M, Pickering R, Kumke K, Nasuda S, Houben A (2011) Loss of centromeric histone H3 (*CENH3*) from centromeres precedes uniparental chromosome elimination in interspecific barley hybrids. Proc Natl Acad Sci U S A 108:E498–E505

Schlegel R, Cakmak I, Torun B, Eker S, Tolay I, Ekiz H, Kalayci M, Braun HJ (1998) Screening for zinc efficiency among wheat relatives and their utilisation for alien gene transfer. Euphytica 100:281–286

Schnable PS, Ware D, Fulton RS, Stein JC, Wei F, Pasternak S, Liang C, Zhang J, Fulton L, Graves TA, Minx P, Reily AD, Courtney L, Kruchowski SS, Tomlinson C, Strong C, Delehaunty K, Fronick C, Courtney B, Rock SM, Belter E, Du F, Kim K, Abbott RM, Cotton M, Levy A, Marchetto P, Ochoa K, Jackson SM, Gillam B, Chen W, Yan L, Higginbotham J, Cardenas M, Waligorski J, Applebaum E, Phelps L, Falcone J, Kanchi K, Thane T, Scimone A, Thane N, Henke J, Wang T, Ruppert J, Shah N, Rotter K, Hodges J, Ingenthron E, Cordes M, Kohlberg S, Sgro J, Delgado B, Mead K, Chinwalla A, Leonard S, Crouse K, Collura K, Kudrna D, Currie J, He R, Angelova A, Rajasekar S, Mueller T, Lomeli R, Scara G, Ko A, Delaney K, Wissotski M, Lopez G, Campos D, Braidotti M, Ashley E, Golser W, Kim H, Lee S, Lin J, Dujmic Z, Kim W, Talag J, Zuccolo A, Fan C, Sebastian A, Kramer M, Spiegel L, Nascimento L, Zutavern T, Miller B, Ambroise C, Muller S, Spooner W, Narechania A, Ren L, Wei S, Kumari S, Faga B, Levy MJ, McMahan L, Van Buren P, Vaughn MW, Ying K, Yeh CT, Emrich SJ, Jia Y, Kalyanaraman A, Hsia AP, Barbazuk WB, Baucom RS, Brutnell TP, Carpita NC, Chaparro C, Chia JM, Deragon JM, Estill JC, Fu Y, Jeddeloh JA, Han Y, Lee H, Li P, Lisch DR, Liu S, Liu Z, Nagel DH, McCann MC, SanMiguel P, Myers AM, Nettleton D, Nguyen J, Penning BW, Ponnala L, Schneider KL, Schwartz DC, Sharma A, Soderlund C, Springer NM, Sun Q, Wang H, Waterman M, Westerman R, Wolfgruber TK, Yang L, Yu Y, Zhang L, Zhou S, Zhu Q, Bennetzen JL, Dawe RK, Jiang J, Jiang N, Presting GG, Wessler SR, Aluru S, Martienssen RA, Clifton SW, McCombie WR, Wing RA, Wilson RK (2009) The B73 maize genome: complexity, diversity, and dynamics. Science 326:1112–1115

Schneider A, Molnár I, Molnár-Láng M (2008) Utilisation of *Aegilops* (goatgrass) species to widen the genetic diversity of cultivated wheat. Euphytica 163:1–19

Schubert I, Shi F, Fuchs J, Endo TR (1998) An efficient screening for terminal deletions and translocations of barley chromosomes added to common wheat. Plant J 14:489–495

Schubert V, Rudnik R, Schubert I (2014) Chromatin associations in *Arabidopsis* interphase nuclei. Front Genet 5:389

Schwarzacher T, Leitch AR, Bennett MD, Heslop-Harrison JS (1989) *In situ* localization of parental genomes in a wide hybrid. Ann Bot 64:315–324

Schwarzacher T, Wang ML, Leitch AR, Moore G, Heslop-Harrison JS, Miller N (1997) Flow cytometric analysis of the chromosomes and stability of a wheat cell-culture line. Theor Appl Genet 94:91–97

Sears ER (1956) The transfer of leaf-rust resistance from *Aegilops umbellulata* to wheat. Brookhaven symposia in biology, 1956. Genet Plant Breed 1956:1–22

Senerchia N, Felber F, Parisod C (2014) Contrasting evolutionary trajectories of multiple retrotransposons following independent allopolyploidy in wild wheats. New Phytol 202:975–985

Sepsi A, Molnár I, Szalay D, Molnár-Láng M (2008) Characterization of a leaf rust-resistant wheat-*Thinopyrum ponticum* partial amphiploid BE-1, using sequential multicolor GISH and FISH. Theor Appl Genet 116:825–834

Sergeeva EM, Afonnikov DA, Koltunova MK, Gusev VD, Miroshnichenko LA, Vrána J, Kubaláková M, Poncet C, Sourdille P, Feuillet C, Doležel J, Salina EA (2014) Common wheat chromosome 5B composition analysis using low-coverage 454 sequencing. Plant Genome 7:2

Seyfarth R, Feuillet C, Schachermayr G, Winzeler M, Keller B (1999) Development of a molecular marker for the adult plant leaf rust resistance gene *Lr35* in wheat. Theor Appl Genet 99:554–560

Shaked H, Kashkush K, Ozkan H, Feldman M, Levy AA (2001) Sequence elimination and cytosine methylation are rapid and reproducible responses of the genome to wide hybridization and allopolyploidy in wheat. Plant Cell 13:1749–1760

Šimková H, Svensson JT, Condamine P, Hřibová E, Suchánková P, Bhat PR, Bartoš J, Šafář J, Close TJ, Doležel J (2008) Coupling amplified DNA from flow-sorted chromosomes to high-density SNP mapping in barley. BMC Genomics 9:294

Soltis DE, Soltis PS, Tate JA (2004) Advances in the study of polyploidy since plant speciation. New Phytol 161:173–191

Stirnweis D, Milani SD, Brunner S, Herren G, Buchmann G, Peditto D, Jordan T, Keller B (2014) Suppression among alleles encoding nucleotide-binding-leucine-rich repeat resistance proteins interferes with resistance in F1 hybrid and allele-pyramided wheat plants. Plant J 79:893–903

Suchánková P, Kubaláková M, Kovářová P, Bartoš J, Číhalíková J, Molnár-Láng M, Endo TR, Doležel J (2006) Dissection of the nuclear genome of barley by chromosome flow sorting. Theor Appl Genet 113:651–659

Szakács É, Molnár-Láng M (2007) Development and molecular cytogenetic identification of new winter wheat – winter barley ("Martonvásári 9 kr1" – "Igri") disomic addition lines. Genome 50:43–50

Szakács É, Molnár-Láng M (2008) Fluorescent *in situ* hybridization polymorphism on the 1RS chromosome arms of cultivated *Secale cereale* species. Cereal Res Commun 36:247–255

Szakács É, Kruppa K, Molnár-Láng M (2013) Analysis of chromosomal polymorphism in barley (*Hordeum vulgare* L. ssp. *vulgare*) and between *H. vulgare* and *H. chilense* using three-color fluorescence *in situ* hybridization (FISH). J Appl Genet 54:427–433

Tanaka T, Kobayashi F, Joshi GP, Onuki R, Sakai H, Kanamori H, Wu J, Simkova H, Nasuda S, Endo TR, Hayakawa K, Doležel J, Ogihara Y, Itoh T, Matsumoto T, Handa H (2014) Next-generation survey sequencing and the molecular organization of wheat chromosome 6B. DNA Res 21:103–114

Tang KS, Hart GE (1975) Use of isozymes as chromosome markers in wheat-rye addition lines and in triticale. Genet Res 26:187–201

Tautz D (1989) Hypervariabflity of simple sequences as a general source for polymorphic DNA markers. Nucleic Acids Res 17:6463–6471

Tayalé A, Parisod C (2013) Natural pathways to polyploidy in plants and consequences for genome reorganization. Cytogenet Genome Res 140:79–96

The International Barley Genome Sequencing Consortium (2012) A physical, genetic and functional sequence assembly of the barley genome. Nature 491:711–716

The International Brachypodium Initiative (2010) Genome sequencing and analysis of the model grass *Brachypodium distachyon*. Nature 463:763–768

Tiwari VK, Wang S, Sehgal S, Vrána J, Friebe B, Kubaláková M, Chhuneja P, Doležel J, Akhunov E, Kalia B, Sabir J, Gill BS (2014) SNP Discovery for mapping alien introgressions in wheat. BMC Genomics 15:273

Valárik M, Bartos J, Kovárová P, Kubaláková M, de Jong JH, Dolezel J (2004) High-resolution FISH on super-stretched flow-sorted plant chromosomes. Plant J 37:940–950

Vos P, Hogers R, Bleeker M, Reijans M, van de Lee T, Hornes M, Frijters A, Pot J, Peleman J, Kuiper M (1995) AFLP: a new technique for DNA fingerprinting. Nucleic Acids Res 23:4407–4414

Vrána J, Kubaláková M, Simková H, Číhalíkovái J, Lysák MA, Dolezel J (2000) Flow sorting of mitotic chromosomes in common wheat (*Triticum aestivum* L). Genetics 156:2033–2041

Wang ML, Leitch AR, Schwarzacher T, Heslop-Harrison JS, Moore G (1992) Construction of a chromosome-enriched *Hpa*ll library from flow-sorted wheat chromosomes. Nucleic Acids Res 20:1897–1901

Wang RR-C, Chen J, Joppa LR (1995) Production and identification of chromosome specific RAPD markers for Langdon durum wheat disomic substitution lines. Crop Sci 35:886

Wang S, Wong D, Forrest K, Allen A, Chao S, Huang BE, Maccaferri M, Salvi S, Milner SG, Cattivelli L, Mastrangelo AM, Whan A, Stephen S, Barker G, Wieseke R, Plieske J, International Wheat Genome Sequencing Consortium, Lillemo M, Mather D, Appels R, Dolferus R, Brown-Guedira G, Korol A, Akhunova AR, Feuillet C, Salse J, Morgante M, Pozniak C, Luo MC, Dvorak J, Morell M, Dubcovsky J, Ganal M, Tuberosa R, Lawley C, Mikoulitch I, Cavanagh C, Edwards KJ, Hayden M, Akhunov E (2014) Characterization of polyploid wheat genomic diversity using a high-density 90 000 single nucleotide polymorphism array. Plant Biotechnol J 12:787–796

Warburton ML, Crossa J, Franco J, Kazi M, Trethowan R, Rajaram S, Pfeiffer W, Zhang P, Dreisigacker S, van Ginkel M (2006) Bringing wild relatives back into the family: recovering genetic diversity in CIMMYT improved wheat germplasm. Euphytica 149:289–301

Waugh R, McLean K, Flavell AJ, Pearce SR, Kumar A, Thomas BBT, Powell W (1997) Genetic distribution of *Bare–1*-like retrotransposable elements in the barley genome revealed by sequence-specific amplification polymorphisms (S-SAP). Mol Gen Genet 253:687–694

Williams JGK, Kubelik AR, Livak KJ, Rafalski JA, Tingey SV (1990) DNA polymorphisms amplified by arbitrary primers are useful as genetic markers. Nucleic Acids Res 18:6531–6535

Wilson AS (1876) Wheat and rye hybrids. Trans Proc Bot Soc (Edinb) 12:286–288

Wulff BBH, Moscou MJ (2014) Strategies for transferring resistance into wheat: from wide crosses to GM cassettes. Front Plant Sci 5:692

Wu Y, Sun Y, Wang X, Lin X, Sun S, Shen K, Wang J, Jiang T, Zhong S, Xu C, Liu B (2015) Transcriptome shock in an interspecific F1 triploid hybrid of *Oryza* revealed by RNA sequencing. J Integr Plant Biol. doi:10.1111/jipb.12357

Xiong LZ, Xu CG, Maroof MAS, Zhang Q (1999) Patterns of cytosine methylation in an elite rice hybrid and its parental lines, detected by a methylation-sensitive amplification polymorphism technique. Mol Gen Genet 261:439–446

Yabe S, Hara T, Ueno M, Enoki H, Kimura T, Nishimura S, Yasui Y, Ohsawa R, Iwata H (2014) Rapid genotyping with DNA micro-arrays for high-density linkage mapping and QTL mapping in common buckwheat (*Fagopyrum esculentum* Moench.). Breed Sci 64:291–299

Yamamoto M, Mukai Y (1989) Application of fluorescence *in situ* hybridization to molecular cytogenetics of wheat. Wheat Inf Serv 69:30–32

Yang X, Cao A, Sun Y, Chen P (2013) Tracing the location of powdery mildew resistance-related gene *Stpk-V* by FISH with a TAC clone in *Triticum aestivum-Haynaldia villosa* alien chromosome lines. Chin Sci Bull 58:4084–4091

Yasui Y, Nasuda S, Matsuoka Y, Kawahara T (2001) The *Au* family, a novel short interspersed element (SINE) from *Aegilops umbellulata*. Theor Appl Genet 102:463–470

Yingshan D, Xiuling B, Yushi L, Mengyuan H, Bao L (2004) Molecular characterization of a cryptic wheat-*Thinopyrum intermedium* translocation line: evidence for genomic instability in nascent allopolyploid and aneuploid lines. Genet Mol Biol 27:237–241

Yoo M-J, Szadkowski E, Wendel JF (2013) Homoeolog expression bias and expression level dominance in allopolyploid cotton. Heredity 110:171–180

Yu H, Deng Z, Xiang C, Tian J (2014) Analysis of diversity and linkage disequilibrium mapping of agronomic traits on B-genome of wheat. J Genomics 2:20–30

Yu J-K, Rota ML, Kantety RV, Sorrells ME (2004) EST derived SSR markers for comparative mapping in wheat and rice. Mol Genet Genomics 271:742–751

Zarco-Hernandez JA, Santiveri F, Michelena A, Javier Peña R (2005) Durum wheat (*Triticum turgidum* L.) carrying the 1BL.1RS chromosomal translocation: agronomic performance and quality characteristics under Mediterranean conditions. Eur J Agron 22:33–43

Zhang H, Jia J, Gale MD, Devos KM (1998) Relationships between the chromosomes of *Aegilops umbellulata* and wheat. Theor Appl Genet 96:69–75

Zhang J, Liu W, Han H, Song L, Bai L, Gao Z, Zhang Y, Yang X, Li X, Gao A, Li L (2015) De novo transcriptome sequencing of *Agropyron cristatum* to identify available gene resources for the enhancement of wheat. Genomics. doi:10.1016/j.ygeno.2015.04.003

Zhang P, Li W, Fellers J, Friebe B, Gill BS (2004a) BAC-FISH in wheat identifies chromosome landmarks consisting of different types of transposable elements. Chromosoma 112:288–299

Zhang P, Li W, Friebe B, Gill BS (2004b) Simultaneous painting of three genomes in hexaploid wheat by BAC-FISH. Genome 47:979–987

ERRATUM TO

Chapter 5
Gametocidal Genes

Takashi R. Endo

M. Molnár-Láng et al. (eds.), *Alien Introgression in Wheat*,
DOI 10.1007/978-3-319-23494-6_5

DOI 10.1007/978-3-319-23494-6_14

The publisher regrets the incorrect figure published in Chapter 5, page 122 in the print and online versions of this book. The correct Figure 5.1 is given below.

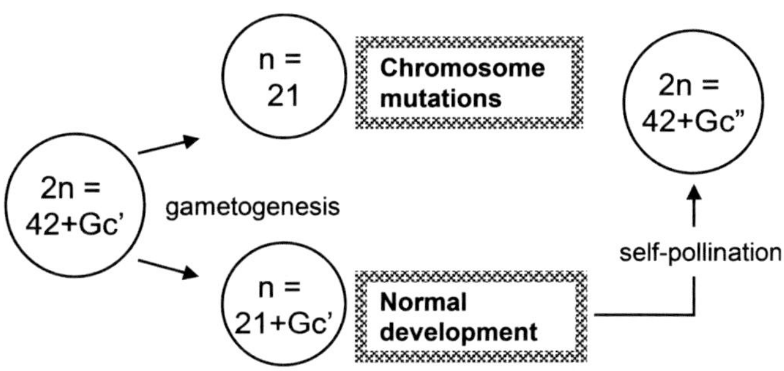

The online version of the original chapter can be found at
http://dx.doi.org/10.1007/978-3-319-23494-6_5

M. Molnár-Láng et al. (eds.), *Alien Introgression in Wheat*,
DOI 10.1007/978-3-319-23494-6_14

Index

M. Molnár-Láng et al. (eds.), *Alien Introgression in Wheat*,
DOI 10.1007/978-3-319-23494-6

Made in the USA
Monee, IL
11 January 2021

2b03d4e7-4110-4a72-9c51-18de16145e33R02